Bernard Valeur

Molecular Fluorescence

Principles and Applications

Related Titles from WILEY-VCH

Broekaert, J. A. C.

Analytical Atomic Spectrometry with Flames and Plasmas

2001. ISBN 3-527-30146-1

Günzler, H. and Williams, A.

Handbook of Analytical Techniques
2 Volumes

2001. ISBN 3-527-30165-8

Feringa, B. L.

Molecular Switches

2001. ISBN 3-527-29965-3

Zander, C.; Keller, R. A. and Enderlein, J.

Single-Molecule Detection in Solution. Methods and Applications

2002. ISBN 3-527-40310-8

Bernard Valeur

Molecular Fluorescence

Principles and Applications

Weinheim – New York – Chichester – Brisbane – Singapore – Toronto

Prof. Dr. Bernard Valeur
Laboratoire de Chimie Générale
Conservatoire National des Arts et Métiers
292 rue Saint-Martin
75141 Paris Cedex 03
France

Cover
The photograph was provided by Prof. R. Clegg (University of Illinois, USA).

1. Auflage 2002
1. Nachdruck, 2002, der 1. Auflage 2002

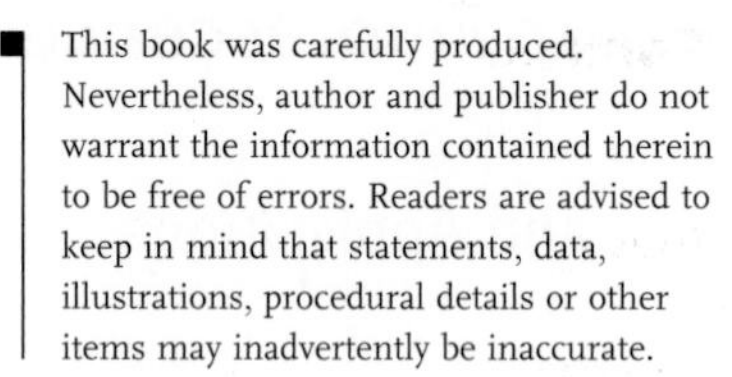
This book was carefully produced. Nevertheless, author and publisher do not warrant the information contained therein to be free of errors. Readers are advised to keep in mind that statements, data, illustrations, procedural details or other items may inadvertently be inaccurate.

Library of Congress Card No.: applied for
A catalogue record for this book is available from the British Library.
Die Deutsche Bibliothek – CIP Cataloguing-in-Publication-Data
A catalogue record for this publication is available from Die Deutsche Bibliothek

Printed in the Federal Republic of Germany.
Printed on acid-free paper.

Typesetting Asco Typesetters, Hongkong
Printing betz-druck gmbh, Darm-stadt
Bookbinding J. Schäffer GmbH&Co. KG, Grünstadt

ISBN 3-527-29919-X

Contents

Preface

This book is intended for students and researchers wishing to gain a deeper understanding of molecular fluorescence, with particular reference to applications in physical, chemical, material, biological and medical sciences.

Fluorescence was first used as an analytical tool to determine the concentrations of various species, either neutral or ionic. When the analyte is fluorescent, direct determination is possible; otherwise, a variety of indirect methods using derivatization, formation of a fluorescent complex or fluorescence quenching have been developed. Fluorescence sensing is the method of choice for the detection of analytes with a very high sensitivity, and often has an outstanding selectivity thanks to specially designed fluorescent molecular sensors. For example, clinical diagnosis based on fluorescence has been the object of extensive development, especially with regard to the design of optodes, i.e. chemical sensors and biosensors based on optical fibers coupled with fluorescent probes (e.g. for measurement of pH, pO_2, pCO_2, potassium, etc. in blood).

Fluorescence is also a powerful tool for investigating the structure and dynamics of matter or living systems at a molecular or supramolecular level. Polymers, solutions of surfactants, solid surfaces, biological membranes, proteins, nucleic acids and living cells are well-known examples of systems in which estimates of local parameters such as polarity, fluidity, order, molecular mobility and electrical potential is possible by means of fluorescent molecules playing the role of probes. The latter can be intrinsic or introduced on purpose. The high sensitivity of fluorimetric methods in conjunction with the specificity of the response of probes to their microenvironment contribute towards the success of this approach. Another factor is the ability of probes to provide information on dynamics of fast phenomena and/or the structural parameters of the system under study.

Progress in instrumentation has considerably improved the sensitivity of fluorescence detection. Advanced fluorescence microscopy techniques allow detection at single molecule level, which opens up new opportunities for the development of fluorescence-based methods or assays in material sciences, biotechnology and in the pharmaceutical industry.

The aim of this book is to give readers an overview of molecular fluorescence, allowing them to understand the fundamental phenomena and the basic techniques, which is a prerequisite for its practical use. The parameters that may affect the characteristics of fluorescence emission are numerous. This is a source of richness but also of complexity. The literature is teeming with examples of erroneous interpretations, due to a lack of knowledge of the basic principles. The reader's attention will be drawn to the many possible pitfalls.

Chapter 1 is an introduction to the field of molecular fluorescence, starting with a short history of fluorescence. In Chapter 2, the various aspects of light absorption (electronic transitions, UV–visible spectrophotometry) are reviewed.

Chapter 3 is devoted to the characteristics of fluorescence emission. Special attention is paid to the different ways of de-excitation of an excited molecule, with emphasis on the time-scales relevant to the photophysical processes – but without considering, at this stage, the possible interactions with other molecules in the excited state. Then, the characteristics of fluorescence (fluorescence quantum yield, lifetime, emission and excitation spectra, Stokes shift) are defined.

The effects of photophysical intermolecular processes on fluorescence emission are described in Chapter 4, which starts with an overview of the de-excitation processes leading to fluorescence quenching of excited molecules. The main excited-state processes are then presented: electron transfer, excimer formation or exciplex formation, proton transfer and energy transfer.

Fluorescence polarization is the subject of Chapter 5. Factors affecting the polarization of fluorescence are described and it is shown how the measurement of emission anisotropy can provide information on fluidity and order parameters.

Chapter 6 deals with fluorescence techniques, with the aim of helping the reader to understand the operating principles of the instrumental set-up he or she utilizes, now or in the future. The section devoted to the sophisticated time-resolved techniques will allow readers to know what they can expect from these techniques, even if they do not yet utilize them. Dialogue with experts in the field, in the course of a collaboration for instance, will be made easier.

The effect of solvent polarity on the emission of fluorescence is examined in Chapter 7, together with the use of fluorescent probes to estimate the polarity of a microenvironment.

Chapter 8 shows how parameters like fluidity, order parameters and molecular mobility can be locally evaluated by means of fluorescent probes.

Chapter 9 is devoted to resonance energy transfer and its applications in the cases of donor–acceptor pairs, assemblies of donor and acceptor, and assemblies of like fluorophores. In particular, the use of resonance energy transfer as a 'spectroscopic ruler', i.e. for the estimation of distances and distance distributions, is presented.

In Chapter 10, fluorescent pH indicators and fluorescent molecular sensors for cations, anions and neutral molecules are described, with an emphasis on design principles in regard to selectivity.

Finally, in Chapter 11 some advanced techniques are briefly described: fluorescence up-conversion, fluorescence microscopy (confocal excitation, two-photon excitation, near-field optics, fluorescence lifetime imaging), fluorescence correlation spectroscopy, and single-molecule fluorescence spectroscopy.

This book is by no means intended to be exhaustive and it should rather be considered as a textbook. Consequently, the bibliography at the end of each chapter has been restricted to a few leading papers, reviews and books in which the readers will find specific references relevant to their subjects of interest.

Fluorescence is presented in this book from the point of view of a physical chemist, with emphasis on the understanding of physical and chemical concepts. Efforts have been made to make this book easily readable by researchers and students from any scientific community. For this purpose, the mathematical developments have been limited to what is strictly necessary for understanding the basic phenomena. Further developments can be found in accompanying boxes for aspects of major conceptual interest. The main equations are framed so that, in a first reading, the intermediate steps can be skipped. The aim of the boxes is also to show illustrations chosen from a variety of fields. Thanks to such a presentation, it is hoped that this book will favor the relationship between various scientific communities, in particular those that are relevant to physicochemical sciences and life sciences.

I am extremely grateful to Professors Elisabeth Bardez and Mario Nuno Berberan-Santos for their very helpful suggestions and constant encouragement. Their critical reading of most chapters of the manuscript was invaluable. The list of colleagues and friends who should be gratefully acknowledged for their advice and encouragement would be too long, and I am afraid I would forget some of them. Special thanks are due to my son, Eric Valeur, for his help in the preparation of the figures and for enjoyable discussions. I wish also to thank Professor Philip Stephens for his help in the translation of French quotations.

Finally, I will never forget that my first steps in fluorescence spectroscopy were guided by Professor Lucien Monnerie; our friendly collaboration for many years was very fruitful. I also learned much from Professor Gregorio Weber during a one-year stay in his laboratory as a postdoctoral fellow; during this wonderful experience, I met outstanding scientists and friends like Dave Jameson, Bill Mantulin, Enrico Gratton and many others. It is a privilege for me to belong to Weber's 'family'.

Paris, May 2001 Bernard Valeur

Prologue

La lumière joue dans notre vie un rôle essentiel: elle intervient dans la plupart de nos activités. Les Grecs de l'Antiquité le savaient bien déjà, eux qui pour dire "mourir" disaient "perdre la lumière".

[Light plays an essential role in our lives: it is an integral part of the majority of our activities. The ancient Greeks, who for "to die" said "to lose the light", were already well aware of this.]

Louis de Broglie, 1941

1
Introduction

From the discovery of the fluorescence of *Lignum Nephriticum* (1965) to fluorescence probing of the structure and dynamics of matter and living systems at a molcular level

... ex arte calcinati, et illuminato aeri seu solis radiis, seu flammae fulgoribus expositi, lucem inde sine calore concipiunt in sese; ...	[... *properly calcinated, and illuminated either by sunlight or flames, they conceive light from themselves without heat;* ...]

Licetus, 1640 *(about the Bologna stone)*

1.1
What is luminescence?

Luminescence is an emission of ultraviolet, visible or infrared photons from an electronically excited species. The word luminescence, which comes from the Latin (*lumen* = light) was first introduced as *luminescenz* by the physicist and science historian Eilhardt Wiedemann in 1888, to describe 'all those phenomena of light which are not solely conditioned by the rise in temperature', as opposed to incandescence. Luminescence is *cold light* whereas incandescence is *hot light*. The various types of luminescence are classified according to the mode of excitation (see Table 1.1).

Luminescent compounds can be of very different kinds:

- organic compounds: aromatic hydrocarbons (naphthalene, anthracene, phenanthrene, pyrene, perylene, etc.), fluorescein, rhodamines, coumarins, oxazines, polyenes, diphenylpolyenes, aminoacids (tryptophan, tyrosine, phenylalanine), etc.
- inorganic compounds: uranyl ion (UO_2^+), lanthanide ions (e.g. Eu^{3+}, Tb^{3+}), doped glasses (e.g. with Nd, Mn, Ce, Sn, Cu, Ag), crystals (ZnS, CdS, ZnSe, CdSe, GaS, GaP, Al_2O_3/Cr^{3+} (ruby)), etc.

Tab. 1.1. The various types of luminescence

Phenomenon	***Mode of excitation***
Photoluminescence (fluorescence, phosphorescence, delayed fluorescence)	Absorption of light (photons)
Radioluminescence	Ionizing radiation (X-rays, α, β, γ)
Cathodoluminescence	Cathode rays (electron beams)
Electroluminescence	Electric field
Thermoluminescence	Heating after prior storage of energy (e.g. radioactive irradiation)
Chemiluminescence	Chemical process (e.g. oxidation)
Bioluminescence	Biochemical process
Triboluminescence	Frictional and electrostatic forces
Sonoluminescence	Ultrasounds

- organometallic compounds: ruthenium complexes (e.g. $Ru(biPy)_3$), complexes with lanthanide ions, complexes with fluorogenic chelating agents (e.g. 8-hydroxy-quinoline, also called oxine), etc.

Fluorescence and *phosphorescence* are particular cases of luminescence (Table 1.1). The mode of excitation is absorption of a photon, which brings the absorbing species into an electronic excited state. The emission of photons accompanying de-excitation is then called *photoluminescence* (fluorescence, phosphorescence or delayed fluorescence), which is one of the possible physical effects resulting from interaction of light with matter, as shown in Figure 1.1.

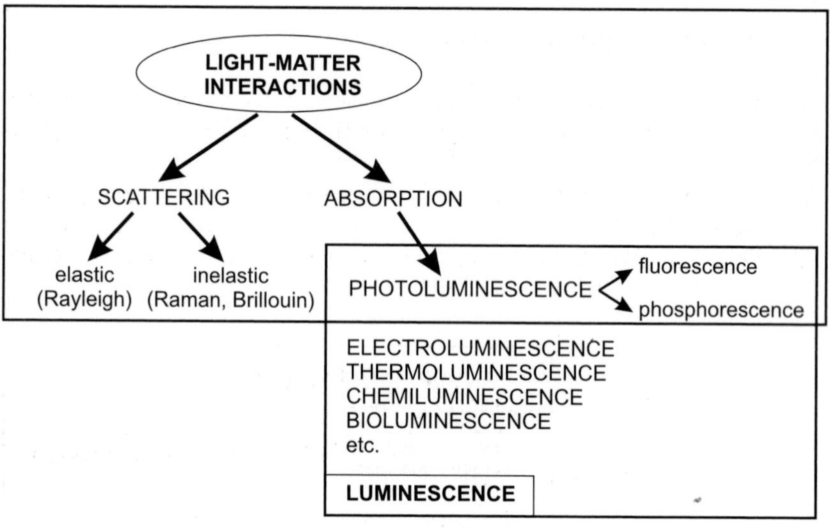

Fig. 1.1. Position of fluorescence and phosphorescence in the frame of light–matter interactions.

Tab. 1.2. Early stages in the history of fluorescence and phosphorescence[a)]

Year	*Scientist*	*Observation or achievement*
1565	N. Monardes	Emission of light by an infusion of wood Lignum Nephriticum (first reported observation of fluorescence)
1602	V. Cascariolo	Emission of light by Bolognese stone (first reported observation of phosphorescence)
1640	Licetus	Study of Bolognese stone. First definition as a non-thermal light emission
1833	D. Brewster	Emission of light by chlorophyll solutions and fluorspar crystals
1845	J. Herschel	Emission of light by quinine sulfate solutions (epipolic dispersion)
1842	E. Becquerel	Emission of light by calcium sulfide upon excitation in the UV. First statement that the emitted light is of longer wavelength than the incident light
1852	G. G. Stokes	Emission of light by quinine sulfate solutions upon excitation in the UV (refrangibility of light)
1853	G. G. Stokes	Introduction of the term fluorescence
1858	E. Becquerel	First phosphoroscope
1867	F. Goppelsröder	First fluorometric analysis (determination of Al(III) by the fluorescence of its morin chelate)
1871	A. Von Baeyer	Synthesis of fluorescein
1888	E. Wiedemann	Introduction of the term luminescence

a) More details can be found in:
Harvey E. N. (1957) *History of Luminescence*, The American Philosophical Society, Philadelphia.
O'Haver T. C. (1978) The Development of Luminescence Spectrometry as an Analytical Tool, *J. Chem. Educ.* **55**, 423–8.

1.2
A brief history of fluorescence and phosphorescence

It is worth giving a brief account of the early stages in the history of fluorescence and phosphorescence (Table 1.2), paying special attention to the origin of these terms.

The term *phosphorescence* comes from the Greek: $\phi\omega\varsigma$ = light (genitive case: $\phi o\tau o\varsigma \rightarrow$ photon) and $\phi o\rho\varepsilon\iota\nu$ = to bear (Scheme 1.1). Therefore, *phosphor* means 'which bears light'. The term *phosphor* has indeed been assigned since the Middle

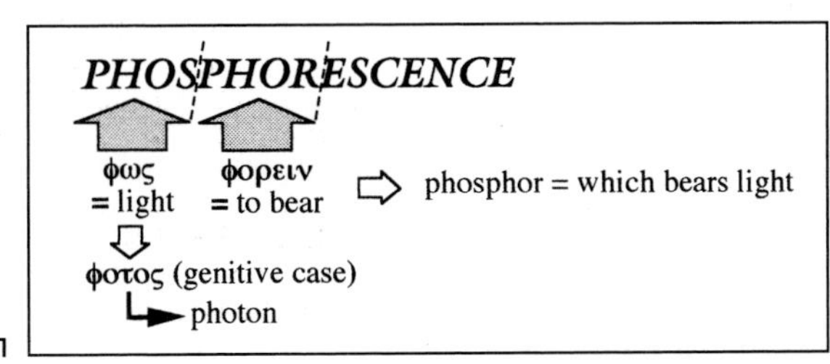

Scheme 1.1

Ages to materials that glow in the dark after exposure to light. There are many examples of minerals reported a long time ago that exhibit this property, and the most famous of them (but not the first one) was the *Bolognian phosphor* discovered by a cobbler from Bologna in 1602, Vincenzo Cascariolo, whose hobby was alchemy. One day he went for a walk in the Monte Paterno area and he picked up some strange heavy stones. After calcination with coal, he observed that these stones glowed in the dark after exposure to light. It was recognized later that the stones contained barium sulfate, which, upon reduction by coal, led to barium sulfide, a phosphorescent compound. Later, the same name *phosphor* was assigned to the element isolated by Brandt in 1677 (despite the fact that it is chemically very different) because, when exposed to air, it burns and emits vapors that glow in the dark.

In contrast to phosphorescence, the etymology of the term *fluorescence* is not at all obvious. It is indeed strange, at first sight, that this term contains *fluor* which is not remarked by its fluorescence! The term *fluorescence* was introduced by Sir George Gabriel Stokes, a physicist and professor of mathematics at Cambridge in the middle of the nineteenth century. Before explaining why Stokes coined this term, it should be recalled that the first reported observation of fluorescence was made by a Spanish physician, Nicolas Monardes, in 1565. He described the wonderful peculiar blue color of an infusion of a wood called *Lignum Nephriticum*. This wood was further investigated by Boyle, Newton and others, but the phenomenon was not understood.

In 1833, David Brewster, a Scottish preacher, reported[1] that a beam of white light passing through an alcoholic extract of leaves (chlorophyll) appears to be red when observed from the side, and he pointed out the similarity with the blue light coming from a light beam passing through fluorspar crystals. In 1845, John Herschel, the famous astronomer, considered that the blue color at the surface of solutions of quinine sulfate and *Lignum Nephriticum* was 'a case of superficial color presented by a homogeneous liquid, internally colorless'. He called this phenomenon *epipolic dispersion*, from the Greek $\varepsilon\pi\iota\pi o\lambda\eta$ = surface[2]. The solutions observed by Herschel were very concentrated so that the majority of the incident light was absorbed and all the blue color appeared to be only at the surface. Herschel used a prism to show that the epipolic dispersion could be observed only upon illumination by the blue end of the spectrum, and not the red end. The crude spectral analysis with the prism revealed blue, green and a small amount of yellow light, but Herschel did not realize that the superficial light was of longer wavelength than the incident light.

The phenomena were reinvestigated by Stokes, who published a famous paper entitled 'On the refrangibility of light' in 1852[3]. He demonstrated that the phenomenon was an emission of light following absorption of light. It is worth describing one of Stokes' experiments, which is spectacular and remarkable for its

1) Brewster D. (1833) *Trans. Roy. Soc. Edinburgh* **12**, 538–45.

2) Herschel J. F. W. (1945) *Phil. Trans.* 143–145 & 147–153.

3) Stokes G. G. (1852) *Phil. Trans.* **142**, 463–562.

Scheme 1.2

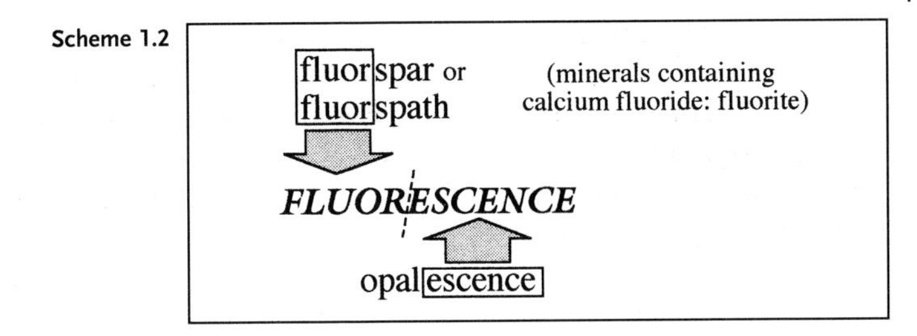

simplicity. Stokes formed the solar spectrum by means of a prism. When he moved a tube filled with a solution of quinine sulfate through the visible part of the spectrum, nothing happened: the solution simply remained transparent. But beyond the violet portion of the spectrum, i.e. in the non-visible zone corresponding to ultraviolet radiations, the solution glowed with a blue light. Stokes wrote: 'It was certainly a curious sight to see the tube instantaneously light up when plunged into the invisible rays; it was literally *darkness visible.*' This experiment provided compelling evidence that there was absorption of light followed by emission of light. Stokes stated that the emitted light is always of longer wavelength than the exciting light. This statement becomes later Stokes' law.

Stokes' paper led Edmond Becquerel, a French physicist[4], to 'réclamation de priorité' for this kind of experiment[5]. In fact, Becquerel published an outstanding paper[6] in 1842 in which he described the light emitted by calcium sulfide deposited on paper when exposed to solar light beyond the violet part of the spectrum. He was the first to state that the emitted light is of longer wavelength than the incident light.

In his first paper[3], Stokes called the observed phenomenon *dispersive reflexion*, but in a footnote, he wrote 'I confess I do not like this term. I am almost inclined to coin a word, and call the appearance *fluorescence*, from fluorspar, as the analogous term *opalescence* is derived from the name of a mineral.' Most of the varieties of fluorspar or fluorspath (minerals containing calcium fluoride (fluorite)) indeed exhibit the property described above. In his second paper[7], Stokes definitely resolved to use the word *fluorescence* (Scheme 1.2).

We understand now why *fluorescence* contains the term *fluor*, but what is the origin of *fluorspar* or *fluorspath* and why are these materials fluorescent? *Spar* (in English) and *spath* (in German) were the names given in the eighteenth century[8] to 'stones' that are more or less transparent and crystallized with a lamellar texture. Because these materials can be easily melted, and some of them can help to melt

4) Edmond Becquerel is the father of Henri Becquerel, who discovered radioactivity. Edmond Becquerel invented the famous phosphoroscope that bears his name. He was Professor at the Museum National d'Histoire Naturelle and at the Conservatoire National des Arts et Métiers in Paris.

5) In *Cosmos* (1854) **3**, 509–10.

6) Becquerel E. (1842) *Annales de Chimie et Physique* (3) **9**, 257–322.

7) Stokes G. G. (1853) *Phil. Trans.* **143**, 385–96.

8) Macquer P. J. (1779) *Dictionnaire de Chymie*, p. 462.

other materials, many mineralogists and metallurgists employed the word *fluor*[9] in order to express some fluidity (*fluere* = to flow in Latin). This is the origin of the name of the element *fluor*, isolated by Moissan in 1886, although there is no direct relationship between the Latin origin of this element and fluidity. Many spaths are known to be colored because of the presence of small amounts of impurities, which explains the fluorescence properties because fluorite itself is not fluorescent. The blue and red fluorescences are due to divalent and trivalent europium ions, respectively. Yttrium and Dysprosium can also be present and yield a yellow fluorescence[10]. Other possible fluorescent impurities may exist but they were found to be difficult to characterize.

The difference between the Stokes and Becquerel experiments described above is that quinine sulfate is fluorescent whereas calcium sulfide is phosphorescent, but both species are relevant to photoluminescence. In the nineteenth century, the distinction between fluorescence and phosphorescence was made on an experimental basis: fluorescence was considered as an emission of light that disappears simultaneously with the end of excitation, whereas in phosphorescence the emitted light persists after the end of excitation. Now we know that in both cases emission lasts longer than excitation and a distinction only based on the duration of emission is not sound because there are long-lived fluorescences (e.g. uranyl salts) and short-lived phosphorescences (e.g. violet luminescence of zinc sulfide). The first theoretical distinction between fluorescence and phosphorescence was provided by Francis Perrin[11]: 'if the molecules pass, between absorption and emission, through a stable or unstable intermediate state and are thus no longer able to reach the emission state without receiving from the medium a certain amount of energy, there is phosphorescence'. Further works clarified the distinction between fluorescence, delayed-fluorescence and phosphorescence[12].

In addition to this clarification, many other major events in the history of fluorescence occurred during the first half of the twentieth century. The most important are reported in Table 1.3, together with the names of the associated scientists. It is remarkable that the period 1918–35 (i.e. less than 20 years) was exceptionally fecund for the understanding of the major experimental and theoretical aspects of fluorescence and phosphorescence.

1.3 Fluorescence and other de-excitation processes of excited molecules

Once a molecule is excited by absorption of a photon, it can return to the ground state with emission of fluorescence, but many other pathways for de-excitation are also possible (Figure 1.2): internal conversion (i.e. direct return to the ground state

9) Macquer P. J. (1779) *Dictionnaire de Chymie*, p. 464.

10) Robbins M. (1994) *Fluorescence. Gems and Minerals under Ulraviolet Light*, Geoscience Press.

11) Perrin F. (1929) Doctoral thesis, Paris; *Annales de Physique* **12**, 2252–4.

12) Nickel B. (1996) Pioneers in Photochemistry. From the Perrin Diagram to the Jablonski Diagram, *EPA Newsletter* **58**, 9–38.

Tab. 1.3. Milestones in the history of fluorescence and phosphorescence during the first half of the twentieth century[a)]

Year	*Scientists*	*Observation or achievement*
1905, 1910	E. L. Nichols and E. Merrit	First fluorescence excitation spectrum of a dye
1907	E. L. Nichols and E. Merrit	Mirror symmetry between absorption and fluorescence spectra
1918	J. Perrin	Photochemical theory of dye fluorescence
1919	Stern and Volmer	Relation for fluorescence quenching
1920	F. Weigert	Discovery of the polarization of the fluorescence emitted by dye solutions
1922	S. J. Vavilov	Excitation-wavelength independence of the fluorescence quantum yield
1923	S. J. Vavilov and W. L. Levshin	First study of the fluorescence polarization of dye solutions
1924	S. J. Vavilov	First determination of fluorescence yield of dye solutions
1924	F. Perrin	Quantitative description of static quenching (active sphere model
1924	F. Perrin	First observation of alpha phosphorescence (E-type delayed fluorescence)
1925	F. Perrin	Theory of fluorescence polarization (influence of viscosity)
1925	W. L. Levshin	Theory of polarized fluorescence and phosphorescence
1925	J. Perrin	Introduction of the term delayed fluorescence Prediction of long-range energy transfer
1926	E. Gaviola	First direct measurement of nanosecond lifetimes by phase fluorometry (instrument built in Pringsheim's laboratory)
1926	F. Perrin	Theory of fluorescence polarization (sphere). Perrin's equation Indirect determination of lifetimes in solution. Comparison with radiative lifetimes
1927	E. Gaviola and P. Pringsheim	Demonstration of resonance energy transfer in solutions
1928	E. Jette and W. West	First photoelectric fluorometer
1929	F. Perrin	Discussion on Jean Perrin's diagram for the explanation of the delayed fluorescence by the intermediate passage through a metastable state First qualitative theory of fluorescence depolarization by resonance energy transfer
1929	J. Perrin and Choucroun	Sensitized dye fluorescence due to energy transfer
1932	F. Perrin	Quantum mechanical theory of long-range energy transfer between atoms
1934	F. Perrin	Theory of fluorescence polarization (ellipsoid)
1935	A. Jablonski	Jablonski's diagram
1944	Lewis and Kasha	Triplet state
1948	Th. Förster	Quantum mechanical theory of dipole–dipole energy transfer

a) More details can be found in the following references:
Nickel B. (1996) From the Perrin Diagram to the Jablonski Diagram. Part 1, *EPA Newsletter* **58**, 9–38.

Tab 1.3 (cont.)

Nickel B. (1997) From the Perrin Diagram to the Jablonski Diagram. Part 2, *EPA Newsletter* **61**, 27–60.
Nickel B. (1998) From Wiedemann's discovery to the Jablonski Diagram *EPA Newsletter* **64**, 19–72.
Berberan-Santos M. N. (2001) Pioneering Contributions of Jean and Francis Perrin to Molecular Fluorescence, in: Valeur B. and Brochon J. C. (Eds), *New Trends in Fluorescence Spectroscopy. Applications to Chemical and Life Sciences*, Springer-Verlag, Berlin, pp. 7–33.

without emission of fluorescence), intersystem crossing (possibly followed by emission of phosphorescence), intramolecular charge transfer and conformational change. These processes will be the subject of Chapter 3. Interactions in the excited state with other molecules may also compete with de-excitation: electron transfer, proton transfer, energy transfer, excimer or exciplex formation. These intermolecular photophysical processes will be described in Chapter 4.

These de-excitation pathways may compete with fluorescence emission if they take place on a time-scale comparable with the average time (lifetime) during which the molecules stay in the excited state. This average time represents the *experimental time window* for observation of dynamic processes. The characteristics of fluorescence (spectrum, quantum yield, lifetime), which are affected by any excited-state process involving interactions of the excited molecule with its close environment, can then provide information on such a microenvironment. It should be noted that some excited-state processes (conformational change, electron transfer, proton transfer, energy transfer, excimer or exciplex formation) may lead to a fluorescent species whose emission can superimpose that of the initially excited molecule. Such an emission should be distinguished from the 'primary' fluorescence arising from the excited molecule.

The success of fluorescence as an investigative tool in studying the structure and dynamics of matter or living systems arises from the high sensitivity of fluoro-

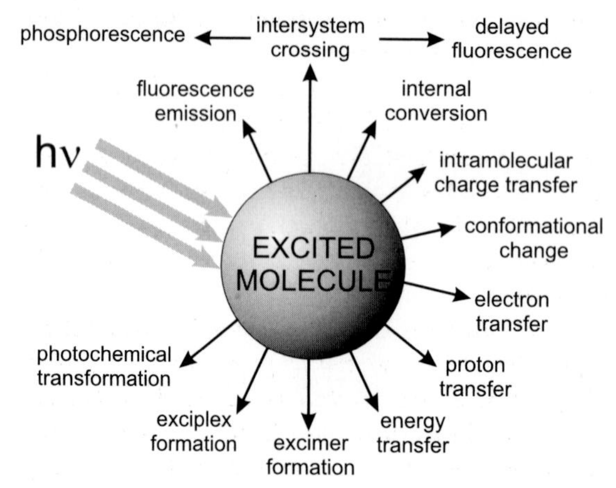

Fig. 1.2. Possible de-excitation pathways of excited molecules.

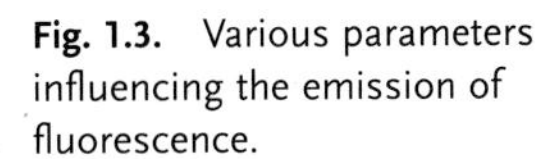
Fig. 1.3. Various parameters influencing the emission of fluorescence.

metric techniques, the specificity of fluorescence characteristics due to the microenvironment of the emitting molecule, and the ability of the latter to provide spatial and temporal information. Figure 1.3 shows the physical and chemical parameters that characterize the microenvironment and can thus affect the fluorescence characteristics of a molecule.

1.4
Fluorescent probes

As a consequence of the strong influence of the surrounding medium on fluorescence emission, fluorescent molecules are currently used as probes for the investigation of physicochemical, biochemical and biological systems. A large part of this book is devoted to the use of so-called *fluorescent probes.*

It is worth recalling that other types of probes are used in practice: for example, radioactive tracers, with their well-known drawback of their radioactivity, and EPR (electronic paramagnetic resonance) probes that provide information mainly on molecular mobility. In contrast to these probes, which are used in rather limited fields of applications, fluorescent probes can offer a wealth of information in various fields, as shown in Table 1.4. The various examples described in this book will demonstrate their outstanding versatility.

Fluorescent probes can be divided into three classes: (i) *intrinsic probes*; (ii) *extrinsic covalently bound probes*; and (iii) *extrinsic associating probes.* Intrinsic probes are ideal but there are only a few examples (e.g. tryptophan in proteins). The advantage of covalently bound probes over the extrinsic associating probes is that the location of the former is known. There are various examples of probes covalently

Tab. 1.4. Information provided by fluorescent probes in various fields

Field	*Information*
Polymers	dynamics of polymer chains; microviscosity; free volume; orientation of chains in stretched samples; miscibility; phase separation; diffusion of species through polymer networks; end-to-end macrocyclization dynamics; monitoring of polymerization; degradation
Solid surfaces	nature of the surface of colloidal silica, clays, zeolites, silica gels, porous Vycor glasses, alumina: rigidity, polarity and modification of surfaces
Surfactant solutions	critical micelle concentration; distribution of reactants among particles; surfactant aggregation numbers; interface properties and polarity; dynamics of surfactant solutions; partition coefficients; phase transitions; influence of additives
Biological membranes	fluidity; order parameters; lipid–protein interactions; translational diffusion; site accessibility; structural changes; membrane potentials; complexes and binding; energy-linked and light-induced changes; effects of additives; location of proteins; lateral organization and dynamics
Vesicles	characterization of the bilayer: microviscosity, order parameters; phase transition; effect of additives; internal pH; permeability
Proteins	binding sites; denaturation; site accessibility; dynamics; distances; conformational transition
Nucleic acids	flexibility; torsion dynamics; helix structure; deformation due to intercalating agents; photocleavage; accessibility; carcinogenesis
Living cells	visualization of membranes, lipids, proteins, DNA, RNA, surface antigens, surface glycoconjugates; membrane dynamics; membrane permeability; membrane potential; intracellular pH; cytoplasmic calcium, sodium, chloride, proton concentration; redox state; enzyme activities; cell–cell and cell–virus interactions; membrane fusion; endocytosis; viability, cell cycle; cytotoxic activity
Fluoroimmunochemistry	fluoroimmunoassays

attached to surfactants, polymer chains, phospholipids, proteins, polynucleotides, etc. A selection of them is presented in Figure 1.4. In particular, the anthroyloxy stearic acids with the anthracene moiety attached in various positions of the paraffinic chain allows micellar systems or bilayers to be probed at various depths.

Protein tagging can be easily achieved by means of labeling reagents having proper functional groups: covalent binding is indeed possible on amino groups (with isothiocyanates, chlorotriazinyl derivatives, hydroxysuccinimido active esters), and on sulfhydryl groups (with iodoacetamido and maleimido functional groups). Fluorescein, rhodamine and erythrosin derivatives with these functional groups are currently used.

Owing to the difficulty of synthesis of molecules or macromolecules with covalently bound specific probes, most of the investigations are carried out with non-

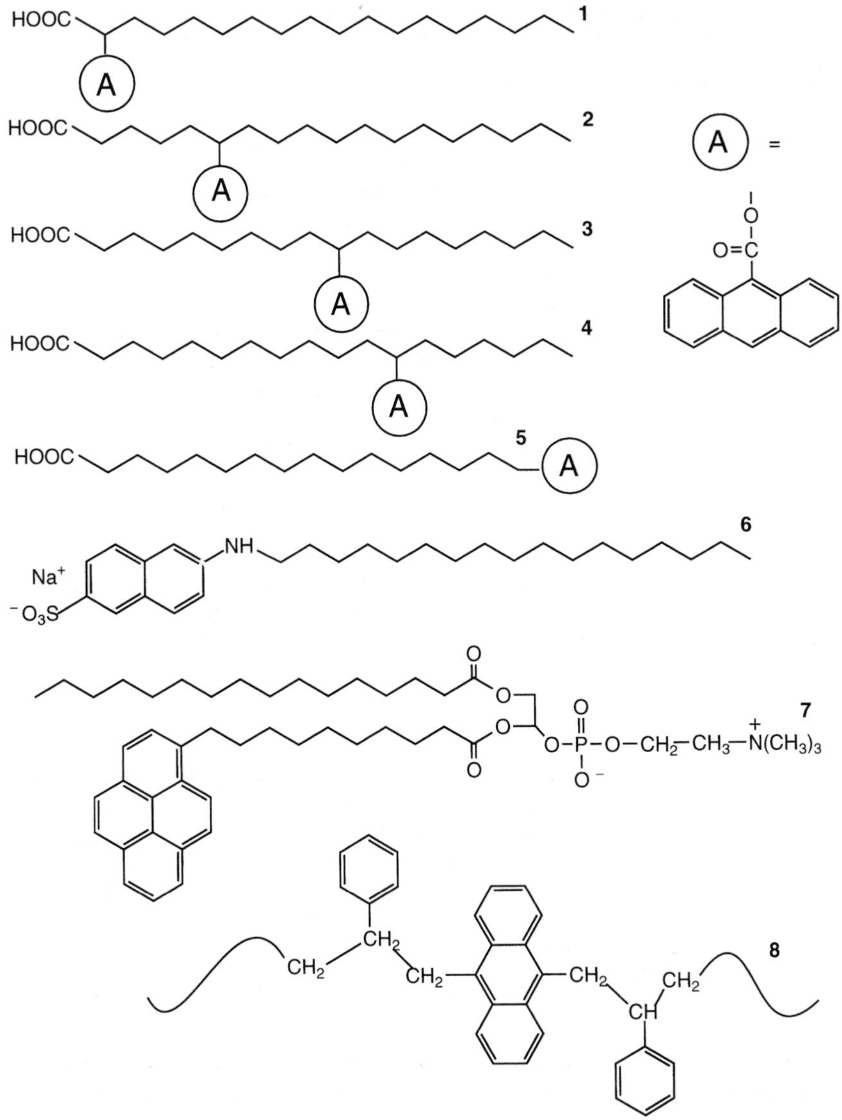

Fig. 1.4. Examples of surfactants, phospholipids and polymers with covalently bound probes. 1: 2-(9-anthroyloxy)stearic acid. 2: 6-(9-anthroyloxy)stearic acid. 3: 10-(9-anthroyloxy)stearic acid. 4: 12-(9-anthroyloxy)stearic acid. 5: (9-anthroyloxy)palmitic acid. 6: 2-(*N*-octadecyl)amino-naphthalene-6-sulfonic acid, sodium salt. 7: 3-palmitoyl-2-(1-pyrenedecanoyl)-L-α-phosphatidylcholine. 8: polystyrene labeled with anthracene.

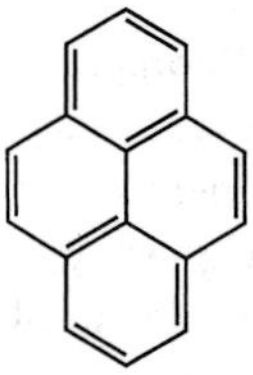

Fig. 1.5. Examples of hydrophobic, hydrophilic and amphiphilic probes. 1: pyrene. 2: 8-hydroxypyrene-1,3,6-trisulfonic acid, trisodium salt (pyranine). 3: 8-alkoxypyrene-1,3,6-trisulfonic acid, trisodium salt. 4: 1-pyrenedodecanoic acid. 5: 1,6-diphenyl-1,2,5-hexatriene (DPH). 6: 1-(4-trimethylammonium-phenyl)-6-phenyl-1,3,5-hexatriene, *p*-toluene sulfonate (TMA–DPH). 7: *cis*-parinaric acid. 8: *trans*-parinaric acid.

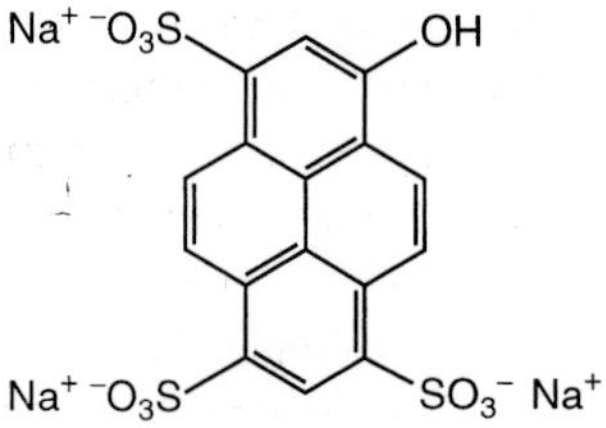

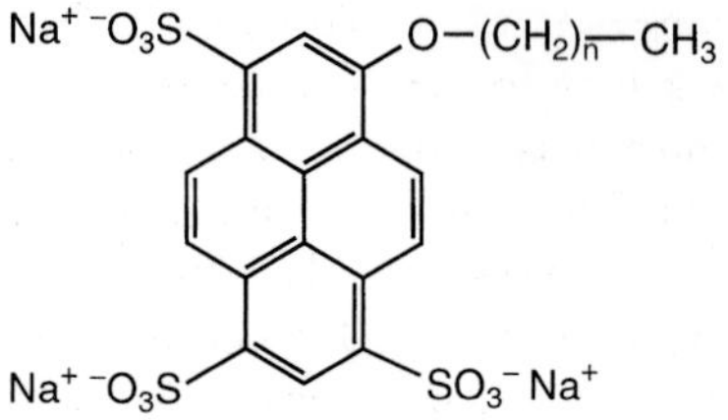

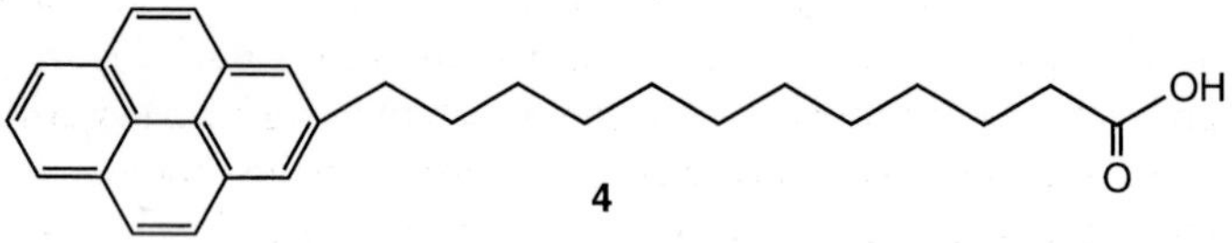

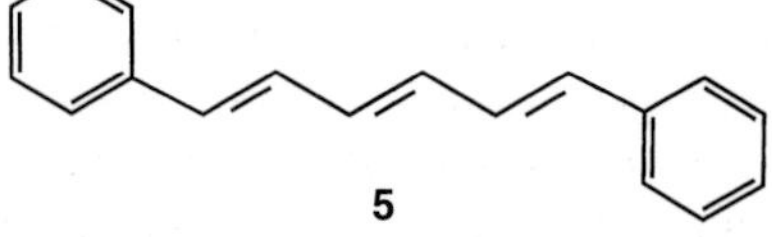

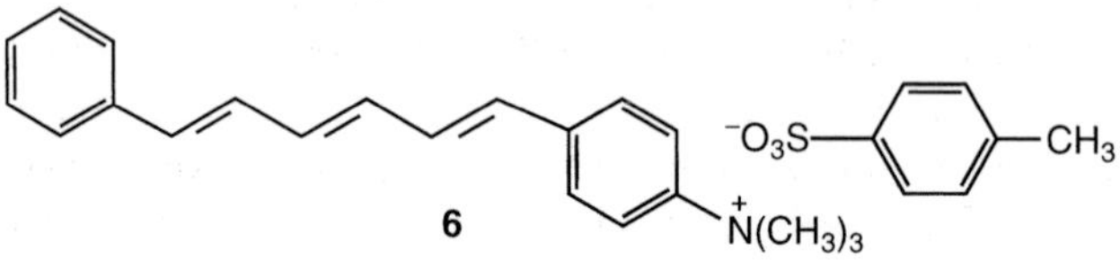

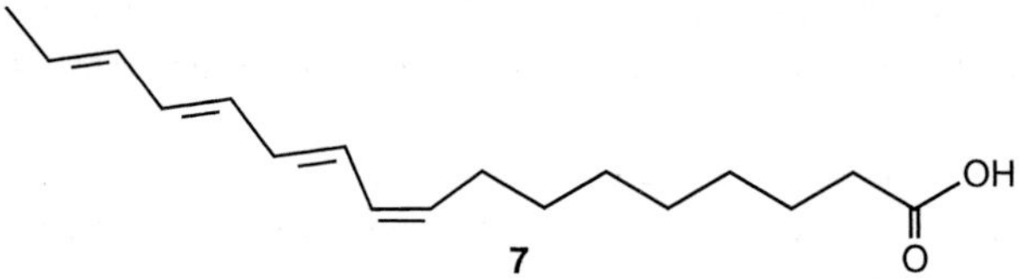

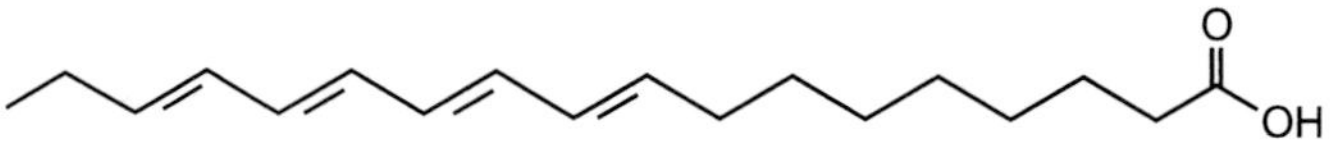

covalently associating probes (class III). The sites of solubilization of extrinsic probes are governed by their chemical nature and the resulting specific interactions that can be established within the region of the system to be probed. The hydrophilic, hydrophobic or amphiphilic character of a probe is essential. Figure 1.5 gives various examples. Pyrene is known as a probe of hydrophobic regions; furthermore its sensitivity to polarity is very useful (see Chapter 7). In contrast, pyranine is very hydrophilic and will be located in hydrophilic aqueous regions;

moreover it is sensitive to pH. If the OH group is replaced by O–$(CH_2)_n$–CH_3, the resulting molecule becomes pH-insensitive and amphiphilic; the fluorophore moiety plays the role of a polar head and thus is located at the surfactant–water interface of systems consisting of amphiphilic molecules (bilayers of membranes and vesicles, micellar systems, etc.). Conversely, the pyrene moiety of pyrenedodecanoic acid is deeply embedded in the hydrophobic part of an organized assembly. 1,6-Diphenyl-1,3,5-hexatriene (DPH) is located in the hydrocarbon region of bilayers of membrane and vesicles, whereas its cationic analog TMA–DPH is anchored with its charged group at the surfactant–water interface. The latter is thus a probe of the upper region of bilayers. *Cis*- and *trans*-parinaric acids are good examples of probes causing minimum spatial perturbation to organized assemblies.

The above examples show that a very important criterion in the choice of a probe is its sensitivity to a particular property of the microenvironment in which it is located (e.g. polarity, acidity, etc.). On the other hand, insensitivity to the chemical nature of the environment is preferable in some cases (e.g. in fluorescence polarization or energy transfer experiments). Environment-insensitive probes are also better suited to fluorescence microscopy and flow cytometry.

A criticism often aimed at the use of extrinsic fluorescent probes is the possible local perturbation induced by the probe itself on the microenvironment to be probed. There are indeed several cases of systems perturbed by fluorescent probes. However, it should be emphasized that many examples of results consistent with those obtained by other techniques can be found in the literature (transition temperature in lipid bilayer, flexibility of polymer chains, etc.). To minimize the perturbation, attention must be paid to the size and shape of the probe with respect to the probed region.

In conclusion, the choice of a fluorescent probe is crucial for obtaining unambiguous interpretations. The major aspects that should be taken into consideration are shown in Figure 1.6.

1.5
Molecular fluorescence as an analytical tool

Analytical techniques based on fluorescence detection are very popular because of their high sensitivity and selectivity, together with the advantages of spatial and temporal resolution, and the possibility of remote sensing using optical fibers.

When an analyte is fluorescent, *direct fluorometric detection* is possible by means of a spectrofluorometer operating at appropriate excitation and observation wavelengths. This is the case for aromatic hydrocarbons (e.g. in crude oils), proteins (e.g. in blood serum, in cow milk), some drugs (e.g. morphine), chlorophylls, etc. Numerous fields of applications have been reported: analysis of air and water pollutants, oils, foods, drugs; monitoring of industrial processes; monitoring of species of clinical relevance; criminology; etc.

However, most ions and molecules are not fluorescent and the main *indirect methods* that are used in this case are the following:

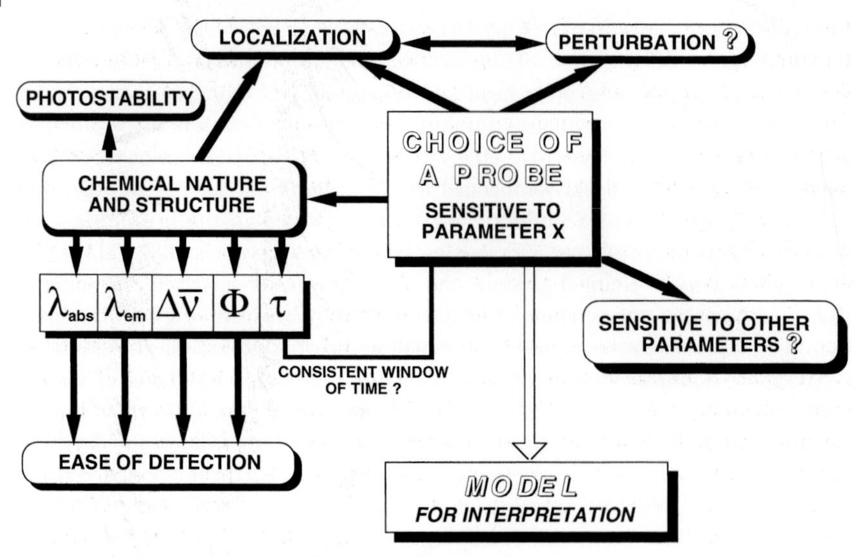

Fig. 1.6. Strategy for the choice of a fluorescent probe. $\Delta\bar{\nu}$, Φ, and τ are the Stokes shift, quantum yield and lifetime, respectively (see definitions in Chapter 3).

- *Derivatization*, i.e. reaction of the analyte with a reagent leading to a fluorescent compound, is often used in conjunction with liquid chromatography with fluorescence detection. This method is currently used in biochemistry and clinical chemistry.
- *Formation of a fluorescent complex* is the basis of most methods of ion and molecule recognition (see Chapter 9).
- *Fluorescence quenching* resulting from the collision of the analyte with a fluorescent compound (see Chapter 4). This method is particularly well suited to the detection of gases such as oxygen (dissolved in water or blood), SO_2, H_2S, ammonia, HCl, Cl_2, chlorocarbons, etc.

Finally, *fluorescence immunoassay* is a method of major importance for biochemical and biomedical applications.

1.6 Ultimate temporal and spatial resolution: femtoseconds, femtoliters, femtomoles and single-molecule detection

The ability of fluorescence to provide temporal information is of major importance. Great progress has been made since the first determination of an excited-state lifetime by Gaviola in 1926 using a phase fluorometer. A time resolution of a few tens of picosecond can easily be achieved in both pulse and phase fluorometries by using high repetition rate picosecond lasers and microchannel plate photomultipliers (see Chapter 6). Such a time resolution is limited by the response of the photomultiplier but not by the width of the laser pulse, which can be as short as 50–100 fs (1 femtosecond = 10^{-15} second) (e.g. with a titanium:sapphire laser). The time resolution can be reduced to a few picoseconds with a streak camera. To get an even better time resolution (100–200 fs), a more recent technique based on *fluorescence up-conversion* has been developed (see Chapter 11).

Regarding spatial resolution, fluorescence microscopy in confocal configuration or with two-photon excitation (see Chapter 11) allows the diffraction limit to be approached, which is approximately half the wavelength of the excitation light (0.2–0.3 μm for visible radiation) with the advantage of three-dimensional resolution. The excitation volume can be as small as 0.1 fL (femtoliter). Compared to conventional fluorometers, this represents a reduction by a factor of 10^{10} of the excitation volume. At high dilution ($\approx 10^{-9}$ M or less), fluorophores entering and leaving such a small volume cause changes in fluorescence intensity. Analysis of these fluctuations (which is the object of *fluorescence correlation spectroscopy*; see Chapter 11) in terms of autocorrelation function can provide information on translational diffusion, flow rates and molecular aggregation. Fluctuations can also be caused by chemical reactions or rotational diffusion. The typical lower limit concentration is ~1 fM (femtomol L^{-1}). The progress of these techniques allows us to study molecular interactions at the unsurpassed sensitivity of *single-molecule detection*.

The diffraction limit can be overcome by using a sub-wavelength light source and by placing the sample very close to this source (i.e. in the near field). The relevant domain is *near-field optics* (as opposed to far-field conventional optics), which has been applied in particular to fluorescence microscopy. This technique, called *near-field scanning optical microscopy* (NSOM), is an outstanding tool in physical, chemical and life sciences for probing the structure of matter or living systems. The resolution is higher than in confocal microscopy, with the additional capability of force mapping of the surface topography, and the advantage of reduced photobleaching. Single molecule detection is of course possible by this technique.

The first optical detection of a single molecule was reported in 1989 by Moerner and Kador, who detected a single pentacene molecule doped into a *p*-terphenyl crystal (at liquid helium temperature) using absorption with a double modulation technique. Fluorescence excitation spectroscopy on a single molecule was demonstrated for the first time by Orrit and Bernard in 1990. The detection of a single fluorescent molecule in solution was achieved not much later. Therefore, Schrödinger's statement (in 1952) has been outspaced by reality: '. . . we never experiment with just one electron or atom or molecule. In thought experiments we sometimes assume we do, this invariably entails ridiculous consequences.'

Single molecule detection offers the possibility of selecting, trapping, sorting, picking, and even manipulating molecules, especially biological macromolecules. Detection and spectroscopy of individual fluorescent molecules thus provide new tools not only in basic research but also in biotechnology and pharmaceutical industries (e.g. drug screening).

1.7
Bibliography

Pioneering books

Bowen E. J. and Wokes F. **(1953)** *Fluorescence of Solutions*, Longmans, Green and Co., London.

Curie M. **(1934)** *Luminescence des Corps Solides*, Presses Universitaires de France, Paris.

Curie M. **(1946)** *Fluorescence et Phosphorescence*, Hermann, Paris.

Dake H. C. and De Ment J. **(1941)** *Fluorescent Light and its Applications*, Chemical Publishing Co., New York.

De Ment J. **(1945)** *Fluorochemistry. A comprehensive Study Embracing the Theory and Applications of Luminescence and Radiation in Physicochemical Science*, Chemical Publishing Co., New York.

Förster T. **(1951)** *Fluoreszenz organischer Verbindungen*, Vandenhoeck and Ruprecht, Göttingen.

Hirschlaff E. **(1939)** *Fluorescence and Phosphorescence*, Chemical Publishing Co., New York.

Perrin F. **(1931)** Fluorescence. Durée Elémentaire d'Emission Lumineuse, Hermann, Paris.

Pringsheim P. **(1921, 1923, 1928)** *Fluorescenz und Phosphorescenz im Lichte der neueren Atomtheorie*, Verlag von Julius Springer, Berlin.

Pringsheim P. **(1949)** *Fluorescence and Phosphorescence*, Interscience, New York.

Pringsheim P. and Vogel M. **(1943)** Luminescence of Liquids and Solids and its Practical Applications, Interscience, New York.

Monographs and edited books after 1960

Baeyens W. R. G, de Keukeleire D. and Korkidis K. (Eds) **(1991)** *Luminescence Techniques in Chemical and Biochemical Analysis*, Marcel Dekker, New York.

Becker R. S. **(1969)** *Theory and Interpretation of Fluorescence and Phosphorescence*, Wiley Interscience, New York.

Beddard G. S. and West M. A. (Eds) **(1981)** *Fluorescent Probes*, Academic Press, London.

Berlman I. B. **(1965, 1971)** *Handbook of Fluorescence Spectra of Aromatic Molecules*, Academic Press, New York.

Birks J. B. **(1970)** *Photophysics of Aromatic Molecules*, Wiley-Interscience, London.

Birks J. B. (Ed.) **(1975)** *Organic Molecular Photophysics*, Vols. 1 and 2, John Wiley & Sons, London.

Bowen E. J. (Ed.) **(1968)** *Luminescence in Chemistry*, Van Nostrand, London.

Chen R. F. and Edelhoch H. (Eds) **(1975, 1976)** *Biochemical Fluorescence. Concepts*, Vols. 1 and 2, Marcel Dekker, New York.

Cundall R. B. and Dale R. E. (Eds) **(1983)** *Time-Resolved Fluorescence Spectroscopy in Biochemistry and Biology*, Plenum Press, New York.

Czarnik A. W. (Ed.) **(1992)** *Fluorescence Chemosensors for Ion and Molecule Recognition*, American Chemical Society, Washington.

Demas J. N. **(1983)** *Excited State Lifetime Measurement*, Academic Press, New York.

Desvergne J.-P. and Czarnik A. W. (Eds) **(1997)** *Chemosensors of Ion and Molecule Recognition*, Kluwer Academic Publishers, Dordrecht.

Dewey G. (Ed.) **(1991)** *Biophysical and Biochemical Aspects of Fluorescence Spectroscopy*, Plenum Press, New York.

Galanin M. D. **(1996)** *Luminescence of Molecules and Crystals*, Cambridge International Science Publishing, Cambridge.

Goldberg M. C. (Ed.) **(1989)** *Luminescence Applications in Biological, Chemical, Environmental, and Hydrological Sciences*, American Chemical Society, Washington.

Guilbault G. (Ed.) **(1973, 1990)** *Practical Fluorescence*, Marcel Dekker, New York (1st edn: 1973; 2nd edn: 1990).

Guillet J. E. **(1985)** *Polymer Photophysics and Photochemistry*, Cambridge University Press, Cambridge, UK.

Hercules D. M. (Ed.) **(1966)** *Fluorescence and Phosphorescence Analysis*, Wiley Interscience, New York.

Jameson D. M. and Reinhart G. D. (Eds)

(1989) *Fluorescent Biomolecules*, Plenum Press, New York.

Krasovitskii B. M. and Bolotin B. M. **(1988)** *Organic Luminescent Materials*, VCH, Weinheim.

Lakowicz J. R. **(1983, 1999)** *Principles of Fluorescence Spectroscopy*, Plenum Press, New York (1st edn, 1983; 2nd edn, 1999).

Lakowicz J. R. (Ed.) *Topics in Fluorescence Spectroscopy*, Plenum Press, New York. Vol. 1: Techniques (1991); Vol. 2: Principles (1991); Vol. 3: Biochemical Applications (1992); Vol. 4: Probe Design and Chemical Sensing (1994); Vol. 5: Non-Linear and Two-Photon-Induced Fluorescence (1997); Vol. 6: Protein Fluorescence (2000).

Lansing Taylor D., Waggoner A. S., Lanni F., Murphy R. F. and Birge R. R. (Eds) **(1986)** *Applications of Fluorescence in the Biomedical Sciences*, Alan R. Liss, New York.

Mielenz K. D. (Ed.) **(1982)** *Measurement of Photoluminescence*, Academic Press, Washington.

Mielenz K. D., Velapodi R. A. and Mavrodineanu R. (Eds) **(1977)** *Standardization in Spectrometry and Luminescence Measurement*, U.S. Dept. Commerce, New York.

Miller J. N. (Ed.) **(1981)** *Standards in Fluorescence Spectrometry*, Ultraviolet Spectrometry Group, Chappman and Hall, London.

Murov S. L., Carmichael I. and Hug G. L. **(1993)** *Handbook of Photochemistry*, 2nd edn, Marcel Dekker, New York.

O'Connor D. V. and Phillips D. **(1984)** *Time-Correlated Single Photon Counting*, Academic Press, London.

Parker C. A. **(1968)** *Photoluminescence of Solutions*, Elsevier, Amsterdam.

Pesce A. J., Rosen C.-G. and Pasby T. L. (Eds) **(1971)** *Fluorescence Spectroscopy*, Marcel Dekker, New York.

Rabek J. F. (Ed.) **(1990)** *Photochemistry and Photophysics*, CRC Press, Boca Raton.

Ramamurthy V. (Ed.) **(1991)** *Photochemistry in Organized and Constrained Media*, VCH, Weinheim.

Rendell D. and Mowthorpe D. (Eds) **(1987)** *Fluorescence and Phosphorescence*, John Wiley and Sons, Chichester.

Rettig W., Strehmel B., Schrader S. and Seifert H. (Eds) **(1999)** *Applied Fluorescence in Chemistry, Biology and Medicine*, Springer, Berlin.

Schenk G. H. **(1973)** *Absorption of Light and Ultraviolet Radiation. Fluorescence and Phosphorescence Emission*, Allyn and Bacon, Boston.

Schulman S. G. **(1977)** *Fluorescence and Phosphorescence Spectroscopy: Physicochemical Principles and Practice*, Pergamon Press, Oxford.

Schulman S. G. (Ed.) *Molecular Luminescence Spectroscopy*, John Wiley and Sons, New York, Part 1 (1985); Part 2 (1988); Part 3 (1993).

Slavik J. (Ed.) **(1996)** *Fluorescence Microscopy and Fluorescent Probes*, Plenum Press, New York.

Suppan P. **(1994)** *Chemistry and Light*, Royal Society of Chemistry, Cambridge.

Turro N. J. **(1978)** *Modern Molecular Photochemistry*, The Benjamin/Cummings Publishing Co., Menlo Park.

Udenfriend S., *Fluorescence Assay in Biology and Medecine*, Academic Press, New York, Vol. 1 (1962); Vol. 2 (1971).

Valeur B. and Brochon J. C. (Eds) **(2001)** *New Trends in Fluorescence Spectroscopy. Applications to Chemical and Life Sciences*, Springer-Verlag, Berlin.

Wolfbeis O. S. (Ed.) **(1993)** Fluorescence Spectroscopy. New Methods and Applications, Springer-Verlag, Berlin.

2
Absorption of UV–visible light

La lumière (...) donne la couleur et l'éclat à toutes les productions de la nature et de l'art; elle multiplie l'univers en le peignant dans les yeux de tout ce qui respire.

[*Light (...) gives color and brilliance to all works of nature and of art; it multiplies the universe by painting it in the eyes of all that breathe.*]

Abbé Nollet, 1783

The aim of this chapter is to recall the basic principles of light absorption by molecules. The reader is referred to more specialized books for further details.

2.1
Types of electronic transitions in polyatomic molecules

An electronic transition consists of the promotion of an electron from an orbital of a molecule in the ground state to an unoccupied orbital by absorption of a photon. The molecule is then said to be in an excited state. Let us recall first the various types of molecular orbitals.

A σ orbital can be formed either from two s atomic orbitals, or from one s and one p atomic orbital, or from two p atomic orbitals having a collinear axis of symmetry. The bond formed in this way is called a σ bond. A π orbital is formed from two p atomic orbitals overlapping laterally. The resulting bond is called a π bond. For example in ethylene ($CH_2{=}CH_2$), the two carbon atoms are linked by one σ and one π bond. Absorption of a photon of appropriate energy can promote one of the π electrons to an antibonding orbital denoted by π^*. The transition is then called $\pi \rightarrow \pi^*$. The promotion of a σ electron requires a much higher energy (absorption in the far UV) and will not be considered here.

A molecule may also possess non-bonding electrons located on heteroatoms such as oxygen or nitrogen. The corresponding molecular orbitals are called n or-

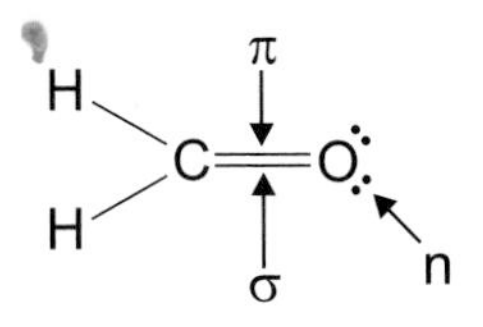

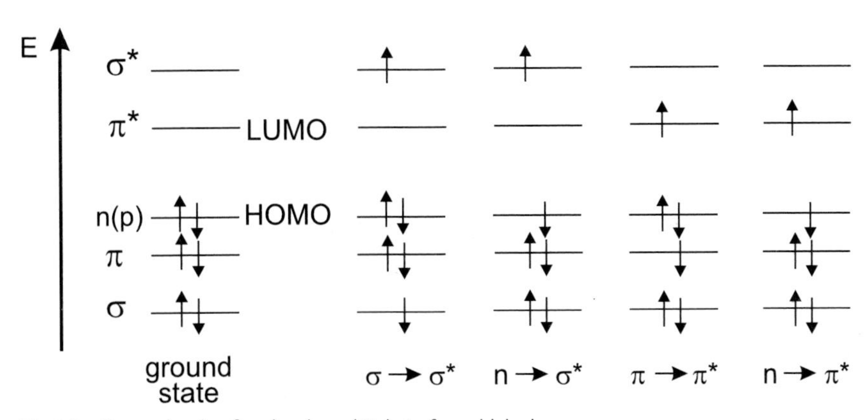

Fig. 2.1. Energy levels of molecular orbitals in formaldehyde (HOMO: Highest Occupied Molecular Orbitals; LUMO: Lowest Unoccupied Molecular Orbitals) and possible electronic transitions.

bitals. Promotion of a non-bonding electron to an antibonding orbital is possible and the associated transition is denoted by $n \rightarrow \pi^*$.

The energy of these electronic transitions is generally in the following order:

$$n \rightarrow \pi^* < \pi \rightarrow \pi^* < n \rightarrow \sigma^* < \sigma \rightarrow \pi^* < \sigma \rightarrow \sigma^*$$

To illustrate these energy levels, Figure 2.1 shows formaldehyde as an example, with all the possible transitions. The $n \rightarrow \pi^*$ transition deserves further attention: upon excitation, an electron is removed from the oxygen atom and goes into the π^* orbital localized half on the carbon atom and half on the oxygen atom. The n–π^* excited state thus has a charge transfer character, as shown by an increase in the dipole moment of about 2 D with respect to the ground state dipole moment of C=O (3 D).

In absorption and fluorescence spectroscopy, two important types of orbitals are considered: the Highest Occupied Molecular Orbitals (HOMO) and the Lowest Unoccupied Molecular Orbitals (LUMO). Both of these refer to the ground state of the molecule. For instance, in formaldehyde, the HOMO is the n orbital and the LUMO is the π^* orbital (see Figure 2.1).

When one of the two electrons of opposite spins (belonging to a molecular orbital of a molecule in the ground state) is promoted to a molecular orbital of higher energy, its spin is in principle unchanged (Section 2.3) so that the total spin quantum number ($S = \Sigma s_i$, with $s_i = +\frac{1}{2}$ or $-\frac{1}{2}$) remains equal to zero. Because the multi-

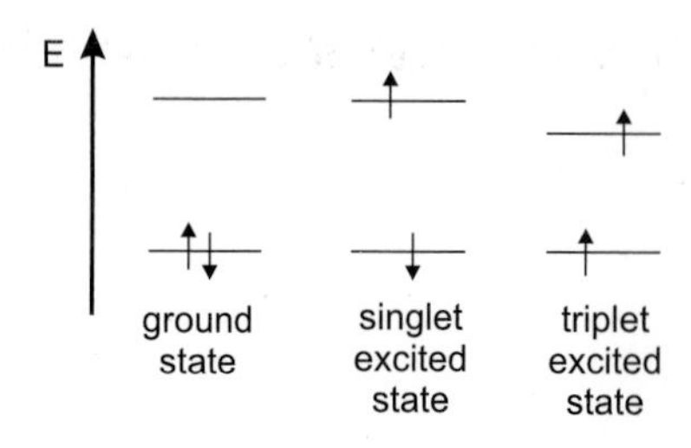

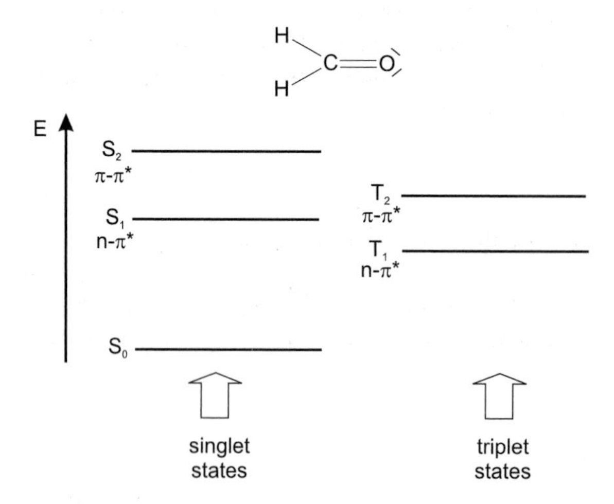

Fig. 2.2. Distinction between singlet and triplet states, using formaldehyde as an example.

plicities of both the ground and excited states ($M = 2S + 1$) is equal to 1, both are called *singlet state* (usually denoted S_0 for the ground state, and $S_1, S_2, \ldots$ for the excited states) (Figure 2.2)[1]. The corresponding transition is called a singlet–singlet transition. It will be shown later that a molecule in a singlet excited state may undergo conversion into a state where the promoted electron has changed its spin; because there are then two electrons with parallel spins, the total spin quantum number is 1 and the multiplicity is 3. Such a state is called a *triplet state* because it corresponds to three states of equal energy. According to Hund's Rule, the triplet state has a lower energy than that of the singlet state of the same configuration.

In a molecule such as formaldehyde, the bonding and non-bonding orbitals are localized (like the bonds) between pairs of atoms. Such a picture of localized orbitals is valid for the σ orbitals of single bonds and for the π orbitals of isolated double bonds, but it is no longer adequate in the case of alternate single and double carbon–carbon bonds (in so-called conjugated systems). In fact, overlap of the π orbitals allows the electrons to be delocalized over the whole system (resonance

1) In some cases, the ground state is not a singlet state, e.g. dioxygen, anion and cation radicals of aromatic molecules.

effect). Butadiene and benzene are the simplest cases of linear and cyclic conjugated systems, respectively.

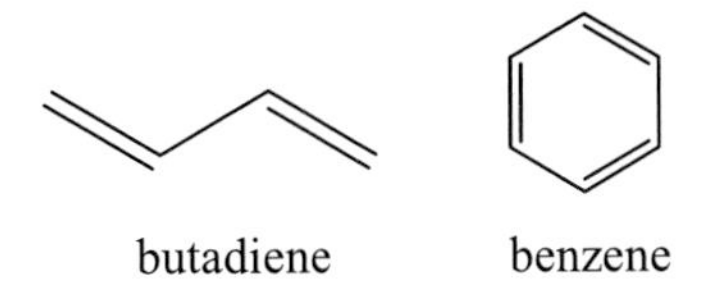

butadiene benzene

Because there is no overlap between the σ and π orbitals, the π electron system can be considered as independent of the σ bonds. It is worth remembering that the greater the extent of the π electron system, the lower the energy of the low-lying $\pi \rightarrow \pi^*$ transition, and consequently, the larger the wavelength of the corresponding absorption band. This rule applies to linear conjugated systems (polyenes) and cyclic conjugated systems (aromatic molecules).

2.2 Probability of transitions. The Beer–Lambert Law. Oscillator strength

Experimentally, the efficiency of light absorption at a wavelength λ by an absorbing medium is characterized by the *absorbance* $A(\lambda)$ or the *transmittance* $T(\lambda)$, defined as

$$A(\lambda) = \log \frac{I_\lambda^0}{I_\lambda} = -\log T(\lambda) \qquad T(\lambda) = \frac{I_\lambda}{I_\lambda^0} \tag{2.1}$$

where I_λ^0 and I_λ are the light intensities of the beams entering and leaving the absorbing medium, respectively[2].

In many cases, the absorbance of a sample follows the *Beer–Lambert Law*

$$A(\lambda) = \log \frac{I_\lambda^0}{I_\lambda} = \varepsilon(\lambda) l c \tag{2.2}$$

where $\varepsilon(\lambda)$ is the *molar (decadic) absorption coefficient* (commonly expressed in L mol^{-1} cm^{-1}), c is the concentration (in mol L^{-1}) of absorbing species and l is the absorption path length (thickness of the absorbing medium) (in cm). Derivation of the Beer–Lambert Law is given in Box 2.1.

2) The term *intensity* is commonly used but is imprecise. According to IUPAC recommendations (see *Pure & Appl. Chem.* **68**, 2223–2286 (1996)), this term should be replaced by the *spectral radiant power* P_λ, i.e. the radiant power at wavelength λ per unit wavelength interval. Radiant power is synonymous with radiant (energy) flux. The SI unit for radiant power is J s^{-1} = W; the SI unit for spectral radiant power is W m^{-1}, but a commonly used unit is W nm^{-1}.

Failure to obey the linear dependence of the absorbance on concentration, according to the Beer–Lambert Law, may be due to aggregate formation at high concentrations or to the presence of other absorbing species.

Various terms for characterizing light absorption can be found in the literature. The recommendations of the International Union of Pure and Applied Chemistry (IUPAC)[3] are very helpful here. In particular, the term optical density, synonymous with absorbance, is not recommended. Also, the term molar absorption coefficient should be used instead of molar extinction coefficient.

The (decadic) *absorption coefficient* $a(\lambda)$ is the absorbance divided by the optical path length, l:

$$a(\lambda) = \frac{A(\lambda)}{l} = \frac{1}{l}\log\frac{I_\lambda^0}{I_\lambda} \quad \text{or} \quad I_\lambda = I_\lambda^0 10^{-a(\lambda)l} \tag{2.3}$$

Physicists usually prefer to use the *Napierian absorption coefficient* $\alpha(\lambda)$

$$\alpha(\lambda) = \frac{1}{l}\ln\frac{I_\lambda^0}{I_\lambda} = a(\lambda)\ln 10 \quad \text{or} \quad I_\lambda = I_\lambda^0 e^{-\alpha(\lambda)l} \tag{2.4}$$

Because absorbance is a dimensionless quantity, the SI unit for a and α is m^{-1}, but cm^{-1} is often used.

Finally, the *molecular absorption cross-section* $\sigma(\lambda)$ characterizes the photon-capture area of a molecule. Operationally, it can be calculated as the (Napierian) absorption coefficient divided by the number N of molecular entities contained in a unit volume of the absorbing medium along the light path:

$$\sigma(\lambda) = \frac{\alpha(\lambda)}{N} \tag{2.5}$$

The relationship between the molecular absorption cross-section and the molar absorption coefficient is described in Box 2.1.

The molar absorption coefficient, $\varepsilon(\lambda)$, expresses the ability of a molecule to absorb light in a given solvent. In the classical theory, molecular absorption of light can be described by considering the molecule as an oscillating dipole, which allows us to introduce a quantity called the *oscillator strength*, which is directly related to the integral of the absorption band as follows:

$$f = 2303\frac{mc_0^2}{N_a \pi e^2 n}\int \varepsilon(\bar{\nu})\,d\bar{\nu} = \frac{4.32 \times 10^{-9}}{n}\int \varepsilon(\bar{\nu})\,d\bar{\nu} \tag{2.6}$$

where m and e are the mass and the charge of an electron, respectively, c_0 is the speed of light, n is the index of refraction, and $\bar{\nu}$ is the wavenumber (in cm^{-1}). f is a dimensionless quantity and values of f are normalized so that its maximum

3) See the *Glossary of Terms Used in Photochemistry* published in *Pure & Appl. Chem.* **68**, 2223–2286 (1996).

Box 2.1 Derivation of the Beer–Lambert Law and comments on its practical use

Derivation of the Beer–Lambert Law from considerations at a molecular scale is more interesting than the classical derivation (stating that the fraction of light absorbed by a thin layer of the solution is proportional to the number of absorbing molecules). Each molecule has an associated photon-capture area, called the molecular absorption cross-section σ, that depends on the wavelength. A thin layer of thickness dl contains dN molecules. dN is given by

$$dN = N_a c S \, dl$$

where S is the cross-section of the incident beam, c is the concentration of the solution and N_a is Avogadro's number. The total absorption cross-section of the thin layer is the sum of all molecular cross-sections, i.e. $\sigma \, dN$. The probability of photon capture is thus $\sigma \, dN/S$ and is simply equal to the fraction of light $(-dI/I)$ absorbed by the thin layer:

$$-\frac{dI}{I} = \frac{\sigma \, dN}{S} = N_a \sigma c \, dl$$

Integration leads to

$$\ln \frac{I_0}{I} = N_a \sigma c l \quad \text{or} \quad \log \frac{I_0}{I} = \frac{1}{2.303} N_a \sigma c l$$

where l is the thickness of the solution. This equation is formally identical to Eq. (2.2) with $\varepsilon = N_a\sigma/2.303$.

The molecular absorption cross-section can then be calculated from the experimental value of ε using the following relation:

$$\sigma = \frac{2.303\varepsilon}{N_a} = 3.825 \times 10^{-19}\varepsilon \text{ (in cm}^2\text{)}$$

Practical use of the Beer–Lambert law deserves attention. In general, the sample is a cuvette containing a solution. The absorbance must be characteristic of the absorbing species only. Therefore, it is important to note that in the Beer–Lambert Law ($A(\lambda) = \log I_0/I = \varepsilon(\lambda)lc$), I_0 is the intensity of the beam entering the solution but not that of the incident beam I_i on the cuvette, and I is the intensity of the beam leaving the solution but not that of the beam I_S leaving the cuvette (see Figure B2.1). In fact, there are some reflections on the cuvette walls and these walls may also absorb light slightly. Moreover, the solvent is assumed to have no contribution, but it may also be partially responsible for a decrease in intensity because of scattering and possible absorption. The contributions of the

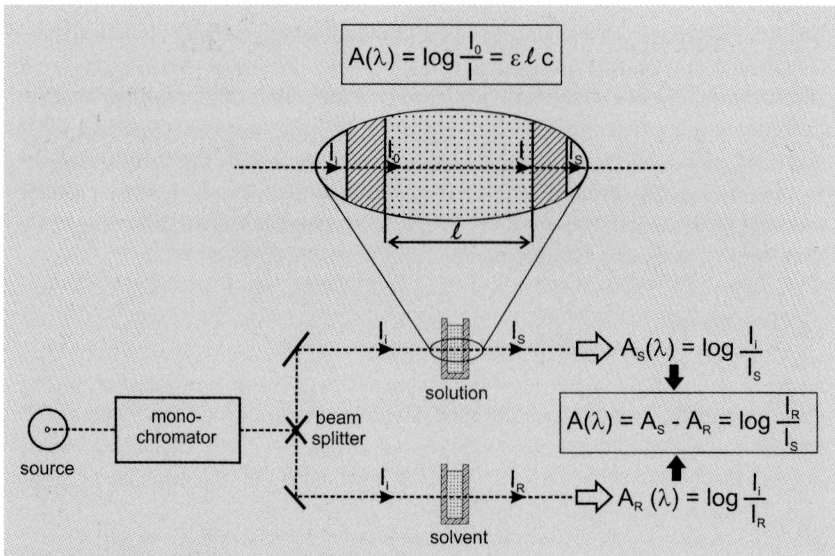

Fig. B2.1. Practical aspects of absorbance measurements.

cuvette walls and the solvent can be taken into account in the following way. The absorbance of the whole sample (including the cuvette walls) is defined as

$$A_{\mathrm{S}}(\lambda) = \log \frac{I_{\mathrm{i}}}{I_{\mathrm{S}}}$$

If the solution is replaced by the solvent, the intensity of the transmitted light is I_{R} and the absorbance becomes

$$A_{\mathrm{R}}(\lambda) = \log \frac{I_{\mathrm{i}}}{I_{\mathrm{R}}}$$

The true absorbance of the solution is then given by

$$A(\lambda) = A_{\mathrm{S}}(\lambda) - A_{\mathrm{R}}(\lambda) = \log \frac{I_{\mathrm{R}}}{I_{\mathrm{S}}}$$

As shown in Figure B2.1, double-beam spectrophotometers automatically record the true absorbance by measuring $\log(I_{\mathrm{R}}/I_{\mathrm{S}})$, thanks to a double compartment containing two cuvettes, one filled with the solution and one filled with the solvent. Because the two cuvettes are never perfectly identical, the baseline of the instrument is first recorded (with both cuvettes filled with the solvent) and stored. Then, the solvent of the sample cuvette is replaced by the solution, and the true absorption spectrum is recorded.

Tab. 2.1. Examples of molar absorption coefficients, ε (at the wavelength corresponding to the maximum of the absorption band of lower energy). Only approximate values are given, because the value of ε slightly depends on the solvent

Compound	ε/L mol^{-1} cm^{-1}	*Compound*	ε/L mol^{-1} cm^{-1}
Benzene	≈200	Acridine	≈12 000
Phenol	≈2 000	Biphenyl	≈16 000
Carbazole	≈4 200	Bianthryl	≈24 000
1-Naphthol	≈5 400	Acridine orange	≈30 000
Indole	≈5 500	Perylene	≈34 000
Fluorene	≈9 000	Eosin Y	≈90 000
Anthracene	≈10 000	Rhodamine B	≈105 000
Quinine sulfate	≈10 000		

value is 1. For $n \rightarrow \pi^*$ transitions, the values of ε are in the order of a few hundreds or less and those of f are no greater than $\sim 10^{-3}$. For $\pi \rightarrow \pi^*$ transitions, the values of ε and f are in principle much higher (except for symmetry-forbidden transitions): f is close to 1 for some compounds, which corresponds to values of ε that are of the order of 10^5. Table 2.1 gives some examples of values of ε.

In the quantum mechanical approach, a *transition moment* is introduced for characterizing the transition between an initial state and a final state (see Box 2.2). The transition moment represents the transient dipole resulting from the displacement of charges during the transition; therefore, it is not strictly a dipole moment.

The concept of transition moment is of major importance for all experiments carried out with polarized light (in particular for fluorescence polarization experiments, see Chapter 5). In most cases, the transition moment can be drawn as a vector in the coordinate system defined by the location of the nuclei of the atoms[4]; therefore, the molecules whose absorption transition moments are parallel to the electric vector of a linearly polarized incident light are preferentially excited. The probability of excitation is proportional to the square of the scalar product of the transition moment and the electric vector. This probability is thus maximum when the two vectors are parallel and zero when they are perpendicular.

For $\pi \rightarrow \pi^*$ transitions of aromatic hydrocarbons, the absorption transition moments are in the plane of the molecule. The direction with respect to the molecular axis depends on the electronic state attained on excitation. For example, in naphthalene and anthracene, the transition moment is oriented along the short axis for the $S_0 \rightarrow S_1$ transition and along the long axis for the $S_0 \rightarrow S_2$ transition. Various examples are shown in Figure 2.3.

4) Note that this is not true for molecules having a particular symmetry, such as benzene (D_{6h}), triphenylene (D_{3h}) and C_{60} (I_h).

Box 2.2 Einstein coefficients. Transition moment. Oscillator strength

Let us consider a molecule and two of its energy levels E_1 and E_2. The Einstein coefficients are defined as follows (Scheme B2.2): B_{12} is the induced absorption coefficient, B_{21} is the induced emission coefficient and A_{21} is the spontaneous emission coefficient.

Scheme B2.2

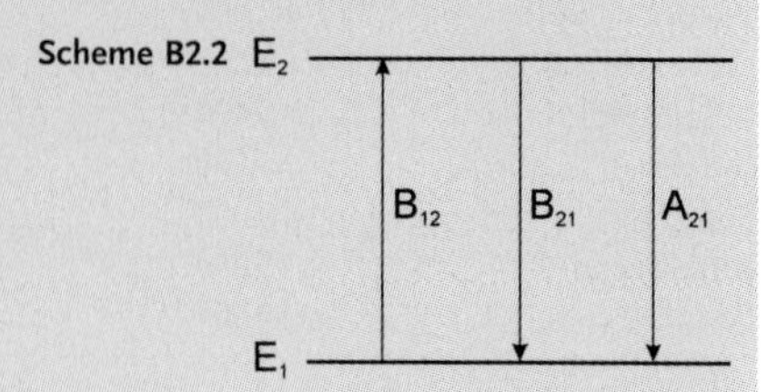

The coefficients for spontaneous and induced emissions will be discussed in Chapter 3 (see Box 3.2).

The rate at which energy is taken up from the incident light is given by

$$\frac{\mathrm{d}P_{12}}{\mathrm{d}t} = B_{12}\rho(\nu)$$

where $\rho(\nu)$ is the energy density incident on the sample at frequency ν. B_{12} appears as the transition rate per unit energy density of the radiation. In the quantum mechanical theory, it is shown that

$$B_{12} = \frac{2}{3}\frac{\pi}{\hbar^2}|\langle\Psi_1|\mathbf{M}|\Psi_2\rangle|^2 \quad (\hbar = h/2\pi)$$

In this expression, $\langle\Psi_1|\mathbf{M}|\Psi_2\rangle = \int \Psi_1\mathbf{M}\Psi_2\,\mathrm{d}\tau$, where Ψ_1 and Ψ_2 are the wavefunctions of states 1 and 2, respectively, $\mathrm{d}\tau$ is over the whole configuration space of the $3N$ coordinates, $\mathbf{M}$ is the dipole moment operator ($\mathbf{M} = \Sigma e r_j$, where r_j is the vector joining the electron j to the origin of a coordinate system linked to the molecule).

It should be noted that the dipole moment of the molecule in its ground state is $\langle\Psi_1|\mathbf{M}|\Psi_1\rangle$. In contrast, the term $\langle\Psi_1|\mathbf{M}|\Psi_2\rangle$ is not strictly a dipole moment because there is a displacement of charges during the transition; it is called *transition moment* $\mathbf{M}_{12}$ that characterizes the transition.

It is interesting that there is a relation between the oscillator strength f (given by Eq. 2.6) and the square of the transition moment integral, which bridges the gap between the classical and quantum mechanical approaches:

$$f = \frac{8\pi^2 m\nu}{3he^2}|\langle\Psi_1|\mathbf{M}|\Psi_2\rangle|^2$$

Note also that the oscillator strength can be related to the coefficient B_{12}:

$$f = \frac{mh\nu}{\pi e^2}B_{12}$$

For more details, see Birks (1970), pp. 48–52.

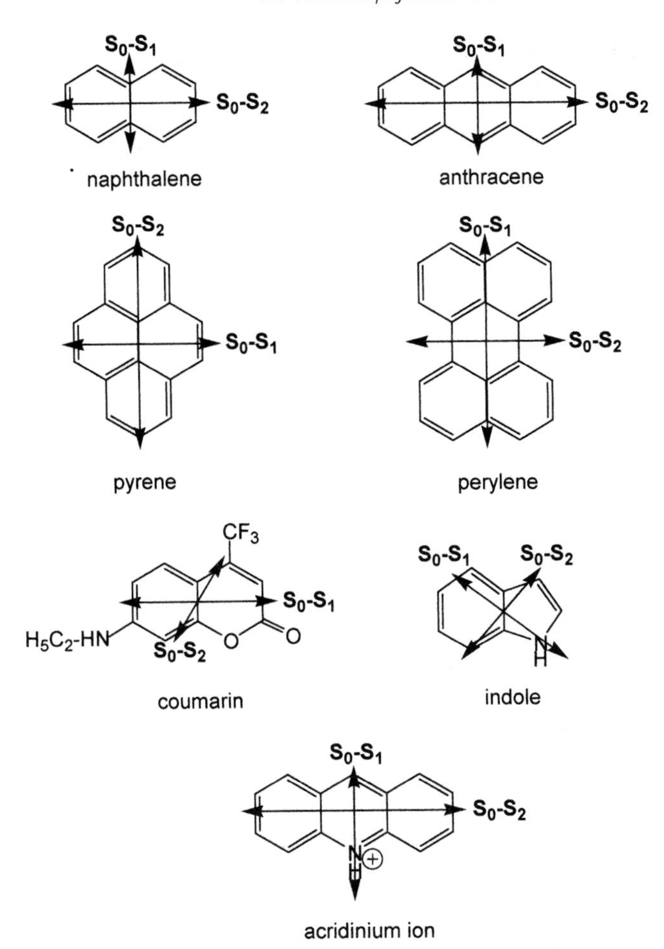

Fig. 2.3. Examples of molecules with their absorption transition moments.

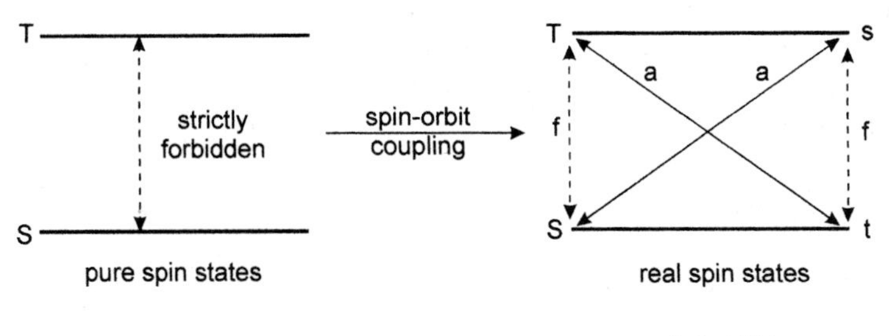

Scheme 2.1

2.3 Selection rules

There are two major selection rules for absorption transitions:

1. **Spin-forbidden transitions.** Transitions between states of different multiplicities are forbidden, i.e. singlet–singlet and triplet–triplet transitions are allowed, but singlet–triplet and triplet–singlet transitions are forbidden. However, there is always a weak interaction between the wavefunctions of different multiplicities via spin–orbit coupling[5]. As a result, a wavefunction for a singlet (or triplet) state always contains a small fraction of a triplet (or singlet) wavefunction $\Psi = \alpha^1\Psi + \beta^3\Psi$; this leads to a small but non-negligible value of the intensity integral during a transition between a singlet state and a triplet state or vice versa (see Scheme 2.1). In spite of their very small molar absorption coefficients, such transitions can be effectively observed.

   Intersystem crossing (i.e. crossing from the first singlet excited state S_1 to the first triplet state T_1) is possible thanks to spin–orbit coupling. The efficiency of this coupling varies with the fourth power of the atomic number, which explains why intersystem crossing is favored by the presence of a heavy atom. Fluorescence quenching by internal heavy atom effect (see Chapter 3) or external heavy atom effect (see Chapter 4) can be explained in this way.
2. **Symmetry-forbidden transitions.** A transition can be forbidden for symmetry reasons. Detailed considerations of symmetry using group theory, and its consequences on transition probabilities, are beyond the scope of this book. It is important to note that a symmetry-forbidden transition can nevertheless be observed because the molecular vibrations cause some departure from perfect symmetry (vibronic coupling). The molar absorption coefficients of these transitions are very small and the corresponding absorption bands exhibit well-defined vibronic bands. This is the case with most $n \rightarrow \pi^*$ transitions in solvents that cannot form hydrogen bonds ($\varepsilon \approx$ 100–1000 L mol^{-1} cm^{-1}).

2.4 The Franck–Condon principle

According to the Born–Oppenheimer approximation, the motions of electrons are much more rapid than those of the nuclei (i.e. the molecular vibrations). Promotion of an electron to an antibonding molecular orbital upon excitation takes about 10^{-15} s, which is very quick compared to the characteristic time for molecular vi-

5) Spin–orbit coupling can be understood in a primitive way by considering the motion of an electron in a Bohr-like orbit. The rotation around the nucleus generates a magnetic moment; moreover, the electron spins about an axis of its own, which generates another magnetic moment. Spin–orbit coupling results from the interaction between these two magnets.

Box 2.3 Classical and quantum mechanical description of the Franck–Condon principle[a)]

Classically, the transition occurs when the distances between nuclei are equal to the equilibrium bond lengths of the molecule in the ground state. While the transition is in progress, the position of the nuclei is unchanged and they do not accelerate. Consequently, the transition terminates where the vertical line intersects the potential energy curve of the lowest excited state, i.e. at the turning point. As soon as the transition is complete, the excited molecule begins to vibrate at an energy corresponding to the intersection.

In the quantum mechanical description (in continuation of Box 2.2), the wavefunction can be described by the product of an electronic wavefunction Ψ and a vibrational wavefunction χ (the rotational contribution can be neglected), so that the probability of transition between an initial state defined by $\Psi_1\chi_a$ and a final state defined by $\Psi_2\chi_b$ is proportional to $|\langle\Psi_1\chi_a|\mathbf{M}|\Psi_2\chi_b\rangle|^2$. Because **M** only depends on the electron coordinates, this expression can be rewritten as the product of two terms $|\langle\Psi_1|\mathbf{M}|\Psi_2\rangle|^2|\langle\chi_a\,|\,\chi_b\rangle|^2$ where the second term is called the Franck–Condon factor. Qualitatively, the transition occurs from the lowest vibrational state of the ground state to the vibrational state of the excited state that it most resembles in terms of vibrational wavefunction.

a) Atkins P. W. and Friedman R. S. (1997) *Molecular Quantum Mechanics*, Oxford University Press, Oxford.

brations (10^{-10}–10^{-12} s). This observation is the basis of the Franck–Condon principle: an electronic transition is most likely to occur without changes in the positions of the nuclei in the molecular entity and its environment (Box 2.3). The resulting state is called a *Franck–Condon state*, and the transition is called *vertical transition*, as illustrated by the energy diagram of Figure 2.4 in which the potential energy curve as a function of the nuclear configuration (internuclear distance in the case of a diatomic molecule) is represented by a Morse function.

At room temperature, most of the molecules are in the lowest vibrational level of the ground state (according to the Boltzmann distribution; see Chapter 3, Box 3.1). In addition to the 'pure' electronic transition called the 0–0 transition, there are several vibronic transitions whose intensities depend on the relative position and shape of the potential energy curves (Figure 2.4).

The width of a band in the absorption spectrum of a chromophore located in a particular microenvironment is a result of two effects: homogeneous and inhomogeneous broadening. Homogeneous broadening is due to the existence of a continuous set of vibrational sublevels in each electronic state. Inhomogeneous broadening results from the fluctuations of the structure of the solvation shell

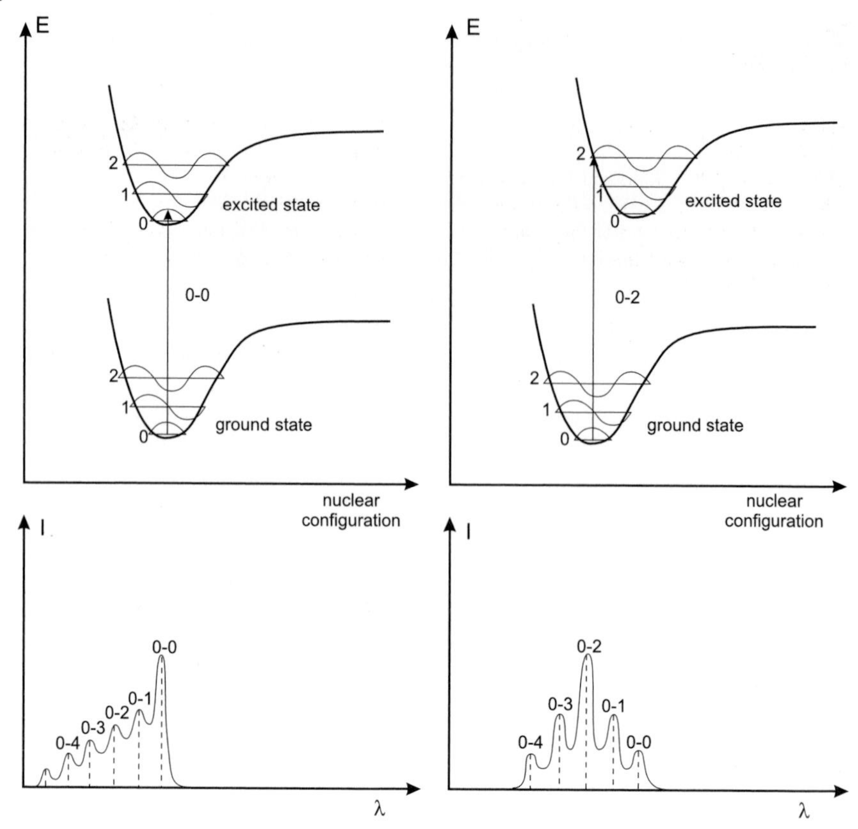

Fig. 2.4. Top: Potential energy diagrams with vertical transitions (Franck–Condon principle). Bottom: shape of the absorption bands; the vertical broken lines represent the absorption lines that are observed for a vapor, whereas broadening of the spectra is expected in solution (solid line).

surrounding the chromophore. Such broadening effects exist also for emission bands in fluorescence spectra and will be discussed in Section 3.5.1.

Shifts in absorption spectra due to the effect of substitution or a change in environment (e.g. solvent) will be discussed in Chapter 3, together with the effects on emission spectra. Note that a shift to longer wavelengths is called a *bathochromic shift* (informally referred to as a *red-shift*). A shift to shorter wavelengths is called a *hypsochromic shift* (informally referred to as a *blue-shift*). An increase in the molar absorption coefficient is called the *hyperchromic effect*, whereas the opposite is the *hypochromic effect*.

2.5 Bibliography

Birks J. B. **(1970)** *Photophysics of Aromatic Molecules*, Wiley, London.

Herzberg G. **(1966)** Molecular Spectra and Molecular Structure. III Electronic Spectra and Electronic Structure of Polyatomic Molecules, Van Nostrand Reinhold Company, New York.

Jaffé H. H. and Orchin M. **(1962)** *Theory and Applications of Ultraviolet Spectroscopy*, John Wiley & Sons, New York.

Turro N. J. **(1978)** *Modern Molecular Photochemistry*, Benjamin/Cummings, Menlo Park, CA.

3
Characteristics of fluorescence emission

> The world of fluorescence
> is a world of beautiful
> color. In the darkness all
> the ordinary colors of our
> daylight world disappear.
> Only the intensely
> glowing hues of
> fluorescent substances
> touched by the ultraviolet
> beam shine out with
> striking clarity.
>
> Sterling Gleason, 1960

This chapter describes the characteristics of the fluorescence emission of an excited molecule in solution. We do not consider here the photophysical processes involving interactions with other molecules (electron transfer, proton transfer, energy transfer, excimer or exciplex formation, etc.). These processes will be examined in Chapter 4.

3.1
Radiative and non-radiative transitions between electronic states

The Perrin–Jablonski diagram (Figure 3.1) is convenient for visualizing in a simple way the possible processes: photon absorption, internal conversion, fluorescence, intersystem crossing, phosphorescence, delayed fluorescence and triplet–triplet transitions. The singlet electronic states are denoted S_0 (fundamental electronic state), $S_1, S_2, \ldots$ and the triplet states, $T_1, T_2, \ldots$. Vibrational levels are associated with each electronic state. It is important to note that absorption is very fast ($\approx 10^{-15}$ s) with respect to all other processes (so that there is no concomitant

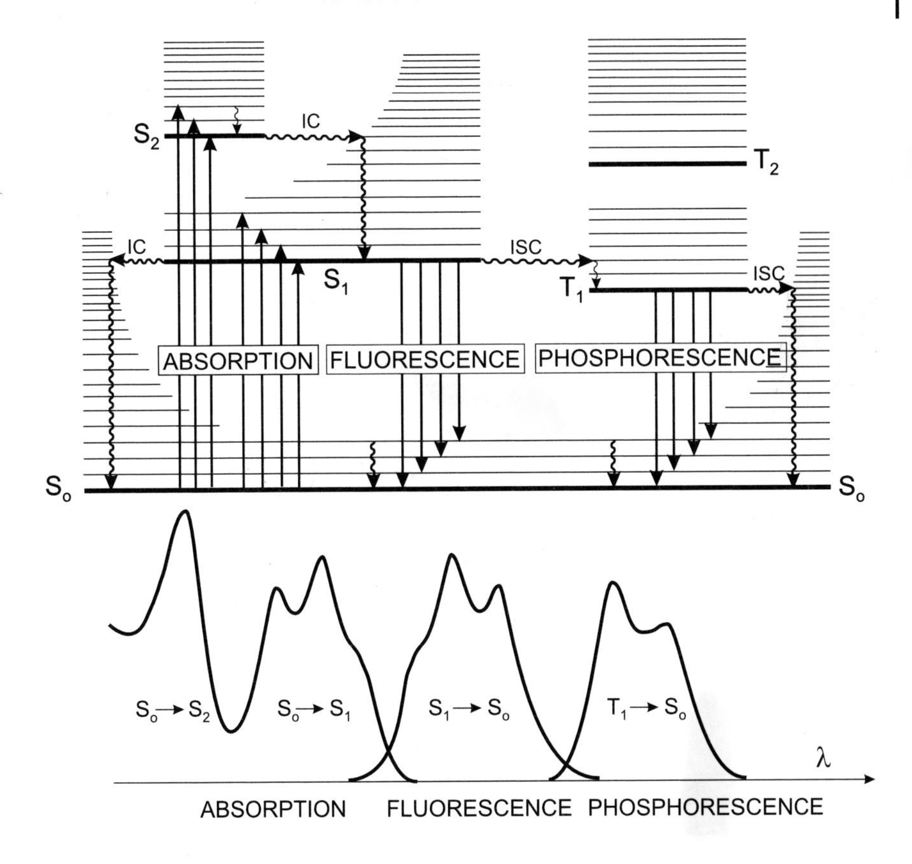

CHARACTERISTIC TIMES

absorption	10^{-15} s
vibrational relaxation	10^{-12}-10^{-10} s
lifetime of the excited state S_1	10^{-10}-10^{-7} s → fluorescence
intersystem crossing	10^{-10}-10^{-8} s
internal conversion	10^{-11}-10^{-9} s
lifetime of the excited state T_1	10^{-6} -1 s → phosphorescence

Fig. 3.1. Perrin–Jablonski diagram and illustration of the relative positions of absorption, fluorescence and phosphorescence spectra.

displacement of the nuclei according to the Franck–Condon principle; see Chapter 2).

The vertical arrows corresponding to absorption start from the 0 (lowest) vibrational energy level of S_0 because the majority of molecules are in this level at room temperature, as shown in Box 3.1. Absorption of a photon can bring a molecule to one of the vibrational levels of $S_1, S_2, \ldots$. The subsequent possible de-excitation processes will now be examined.

Box 3.1 Relative populations of molecules in the vibrational energy levels according to the Boltzmann Law

For some aromatic hydrocarbons such as naphthalene, anthracene and perylene, the absorption and fluorescence spectra exhibit vibrational bands. The energy spacing between the vibrational levels and the Franck–Condon factors (see Chapter 2) that determine the relative intensities of the vibronic bands are similar in S_0 and S_1 so that the emission spectrum often appears to be symmetrical to the absorption spectrum ('mirror image' rule), as illustrated in Figure B3.1.

The ratio of the numbers of molecules N_1 and N_0 in the 1 and 0 vibrational levels of energy E_1 and E_0, respectively, is given by the Boltzmann Law:

$$N_1/N_0 = \exp[-(E_1 - E_0)/kT]$$

where k is the Boltzmann constant ($k = 1.3807 \times 10^{-23}$ J K^{-1}) and T is the absolute temperature.

For instance, the absorption and emission spectra of anthracene show a wavenumber spacing of about 1400 cm^{-1}, i.e. an energy spacing of 2.8×10^{-20} J, between the 0 and 1 vibrational levels. In this case, the ratio N_1/N_0 at room temperature (298 K) is about 0.001.

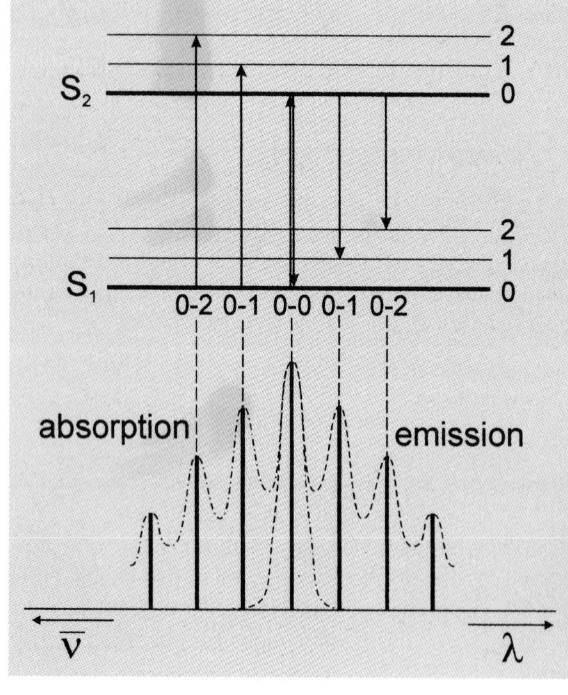

Fig. B3.1. Illustration of the vibrational bands in the absorption and fluorescence spectra of aromatic hydrocarbons. Broadening of the bands will be explained in Section 3.5.1.

However, it should be noted that most fluorescent molecules exhibit broad and structureless absorption and emission bands, which means that each electronic state consists of an almost continuous manifold of vibrational levels. If the energy difference between the 0 and 1 vibrational levels of S_0 (and S_1) is, for instance, only about 500 cm^{-1}, the ratio N_1/N_0 becomes about 0.09. Consequently, excitation can then occur from a vibrationally excited level of the S_0 state. This explains why the absorption spectrum can partially overlap the fluorescence spectrum (see Section 3.1.2).

Finally, in all cases, the energy gap between S_0 and S_1 is of course much larger than between the vibrational levels, so the probability of finding a molecule in S_1 at room temperature as a result of thermal energy is nearly zero ($E_{S1} - E_{S0} \approx 4 \times 10^{-19}$ J, compared with $kT \approx 4 \times 10^{-21}$ J.)

3.1.1 Internal conversion

Internal conversion is a non-radiative transition between two electronic states of the same spin multiplicity. In solution, this process is followed by a vibrational relaxation towards the lowest vibrational level of the final electronic state. The excess vibrational energy can be indeed transferred to the solvent during collisions of the excited molecule with the surrounding solvent molecules.

When a molecule is excited to an energy level higher than the lowest vibrational level of the first electronic state, vibrational relaxation (and internal conversion if the singlet excited state is higher than S_1) leads the excited molecule towards the 0 vibrational level of the S_1 singlet state with a time-scale of 10^{-13}–10^{-11} s.

From S_1, internal conversion to S_0 is possible but is less efficient than conversion from S_2 to S_1, because of the much larger energy gap between S_1 and S_0[1]. Therefore, internal conversion from S_1 to S_0 can compete with emission of photons (fluorescence) and intersystem crossing to the triplet state from which emission of photons (phosphorescence) can possibly be observed.

3.1.2 Fluorescence

Emission of photons accompanying the $S_1 \rightarrow S_0$ relaxation is called *fluorescence*. It should be emphasized that, apart from a few exceptions[2], fluorescence emission occurs from S_1 and therefore its characteristics (except polarization) do not depend

1) The smaller the energy gap between the initial and final electronic states, the larger the efficiency of internal conversion.

2) For instance, emission from S_2 in the case of azulene; simultaneous emission from S_1 and S_2 in the case of indole in some solvents.

on the excitation wavelength (provided of course that only one species exists in the ground state).

The 0–0 transition is usually the same for absorption and fluorescence. However, the fluorescence spectrum is located at higher wavelengths (lower energy) than the absorption spectrum because of the energy loss in the excited state due to vibrational relaxation (Figure 3.1). According to the Stokes Rule (an empirical observation pre-dating the Perrin–Jablonski diagram), the wavelength of a fluorescence emission should always be higher than that of absorption. However in most cases, the absorption spectrum partly overlaps the fluorescence spectrum, i.e. a fraction of light is emitted at shorter wavelengths than the absorbed light. Such an observation seems to be, at first sight, in contradiction to the principle of energy conservation. However, such an 'energy defect' is compensated for (as stated by Einstein for the first time) by the fact that at room temperature, a small fraction of molecules is in a vibrational level higher than level 0 (distribution among the energy levels fulfilling the Boltzmann Law; see Box 3.1) in the ground state as well as in the excited state. At low temperature, this departure from the Stokes Law should disappear.

In general, the differences between the vibrational levels are similar in the ground and excited states, so that the fluorescence spectrum often resembles the first absorption band ('mirror image' rule). The gap (expressed in wavenumbers) between the maximum of the first absorption band and the maximum of fluorescence is called the *Stokes shift*.

It should be noted that emission of a photon is as fast as absorption of a photon ($\approx 10^{-15}$ s). However, excited molecules stay in the S_1 state for a certain time (a few tens of picoseconds to a few hundreds of nanoseconds, depending on the type of molecule and the medium) before emitting a photon or undergoing other de-excitation processes (internal conversion, intersystem crossing). Thus, after excitation of a population of molecules by a very short pulse of light, the fluorescence intensity decreases exponentially with a characteristic time, reflecting the average lifetime of the molecules in the S_1 excited state (excited-state lifetime; see Section 3.2.1). Such an intensity decay is formally comparable with a radioactive decay that is also exponential, with a characteristic time, called the *radioactive period*, reflecting the average lifetime of a radioelement before disintegration.

The emission of fluorescence photons just described is a spontaneous process. Under certain conditions, stimulated emission can occur (e.g. dye lasers) (see Box 3.2).

3.1.3
Intersystem crossing and subsequent processes

A third possible de-excitation process from S_1 is intersystem crossing toward the T_1 triplet state followed by other processes, according to Scheme 3.1.

Box 3.2 Spontaneous and stimulated emissions

The Einstein coefficients characterize the probability of transition of a molecule between two energy levels E_1 and E_2 (Scheme B3.2). B_{12} is the *induced absorption coefficient* (see Chapter 2), B_{21} is the *induced emission coefficient* and A_{21} is the *spontaneous emission coefficient.* The emission-induced process $E_2 \to E_1$ occurs at exactly the same rate as the absorption-induced process $E_1 \to E_2$, so that $B_{12} = B_{21}$.

Scheme B3.2

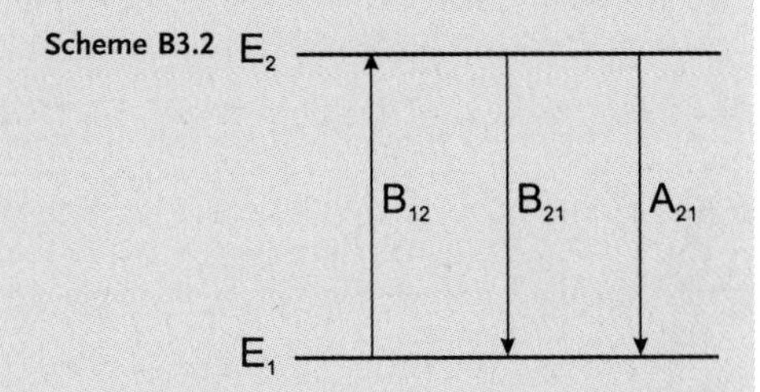

The number of molecules in states 1 and 2 is N_1 and N_2, respectively. These numbers must satisfy the Boltzmann Law:

$$\frac{N_1}{N_2} = \exp[-(E_1 - E_2)/kT] = \exp(+h\nu/kT)$$

where h is Planck's constant.

The rate of absorption from state 1 to state 2 is $N_1 B_{12}\rho(\nu)$, where $\rho(\nu)$ is the energy density incident on the sample at frequency ν. The rate of emission from state 2 to state 1 is $N_2[A_{21} + B_{21}\rho(\nu)]$. At equilibrium, these two rates are equal, hence

$$\frac{N_1}{N_2} = \frac{B_{21}\rho(\nu) + A_{21}}{B_{12}\rho(\nu)} = 1 + \frac{A_{21}}{B_{12}\rho(\nu)}$$

The radiation density $\rho(\nu)$ is given by Planck's black body radiation law:

$$\rho(\nu) = \frac{8\pi h\nu^3}{c^3 \exp(+h\nu/kT) - 1}$$

The three equations above lead to

$$A_{21} = \frac{8\pi h\nu^3}{c^3} B_{21}$$

Note that the ratio A_{21}/B_{21} is proportional to the cube of the frequency. For this reason, while in the visible region essentially all emission is spontaneous for the usual radiation levels, the same is not true for longer wavelengths (e.g. radiofrequencies), where spontaneous emission is negligible.

The condition for observing induced emission is that the population of the first singlet state S_1 is larger than that of S_0, which is far from the case at room temperature because of the Boltzmann distribution (see above). An inversion of population (i.e. $N_{S1} > N_{S0}$) is thus required. For a four-level system inversion can be achieved using optical pumping by an intense light source (flash lamps or lasers); dye lasers work in this way. Alternatively, electrical discharge in a gas (gas lasers, copper vapor lasers) can be used.

In contrast to spontaneous emission, induced emission (also called *stimulated emission*) is coherent, i.e. all emitted photons have the same physical characteristics – they have the same direction, the same phase and the same polarization. These properties are characteristic of laser emission (L.A.S.E.R. = Light Amplification by Stimulated Emission of Radiation). The term *induced emission* comes from the fact that de-excitation is triggered by the interaction of an incident photon with an excited atom or molecule, which induces emission of photons having the same characteristics as those of the incident photon.

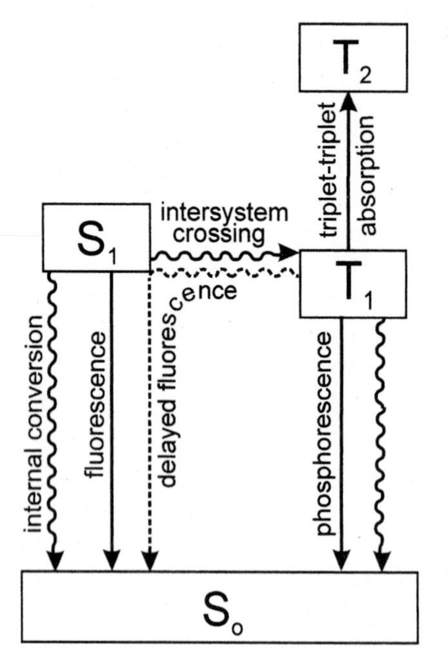

Scheme 3.1

3.1.3.1 Intersystem crossing

Intersystem crossing[3)] is a non-radiative transition between two isoenergetic vibrational levels belonging to electronic states of different multiplicities. For example, an excited molecule in the 0 vibrational level of the S_1 state can move to the isoenergetic vibrational level of the T_n triplet state; then vibrational relaxation brings it into the lowest vibrational level of T_1. Intersystem crossing may be fast enough (10^{-7}–10^{-9} s) to compete with other pathways of de-excitation from S_1 (fluorescence and internal conversion $S_1 \rightarrow S_0$).

Crossing between states of different multiplicity is in principle forbidden, but spin–orbit coupling (i.e. coupling between the orbital magnetic moment and the spin magnetic moment) (see Chapter 2) can be large enough to make it possible. The probability of intersystem crossing depends on the singlet and triplet states involved. If the transition $S_0 \rightarrow S_1$ is of $n \rightarrow \pi^*$ type for instance, intersystem crossing is often efficient. It should also be noted that the presence of heavy atoms (i.e. whose atomic number is large, for example Br, Pb) increases spin–orbit coupling and thus favors intersystem crossing.

3.1.3.2 Phosphorescence versus non-radiative de-excitation

In solution at room temperature, non-radiative de-excitation from the triplet state T_1, is predominant over radiative de-excitation called *phosphorescence*. In fact, the transition $T_1 \rightarrow S_0$ is forbidden (but it can be observed because of spin–orbit coupling), and the radiative rate constant is thus very low. During such a slow process, the numerous collisions with solvent molecules favor intersystem crossing and vibrational relaxation in S_0.

On the contrary, at low temperatures and/or in a rigid medium, phosphorescence can be observed. The lifetime of the triplet state may, under these conditions, be long enough to observe phosphorescence on a time-scale up to seconds, even minutes or more.

The phosphorescence spectrum is located at wavelengths higher than the fluorescence spectrum (Figure 3.1) because the energy of the lowest vibrational level of the triplet state T_1 is lower than that of the singlet state S_1.

3.1.3.3 Delayed fluorescence

Thermally activated delayed fluorescence Reverse intersystem crossing $T_1 \rightarrow S_1$ can occur when the energy difference between S_1 and T_1 is small and when the lifetime of T_1 is long enough. This results in emission with the same spectral distribution as normal fluorescence but with a much longer decay time constant because the molecules stay in the triplet state before emitting from S_1. This fluorescence emission is thermally activated; consequently, its efficiency increases with increasing tempera-

3) The word 'crossing' comes from the fact that the intersection between the potential energy surfaces corresponding to the S_1 and T_n states allows a molecule to cross from the S_1 state to the T_n state. The smaller the difference between the crossing point of these two surfaces and the minimum energy of the S_1 state, the more likely the crossing. If the difference in energy between the S_1 and T_1 states is small, molecules may return to the S_1 state. The subsequent emission from this state is called *delayed fluorescence* (see section 3.1.3.3).

ture. It is also called *delayed fluorescence of E-type* because it was observed for the first time with eosin. It does not normally occur in aromatic hydrocarbons because of the relatively large difference in energy between S_1 and T_1. In contrast, delayed fluorescence is very efficient in fullerenes.

Triplet–triplet annihilation In concentrated solutions, a collision between two molecules in the T_1 state can provide enough energy to allow one of them to return to the S_1 state. Such a triplet–triplet annihilation thus leads to a delayed fluorescence emission (also called *delayed fluorescence of P-type* because it was observed for the first time with pyrene). The decay time constant of the delayed fluorescence process is half the lifetime of the triplet state in dilute solution, and the intensity has a characteristic quadratic dependence with excitation light intensity.

3.1.3.4 Triplet–triplet transitions

Once a molecule is excited and reaches triplet state T_1, it can absorb another photon at a different wavelength because triplet–triplet transitions are spin allowed. These transitions can be observed provided that the population of molecules in the triplet state is large enough, which can be achieved by illumination with an intense pulse of light.

3.2 Lifetimes and quantum yields

3.2.1 Excited-state lifetimes

The rate constants for the various processes will be denoted as follows (see Scheme 3.2):

k_r^S: rate constant for radiative deactivation $S_1 \rightarrow S_0$ with emission of fluorescence.
k_{ic}^S: rate constant for internal conversion $S_1 \rightarrow S_0$.
k_{isc}: rate constant for intersystem crossing.

Regarding the two latter non-radiative pathways of de-excitation from S_1, it is convenient to introduce the overall non-radiative rate constant k_{nr}^S such that $k_{nr}^S = k_{ic}^S + k_{isc}$.

For deactivation from T_1, we have
k_r^T: rate constant for radiative deactivation $T_1 \rightarrow S_0$ with emission of phosphorescence.
k_{nr}^T: rate constant for non-radiative deactivation (intersystem crossing) $T_1 \rightarrow S_0$.

De-excitation processes resulting from intermolecular interactions are not considered in this chapter; they will be described in Chapter 4.

Let us consider a dilute solution of a fluorescent species A whose concentration

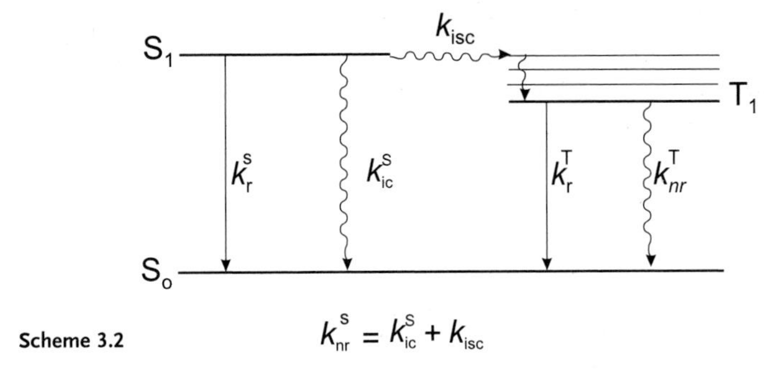

Scheme 3.2

is [A] (in mol L^{-1}). A very short pulse of light[4] (i.e. whose duration is short with respect to the reciprocal of the involved rate constants) at time 0 brings a certain number of molecules A to the S_1 excited state by absorption of photons. These excited molecules then return to S_0, either radiatively or non-radiatively, or undergo intersystem crossing. As in classical chemical kinetics, the rate of disappearance of excited molecules is expressed by the following differential equation:

$$-\frac{d[^1A^*]}{dt} = (k_r^S + k_{nr}^S)[^1A^*] \tag{3.1}$$

Integration of this equation yields the time evolution of the concentration of excited molecules $[^1A^*]$. Let $[^1A^*]_0$ be the concentration of excited molecules at time 0 resulting from pulse light excitation. Integration leads to

$$[^1A^*] = [^1A^*]_0 \exp\left(-\frac{t}{\tau_S}\right) \tag{3.2}$$

where τ_S, the lifetime of excited state S_1, is given by

$$\boxed{\tau_S = \frac{1}{k_r^S + k_{nr}^S}} \tag{3.3}$$

The fluorescence intensity is defined as the amount of photons (in mol, or its equivalent, in einsteins; 1 einstein = 1 mole of photons) emitted per unit time (s) and per unit volume of solution (liter: L) according to

$$A^* \xrightarrow{k_r^S} A + \text{photon}$$

The fluorescence intensity i_F at time t after excitation by a very short pulse of light at time 0 is proportional, at any time, to the instantaneous concentration of molecules still excited $[^1A^*]$; the proportionality factor is the rate constant for radiative

4) Strictly speaking, the light pulse is a δ-function (Dirac). The response of the system in terms of fluorescence intensity will thus be called a δ-pulse response.

de-excitation k_r^S:

$$i_F(t) = k_r^S[^1A^*] = k_r^S[^1A^*]_0 \exp\left(-\frac{t}{\tau_S}\right) \tag{3.4}$$

$i_F(t)$, the *δ-pulse response* of the system, decreases according to a single exponential.

It should be emphasized that, in any practical measurement of fluorescence intensity, the measured quantity is proportional to i_F, the proportionality factor depending on instrumental conditions (see Chapter 6). The 'measured' fluorescence intensity will be denoted I_F. It will be helpful to keep in mind that the numerical value of I_F is obtained on an arbitrary scale, depending on the experimental settings.

If the only way of de-excitation from S_1 to S_0 was fluorescence emission, the lifetime would be $1/k_r^S$: this is called the *radiative lifetime* (in preference to *natural lifetime*) and denoted by τ_r[5]. The radiative lifetime can be theoretically calculated from the absorption and fluorescence spectra using the Strickler–Berg relation[6].

The lifetime of a homogeneous population of fluorophores is very often independent of the excitation wavelength as the emission spectrum (but there are some exceptions). In fact, internal conversion and vibrational relaxation are always very fast in solution and emission arises from the lowest vibrational level of state S_1.

The fluorescence decay time τ_S is one of the most important characteristics of a fluorescent molecule because it defines the time window of observation of dynamic phenomena. As illustrated in Figure 3.2, no accurate information on the rate of phenomena occurring at time-scales shorter than about $\tau/100$ ('private life' of the molecule) or longer than about 10τ ('death' of the molecule) can be obtained, whereas at intermediate times ('public life' of the molecule) the time evolution of phenomena can be followed. It is interesting to note that a similar situation is found in the use of radioisotopes for dating: the period (i.e. the time constant of the exponential radioactive decay) must be of the same order of magnitude as the age of the object to be dated (Figure 3.2).

Following a δ-pulse excitation, a fraction of excited molecules can reach the triplet state, from which they return to the ground state either radiatively or non-radiatively. The concentration of molecules in the triplet state decays exponentially with a time constant τ_T representing the lifetime of the triplet state

5) It is interesting to note that for a resonant transition (i.e. coinciding absorption and emission frequencies), the reciprocal of the radiative lifetime is equal to the Einstein coefficient A_{21} for spontaneous emission (see Box 3.2).

6) The Strickler–Berg relation (*J. Chem. Phys.* 37, 814 (1962)) is:

$$\frac{1}{\tau_r} = \frac{8\pi \times 230 c_0 n^2}{N_a} \frac{\int F_{\bar{\nu}}(\bar{\nu}_F)\,d\bar{\nu}_F}{\int \bar{\nu}_F^{-3} F_{\bar{\nu}}(\bar{\nu}_F)\,d\bar{\nu}_F} \int \frac{\varepsilon(\bar{\nu}_A)\,d\bar{\nu}_A}{\bar{\nu}_A}$$

$$= 2.88 \times 10^{-9} n^2 \frac{\int F_{\bar{\nu}}(\bar{\nu}_F)\,d\bar{\nu}_F}{\int \bar{\nu}_F^{-3} F_{\bar{\nu}}(\bar{\nu}_F)\,d\bar{\nu}_F} \int \frac{\varepsilon(\bar{\nu}_A)\,d\bar{\nu}_A}{\bar{\nu}_A}$$

where n is the index of refraction, c_0 is the speed of light, ε is the molar absorption coefficient, and $F_{\bar{\nu}}(\bar{\nu}_F)$ is defined by Eq. 3.21 (see below). The Strickler–Berg equation yields values of τ_r that are often in agreement with the experimental ones, but it fails in a number of cases, especially when the interactions with the solvent cannot be ignored and when there is a change in the excited-state geometry. An important consequence of this equation is that the lower the molar absorption coefficient, the longer the radiative lifetime, i.e. the lower the rate of the radiative process.

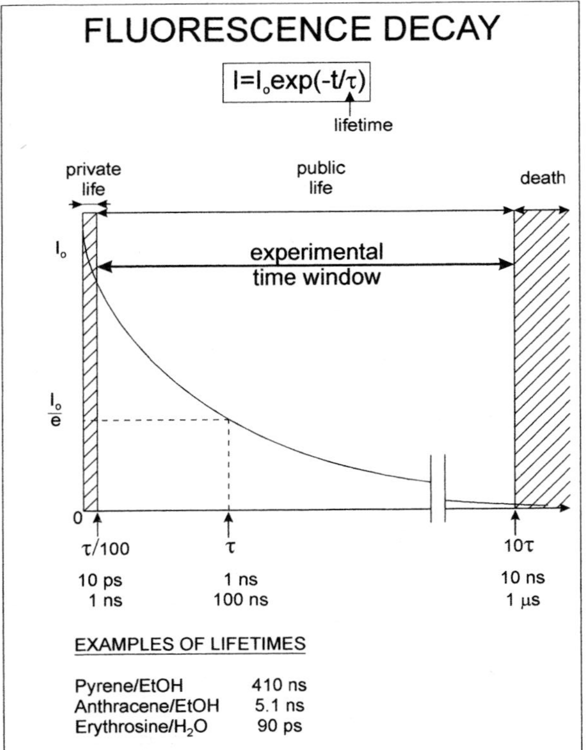

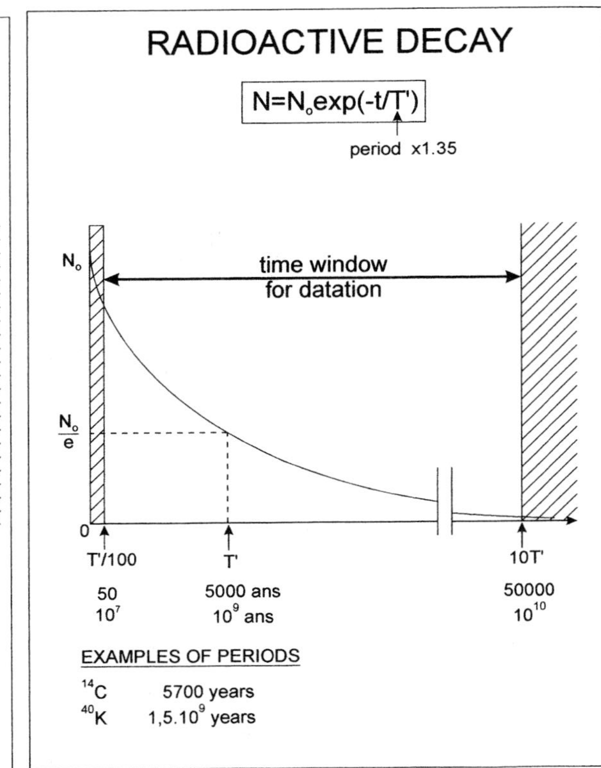

Fig. 3.2. Decay of fluorescence intensity and analogy with radioactive decay. Note that the lifetime τ is the time needed for the concentration of molecular entities to decrease to 1/e of its original value, whereas the radioactive period T is the time needed for the number of radioactive entities to decrease to $\frac{1}{2}$ of its original value. Therefore, T' (the decay time constant equivalent to the lifetime) is equal to $1.35T$.

$$\tau_T = \frac{1}{k_r^T + k_{nr}^T} \tag{3.5}$$

For organic molecules, the lifetime of the singlet state ranges from tens of picoseconds to hundreds of nanoseconds, whereas the triplet lifetime is much longer (microseconds to seconds). However, such a difference cannot be used to make a distinction between fluorescence and phosphorescence because some inorganic compounds (for instance, uranyl ion) or organometallic compounds may have a long lifetime.

Monitoring of phosphorescence or delayed fluorescence enables us to study much slower phenomena. Examples of lifetimes are given in Table 3.1.

3.2.2
Quantum yields

The *fluorescence quantum yield* Φ_F is the fraction of excited molecules that return to the ground state S_0 with emission of fluorescence photons:

$$\Phi_F = \frac{k_r^S}{k_r^S + k_{nr}^S} = k_r^S \tau_S \tag{3.6}$$

In other words, the fluorescence quantum yield is the ratio of the number of emitted photons (over the whole duration of the decay) to the number of absorbed photons. According to Eq. (3.4), the ratio of the δ-pulse response $i_F(t)$ and the number of absorbed photons is given by

$$\frac{i_F(t)}{[^1A^*]_0} = k_r^S \exp\left(-\frac{t}{\tau_S}\right) \tag{3.7}$$

and integration of this relation over the whole duration of the decay (mathematically from 0 to infinity) yields Φ_F:

$$\frac{1}{[^1A^*]_0} \int_0^{\infty} i_F(t)\,dt = k_r^S \tau_S = \Phi_F \tag{3.8}$$

The quantum yields of intersystem crossing (Φ_{isc}) and phosphorescence (Φ_P) are given by

$$\Phi_{isc} = \frac{k_{isc}}{k_r^S + k_{nr}^S} = k_{isc} \tau_S \tag{3.9}$$

$$\Phi_P = \frac{k_r^T}{k_r^T + k_{nr}^T} \Phi_{isc} \tag{3.10}$$

Using the radiative lifetime, as previously defined, the fluorescence quantum yield can also be written as

Tab. 3.1. Quantum yields and lifetimes of some aromatic hydrocarbons

Compound	*Formula*	*Solvent (temp.)*	Φ_F	τ_S (ns)	Φ_{isc}	Φ_P	τ_T (s)
Benzene		Ethanol (293 K)	0.04	31			
		EPA[a] (77 K)				0.17	7.0
Naphthalene	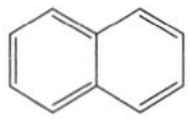	Ethanol (293 K)	0.21	2.7	0.79		
		Cyclohexane (293 K)	0.19	96			
		EPA (77 K)				0.06	2.6
Anthracene	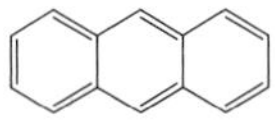	Ethanol (293 K)	0.27	5.1	0.72		
		Cyclohexane (293 K)	0.30	5.24			0.09
		EPA (77 K)					
Perylene	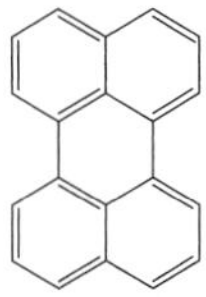	n-Hexane	0.98		0.02		
		Cyclohexane (293 K)	0.78	6			
Pyrene	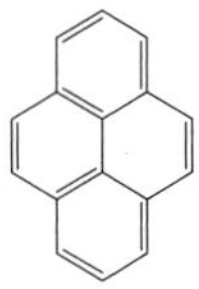	Ethanol (293 K)	0.65	410	0.35		
		Cyclohexane (293 K)	0.65	450			
Phenanthrene	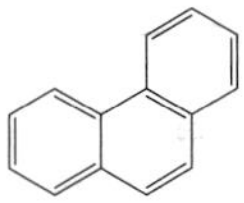	Ethanol (293 K)	0.13		0.85		
		n-Heptane (293 K)	0.16	0.60			
		EPA (77 K)				0.31	3.3
		Polymer film	0.12		0.88		0.11

a) EPA: mixture of ethanol, isopentane, diethyl ether 2:5:5 v/v/v

$$\Phi_F = \frac{\tau_S}{\tau_r} \tag{3.11}$$

Following an external perturbation, the fluorescence quantum yield can remain proportional to the lifetime of the excited state (e.g. in the case of dynamic quenching (see Chapter 4), variation in temperature, etc.). However, such a proportionality may not be valid if de-excitation pathways – different from those described above – result from interactions with other molecules. A typical case where the fluorescence quantum yield is affected without any change in excited-state lifetime is the formation of a ground-state complex that is non-fluorescent (static quenching; see Chapter 4).

Examples of fluorescence quantum yields and lifetimes of aromatic hydrocarbons are reported in Table 3.1[7].

7) It is interesting to note that when the fluorescence quantum yield and the excited-state lifetime of a fluorophore are measured under the same conditions, the non-radiative and radiative rate constants can be easily calculated by means of the following relations:

$$k_r^S = \frac{\Phi_F}{\tau_S} \qquad k_{nr}^S = \frac{1}{\tau_S}(1 - \Phi_F)$$

It is well known that dioxygen quenches fluorescence (and phosphorescence) (see Chapter 4), but its effect on quantum yields and lifetimes strongly depends on the nature of the compound and the medium. Oxygen quenching is a collisional process and is therefore diffusion-controlled. Consequently, compounds of long lifetime, such as naphthalene and pyrene, are particularly sensitive to the presence of oxygen. Moreover, oxygen quenching is less efficient in media of high viscosity. Oxygen quenching can be avoided by bubbling nitrogen or argon in the solution; however the most efficient method (used particularly in phosphorescence studies) is to perform a number of freeze–pump-thaw cycles.

3.2.3 Effect of temperature

Generally, an increase in temperature results in a decrease in the fluorescence quantum yield and the lifetime because the non-radiative processes related to thermal agitation (collisions with solvent molecules, intramolecular vibrations and rotations, etc.) are more efficient at higher temperatures. Experiments are often in good agreement with the empirical linear variation of $\ln(1/\Phi_F - 1)$ versus $1/T$.

As mentioned above, phosphorescence is observed only under certain conditions because the triplet states are very efficiently deactivated by collisions with solvent molecules (or oxygen and impurities) because their lifetime is long. These effects can be reduced and may even disappear when the molecules are in a frozen solvent, or in a rigid matrix (e.g. polymer) at room temperature. The increase in phosphorescence quantum yield by cooling can reach a factor of 10^3, whereas this factor is generally no larger than 10 or so for fluorescence quantum yield.

In conclusion, lifetimes and quantum yields are characteristics of major importance. Obviously, the larger the fluorescence quantum yield, the easier it is to observe a fluorescent compound, especially a fluorescent probe. It should be emphasized that, in the condensed phase, many parameters can affect the quantum yields and lifetimes: temperature, pH, polarity, viscosity, hydrogen bonding, presence of quenchers, etc. Attention should be paid to possible erroneous interpretation arising from the simultaneous effects of several factors (for instance, changes in viscosity due to a variation in temperature).

3.3 Emission and excitation spectra

3.3.1 Steady-state fluorescence intensity

Emission and excitation spectra are recorded using a spectrofluorometer (see Chapter 6). The light source is a lamp emitting a constant photon flow, i.e. a constant amount of photons per unit time, whatever their energy. Let us denote by N_0 the constant amount of incident photons entering, during a given time, a unit volume of the sample where the fluorophore concentration is [A] (N_0 and [A] in mol L^{-1}). αN_0 represents the amount of absorbed photons per unit volume involved in the excitation process

$$^1\mathrm{A} + h\nu \xrightarrow{k_a} {}^1\mathrm{A}^*$$

Let us recall that the pseudo-first order rate constant for this process is very large ($k_a \approx 10^{15}\ \mathrm{s}^{-1}$) whereas the subsequent steps of de-excitation occur with much lower rate constants (k_r^S and $k_{nr}^S \approx 10^7$–$10^{10}\ \mathrm{s}^{-1}$), according to

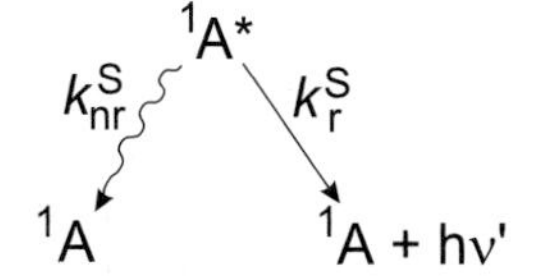

Under continuous illumination, the concentration $[^1\mathrm{A}^*]$ remains constant, which means that $^1\mathrm{A}^*$ is in a steady state. Measurements under these conditions are then called *steady-state measurements.*

The rate of change of $[^1\mathrm{A}^*]$ is equal to zero:

$$\frac{d[^1\mathrm{A}^*]}{dt} = 0 = k_a \alpha N_0 - (k_r^S + k_{nr}^S)[^1\mathrm{A}^*] \tag{3.12}$$

$k_a \alpha N_0$ represents the amount of absorbed photons per unit volume and per unit time. It can be rewritten as αI_0 where I_0 represents the intensity of the incident light (in moles of photons per liter and per second).

The constant concentration $[^1\mathrm{A}^*]$ is given by

$$[^1\mathrm{A}^*] = \frac{\alpha I_0}{k_r^S + k_{nr}^S} \tag{3.13}$$

The amount of fluorescence photons emitted per unit time and per unit volume, i.e. the *steady-state fluorescence intensity*, is then given by

$$i_F = k_r^S[^1\mathrm{A}^*] = \alpha I_0 \frac{k_r^S}{k_r^S + k_{nr}^S} = \alpha I_0 \Phi_F \tag{3.14}$$

This expression shows that the steady-state fluorescence intensity per absorbed photon $i_F/\alpha I_0$ is the fluorescence quantum yield[8].

8) It is worth noting that integration of the δ-pulse response $i_F(t)$ of the fluorescence intensity over the whole duration of the decay (Eq. 3.8) yields

$$\int_0^\infty i_F(t)\,dt = k_r^S \tau_S [^1\mathrm{A}^*]_0 = [^1\mathrm{A}^*]_0 \Phi_F$$

This quantity is the total amount of photons emitted per unit volume under steady-state conditions which, divided by time, yields the above expression for i_F. An exciting light of constant intensity can then be considered as an infinite sum of infinitely short light pulses.

3.3.2
Emission spectra

We have so far considered all emitted photons, whatever their energy. We now focus our attention on the energy distribution of the emitted photons. With this in mind, it is convenient to express the steady-state fluorescence intensity per absorbed photon as a function of the wavelength of the emitted photons, denoted by $F_\lambda(\lambda_F)$ (in m^{-1} or nm^{-1}) and satisfying the relationship

$$\int_0^\infty F_\lambda(\lambda_F)\,d\lambda_F = \Phi_F \tag{3.15}$$

where Φ_F is the fluorescence quantum yield defined above. $F_\lambda(\lambda_F)$ represents the *fluorescence spectrum* or *emission spectrum*: it reflects the distribution of the probability of the various transitions from the lowest vibrational level of S_1 to the various vibrational levels of S_0. The emission spectrum is characteristic of a given compound.

In practice, the steady-state fluorescence intensity $I_F(\lambda_F)$ measured at wavelength λ_F (selected by a monochromator with a certain wavelength bandpass $\Delta\lambda_F$) is proportional to $F_\lambda(\lambda_F)$ and to the number of photons absorbed at the excitation wavelength λ_E (selected by a monochromator). It is convenient to replace this number of photons by the absorbed intensity $I_A(\lambda_E)$, defined as the difference between the intensity of the incident light $I_0(\lambda_E)$ and the intensity of the transmitted light $I_T(\lambda_E)$:

$$I_A(\lambda_E) = I_0(\lambda_E) - I_T(\lambda_E) \tag{3.16}$$

The fluorescence intensity can thus be written as

$$I_F(\lambda_E, \lambda_F) = kF_\lambda(\lambda_F)I_A(\lambda_E) \tag{3.17}$$

The proportionality factor k depends on several parameters, in particular on the optical configuration for observation (i.e. the solid angle through which the instrument collects fluorescence, which is in fact emitted in all directions) and on the bandwidth of the monochromators (i.e. the entrance and exit widths; see Chapter 5).

Furthermore, the intensity of the transmitted light can be expressed using the Beer–Lambert law (see Chapter 2):

$$I_T(\lambda_E) = I_0(\lambda_E)\exp[-2.3\varepsilon(\lambda_E)lc] \tag{3.18}$$

where $\varepsilon(\lambda_E)$ denotes the molar absorption coefficient of the fluorophore at wavelength λ_E (in L mol^{-1} cm^{-1}), l the optical path in the sample (in cm) and c the concentration (in mol L^{-1}). The quantity $\varepsilon(\lambda_E)lc$ represents the absorbance $A(\lambda_E)$ at wavelength λ_E.

Tab. 3.2. Deviation from linearity in the relation between fluorescence intensity and concentration for various absorbances

Absorbance	*Deviation (%)*
10^{-3}	0.1
10^{-2}	1.1
0.05	5.5
0.10	10.6
0.20	19.9

Equations (3.16)–(3.18) lead to

$$I_F(\lambda_E, \lambda_F) = kF_\lambda(\lambda_F)I_0(\lambda_E)\{1 - \exp[-2.3\varepsilon(\lambda_E)lc]\} \tag{3.19}$$

In practice, measurement of the variations in I_F as a function of wavelength λ_F, for a fixed excitation wavelength λ_E, reflects the variations in $F_\lambda(\lambda_F)$ and thus provides the fluorescence spectrum[9]. Because the proportionality factor k is generally unknown, the numerical value of the measured intensity I_F has no meaning, and generally speaking, I_F is expressed in arbitrary units. In the case of low concentrations, the following expansion can be used in Eq. (3.17):

$$1 - \exp(2.3\varepsilon lc) = 2.3\varepsilon lc - \frac{1}{2}(2.3\varepsilon lc)^2 + \cdots$$

In highly diluted solutions, the terms of higher order become negligible. By keeping only the first term, we obtain

$$I_F(\lambda_E, \lambda_F) \cong kF_\lambda(\lambda_F)I_0(\lambda_E)[2.3\varepsilon(\lambda_E)lc] = 2.3kF_\lambda(\lambda_F)I_0(\lambda_E)A(\lambda_E) \tag{3.20}$$

This relation shows that the fluorescence intensity is proportional to the concentration only for low absorbances. Deviation from a linear variation increases with increasing absorbance (Table 3.2).

Moreover, when the concentration of fluorescent compound is high, inner filter effects reduce the fluorescence intensity depending on the observation conditions (see Chapter 6). In particular, the photons emitted at wavelengths corresponding to the overlap between the absorption and emission spectra can be reabsorbed (radiative transfer). Consequently, when fluorometry is used for a quantitative evaluation of the concentration of a species, it should be kept in mind that the fluorescence intensity is proportional to the concentration only for diluted solutions.

Fluorometry is a very sensitive technique – up to 1000 times more sensitive than spectrophotometry. This is because the fluorescence intensity is measured above a

9) It will be shown in Chapter 6 that k depends on the wavelength because the transmission of the monochromator and the sensitivity of the detector are wavelength-dependent. Therefore, correction of spectra is necessary for quantitative measurements.

low background level whereas in the measurement of low absorbances, two large signals that are slightly different are compared. Thanks to outstanding progress in instrumentation, it is now possible in some cases to even detect a single fluorescent molecule (see Chapter 11).

The fluorescence spectrum of a compound may be used in some cases for the identification of species, especially when the spectrum exhibits vibronic bands (e.g. in the case of aromatic hydrocarbons), but the spectra of most fluorescent probes (in the condensed phase) exhibit broad bands.

Equations (3.15)–(3.20) have been written using wavelengths, but they could also have been written using wavenumbers. For example, the integral in Eq. (3.15) is found written in some books (e.g. Birks, 1969, 1973) using wavenumbers instead of wavelengths:

$$\int_{\infty}^{0} F_{\bar{\nu}}(\bar{\nu}_F)\,\mathrm{d}\bar{\nu}_F = \Phi_F \tag{3.21}$$

where $F_{\bar{\nu}}(\bar{\nu}_F)$ is the fluorescence intensity per unit wavenumber.

Comments should be made here on the theoretical equivalence between Eqs (3.15) and (3.21). The fluorescence quantum yield Φ_F, i.e. the number of photons emitted over the whole fluorescence spectrum divided by the number of absorbed photons, must of course be independent of the representation of the fluorescence spectrum in the wavelength scale (Eq. 3.15) or the wavenumber scale (Eq. 3.21):

$$\Phi_F = \int_{0}^{\infty} F_{\lambda}(\lambda_F)\,\mathrm{d}\lambda_F = \int_{\infty}^{0} F_{\bar{\nu}}(\bar{\nu}_F)\,\mathrm{d}\bar{\nu}_F \tag{3.22}$$

However, as shown in Box 3.3, it should be emphasized that $F_{\lambda}(\lambda_F)$ is not equal to $F_{\bar{\nu}}(\bar{\nu}_F)$ and this has practical consequences.

From the theoretical point of view, the important consequence of Eq. (3.22) is that the conversion of an integral from the wavenumber form to the wavelength form simply consists of replacing $F_{\bar{\nu}}(\bar{\nu}_F)$ by $F_{\lambda}(\lambda_F)$, and $\mathrm{d}\bar{\nu}_F$ by $\mathrm{d}\lambda_F$. However, from the practical point of view, because all spectrofluorometers are equipped with grating monochromators, calculation of the integral must be done with the wavelength form. The fluorescence spectrum is then recorded on a linear wavelength scale at constant wavelength bandpass $\Delta\lambda_F$ (which is the integration step) (see Box 3.3).

3.3.3
Excitation spectra

The variations in fluorescence intensity as a function of the excitation wavelength λ_E for a fixed observation wavelength λ_F represents the *excitation spectrum.* According to Eq. (3.20), these variations reflect the evolution of the product $I_0(\lambda_E)A(\lambda_E)$. If we can compensate for the wavelength dependence of the incident light (see Chapter 6), the sole term to be taken into consideration is $A(\lambda_E)$, which represents the absorption spectrum. The corrected excitation spectrum is thus identical in

Box 3.3 Determination of fluorescence quantum yields from fluorescence spectra: wavelength scale or wavenumber scale?

Fluorescence quantum yields are usually determined by integration of the fluorescence spectrum (and subsequent normalization using a standard of known fluorescence quantum yield in order to get rid of the instrumental factor k appearing in Eqs 3.17 or 3.18; see Chapter 6). In practice, attention should be paid to the method of integration.

When the emission monochromator of the spectrofluorometer is set at a certain wavelength λ_F with a bandpass $\Delta\lambda_F$, the reading is proportional to the number of photons emitted in the *wavelength range* from λ_F to $\lambda_F + \Delta\lambda_F$, or in the corresponding *wavenumber range* from to $1/\lambda_F$ to $1/(\lambda_F + \Delta\lambda_F)$. The number of detected photons satisfies the relationship:

$$F_\lambda(\lambda_F)\Delta\lambda_F = F_{\bar{\nu}}(\bar{\nu}_F)\Delta\bar{\nu}_F$$

where $\Delta\bar{\nu}_F$ must be positive and is thus defined as $1/\lambda_F - 1/(\lambda_F + \Delta\lambda_F)$. Hence,

$$F_\lambda(\lambda_F) = F_{\bar{\nu}}(\bar{\nu}_F)\frac{1}{\lambda_F(\lambda_F + \Delta\lambda_F)}$$

Because $\Delta\lambda_F \ll \lambda_F$, this equation becomes

$$F_\lambda(\lambda_F)\lambda_F^2 = F_{\bar{\nu}}(\bar{\nu}_F)$$

which clearly shows that $F_\lambda(\lambda_F)$ is not equal to $F_{\bar{\nu}}(\bar{\nu}_F)$.

If a grating monochromator is used to record the fluorescence spectrum, the bandpass $\Delta\lambda_F$ is constant and the scale is linear in wavelength. Therefore, after correction of the spectrum (see Chapter 6), the integral should be calculated with the wavelength form (Eq. 3.13). Some workers convert the wavelength scale into the wavenumber scale before integration, but this procedure is incorrect. In some cases, the consequence might be that the quantum yield calculated by comparison with a fluorescent standard is found to be greater than 1!

It should be noted that no such difficulty appears with the integral of an absorption spectrum because the absorption coefficient is proportional to the logarithm of a ratio of intensities, so that $\varepsilon(\lambda) = \varepsilon(\bar{\nu})$. For instance, in the calculation of an oscillator strength (defined in Chapter 2), integration can be done either in the wavelength scale or in the wavenumber scale.

shape to the absorption spectrum, provided that there is a single species in the ground state. In contrast, when several species are present, or when a sole species exists in different forms in the ground state (aggregates, complexes, tautomeric forms, etc.), the excitation and absorption spectra are no longer superimposable. Comparison of absorption and emission spectra often provides useful information.

3.3.4
Stokes shift

The *Stokes shift* is the gap between the maximum of the first absorption band and the maximum of the fluorescence spectrum (expressed in wavenumbers), $\Delta\bar{\nu} = \bar{\nu}_a - \bar{\nu}_f$ (Figure 3.3).

This important parameter can provide information on the excited states. For instance, when the dipole moment of a fluorescent molecule is higher in the excited state than in the ground state, the Stokes shift increases with solvent polarity. The consequences of this in the estimation of polarity using fluorescent polarity probes is discussed in Chapter 7.

From a practical point of view, the detection of a fluorescent species is of course easier when the Stokes shift is larger.

3.4
Effects of molecular structure on fluorescence

3.4.1
Extent of π-electron system. Nature of the lowest-lying transition

Most fluorescent compounds are aromatic. A few highly unsaturated aliphatic compounds are also fluorescent. Generally speaking, an increase in the extent of the π-electron system (i.e. degree of conjugation) leads to a shift of the absorption and fluorescence spectra to longer wavelengths and to an increase in the fluorescence quantum yield. This simple rule is illustrated by the series of linear aromatic hydrocarbons: naphthalene, anthracene, naphthacene and pentacene emit fluorescence in the ultraviolet, blue, green and red, respectively.

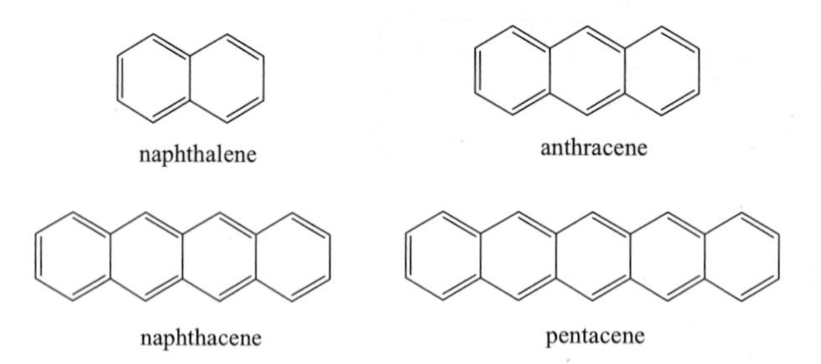

The lowest-lying transitions of aromatic hydrocarbons are of $\pi \rightarrow \pi^*$ type, which are characterized by high molar absorption coefficients and relatively high fluorescence quantum yields. When a heteroatom is involved in the π-system, an $n \rightarrow \pi^*$ transition may be the lowest-lying transition. Such transitions are characterized by molar absorption coefficients that are at least a factor of 10^2 smaller than those of $\pi \rightarrow \pi^*$ transitions. Therefore, according to the Strickler–Berg equation (Section 3.2.1), the radiative lifetime τ_r is at least 100 times longer than that of low-lying

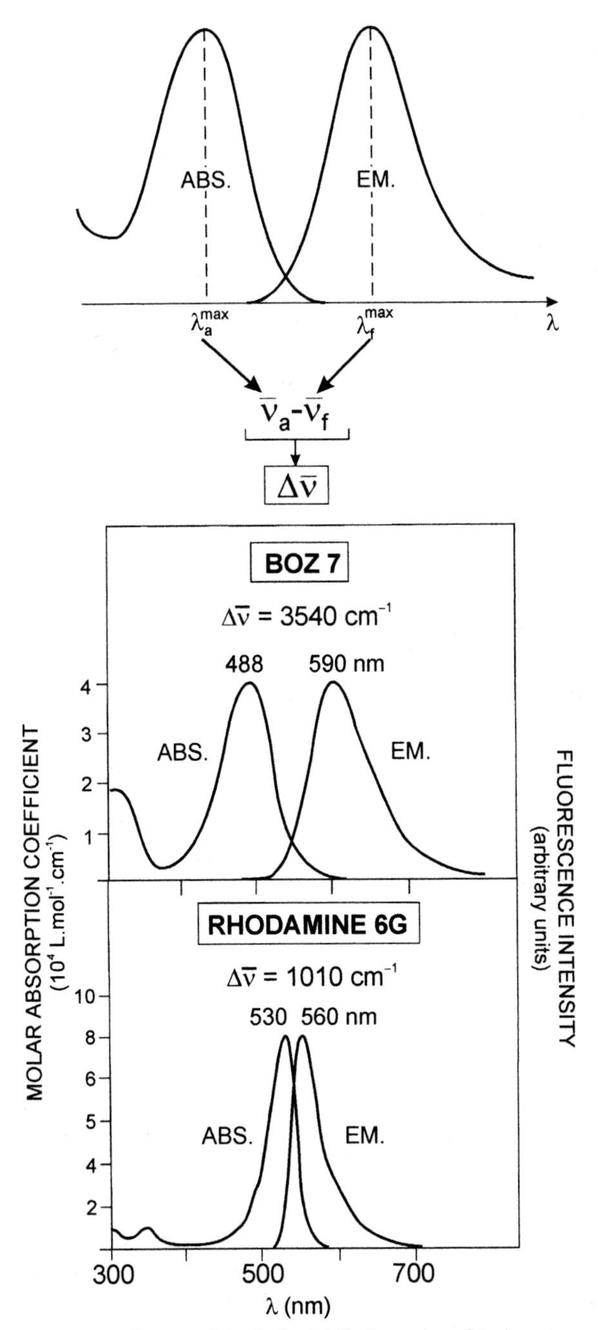

Fig. 3.3. Definition of the Stokes shift. Examples of Stokes shift: benzoxazinone derivative (BOZ 7) and rhodamine 6G.

Tab. 3.3. Heavy atom effect on emissive properties of naphthalene (from Wehry, 1990)

	Φ_F	k_{isc}/s^{-1}	Φ_P	τ_T/s
Naphthalene	0.55	1.6×10^6	0.051	2.3
1-Fluoronaphthalene	0.84	5.7×10^5	0.056	1.5
1-Chloronaphthalene	0.058	4.9×10^7	0.30	0.29
1-Bromonaphthalene	0.0016	1.9×10^9	0.27	0.02
1-Iodonaphthalene	<0.0005	$>6 \times 10^9$	0.38	0.002

$\pi \rightarrow \pi^*$ transitions, i.e. about 10^{-6} s. Such a slow process cannot compete with the dominant non-radiative processes. This explains the low fluorescence quantum yields of many molecules in which the lowest excited state is n–π^* in nature. This is the case for most azo compounds and some compounds containing carbonyl groups and nitrogen heterocycles (with pyridine-type nitrogens) (see Sections 3.4.2.3 and 3.4.3).

3.4.2
Substituted aromatic hydrocarbons

The effect of substituents on the fluorescence characteristics of aromatic hydrocarbons is quite complex and generalizations should be made with caution. Both the nature and position of a substituent can alter these characteristics.

3.4.2.1 Internal heavy atom effect

In general, the presence of heavy atoms as substituents of aromatic molecules (e.g. Br, I) results in fluorescence quenching (*internal heavy atom effect*) because of the increased probability of intersystem crossing. In fact, intersystem crossing is favored by spin–orbit coupling whose efficiency has a Z^4 dependence (Z is the atomic number). Table 3.3 exemplifies this effect.

However, the heavy atom effect can be small for some aromatic hydrocarbons if: (i) the fluorescence quantum yield is large so that de-excitation by fluorescence emission dominates all other de-excitation processes; (ii) the fluorescence quantum yield is very low so that the increase in efficiency of intersystem crossing is relatively small; (iii) there is no triplet state energetically close to the fluorescing state (e.g. perylene)[10].

3.4.2.2 Electron-donating substituents: –OH, –OR, –NH_2, –NHR, –NR_2

In general, substitution with electron-donating groups induces an increase in the molar absorption coefficient and a shift in both absorption and fluorescence spec-

10) The energy difference between S_1 and the accepting triplet state has to be small enough for intersystem crossing to compete effectively with fluorescence. This is not the case with perylene, as demonstrated by the fact that its fluorescence quantum yield is nearly identical to those of bromo-substituted perylenes (Dreeskamp H., Koch E. and Zander M. (1975) *Chem. Phys. Lett.* **31**, 251).

tra. Moreover, these spectra are broad and often structureless compared to the parent aromatic hydrocarbons (e.g. 1- and 2-naphthol compared to naphthalene).

The presence of lone pairs of electrons on the oxygen and nitrogen atoms does not change the π–π^* nature of the transitions of the parent molecule. These lone pairs are indeed involved directly in π bonding with the aromatic system, in contrast to the lone pairs of electrons of carbonyl substituents (see Section 3.4.2.3) or heterocyclic nitrogen (see Section 3.4.3). To make the distinction between the two types of lone pairs, Kasha and Rawls suggested using the term *l orbital* for the lone-pair orbitals of aromatic amines and phenols. A significant intramolecular charge transfer character of the relevant transitions is expected for planar aromatic amines and phenol; this is confirmed by the fact that the fluorescence spectra are broad and structureless.

If, for steric reasons, the $-NH_2$ group is twisted out of the plane of the aromatic ring, the degree of conjugation of the l orbitals is decreased, but the transitions corresponding to the promotion of an electron from an l orbital to a π^* orbital are still different (in particular more intense) to the $n \rightarrow \pi^*$ transitions involving the lone pairs of a carbonyl or nitro group. Departure from coplanarity with the aromatic ring is also pronounced with $-OR$ substituents, whereas an $-OH$ group is nearly coplanar.

The absorption and emission characteristics of phenols and aromatic amines are pH-dependent. These aspects will be discussed in Section 4.5.

3.4.2.3 **Electron-withdrawing substituents: carbonyl and nitro compounds**

The fluorescence properties of aromatic carbonyl compounds are complex and often difficult to predict.

Many aromatic aldehydes and ketones (e.g. benzophenone, anthrone, 1- and 2-naphthaldehyde) have a low-lying n–π^* excited state and thus exhibit low fluorescence quantum yields, as explained above. The dominant de-excitation pathway is intersystem crossing (whose efficiency has been found to be close to 1 for benzophenone).

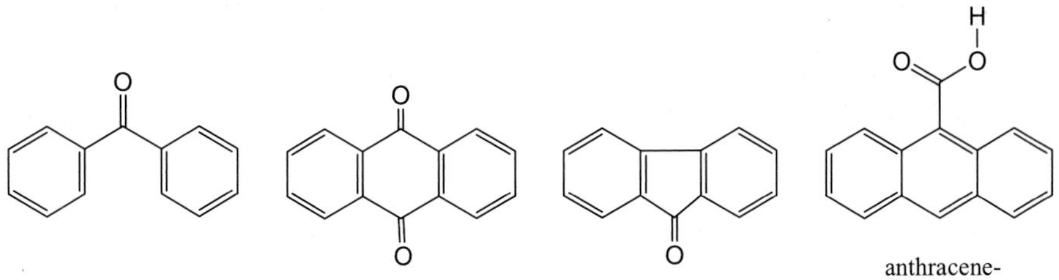

benzophenone | anthrone | fluorenone | anthracene-9-carboxylic acid

Some aromatic carbonyl compounds have a low-lying π–π^* excited state and thus have a reasonable quantum yield (e.g. 0.12 for fluorenone in ethanol at 77 K and 0.01 at room temperature). However, if an n–π^* state lies only slightly higher in energy, the fluorescence quantum yield strongly depends on the polarity of the solvent (*proximity effect*). In fact, in some solvents, the energy of the n–π^* state can become lower than that of the π–π^* state. When the polarity and the hydrogen bonding power of the solvent increases, the n–π^* state shifts to higher energy whereas the π–π^* state shifts to lower energy. Therefore, intense fluorescence can be observed in polar solvents and weak fluorescence in nonpolar solvents (e.g. xanthone).

When an aromatic molecule has a carboxylic group as a substituent, photophysical effects due to conformational changes can be observed. For instance, anthracene-9-carboxylic acid exhibits a broad fluorescence spectrum deprived of apparent vibronic bands, in contrast to its absorption spectrum and to both absorption and fluorescence spectra of the conjugate base (similar to the anthracene spectrum). Such a difference between the fluorescence spectra of the acidic and basic forms can be explained in terms of conformation of the carboxylate group $-COO^-$, which should be almost perpendicular to the ring so that the π system of the anthracene ring is only slightly perturbed. On the contrary, the carboxylic group $-COOH$ may be in a position close to the coplanarity of the ring; the resulting interaction induces an intramolecular charge-transfer character to the $\pi \to \pi^*$ transition. Charge-transfer fluorescence bands are indeed usually broad and structureless. However, because the absorption spectrum of the acidic form exhibits vibronic bands, the rotation of the $-COOH$ is likely to be photoinduced.

In general, the fluorescence of aromatic hydrocarbons possessing an $-NO_2$ substituent is not detectable. The existence of a low-lying $n \to \pi^*$ transition explains the efficient intersystem crossing process (e.g. for 2-nitronaphthalene, the quantum yield for intersystem crossing is 0.83 in benzene solution at room temperature). Many nitroaromatics are indeed phosphorescent. However, in some cases, the quantum yield for intersystem crossing is significantly less than 1. Therefore, the absence of detectable fluorescence is likely to be due to a high rate of $S_1 \to S_0$ internal conversion, which may be related to the considerable charge-transfer character of the excited state, as a result of the strong electron-withdrawing power of the $-NO_2$ group.

It should be mentioned that many nitroaromatics undergo photodegradation. For instance, 9-nitroanthracene is transformed into anthraquinone upon illumination.

3.4.2.4 **Sulfonates**

The solubility in water of many fluorophores is achieved by grafting sulfonate groups. Fortunately, these groups only slightly affect the fluorescence characteristics of the parent molecule. In general, there is a small red-shift of the fluorescence spectrum, whose vibrational structure is somewhat blurred, and the fluorescence quantum yield is slightly decreased.

Finally, it should be emphasized that the fluorescence characteristics of aromatic hydrocarbons containing more than one substituent are difficult to predict. These effects cannot be simply extrapolated from those of individual substituents. For instance, in spite of the presence of a nitro group, o- and m-nitroaniline and 3-nitro-*N*,*N*-dimethylaniline exhibit fluorescence.

3.4.3
Heterocyclic compounds

Compounds called *azarenes* containing one or more heterocyclic nitrogen atoms (e.g. pyridine, quinoline, acridine) have low-lying n $\rightarrow \pi^*$ transitions, which explains their relatively low fluorescence quantum yields in hydrocarbons.

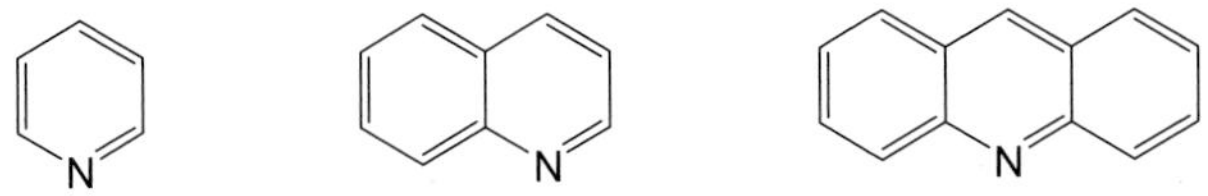

pyridine quinoline acridine

However, the fluorescence characteristics of these compounds are strongly solvent-dependent. In protic solvents such as alcohols, hydrogen bonds can be formed between the nitrogen atoms and the solvent molecules. This results in an inversion of the lowest-lying n–π^* and π–π^* states. As the lowest-lying transition becomes of $\pi \rightarrow \pi^*$ character in these solvents, the fluorescence quantum yield is much higher than in hydrocarbon solvents.

It should be noted that the electron density on the nitrogen atom is reduced on excitation, so that the ability to form hydrogen bonds is lower in the excited state. The ground state is thus more stabilized by hydrogen bonding than the excited state. This results in a red-shift of the spectra when going from nonpolar solvents to hydrogen bond donating solvents.

In some heterocyclic compounds such as acridine, the n–π^* absorption band is difficult to distinguish from the much more intense π–π^* absorption bands.

There are many interesting derivatives of quinoline and acridine obtained by substitution. In particular, 8-hydroxyquinoline (oxine) is the second complexing agent in importance after EDTA. Sulfonation in position 5 leads to a compound which is soluble in water and that exhibits outstanding fluorogenic properties (i.e. fluorescence enhancement) on complexation with metal ions (e.g. aluminum).

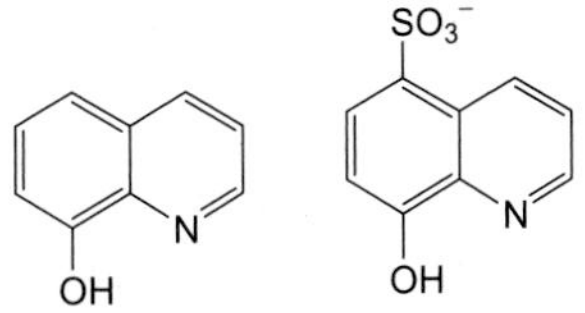

8-hydroxyquinoline and 5-sulfonate derivative

When a heteronitrogen is singly bonded to carbon atoms in a heterocycle, as in pyrrole rings (e.g. indole, carbazole), the transitions involving the non-bonding electrons have properties similar to those of $\pi \rightarrow \pi^*$ transitions. In fact, the non-bonding orbital is perpendicular to the plane of the ring, which allows it to overlap the π orbitals on the adjacent carbon atoms. The relatively high fluorescence quantum yield of carbazole and indole can be explained in this way. Tryptophan is an important derivative of indole, whose photophysical properties have been extensively studied because of its importance in fluorescence investigations of proteins.

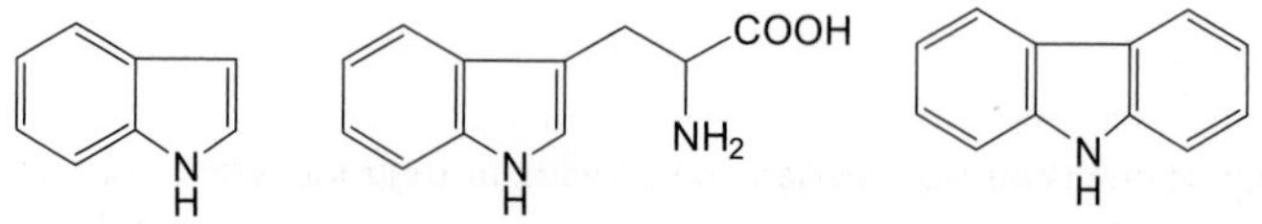

indole   tryptophan   carbazole

The properties of the related heterocycles containing oxygen and sulfur (e.g. dibenzofuran, dibenzothiophene) can be similarly interpreted.

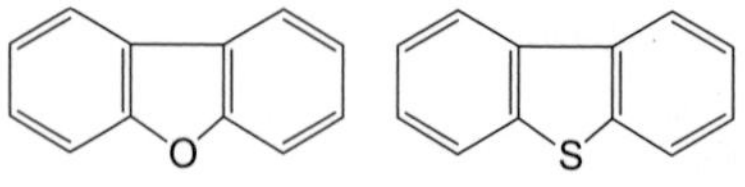

dibenzofuran   dibenzothiophene

Many fluorophores of practical interest are heterocyclic: *coumarins*, *rhodamines*, *pyronines*, *fluoresceins*, *oxazines*, etc. A few examples will now be presented.

Coumarin itself has a poor quantum yield, but appropriate substitution leads to fluorescent compounds emitting in the blue–green region (400–550 nm). Substitution in position 4 by a methyl group leads to *umbelliferone*. *7-Hydroxycoumarins* are very sensitive to pH. For example, 4-methyl-7-hydroxycoumarin (4-methylumbelliferone) can be used as a fluorescent pH probe (see Chapter 10).

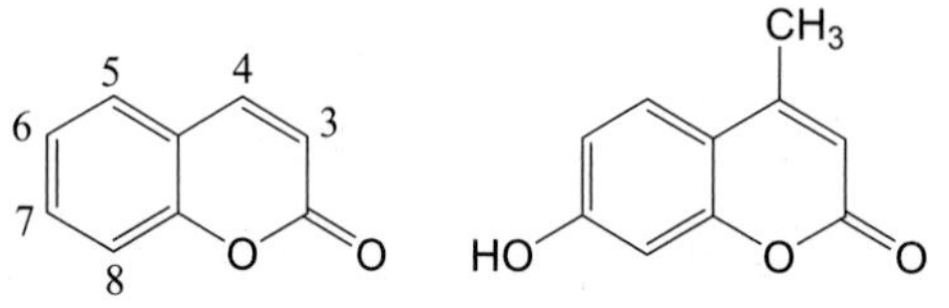

coumarin   4-methylumbelliferone

7-Aminocoumarins are of particular interest because they possess an electron-donating group (amino group) conjugated to an electron-withdrawing group (carbonyl group). This results in a photoinduced charge transfer (see Section 3.4.4). *Aminoquinoxalinones* and *aminobenzoxazinones* exhibit similar properties.

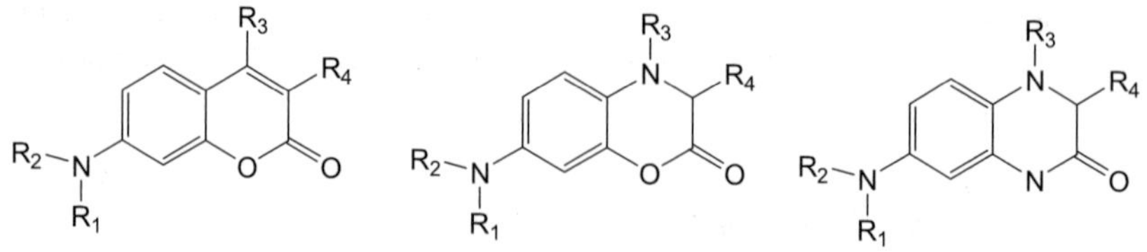

R_1, R_2 = H, alkyl R_3 = H, CH_3, CF_3...
R_4 = H, COOH, $COOC_2H_5$...,

7-aminocoumarins   7-aminobenzoxazinones   7-aminoquinoxalinones

The two well-known classes of highly fluorescent dyes, *rhodamines* and *fluoresceins*, are derivatives of *xanthene*.

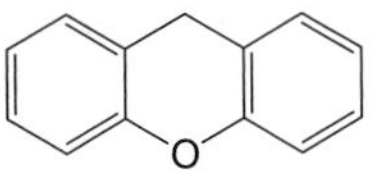

xanthene

Rhodamines (e.g. rhodamine 6G, rhodamine B) were among the first fluorescent dyes to be used as laser dyes. In contrast to coumarins, their absorption and emission spectra are quite narrow and the Stokes shift is small. They emit fluorescence in the range 500–700 nm.

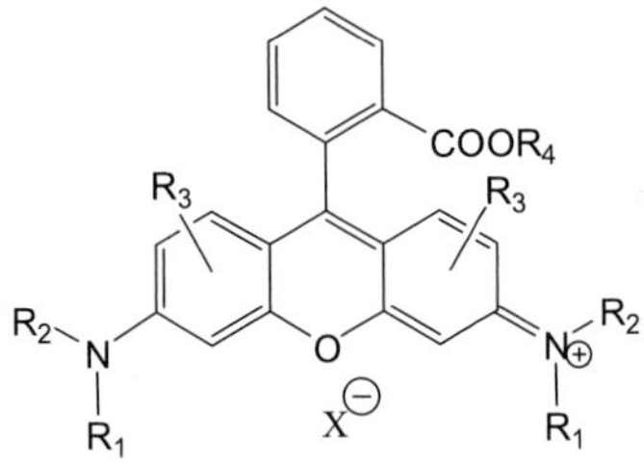

$R_1, R_2, R_3, R_4 = H$, alkyl

rhodamines

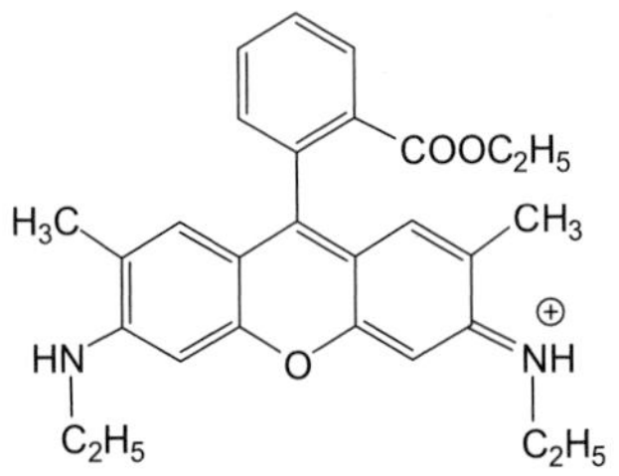

rhodamine 6G

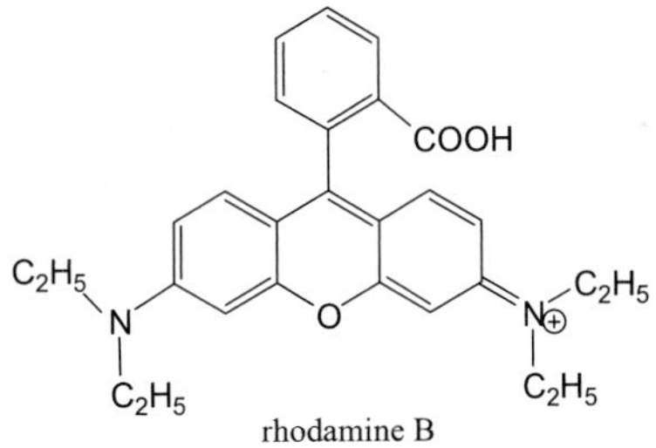

rhodamine B

It is worth noting that the carboxyphenyl group of rhodamines is only slightly involved in the conjugation of the π electron system because it is almost perpendicular to the xanthenic ring for steric reasons. Thus, replacement of this group by a hydrogen atom should not greatly affect the photophysical properties. In fact, the resulting compounds, called *pyronines*, exhibit almost identical properties to those of the corresponding rhodamines: their absorption and emission spectra are shifted by 1–3 nm with respect to the rhodamines, and the Stokes shift is slightly smaller.

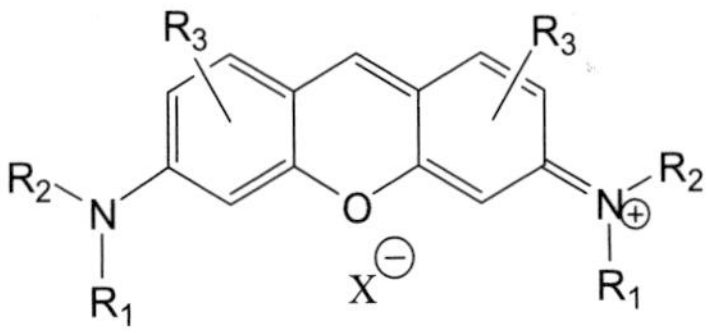

$R_1, R_2, R_3, R_4 = H$, alkyl

pyronines

The second family of xanthene dyes is *fluorescein* and its derivatives. Fluorescein itself is only slightly fluorescent in alcohol solution. In contrast, the alkali salt obtained by addition of alkali exhibits the well-known yellow–green fluorescence characteristic of the fluorescein dianion (uranin). Fluorescein and its derivatives, e.g. *eosin Y* and *erythrosin Y*, are known to be very sensitive to pH and can thus be used as pH fluorescent probes (see Chapter 10).

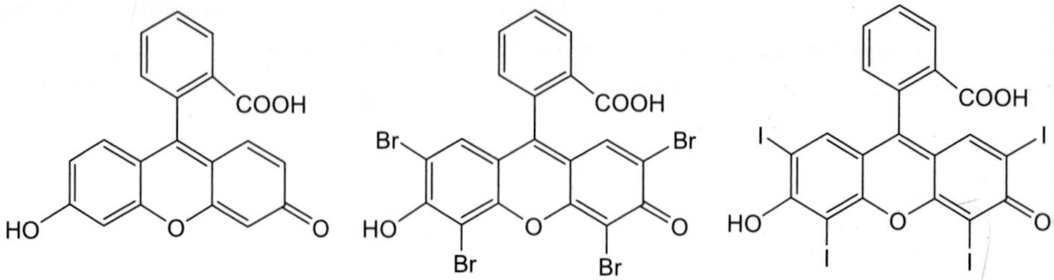

fluorescein   eosin Y   erythrosin Y

In pyronines, replacement of the carbon atom of the central ring (opposite to the oxygen atom) by a nitrogen atom produces *oxazine dyes*, which are used as laser dyes emitting in the range 600–750 nm.

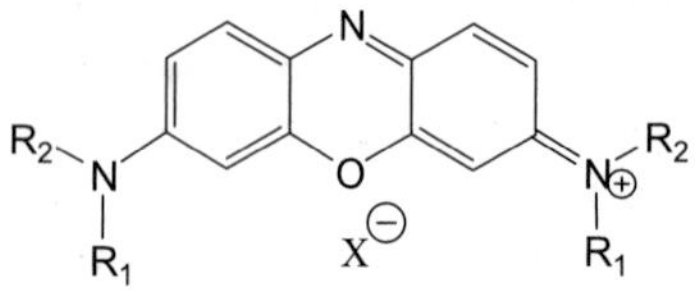

$R_1, R_2 = H$, alkyl

oxazines

3.4.4
Compounds undergoing photoinduced intramolecular charge transfer (ICT) and internal rotation

Excitation of a fluorophore induces the motion of an electron from one orbital to another. If the initial and final orbitals are separated in space, the electronic transition is accompanied by an almost instantaneous change in the dipole moment of the fluorophore. When the latter possesses an electron-donating group (e.g. $-NH_2$, $-NMe_2$, $-CH_3O$) conjugated to an electron-withdrawing group (e.g. >C=O, –CN), the increase in dipole moment can be quite large. Consequently, the excited state reached upon excitation (called the *Franck–Condon state* or *locally excited state*, LE) is not in equilibrium with the surrounding solvent molecules if the latter are polar. In a medium that is sufficiently fluid, the solvent molecules rotate during the lifetime of the excited state until the solvation shell is in thermodynamic equilibrium with the fluorophore. A relaxed *intramolecular charge transfer (ICT) state* is then reached.

Such a *solvent relaxation* explains the increase in the red-shift of the fluorescence spectrum as the polarity of the solvent increases. The effect of polarity on fluorescence emission will be further discussed in Chapter 7, together with polarity probes. Moreover, when a cation receptor is linked to an intramolecular charge transfer fluorophore so that the bound cation can interact with either the donor group or the acceptor group, the ICT is perturbed; the consequent changes in photophysical properties of the fluorophore can be used for sensing cations (see Section 10.3.3).

Relaxation towards an ICT state may be accompanied by internal rotation within the fluorophore. The prime example of great interest is 4-*N*,*N*-dimethylamino-benzonitrile (DMABN).

H_3C / H_3C — N — (phenyl) — CN

DMABN

This molecule has been the object of many studies because, in spite of its simplicity, it exhibits dual fluorescence in polar solvents. This intriguing phenomenon can be explained in the following way (Lippert et al., 1987). In the ground state, the molecule is almost planar, which corresponds to the maximum conjugation between the dimethylamino group and the phenyl ring. According to the Franck–Condon principle, the locally excited state (LE) is still planar, but solvent relaxation takes place with a concomitant rotation of the dimethylamino group until it is twisted at right angles and the conjugation is lost. In the resulting TICT (Twisted Intramolecular Charge Transfer) state, stabilized by the polar solvent molecules, there is a total charge separation between the dimethylamino group and the cyanophenyl moiety (Scheme 3.3).

In addition to the fluorescence band due to emission from the LE state ('normal' band), an emission band corresponding to emission from the TICT state is observed at higher wavelengths ('anomalous' band) (Figure 3.4).

The twisting assumption can be demonstrated by comparing the fluorescence characteristics of the bridged model compounds **2** and **3** with those of DMABN (**1**) in a polar solvent: no twist is possible in compound **2** and LE fluorescence is solely

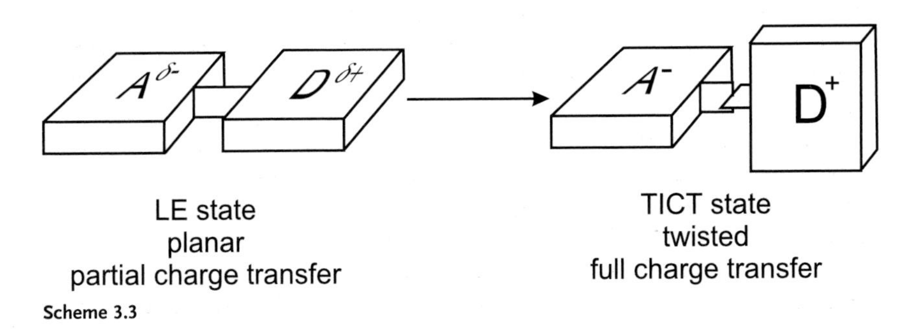

Scheme 3.3

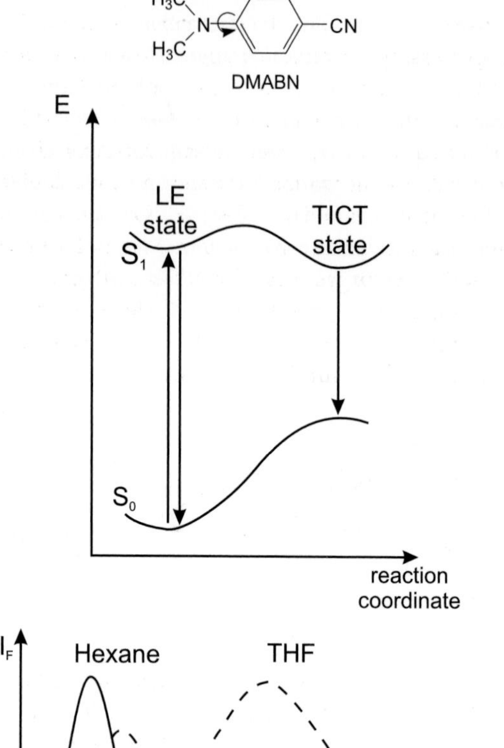

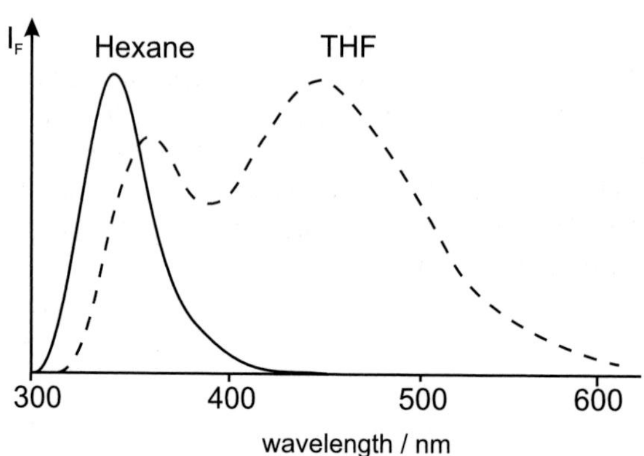

Fig. 3.4. Potential energy diagram of DMABN (top); the reaction coordinate contains both solvent relaxation and rotation of the dimethylamino group. Room temperature fluorescence spectrum in hexane and tetrahydrofurane (bottom) (adapted from Lippert et al., 1987).

observed; the twisted compound **3** exhibits only the TICT fluorescence band. In contrast, DMABN shows both LE and TICT fluorescence.

1   **2**   **3**

Internal rotation, accompanying solvent relaxation or not, can occur in many fluorophores, but dual fluorescence and right angle twist (required for full charge separation) are exceptional. The existence of TICT states (fluorescent or not) has often been invoked (sometimes abusively!) to interpret the photophysical properties of fluorophores, especially those possessing an anilino moiety. When there are several possibilities for internal rotation in an excited molecule, interpretation of the photophysical properties becomes difficult. For instance, in the case of DCM (a well-known merocyanine widely used as a laser dye), several rotations about single bonds are possible. Moreover, photoisomerization (twisting about the double bond) can occur; this process is the main non-radiative de-excitation pathway in nonpolar solvents but it is quite inefficient in polar solvents where the ICT state is stabilized. A very efficient ICT can indeed occur from the dimethylamino group to the dicyanomethylene group, as revealed by the much higer dipole moment of DCM in the excited state than in the ground state (the difference is about 20 D). Donor–acceptor stilbenes such as DCS exhibit a similar behavior.

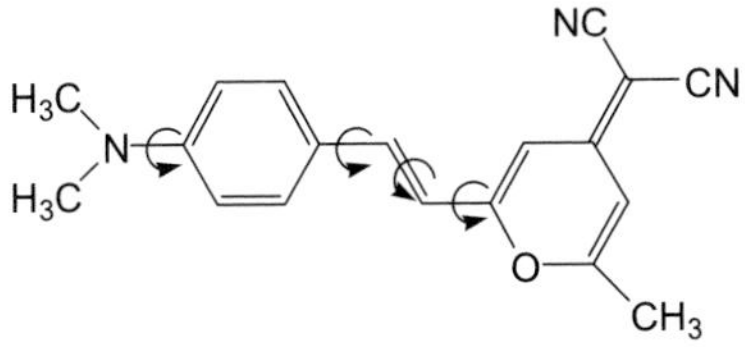

NC CN H3C N H3C O CH3

DCM

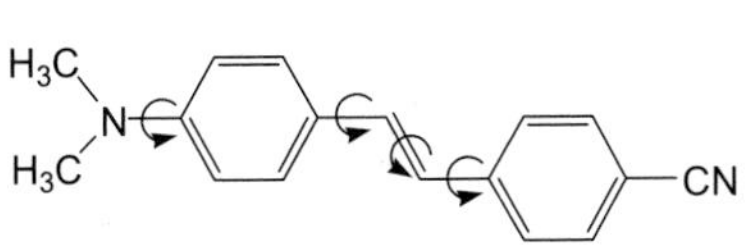

H3C N H3C CN

DCS

Intramolecular charge transfer and internal rotation can also occur in nonpolar and highly symmetric molecules. An outstanding example is 9,9′-bianthryl. This compound in fact exhibits dual fluorescence and the band located at higher wavelengths exhibits a large red-shift as the solvent polarity increases, which is indicative of a highly dipolar character (TICT state). Thus, the high symmetry of bianthryl is broken when the charge transfer state is formed.

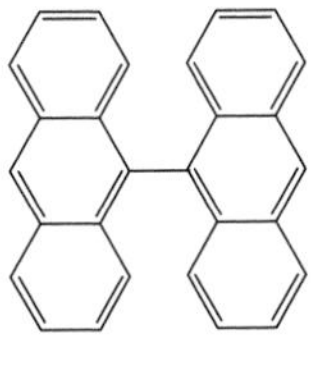

9,9'-bianthryl

Triphenylmethane dyes such as *malachite green* are short-chain cyanine dyes possessing two equivalent resonance structures with a charged and an uncharged nitrogen atom. Consequently, they are symmetric with respect to charge distribution. The photophysics of these dyes is quite complex. Their fluorescence quantum yields and decays are very sensitive to the solvent viscosity, which can be explained by the rotational diffusion of the phenyl rings occurring along a barrierless potential, with a non-radiative decay rate that depends on the twist angle. The fluorescence quantum yield is thus dependent on solvent viscosity but the internal rotations do not reflect the macroscopic viscosity of the solvent because the free volume effects are important. This point will be discussed in Chapter 8.

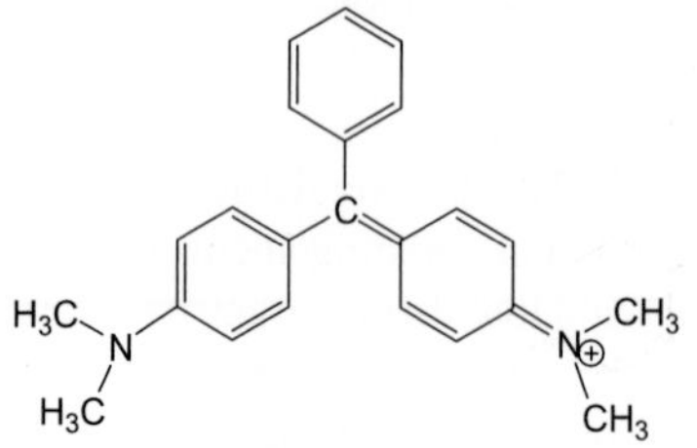

malachite green

The substituted diphenylmethane dye, *auramine O*, is weakly fluorescent in fluid solvents but highly fluorescent in viscous or rigid media. It was originally used to probe the viscosity of viscous polymeric samples. Such a strong dependence on solvent viscosity can be explained in the same way as for triphenylmethane dyes.

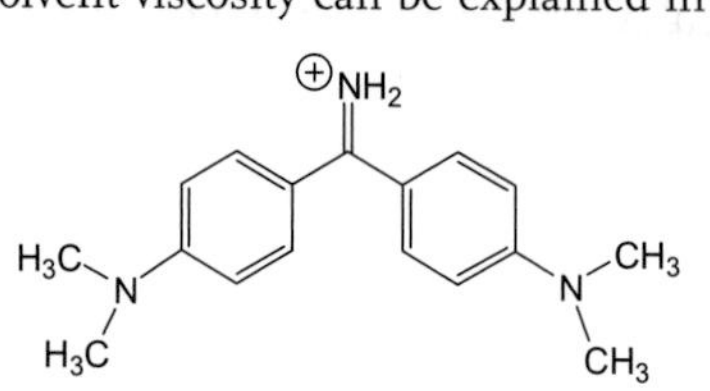

auramine O

As a general rule, internal rotations often provide additional channels for non-radiative de-excitation. Bridging the triphenylmethane structure with an oxygen atom reduces the possibilities of internal rotation. The resulting compounds are *rhodamines* with much higher fluorescence quantum yields (see Section 3.4.3). Further reduction of internal rotation can be achieved via inclusion of the nitrogen atoms in a julolidyl ring, e.g. in rhodamine 101. Consequently, the fluorescence quantum yield of rhodamine 101 is higher (0.92 in ethanol) than that of rhodamine B (0.54 in ethanol)[11]. Moreover, in contrast to rhodamine B, the fluorescence quantum yield of rhodamine 101 is almost independent of temperature, and so it is preferred as a quantum counter in spectrofluorometers.

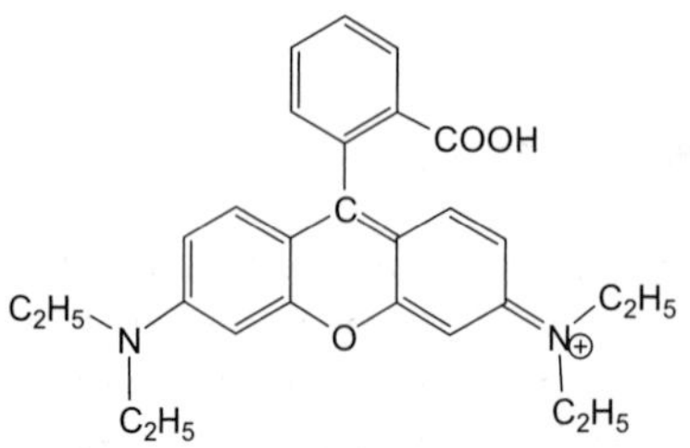

rhodamine B

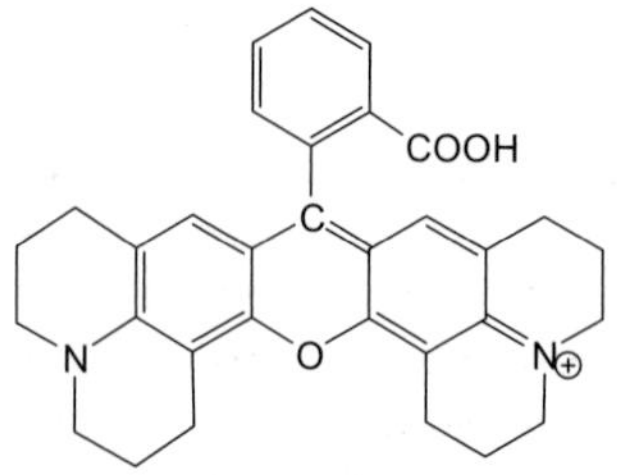

rhodamine 101

11) These compounds are sensitive to pH because of the presence of the carboxylic group. The given fluorescence quantum yields are those of the acidic form. There is of course no pH sensitivity exhibited by the ester derivatives (e.g. rhodamine 6G).

Rhodamine differs from malachite green in the same way as fluorescein differs from phenolphthalein. The latter is known to be non-fluorescent.

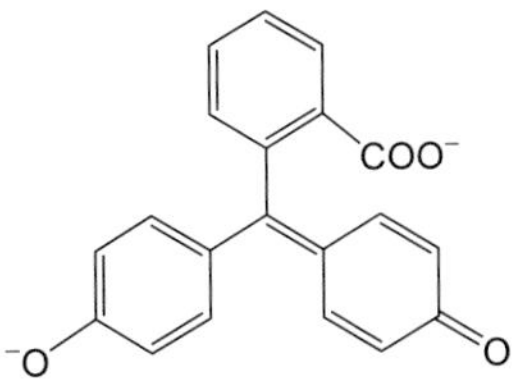

phenolphthalein (basic form)

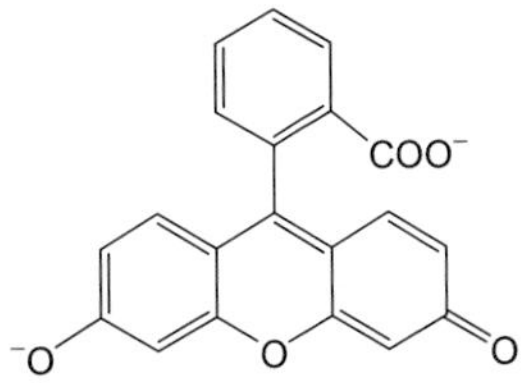

fluorescein dianion

The photophysics of fluorophores undergoing photoinduced charge transfer and/or internal rotation(s) is often complex. Time-resolved fluorescence experiments, transient absorption spectroscopy measurements, quantum chemical calculations, and comparison with model molecules are helpful in understanding their complex photophysical behavior.

3.5 Environmental factors affecting fluorescence

In Section 3.4, structural effects were often discussed in conjunction with the nature of the solvent. As emphasized in the introduction to this book, the fluorescence emitted by most molecules is indeed extremely sensitive to their microenvironment (see Figure 1.3), which explains the extensive use of fluorescent probes. The effects of solvent polarity, viscosity and acidity deserves much attention because these effects are the basis of fluorescence probing of these microenvironmental characteristics and so, later chapters of this book are devoted to these aspects. The effects of polarity and viscosity on fluorescence characteristics in fluid media and the relevant applications are presented in Chapters 7 and 8, respectively. The effect of acidity is discussed in Sections 4.5 and 10.2. This section is thus mainly devoted to rigid matrices or very viscous media, and gases.

3.5.1 Homogeneous and inhomogeneous band broadening. Red-edge effects

The width of a band in the absorption or emission spectrum of a fluorophore located in a particular microenvironment is a result of two effects: *homogeneous* and *inhomogeneous broadening*. Homogeneous broadening is due to the existence of a continuous set of vibrational sublevels in each electronic state. Absorption and emission spectra of moderately large and rigid fluorophores in solution could therefore be almost structureless at room temperature. However, in some cases, many of the vibrational modes are not active, neither in absorption nor in emission, so that a clear vibrational structure is observed (e.g. naphthalene, pyrene).

The second cause of broadening of electronic spectra is the fluctuations in the structure of the solvation shell surrounding the fluorophore. The distribution of solute–solvent configurations and the consequent variation in the local electric field leads to a statistical distribution of the energies of the electronic transitions. This phenomenon is called *inhomogeneous broadening* (for a review see Nemkovich et al., 1991).

In most cases, the extent of inhomogeneous broadening is much greater than that of homogeneous broadening. When interactions with the surrounding molecules are strong, and many configurations are possible, the spectra may become completely blurred. Such spectra are of limited value in analytical fluorescence spectroscopy, especially when the samples contain several compounds whose fluorescence spectra overlap. However, there are several ways to reduce the effect of inhomogeneous broadening for analytical purposes (see Section 3.5.2).

In polar rigid media such as frozen solutions or polymer matrices, inhomogeneous broadening is reduced but still exists, and is responsible for *red-edge effects*, i.e. specific effects that are observed when the fluorophores are excited on the red-edge of the absorption spectrum. Red-edge excitation selects the 'hot' molecules, i.e. those that absorb from vibrational levels above that of the lowest vibrational level of the ground state. The corresponding fluorescence spectrum is red-shifted with respect to the fluorescence spectrum observed upon excitation in the bulk of the absorption spectrum. In liquid solutions, the inhomogeneous broadening becomes dynamic, and the red-shift disappears because a dynamic equilibrium exists among the various solvation sites, but it is still observable if the solvent reorientation relaxation competes with the fluorescence decay.

An excitation-wavelength dependence at the longwave edge of the absorption spectrum has been observed not only for spectral displacement but also for other parameters such as lifetime, quantum yield and apparent rotational rate. Applications to the investigation of polymer rigidity and/or free volume, and to the study of biological systems and excited-state reactions have been developed.

Finally, there is a specific red-edge effect related to non-radiative energy transfer between a donor fluorophore whose emission spectrum overlaps the absorption spectrum of an acceptor fluorophore: in rigid polar solutions, there is a lack of energy transfer upon excitation at the red-edge. This effect, called *Weber's effect*, will be described in Section 9.4.3.

3.5.2
Solid matrices at low temperature

For applications to chemical analysis, a high spectral resolution is desirable and can be achieved using solid matrices at low temperature. In fact, under these conditions, the degree of microenvironmental heterogeneity (expressed as the number of solvation sub-classes) is reduced, which results in decreased inhomogeneous broadening. Moreover, it is advantageous to use laser excitation whose linewidth is less than the width of the inhomogeneously broadened absorption band. Selective excitation of individual compounds in complex mixtures is then possible (Wehry and Mamantov, 1981; Hofstraat et al., 1988).

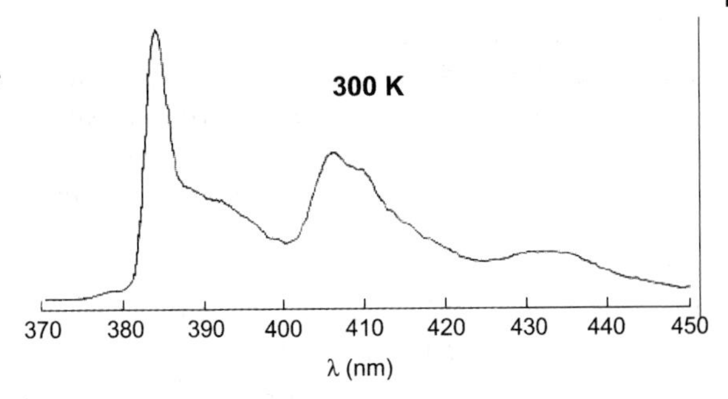

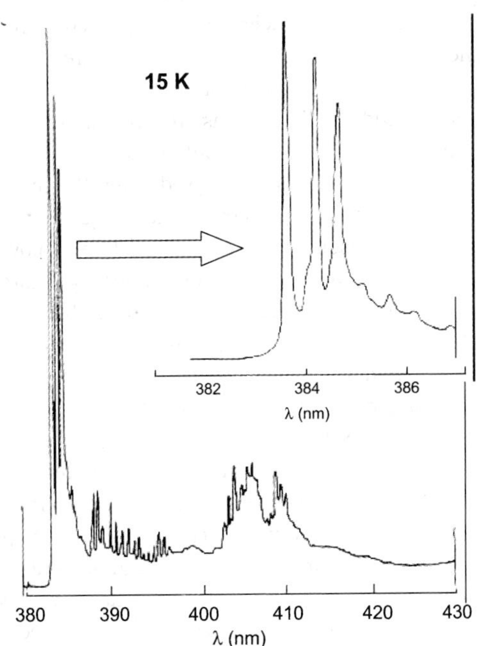

Fig. 3.5. Fluorescence spectrum of benz[a]anthracene in *n*-heptane liquid solution at 300 K (top) and in *n*-heptane frozen solution at 15 K (bottom). Concentration: 2×10^{-5} M. Excitation wavelength: 329.2 nm (adapted from Wehry and Mamantov, 1981).

Shpol'skii spectroscopy Shpol'skii matrices are formed by freezing a solution of fluorophores in *n*-alkanes (provided that the solubility in these solvents is high enough). Rapid immersion in liquid nitrogen or helium leads to a polycrystalline solid matrix. The solute molecules are in very similar microenvironments and inhomogeneous broadening is therefore considerably reduced (Figure 3.5). The spectral lines are thus very narrow (≈ 10 cm^{-1} or less) and Shpol'skii spectra can serve as 'fingerprints' of complex mixtures. Identification of trace amounts of polynuclear aromatic hydrocarbons (PAHs) in complex samples is a great success of this technique.

Matrix isolation spectroscopy In this technique, the fluorophore in the solid or liquid form is vaporized, and the resulting gaseous sample is mixed with a large excess of an inert gas (rare gases – mainly argon and in some cases xenon or krypton or nitrogen). The mixture is deposited on a solid surface at cryogenic temperatures (15 K or less) obtained with a helium bath cryostat. Aggregation effects are avoided in such matrices, in contrast to solid samples prepared by freezing of liquid solutions.

In argon or nitrogen matrices, the spectral resolution (bandwidth $\approx 100\ cm^{-1}$) is not as high as in Shpol'skii frozen solutions (bandwidth $\approx 10\ cm^{-1}$ or less). In the case of complex mixtures requiring high resolution, it is thus preferable to use Shpol'skii solvents (n-alkanes and n-perfluoroalkanes).

Site-selection spectroscopy Maximum selectivity in frozen solutions or vapor-deposited matrices is achieved by using exciting light whose bandwidth (0.01–0.1 cm^{-1}) is less than that of the inhomogeneously broadened absorption band. Lasers are optimal in this respect. The spectral bandwidths can then be minimized by selective excitation only of those fluorophores that are located in very similar matrix sites. The temperature should be very low (5 K or less). The techniques based on this principle are called in the literature *site-selection spectroscopy, fluorescence line narrowing* or *energy-selection spectroscopy*. The solvent (3-methylpentane, ethanol–methanol mixtures, EPA (mixture of ethanol, isopentane and diethyl ether)) should form a clear glass in order to avoid distortion of the spectrum by scatter from cracks.

Various compounds have been analyzed by site-selection spectroscopy. An interesting application is the identification of loci of damage in DNA resulting from reactions with carcinogens such as polycyclic aromatic hydrocarbons.

3.5.3 Fluorescence in supersonic jets

Although this book is devoted to molecular fluorescence in condensed phases, it is worth mentioning the relevance of fluorescence spectroscopy in supersonic jets (Ito et al., 1988). A gas expanded through an orifice from a high-pressure region into a vacuum is cooled by the well-known Joule–Thomson effect. During expansion, collisions between the gas molecules lead to a dramatic decrease in their translational velocities. Translational temperatures of 1 K or less can be attained in this way. The *supersonic jet technique* is an alternative low-temperature approach to the solid-phase methods described in Section 3.5.2; all of them have a common aim of improving the spectral resolution.

In a supersonic jet, the fluorescence spectra are virtually free of environmental perturbations (in contrast to condensed-phase samples) and can thus provide information on isolated solute molecules. Moreover, van der Waals complexes with other solute molecules can be studied, which is of great fundamental interest.

3.6 Bibliography

Birks J. B. **(1969)** *Photophysics of Aromatic Molecules*, Wiley, London.

Birks J. B. (Ed.) **(1973)** *Organic Molecular Photophysics*, Wiley, London.

Guilbault G. G. (Ed.) **(1990)** *Practical Fluorescence*, Marcel Dekker, New York.

Hofstraat J. W., Gooijer C. and Velthorst N. H. (1988) Highly resolved molecular Luminescence Spectroscopy, In: Schulman S. G., Ed., *Molecular Luminescence Spectroscopy*, Part 2, John Wiley and Sons, New York. pp. 283–400.

Ito M., Ebata T. and Mikami N. **(1988)** Laser Spectroscopy of Large Polyatomic Molecules in Supersonic Jets, *Ann. Rev. Phys. Chem.* 39, 123–147.

Lippert E., Rettig W., Bonacic-Koutecky V., Heisel F. and Miehé, J. A. **(1987)** Photophysics of Internal Twisting, *Adv. Chem. Phys.* 68, 1–173.

Nemkovich N. A., Rubinov A. N. and Tomin I. T. **(1991)** Inhomogeneous Broadening of Electronic Spectra of Dye Molecules in Solutions, in: Lakowicz J. R. (Ed.), *Topics in Fluorescence Spectroscopy, Vol. 2, Principles*, Plenum Press, New York, pp. 367–428.

Parker C. A. **(1968)** *Photoluminescence of Solutions*, Elsevier, Amsterdam.

Rettig W. **(1994)** Photoinduced Charge Separation via Twisted Intramolecular Charge Transfer States, *Topics Curr. Chem.* 169, 253–299.

Turro N. J. **(1978)** *Modern Molecular Photochemistry*, Benjamin/Cummings, Menlo Park, CA.

Wehry E. L. **(1990)** Effects of Molecular Structure on Fluorescence and Phosphorescence, in: Guilbault G. G. (Ed.), *Practical Fluorescence*, Marcel Dekker, New York, pp. 75–125.

Wehry E. L. and Mamantov G. **(1981)** Low-Temperature Fluorometric Techniques and Their Application to Analytical Chemistry, in: Wehry E. L. (Ed.), *Modern Fluorescence Spectroscopy*, Vol. 4, Plenum Press, New York, pp. 193–250.

4
Effects of intermolecular photophysical processes on fluorescence emission

> Luminescence research has a long history, full of splendor and surprise, and a bright future, promising variegated applications to probe molecules and crystals, to visualize atomic phenomena, ...
>
> H. J. Queisser, 1981

4.1
Introduction

In Chapter 3 we described the intrinsic pathways of de-excitation of a molecule M^*; the sum k_M of the rate constants for these processes is equal to the reciprocal of the excited-state lifetime τ_0[1]:

$$k_M = k_r + k_{ic} + k_{isc} = k_r + k_{nr} = 1/\tau_0 \tag{4.1}$$

As outlined in Chapter 3, this lifetime is the experimental time window through which processes of similar duration, and competing with the intrinsic de-excitation, can be observed. Most of these processes involve interaction of an excited molecule M^* with another molecule Q according to Scheme 4.1, where k_q represents the *observed rate constant for the bimolecular process.*

The main intermolecular photophysical processes responsible for de-excitation of molecules are presented in Table 4.1. It is interesting to note that most of them

1) Because this chapter deals only with fluorescence (unless otherwise stated), the subscript F and the superscript S used in Chapter 3 are omitted. The subscript 0 (and in some cases, the superscript 0) for fluorescence intensity, quantum yield and lifetime will refer to these characteristics in the absence of intermolecular processes.

Scheme 4.1

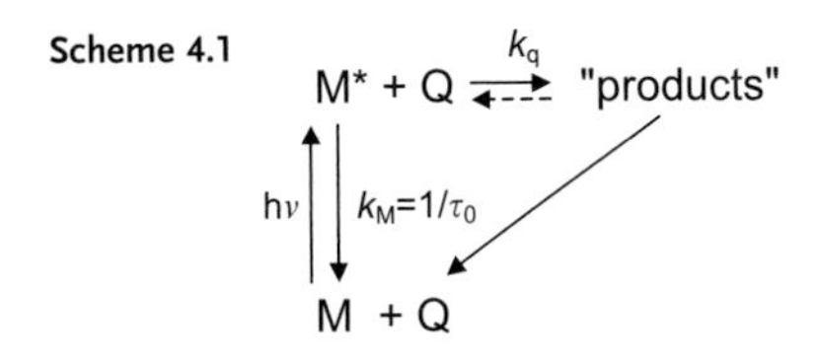

Tab. 4.1. Main photophysical processes responsible for fluorescence quenching

Photophysical process	*$M^* + Q \rightarrow$ products*	*Donor*	*Acceptor*
Collision with a heavy atom (e.g., I^-, Br in CBr_4)[a] or a paramagnetic species (e.g. O_2, NO)[b]	$M^* + Q \rightarrow M + Q + \text{heat}$		
Electron transfer	${}^1D^* + A \rightarrow D^{\cdot+} + A^{\cdot-}$	${}^1D^*$	A
	${}^1A^* + D \rightarrow A^{\cdot-} + D^{\cdot+}$	D	${}^1A^*$
Excimer formation	${}^1M^* + {}^1M \rightarrow {}^1(MM)^*$		
Exciplex formation	${}^1D^* + A \rightarrow {}^1(DA)^*$	${}^1D^*$	A
	${}^1A^* + D \rightarrow {}^1(DA)^*$	D	${}^1A^*$
Proton transfer	$AH^* + B \rightarrow A^{-*} + BH^+$	AH^*	B
	$B^* + AH \rightarrow BH^{+*} + A^-$	AH	B^*
Energy transfer	${}^1D^* + {}^1A \rightarrow {}^1D + {}^1A^*$	${}^1D^*$	${}^1A$
	${}^3D^* + {}^1A \rightarrow {}^1D + {}^3A^*$	${}^3D^*$	${}^1A$
	${}^3D^* + {}^3A \rightarrow {}^1D + {}^3A^*$	${}^3D^*$	${}^3A$
	${}^1M^* + {}^1M \rightarrow {}^1M + {}^1M^*$	${}^1M^*$	${}^1M$

a) As in the case of the intramolecular heavy atom effect (see Chapter 3), intersystem crossing is favored by collision with heavy atoms.

b) Molecular oxygen is a well-known quencher of fluorescence. It deserves special attention because it is ubiquitous in solutions. Its ground state is a triplet state (denoted ${}^3O_2$ or ${}^3\Sigma$) and it has two low-lying singlet states (${}^1\Delta$ and ${}^1\Sigma$). Quenching of singlet or triplet states via energy transfer to produce ${}^1\Delta$ oxygen is thus possible:

$$ {}^1A^* + {}^3O_2({}^3\Sigma) \rightarrow {}^1A + {}^1O_2({}^1\Delta) \quad \text{or} $$

$$ {}^3A^* + {}^3O_2({}^3\Sigma) \rightarrow {}^3A + {}^1O_2({}^1\Delta) $$

'Chemical' quenching by formation of a complex A^*O_2 may also be involved in oxygen quenching.

The contribution of oxygen quenching to the decay of an excited state can be expressed by a quenching term $k_q[O_2]$ to be added to the rate constants of de-excitation. Under atmospheric pressure, the concentration of oxygen in most solvents is 10^{-3}–10^{-4} mol L^{-1}. Therefore, if $k_q \sim 10^{10}$ L mol^{-1} s^{-1}, the lifetime of A^* in the presence of O_2 cannot be more than 10^{-6}–10^{-7} s in air-saturated solutions. The longer the lifetime in the absence of O_2, the stronger the sensitivity to the presence of oxygen, and for some applications, solutions must be degassed by bubbling with nitrogen (or argon), or by the freeze–pump–thaw technique.

involve a fast transfer process from a donor to an acceptor: electron transfer, proton transfer or energy transfer.

The fluorescence characteristics (decay time and/or fluorescence quantum yield) of M^* are affected by the presence of Q as a result of competition between the intrinsic de-excitation and these intermolecular processes:

(i) after excitation by a light pulse, the excited-state M^* population, and consequently the fluorescence intensity, decreases more rapidly than in the absence of excited-state interaction with Q.
(ii) the fluorescence quantum yield is decreased. The loss of fluorescence intensity is called *fluorescence quenching* whatever the nature of the competing intermolecular process and even if this process leads to a fluorescent species (the word quenching applies only to the initially excited molecule).

Analysis of the observed phenomena provides much information on the surroundings of a fluorescent molecule, either quantitatively (if a kinetic model of the competitive processes can be developed so that a kinetic analysis is possible), or at least qualitatively. Special attention will be paid in this chapter to the case where fluorescent molecules are used for probing matter or living systems.

This chapter is restricted to *intermolecular photophysical processes*[2]. Intramolecular excited-state processes will not be considered here, but it should be noted that they can also affect the fluorescence characteristics: intramolecular charge transfer, internal rotation (e.g. formation of twisted charge transfer states), intramolecular proton transfer, etc.

Photochemical de-excitation (i.e. de-excitation resulting from organic photochemical reactions implying bond breaking and formation of new bonds so that the ground state of M is not recovered) is beyond the scope of this book.

4.2 Overview of the intermolecular de-excitation processes of excited molecules leading to fluorescence quenching

4.2.1 Phenomenological approach

Several cases can be identified in which an intermolecular process involving M^* and Q is in competition with intrinsic de-excitation of M^*.

- *Case A*: Q is in large excess so that there is a high probability that M^* and Q, at the time of excitation, are at a distance where the interaction is significant. Then, no mutual approach during the excited-state lifetime is required.

2) The word 'photophysical' implies that, after completion of all de-excitation processes, the ground state of M is recovered unaltered.

When the probability of finding a quencher molecule within the encounter distance with a molecule M^* is less than 1, this situation is relevant to *static quenching* (see Section 4.2.3).

When this probability is equal to 1 (uniform concentration), the 'reaction' is of pseudo-first order. This is the case, for example, in *photoinduced proton transfer* in aqueous solutions from an excited acid M^* ($\equiv AH^*$) (see Section 4.5): M^* is always within the encounter distance with a water molecule acting as a proton acceptor, and thus proton transfer occurs effectively according to a unimolecular process. This is also the case of *photoinduced electron transfer* in aniline or its derivatives as solvents: an excited acceptor is always in the vicinity of an aniline molecule as an electron donor. In both cases, the excited-state reaction occurs under non-diffusive conditions and is of pseudo-first order.

- ***Case B***: Q is not in large excess. However, mutual approach of M^* and Q is not possible during the excited-state lifetime (because the lifetime is too short, or the medium is too viscous). Quenching can then occur only if the interaction is significant at distances longer than the collisional distance. This is the case for *long-range non-radiative energy transfer* that can occur between molecules at distances up to ~80 Å (see Section 4.6.3 and Chapter 9).
- ***Case C***: Q is not in large excess and mutual approach of M^* and Q is possible during the excited-state lifetime. The bimolecular excited-state process is then *diffusion-controlled*. This type of quenching is called *dynamic quenching* (see Section 4.2.2). At high concentrations of Q, static quenching may occur in addition to dynamic quenching (see Section 4.2.4).

Case C is illustrated in Scheme 4.2, where k_1 is the diffusional second-order rate constant for the formation of the pair $(M^* \ldots Q)$ from separated M^* and Q, k_{-1} is the first-order backward rate constant for this step, k_R is the first-order rate constant for the reaction of $(M^* \ldots Q)$ to form 'products'[3], and k_M $(= 1/\tau_0)$ is the rate constant for intrinsic de-excitation of M^*. If the interaction between M^* and Q is weak, the fluorescence characteristics of the pair $M^* \ldots Q$ are the same as those of M^* (Scheme 4.2).

If the fluorescence of the pair is different from that of M^* (e.g. in the case of the formation of an exciplex $(MQ)^*$, which may be an intermediate in electron transfer reactions; see Sections 4.3 and 4.4.2), a different rate constant k_{MQ} $(= 1/\tau_0')$ for intrinsic de-excitation must be introduced (Scheme 4.3).

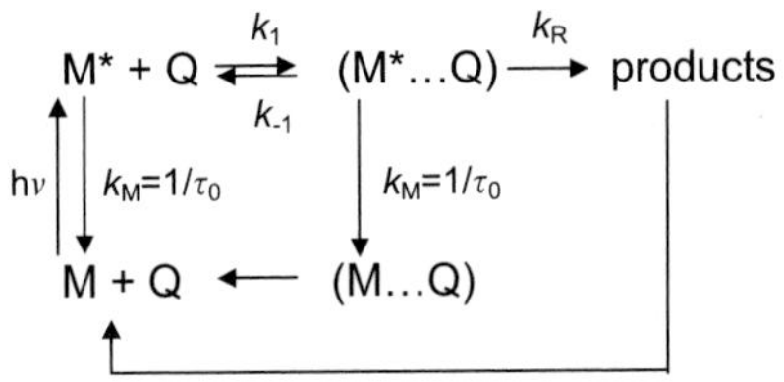

$$M^* + Q \underset{k_{-1}}{\overset{k_1}{\rightleftharpoons}} (M^* \ldots Q) \xrightarrow{k_R} \text{products}$$

$h\nu$ (M + Q → M^* + Q); $k_M = 1/\tau_0$ (M^* + Q → M + Q); $k_M = 1/\tau_0$ ($(M^* \ldots Q)$ → $(M \ldots Q)$); $(M \ldots Q)$ → M + Q; products → M + Q

Scheme 4.2

3) For the sake of clarity, it is assumed that there is no back-reaction from the products to the pair.

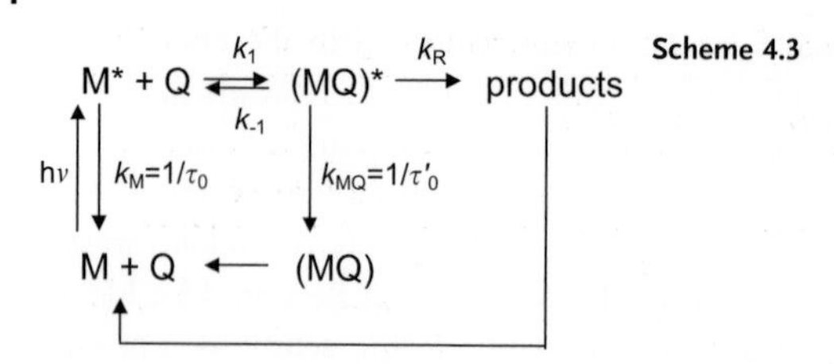

Scheme 4.3

If Q is identical to M (self-quenching without formation of products), $(MM)^*$ is called an *excimer* (excited dimer) with its own fluorescence spectrum and lifetime (Section 4.4.1).

A kinetic analysis of Schemes 4.2 and 4.3 shows that, in all cases, it is valid to consider a pseudo-first order rate constant $k_q[Q]$ and that three main cases should be considered:

1. $k_R \gg k_1[Q], k_{-1}, 1/\tau_0$: the reaction is *diffusion-limited*; the observed rate constant for quenching, k_q, is equal to the diffusional rate constant k_1.
2. $k_R \ll k_1[Q], k_{-1} \gg 1/\tau_0$: the equilibrium is reached prior to the formation of products.
3. k_R is of the same order of magnitude or is smaller than the other rate constants: k_q is smaller than k_1 and can be written as

$$k_q = pk_1 \tag{4.2}$$

where p is the probability of reaction for the encounter pair (often called efficiency), which can be expressed as a function of the rate constants (Eftink and Ghiron, 1981). For instance, p is close to 1 for oxygen, acrylamide and I^-, whereas it is less than 1 for succinimide, Br^- and IO_3^-. Examples of p and k_q values are given in Tables 4.2 and 4.3.

Tab. 4.2. Examples of p and k_q values for indole in the presence of various quenchers (from Eftink, 1991b)

Quencher	p	$k_q/(10^9\ L\ mol^{-1}\ s^{-1})$
Oxygen	1.0	12.3
Acrylamide	1.0	7.1
Succinimide	0.7	4.8
I^-	1.0	6.4
Br^-	0.04	0.2
IO_3^-	0.7	4
Pyridinium-HCl	1.0	9.4
Tl^+	1.0	9.2
Cs^+	0.2	1.1

Tab. 4.3. Examples of k_q values for various fluorophore–quencher pairs (from Eftink, 1991b)

Fluorophore	$k_q/(10^9\ \mathrm{L\ mol^{-1}\ s^{-1}})$			
	O_2	*Acrylamide*	*Succinimide*	I^-
Tyrosine	12	7.0	5.3	
Fluorescein		0	0.2	1.5
Anthracene-9-carboxylic acid	7.7	2.4	0.1	
PRODAN		0	0	2.8
DENS	7.7	3.2	0.1	0.03
Eosin Y		0	0	2.2
Pyrenebutyric acid	10	3.4	0.06	
Carbazole		7.0	5.4	

Abbreviations: PRODAN: 6-propionyl-2-(dimethylamino)naphthalene;
DENS: 6-diethylaminonaphthalene-1-sulfonic acid.

It should be emphasized that *for diffusion-controlled reactions, the observed rate constant for quenching is time-dependent.* In fact, the excited fluorophores M^* that are at a short distance from a quencher Q at the time of excitation react, on average, at shorter times than those that are more distant, because mutual approach requires a longer time before reaction occurs. The important consequence of this is that the very beginning of fluorescence decay curves following pulse excitation will be affected. Such transient effects are not significant for moderate concentrations of quenchers in fluid solvents but they are noticeable at larger quencher concentrations and/or in viscous media. In steady-state experiments, it will be demonstrated that the consequence of these transient effects is a departure from the well known and widely used Stern–Volmer plot. Stern–Volmer kinetics that ignore the transient effects will be presented first.

4.2.2
Dynamic quenching

4.2.2.1 Stern–Volmer kinetics

As a first approach, the experimental quenching rate constant k_q is assumed to be time-independent. According to the simplified Scheme 4.1, the time evolution of the concentration of M^* following a δ-pulse excitation obeys the following differential equation:

$$\frac{d[M^*]}{dt} = -(k_M + k_q[Q])[M^*] = -(1/\tau_0 + k_q[Q])[M^*] \tag{4.3}$$

Integration of this differential equation with the initial condition $[M^*] = [M^*]_0$ at $t = 0$ yields:

$$[M^*] = [M^*]_0 \exp\{-(1/\tau_0 + k_q[Q])t\} \tag{4.4}$$

The fluorescence intensity is proportional to the concentration of M^* and is given by

$$\begin{aligned} i(t) &= k_r[M^*] = k_r[M^*]_0 \exp\{-(1/\tau_0 + k_q[Q])t\} \\ &= i(0) \exp\{-(1/\tau_0 + k_q[Q])t\} \end{aligned} \tag{4.5}$$

where k_r is the radiative rate constant of M^*. The fluorescence decay is thus a single exponential whose time constant is

$$\tau = \frac{1}{\dfrac{1}{\tau_0} + k_q[Q]} = \frac{\tau_0}{1 + k_q\tau_0[Q]} \tag{4.6}$$

Hence,

$$\boxed{\frac{\tau_0}{\tau} = 1 + k_q\tau_0[Q]} \tag{4.7}$$

Time-resolved experiments in the absence and presence of quencher allow us to check whether the fluorescence decay is in fact a single exponential, and provide directly the value of k_q. The fluorescence quantum yield in the presence of quencher is

$$\Phi = \frac{k_r}{k_r + k_{nr} + k_q[Q]} = \frac{k_r}{1/\tau_0 + k_q[Q]} \tag{4.8}$$

whereas, in the absence of quencher, it is given by

$$\Phi_0 = k_r\tau_0 \tag{4.9}$$

Equations (4.8) and (4.9) lead to the *Stern–Volmer relation*:

$$\boxed{\frac{\Phi_0}{\Phi} = \frac{I_0}{I} = 1 + k_q\tau_0[Q] = 1 + K_{SV}[Q]} \tag{4.10}$$

where I_0 and I are the steady-state fluorescence intensities (for a couple of wavelengths λ_E and λ_F) in the absence and in the presence of quencher, respectively. $K_{SV} = k_q\tau_0$ is the Stern–Volmer constant. Generally, the ratio I_0/I is plotted against the quencher concentration (Stern–Volmer plot). If the variation is found to be linear, the slope gives the Stern–Volmer constant. Then, k_q can be calculated if the excited-state lifetime in the absence of quencher is known.

Two cases can be identified:

- If the bimolecular process is not diffusion-limited: $k_q = pk_1$, where p is the efficiency of the 'reaction' and k_1 is the diffusional rate constant.
- If the bimolecular process is diffusion-limited: k_q is identical to the diffusional rate constant k_1, which can be written in the following simplified form (proposed for the first time by Smoluchowski):

$$k_1 = 4\pi N R_c D \quad (\text{in L mol}^{-1}\ \text{s}^{-1}) \tag{4.11}$$

where R_c is the distance of closest approach (in cm), D is the mutual diffusion coefficient (in $\text{cm}^2\ \text{s}^{-1}$), and N is equal to $N_a/1000$[4], N_a being Avogadro's number. The distance of closest approach is generally taken as the sum of the radii of the two molecules (R_M for the fluorophore and R_Q for the quencher). The mutual diffusion coefficient D is the sum of the translational diffusion coefficients of the two species, D_M and D_Q, which can be expressed by the Stokes–Einstein relation[5]

$$D = D_M + D_Q = \frac{kT}{f\pi\eta}\left(\frac{1}{R_M} + \frac{1}{R_Q}\right) \tag{4.12}$$

where k is Boltzmann's constant, η is the viscosity of the medium, f is a coefficient that is equal to 6 for 'stick' boundary conditions and 4 for 'slip' boundary conditions.

The diffusion coefficient of molecules in most solvents at room temperature is generally of the order of $10^{-5}\ \text{cm}^2\ \text{s}^{-1}$. In liquids, k_1 is about 10^9–10^{10} $\text{L mol}^{-1}\ \text{s}^{-1}$.

If R_M and R_Q are of the same order, the diffusional rate constant is approximately equal to $8RT/3\eta$.

4.2.2.2 Transient effects

In reality, the diffusional rate constant is time-dependent, as explained at the end of Section 4.2.1, and should be written as $k_1(t)$. Several models have been developed to express the time-dependent rate constant (see Box 4.1). For instance, in Smoluchowski's theory, $k_1(t)$ is given by

4) Because R_c is expressed in cm and D in $\text{cm}^2\ \text{s}^{-1}$, the factor $N_a/1000$ accounts for the conversion from molecules per cm^3 to moles per dm^3 (L).

5) The Stokes–Einstein relation is only valid for a rigid sphere that is large compared to the molecular dimensions, moving in a homogeneous Newtonian fluid, and obeying the Stokes hydrodynamic law. Therefore, the use of this relation is questionable when the size of the moving molecules is comparable to that of the surrounding molecules forming the microenvironment. This point will be discussed in detail in Chapter 8 dealing with the use of fluorescent probes to estimate the fluidity of a medium.

Box 4.1 Theories of diffusion-controlled reactions

The aim of this box is to present the most important expressions for the time-dependent rate constant $k_1(t)$ that have been obtained for a diffusion-controlled reaction between A and B.

$$\mathrm{A} + \mathrm{B} \underset{k_{-1}}{\overset{k_1(t)}{\rightleftharpoons}} (\mathrm{AB}) \xrightarrow{k_\mathrm{R}} \text{products}$$

The δ-pulse response of the fluorescence intensity can then be obtained by introducing the relation giving $k_1(t)$ into Eq. (4.14) and by analytical or numerical integration of this equation.

Smoluchowski's theory[a]

Smoluchowski, who worked on the rate of coagulation of colloidal particles, was a pioneer in the development of the theory of diffusion-controlled reactions. His theory is based on the assumption that the probability of reaction is equal to 1 when A and B are at the distance of closest approach (R_c) ('absorbing boundary condition'), which corresponds to an infinite value of the intrinsic rate constant k_R. The rate constant k_{-1} for the dissociation of the encounter pair can thus be ignored. As a result of this boundary condition, the concentration of B is equal to zero on the surface of a sphere of radius R_c, and consequently, there is a concentration gradient of B. The rate constant for reaction $k_1(t)$ can be obtained from the flux of B, in the concentration gradient, through the surface of contact with A. This flux depends on the radial distribution function of B, $p(r,t)$, which is a solution of Fick's equation

$$\frac{\partial p(r,t)}{\partial t} = D\nabla^2 p(r,t) \tag{B4.1.1}$$

where r is the distance between A and B, and D is their mutual diffusion coefficient (given by Eq. 4.12). The initial distribution of B is assumed to be uniform.

Under these conditions, the solution of Eq. (B4.1.1) is the Smoluchowski relation:

$$k_1^\mathrm{S}(t) = 4\pi N R_\mathrm{c} D\left[1 + \frac{R_\mathrm{c}}{(\pi D t)^{1/2}}\right] \tag{B4.1.2}$$

where N is Avogadro's number divided by 1000.

Collins–Kimball's theory[b]

In contrast to Smoluchowski's theory, the rate constant for the intrinsic reaction has a finite value k_R when A and B are at the distance of closest approach

$(r = R_c)$, but it is equal to zero for larger distances. It is assumed that k_R is proportional to the probability that a molecule B is located at a distance from A between R_c and $R_c + \delta r$. Assuming that the rate constant k_{-1} for the dissociation of the encounter pair can still be ignored, the resolution of the diffusion equation (B4.1.1) yields

$$k_1^{SCK}(t) = \frac{\beta k_R}{\beta + k_R}\left\{1 + \frac{k_R}{\beta}\exp(\alpha^2 Dt)\,\mathrm{erfc}(\alpha\sqrt{Dt})\right\} \tag{B4.1.3}$$

where $\alpha = (\beta + k_R)/R_c\beta$ and $\beta = 4\pi N R_c D$. erfc is the complementary error function.

Time-resolved fluorescence experiments carried out with 1,2-benzanthracene quenched by CBr_4 in propane-1,2-diol show a better fit with the Collins–Kimball equation than with the Smoluchowski equation.

Cases of distance-dependent rate constants

1. Exponential distance dependence

In the Collins–Kimball theory, the rate constant for the reaction was assumed to be distance-independent. Further refinement proposed by Wilemski and Fixman[c] consists of considering that the reaction rate constant has an *exponential dependence on distance*, which is indeed predicted for electron transfer reactions and energy transfer via electron exchange (see Dexter's formula in Section 4.6.3). The rate constant can thus be written in the following form:

$$k_1(r) = k_R \exp\left(-\frac{r - R_c}{r_e}\right) \tag{B4.1.4}$$

The distance dependence is characterized by the parameter r_e, which is in the range 0.5–2 Å. The diffusion equation (B4.1.2) must be modified by adding a distance-dependent sink term $k(r)$

$$\frac{\partial p(r,t)}{\partial t} = D\nabla^2 p(r,t) - k(r)p(r,t) \tag{B4.1.5}$$

The time-dependent rate constant can be written as

$$k_1^{WF}(t) = 4\pi \int_{R_c}^{\infty} k(r)p(r,t)r^2\,\mathrm{d}r \tag{B4.1.6}$$

For instance, very satisfactory fits of the experimental decay curves of coumarine 1 in the presence of aniline or *N,N*-dimethylaniline as quenchers were observed by Shannon and Eads[d] (with $r_e = 0.5$ Å and $R_c = 5.5$ Å).

2. Distance dependence in $1/r^6$

According to Förster's theory of non-radiative energy transfer via dipole–dipole interaction (see Section 4.6.3), the distance dependence of the rate constant can be written as $k^{dd}(r) = a/r^6$. Under conditions where the effects of diffusion are significant (see Section 9.3.1), the diffusion equation (B4.1.5) must be solved with the added sink term depending on r^{-6}. Gösele et al.[e] obtained an approximate solution based on interpolation between the known solutions at early and long times and they proposed writing it in a form resembling the Smoluchowski equation (B4.1.2)

$$k_1^G(t) = 4\pi N R_{eff} D\left[1 + \frac{R_{eff}}{(\pi D t)^{1/2}}\right] \quad \text{(B4.1.7)}$$

where R_{eff} is defined by

$$R_{eff} = 0.676\left(\frac{R_0^6}{\tau_D^0 D}\right)^{1/4} \quad \text{(B4.1.8)}$$

R_0 is the Förster critical radius (defined in Section 4.6.3), and τ_D^0 is the excited-state lifetime of the donor in the absence of transfer.

Butler and Pilling[f] calculated an exact numerical solution of the diffusion equation. They showed that the interpolation formula proposed by Gösele et al.[e] reproduces the numerical solution with high precision.

a) Smoluchowski M. V. (1927) *Z. Phys. Chem.* **92**, 129.
b) Collins F. C. and Kimball G. E. (1949) *J. Colloid. Sci.* **4**, 425–37.
c) Wilemski G. and Fixman M. (1973) *J. Chem. Phys.* **58**, 4009–19.
d) Shannon C. F. and Eads D. D. (1995) *J. Chem. Phys.* **103**, 5208–23.
e) Gösele U., Hauser M., Klein U. K. A and Frey R. (1975) *Chem. Phys. Lett.* **34**, 519–22.
f) Butler P. R. and Pilling M. J. (1979) *Chem. Phys.* **41**, 239–43.

$$k_1(t) = 4\pi N R_c D\left[1 + \frac{R_c}{(\pi D t)^{1/2}}\right] \quad (4.13)$$

Consequently, in the case of diffusion-limited quenching, the δ-pulse response of the fluorescence intensity can be calculated by means of the following equation, which replaces Eq. (4.5):

$$i(t) = i(0)\exp\left\{1/\tau_0 + [\mathrm{Q}]\int_0^t k_1(t')\,dt'\right\} \quad (4.14)$$

hence

$$i(t) = i(0)\exp(-at - 2b\sqrt{t}) \tag{4.15}$$

where

$$a = \frac{1}{\tau_0} + 4\pi N R_c D[\mathrm{Q}] \tag{4.16}$$

$$b = 4\sqrt{\pi D} N R_c^2 [\mathrm{Q}] \tag{4.17}$$

It should be noted that, in media of low viscosity, the transient term is significant only at short times ($<$ 100 ps at viscosities similar to that of water) and can be neglected, whereas in viscous media, this term cannot be ignored (the fluorescence decay is then no longer a single exponential).

Under continuous illumination, the steady-state intensity can be easily calculated by considering a light of constant intensity as an infinite sum of infinitely short light pulses. It is thus simply obtained by integration of the δ-pulse response. The ratio of the intensities in the absence and in the presence of quencher is then given by

$$\frac{I_0}{I} = \frac{1 + 4\pi N R_c D \tau_0 [\mathrm{Q}]}{Y} \tag{4.18}$$

where

$$Y = \frac{b\sqrt{\pi}}{\sqrt{a}} \exp(b^2/a)\,\mathrm{erfc}(b/\sqrt{a}) \tag{4.19}$$

with a and b given by Eqs (4.16) and (4.17), respectively. erfc is the complementary error function:

$$\mathrm{erfc}(x) = \frac{2}{\sqrt{\pi}} \int_x^{\infty} \mathrm{e}^{-u^2}\, du$$

Equation (4.18) can be compared to the Stern–Volmer relation (4.10). The multiplying factor Y^{-1} accounts for the transient term and leads to a slight upward curvature of the Stern–Volmer plot.

At long times, the second term in Eq. (4.13) is negligible so that the decay becomes exponential with a decay time τ given by

$$\frac{\tau_0}{\tau} = 1 + 4\pi N R_c D \tau_0 [\mathrm{Q}] \tag{4.20}$$

This relation is identical to Eq. (4.7) with $k_q = 4\pi N R_c D$.

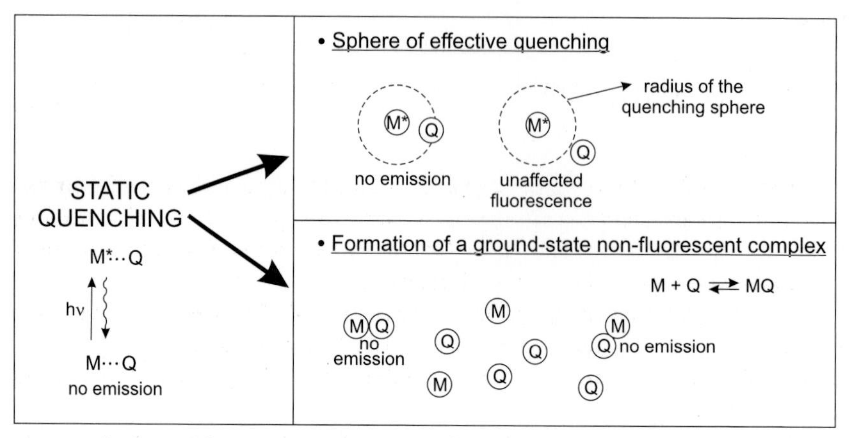

Fig. 4.1. Illustration of static quenching.

4.2.3
Static quenching

The term *static quenching* implies either the existence of a sphere of effective quenching or the formation of a ground-state non-fluorescent complex (Figure 4.1) (Case A of Section 4.2.1).

4.2.3.1 Sphere of effective quenching

When M^* and Q cannot change their positions in space relative to one another during the excited-state lifetime of M^* (i.e. in viscous media or rigid matrices), Perrin proposed a model in which quenching of a fluorophore is assumed to be complete if a quencher molecule Q is located inside a sphere (called the *sphere of effective quenching, active sphere* or *quenching sphere*) of volume V_q surrounding the fluorophore M. If a quencher is outside the active sphere, it has no effect at all on M. Therefore, the fluorescence intensity of the solution is decreased by addition of Q, but the fluorescence decay after pulse excitation is unaffected.

The probability that n quenchers reside within this volume is assumed to obey a Poisson distribution:

$$P_n = \frac{\langle n \rangle^n}{n!} \exp(-\langle n \rangle) \tag{4.21}$$

where $\langle n \rangle$ is the mean number of quenchers in the volume V_q: $\langle n \rangle = V_q N_a[Q]$ (V_q is expressed in L, [Q] in mol L^{-1}, N_a is Avogadro's number). The probability that there is no quencher in this volume is

$$P_0 = \exp(-\langle n \rangle) = \exp(-V_q N_a[Q]) \tag{4.22}$$

Therefore, because the emission intensity is proportional to P_0, Perrin's model leads to

$$\frac{I_0}{I} = \exp(V_q N_a[Q]) \tag{4.23}$$

In contrast to the Stern–Volmer equation (4.10), the ratio I_0/I is not linear and shows an upward curvature at high quencher concentrations. At low concentrations, $\exp(V_q N_a[Q]) \approx 1 + V_q N_a[Q]$, so that the concentration dependence is almost linear (as in the case of the Stern–Volmer plot).

A plot of $\ln(I_0/I)$ versus [Q] yields V_q. The values of $V_q N_a$ are often found to be in the range of 1–3 L mol^{-1}. This corresponds to a quenching sphere radius of about 10 Å, which is somewhat larger than the van der Waals contact distance between M and Q.

Perrin's model has been used in particular for the interpretation of non-radiative energy transfer in rigid media (see Chapter 9).

4.2.3.2 Formation of a ground-state non-fluorescent complex

Let us consider the formation of a non-fluorescent 1:1 complex according to the equilibrium

$$M + Q \rightleftharpoons MQ$$

The excited-state lifetime of the uncomplexed fluorophore M is unaffected, in contrast to dynamic quenching. The fluorescence intensity of the solution decreases upon addition of Q, but the fluorescence decay after pulse excitation is unaffected. Quinones, hydroquinones, purines and pyrimidines are well-known examples of molecules responsible for static quenching.

Using the relation for the stability constant of the complex

$$K_S = \frac{[MQ]}{[M][Q]} \tag{4.24}$$

and the mass conservation law (where $[M]_0$ is the total concentration of M)

$$[M]_0 = [M] + [MQ] \tag{4.25}$$

we obtain the fraction of uncomplexed fluorophores:

$$\frac{[M]}{[M]_0} = \frac{1}{1 + K_S[Q]} \tag{4.26}$$

Considering that the fluorescence intensities are proportional to the concentrations (which is valid only in dilute solutions), this relationship can be rewritten as

$$\frac{I_0}{I} = 1 + K_S[Q] \tag{4.27}$$

A linear relationship is thus obtained, as in the case of the Stern–Volmer plot (Eq. 4.10), but there is no change in excited-state lifetime for static quenching, whereas in the case of dynamic quenching the ratio I_0/I is proportional to the ratio τ_0/τ of the lifetimes.

In some cases, evidence for the formation of a complex can be obtained (e.g. changes in the absorption spectrum upon complexation), but in the absence of such evidence, the interaction is likely to be non-specific and the model of an effective sphere of quenching is more appropriate. A nonlinear variation of I_0/I is predicted in the latter case, but at low quencher concentration, $\exp(V_q N_a[Q]) \approx 1 + V_q N_a[Q]$.

The above considerations can be generalized to complexes of the type M^*Q_n ($n > 1$). The probability that a molecule M^* is in contact with n quencher molecules can be approximately expressed by the Poisson distribution (Eq. 4.21). Perrin's equation (4.23) is then found again.

An interesting application of static quenching is the determination of micellar aggregation numbers (see Box 4.2).

4.2.4
Simultaneous dynamic and static quenching

Static and dynamic quenching may occur simultaneously, resulting in a deviation of the plot of I_0/I against [Q] from linearity.

Let us consider first the case of static quenching by formation of a non-fluorescent complex. The ratio I_0/I obtained for dynamic quenching must be multiplied by the fraction of fluorescent molecules (i.e. uncomplexed)

$$\frac{I}{I_0} = \left[\frac{I}{I_0}\right]_{\text{dyn}} \times \frac{[\mathrm{M}]}{[\mathrm{M}]_0} \tag{4.28}$$

Using Eqs (4.10) and (4.26), the ratio I_0/I can be written as

$$\boxed{\begin{aligned}\frac{I_0}{I} &= (1 + K_{SV}[Q])(1 + K_S[Q]) \\ &= 1 + (K_{SV} + K_S)[Q] + K_{SV}K_S[Q]^2\end{aligned}} \tag{4.29}$$

An upward curvature is thus observed. K_{SV} and K_S can be determined by curve fitting using Eq. (4.29), or alternatively from the plot of $(I_0/I - 1)/[Q]$ against [Q], which should be linear.

Alternatively, using the sphere of effective quenching model, we obtain the following relation instead of Eq. (4.29)

$$\boxed{\frac{I_0}{I} = (1 + K_{SV}[Q])\exp(V_q N_a[Q])} \tag{4.30}$$

Box 4.2 Determination of micellar aggregation numbers by means of fluorescence quenching[a)]

Method I: Static quenching by totally micellized quenchers
Let us consider a fluorescent probe and a quencher that are soluble only in the micellar pseudo-phase. If the quenching is static, fluorescence is observed only from micelles devoid of quenchers. Assuming a Poissonian distribution of the quencher molecules, the probability that a micelle contains no quencher is given by Eq. (4.22), so that the relationship between the fluorescence intensity and the mean occupancy number $\langle n \rangle$ is

$$\ln\left(\frac{I_0}{I}\right) = \langle n \rangle \tag{B4.2.1}$$

$\langle n \rangle$ is related to the micellar aggregation number N_{ag} by the following relation:

$$\langle n \rangle = \frac{[\mathrm{Q}]}{[\mathrm{Mic}]} = \frac{[\mathrm{Q}]N_{ag}}{[\mathrm{Surf}] - [\mathrm{CMC}]} \tag{B4.2.2}$$

where [Q] is the total concentration of quencher, [Mic] is the concentration of micelles, [Surf] is the total concentration of surfactant and [CMC] is the critical micellar concentration. N_{ag} can be calculated from this relation, when all the concentrations are known.

Static quenching by totally micellized quenchers provides a simple steady-state method for the determination of N_{ag}. This method was originally employed with $Ru(bpy)_3^{2+}$ as a fluorophore and 2-methylanthracene as a quencher in SDS (sodium dodecylsulfate) micelles. It is no longer applicable when the contribution of dynamic quenching is not negligible. The validity can be checked by time-resolved measurements: the fluorescence decay should indeed be a single exponential for pure static quenching. Otherwise, the relations given in Method II should be applied.

Method II: Dynamic quenching by totally micellized immobile quenchers
It is assumed that the probability of quenching of a fluorescent probe in a given micelle is proportional to the number of quenchers residing in this micelle. The rate constant for de-excitation of a probe in a micelle containing n quencher molecules is given by

$$k_n = 1/\tau_0 + n k_q \tag{B4.2.3}$$

where k_q is the first-order rate constant for the quenching by one quencher molecule (the intramicellar quenching process is assumed to be a first-order process, as for intramolecular processes).

Assuming a Poissonian distribution, the probability P_n that a micelle contains n quenchers is given by Eq. (4.21). The observed fluorescence intensity following δ-pulse excitation is obtained by summing the contributions from micelles with different numbers of quenchers:

$$\begin{aligned} i(t) &= i(0) \sum_{n=0} P_n \exp(-k_n t) \\ &= i(0) \sum_{n=0} \frac{\langle n \rangle^{\mathrm{n}}}{\mathrm{n}!} \exp(-\langle n \rangle) \exp(-t/\tau_0 + n k_q t) \\ &= i(0) \exp\{-t/\tau_0 + \langle n \rangle [\exp(-k_q t) - 1]\} \end{aligned} \quad \text{(B4.2.4)}$$

At long times, this equation becomes single-exponential:

$$i(t) = i(0) \exp(-\langle n \rangle) \exp(-t/\tau_0) \quad \text{(B4.2.5)}$$

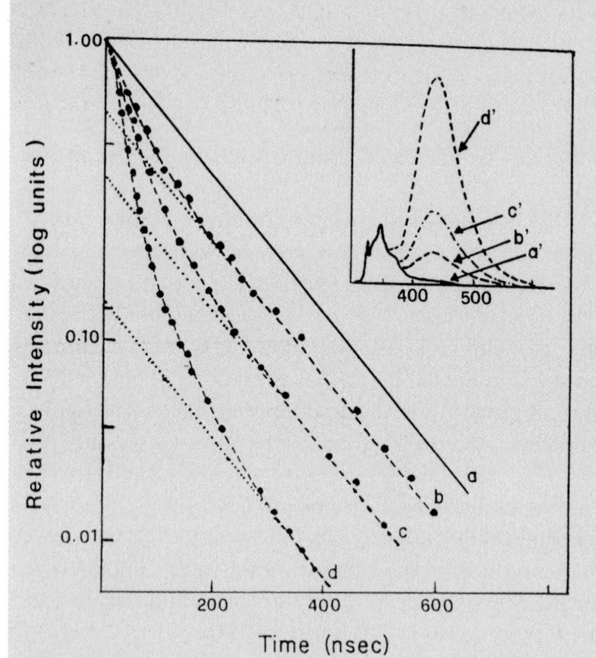

Fig. B4.2.1. Fluorescence decay curves for pyrene monomers in cetyltrimethylammonium (CTAC) micellar solutions (10^{-2} M) at various pyrene concentrations: (a) 7.5×10^{-6} M, (b) 5.2×10^{-5} M, (c) 1.04×10^{-4} M, (d) 2.08×10^{-4} M. The closed circles are the experimental points and the broken lines are the fitted curves according to Eq. (B4.2.4). The dotted lines correspond to Eq. (B4.2.5). Insert: Steady-state fluorescence spectra of corresponding solutions normalized to the monomer emission (reproduced with permission from Atik et al., 1979[b]).

In a logarithmic representation, the slope at long times is the same as in the absence of quencher, and extrapolation to time 0 yields $\langle n \rangle$, from which N_{ag} can be calculated as in Method I.

Quenching of pyrene by excimer formation ($Py^* + Py \rightarrow (PyPy)^* \rightarrow 2Py$) (see Section 4.4.1) is widely used for the determination of micellar aggregation numbers for new surfactant systems. An example is given in Figure B4.2.1.

Fluorescence quenching studies in micellar systems provide quantitative information not only on the aggregation number but also on counterion binding and on the effect of additives on the micellization process. The solubilizing process (partition coefficients between the aqueous phase and the micellar pseudo-phase, entry and exit rates of solutes) can also be characterized by fluorescence quenching.

a) Kalyanasundaran K. (1987) *Photochemistry in Microheterogeneous Systems*, Academic Press, Orlando, Chapter 2.

b) Atik S. S., Nam M. and Singer L. A. (1979) *Chem. Phys. Lett.* **67**, 75.

For example, this relation has been successfully used to describe oxygen quenching of perylene in dodecane at high oxygen pressure.

It should be recalled that an upward curvature can also be due to transient effects that may superimpose the effects of static quenching. The following general relation can then be used

$$\frac{I_0}{I} = (1 + K_{SV}[Q])\frac{\exp(V_q N_a[Q])}{Y} \qquad (4.31)$$

where Y is given by Eq. (4.19).

Figure 4.2 summarizes the various cases of quenching, together with the possible origins of a departure from a linear Stern–Volmer plot.

It should be emphasized that time-resolved experiments are required for unambiguous assignment of the dynamic and static quenching constants.

4.2.5
Quenching of heterogeneously emitting systems

When a system contains a fluorophore in different environments (e.g. a fluorophore embedded in microheterogeneous materials such as sol–gel matrices, polymers, etc.) or more than one fluorophore (e.g. different tryptophanyl residues of a protein), the preceding relations must be modified. If dynamic quenching is predominant, the Stern–Volmer relation should be rewritten as

$$\frac{I}{I_0} = \sum_{i=1}^{n} \frac{f_i}{1 + K_{SV,i}[Q]} \qquad (4.32)$$

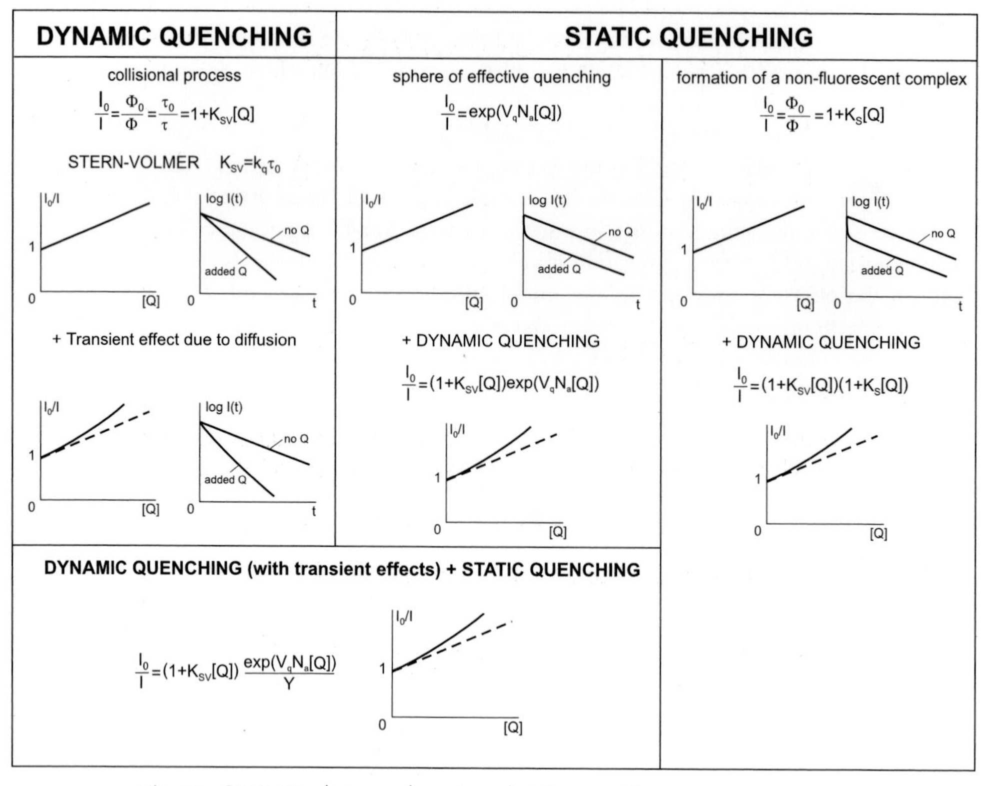

Fig. 4.2. Distinction between dynamic and static quenching.

where $K_{SV,i}$ is the Stern–Volmer constant for the ith species and f_i is the fractional contribution of the ith species to the total fluorescence intensity for a given couple of selected excitation and observation wavelengths. If the values of $K_{SV,i}$ and f_i are quite different, a downward curvature of I_0/I is observed.

In the case of additional static quenching, Eq. (4.32) becomes

$$\frac{I}{I_0} = \sum_{i=1}^{n} \frac{f_i}{(1 + K_{SV,i}[Q]) \exp(V_{q,i} N[Q])} \tag{4.33}$$

4.3 Photoinduced electron transfer

Photoinduced electron transfer (PET) is often responsible for fluorescence quenching. This process is involved in many organic photochemical reactions. It plays a major role in photosynthesis and in artificial systems for the conversion of solar energy based on photoinduced charge separation. Fluorescence quenching experiments provide a useful insight into the electron transfer processes occurring in these systems.

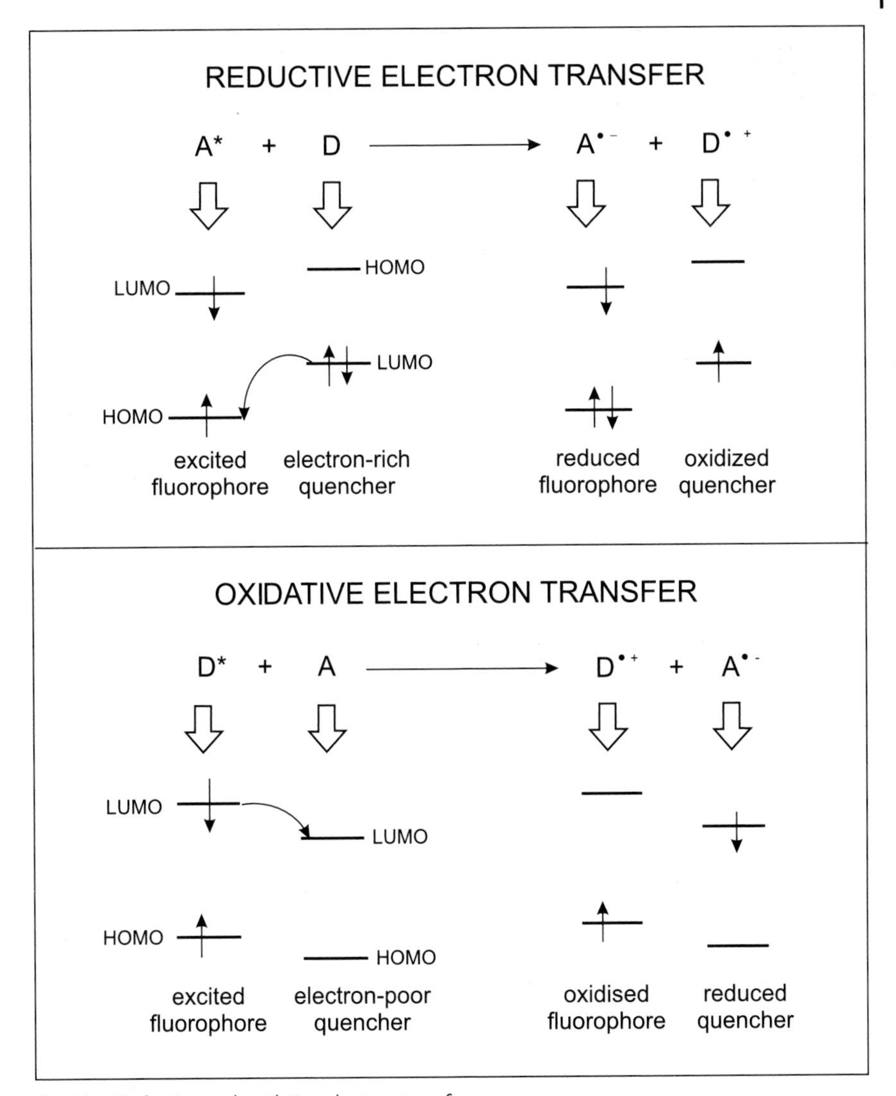

Fig. 4.3. Reductive and oxidative electron transfers.

The oxidative and reductive properties of molecules can be enhanced in the excited state. Oxidative and reductive electron transfer processes according to the following reactions:

$$^{1}D^{*} + A \rightarrow D^{\cdot +} + A^{\cdot -}$$

$$^{1}A^{*} + D \rightarrow A^{\cdot -} + D^{\cdot +}$$

are schematically illustrated in Figure 4.3. Examples of donor and acceptor molecules are given in Figure 4.4.

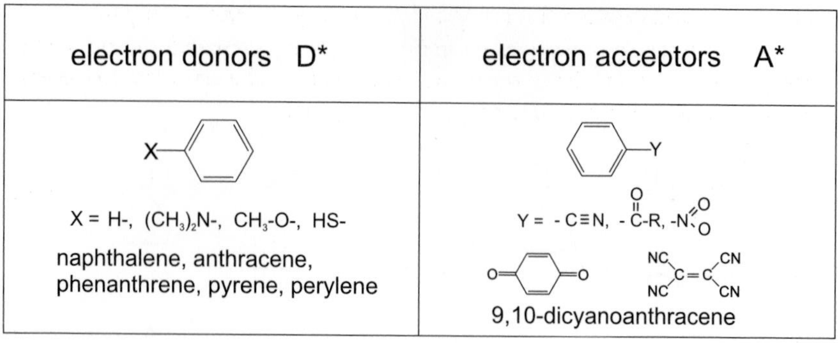

electron donors D*	electron acceptors A*
X = H-, $(CH_3)_2N$-, CH_3-O-, HS- naphthalene, anthracene, phenanthrene, pyrene, perylene	Y = -C≡N, -C(=O)-R, $-NO_2$ 9,10-dicyanoanthracene

Fig. 4.4. Examples of electron donors and acceptors in the excited state.

In the gas phase, the variations in standard free enthalpy ΔG^0 for the above reactions can be expressed using the redox potentials E^0 and the excitation energy ΔE_{00}, i.e. the difference in energy between the lowest vibrational levels of the excited state and the ground state:

$$\Delta G^0 = E^0_{D^{\bullet+}/D^*} - E^0_{A/A^{\bullet-}} = E^0_{D^{\bullet+}/D} - E^0_{A/A^{\bullet-}} - \Delta E_{00}(D) \tag{4.34}$$

$$\Delta G^0 = E^0_{D^{\bullet+}/D} - E^0_{A^*/A^{\bullet-}} = E^0_{D^{\bullet+}/D} - E^0_{A/A^{\bullet-}} - \Delta E_{00}(A) \tag{4.35}$$

These equations are called Rehm–Weller equations. If the redox potentials are expressed in volts, ΔG^0 is then given in volts. Conversion into J mol^{-1} requires multiplication by the Faraday constant ($F = 96\,500$ C mol^{-1}).

In solution, two terms need to be added to take into account the solvation effect (enthalpic term ΔH_{solv}) and the Coulombic energy of the formed ion pair:

$$\Delta G^0 = E^0_{D^{\bullet+}/D} - E^0_{A/A^{\bullet-}} - \Delta E_{00}(D) - \Delta H_{solv} - \frac{e^2}{4\pi\varepsilon r} \tag{4.36}$$

$$\Delta G^0 = E^0_{D^{\bullet+}/D} - E^0_{A/A^{\bullet-}} - \Delta E_{00}(A) - \Delta H_{solv} - \frac{e^2}{4\pi\varepsilon r} \tag{4.37}$$

where e is the electron charge, ε is the dielectric constant of the solvent and r is the distance between the two ions.

The redox potentials can be determined by electrochemical measurements or by theoretical calculations using the energy levels of the LUMO (lowest unoccupied molecular orbital) and the HOMO (highest occupied molecular orbital).

In solution, electron transfer reactions can be described by Schemes 4.2 or 4.3. If the reaction is not diffusion-limited, the reaction rate k_R, denoted here k_{ET} for electron transfer, can be determined. Two cases are possible.

- If the interaction between the donor and acceptor in the encounter pair is strong (Scheme 4.3), this encounter pair $(DA)^*$ is called an 'exciplex' (see Section 4.4).

- If the interaction between the donor and acceptor in the encounter pair ($D^*\ldots A$) is weak (Scheme 4.2), the rate constant k_{ET} can be estimated by the Marcus theory. This theory predicts a quadratic dependence of the activation free energy ΔG^* versus ΔG^0 (standard free energy of the reaction).

$$\Delta G^* = \frac{(\Delta G^0 + \lambda)^2}{4\lambda} \tag{4.38}$$

In this equation, λ is the total reorganization energy given by

$$\lambda = \lambda_{in} + \lambda_s \tag{4.39}$$

where λ_{in} is the contribution due to changes in the intramolecular bond length and bond angle in the donor and acceptor during electron transfer, and λ_s is the contribution of the solvent reorganization.

In the framework of the collision theory, k_{ET} can then be evaluated by means of the usual relation

$$k_{ET} = Z\exp(-\Delta G^*/RT) \tag{4.40}$$

where Z is the frequency of collisions and ΔG^* is the free energy of activation for the ET process.

Because of the quadratic dependence, the variation of $\ln(k_{ET})$ versus ΔG^0 is expected to be a parabol whose maximum corresponds to $\Delta G^0 = 0$ (Figure 4.5). Beyond the maximum ($\Delta G^0 > 0$), k_{ET} decreases when ΔG^0 increases (normal region), whereas below the maximum ($\Delta G^0 < 0$), the inverse behavior is expected (Marcus' inverse region).

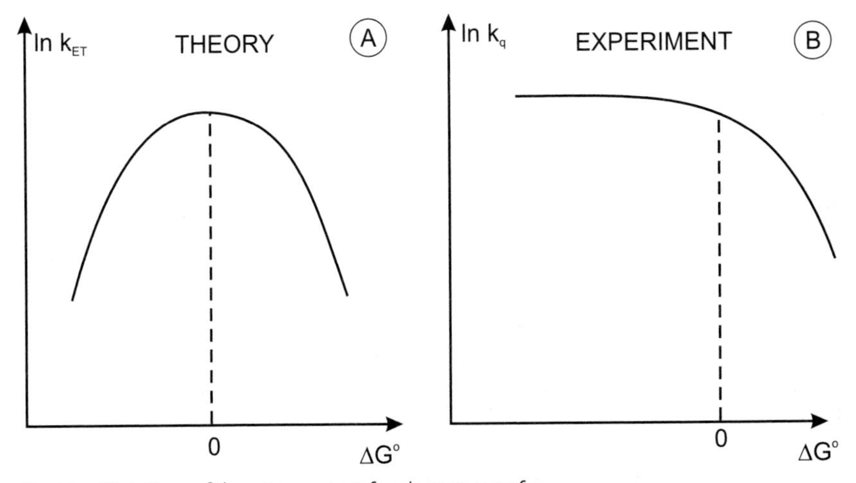

Fig. 4.5. Variations of the rate constant for electron transfer versus ΔG^0 according to the Marcus theory.

The theoretical prediction has been confirmed in the normal region by fluorescence quenching experiments in which the observed rate constant k_q is directly representative of the rate constant k_{ET} for electron transfer. Conversely, the inverse region has not been observed for intermolecular ET because in the inverse region, where $\Delta G^0 \ll 0$, k_{ET} is so large (according to Eqs 4.40 and 4.38) that the reaction is diffusion-limited ($k_R = k_{ET} \gg k_1$ in Scheme 4.2); in other words, $k_q = k_1$ so that the Marcus theory is irrelevant. Consequently, when ΔG^0 becomes more and more negative, k_q reaches a plateau corresponding to the diffusional rate constant k_1 (Figure 4.5B), whereas a decrease is predicted by the Marcus theory.

Nevertheless, the inverse region is observed in the particular case where the electron donor and the electron acceptor are held apart by a bridge (e.g. porphyrins covalently linked to quinones).

4.4
Formation of excimers and exciplexes

Excimers are dimers in the excited state (the term excimer results from the contraction of '**exci**ted di**mer**'). They are formed by collision between an excited molecule and an identical unexcited molecule:

$$^1M^* + {}^1M \rightleftarrows {}^1(MM)^*$$

The symbolic representation $(MM)^*$ shows that the excitation energy is delocalized over the two moieties (as in an excitonic interaction described in Section 4.6).

Exciplexes are excited-state complexes (the term exciplex comes from '**exci**ted com**plex**'). They are formed by collision of an excited molecule (electron donor or acceptor) with an unlike unexcited molecule (electron acceptor or donor):

$$^1D^* + A \rightleftarrows {}^1(DA)^*$$

$$^1A^* + D \rightleftarrows {}^1(DA)^*$$

The formation of excimers and exciplexes are diffusion-controlled processes. The photophysical effects are thus detected at relatively high concentrations of the species so that a sufficient number of collisions can occur during the excited-state lifetime. Temperature and viscosity are of course important parameters.

4.4.1
Excimers

Many aromatic hydrocarbons such as naphthalene or pyrene can form excimers. The fluorescence band corresponding to an excimer is located at wavelengths higher than that of the monomer and does not show vibronic bands (see Figure 4.6 and the example of pyrene in Figure 4.7).

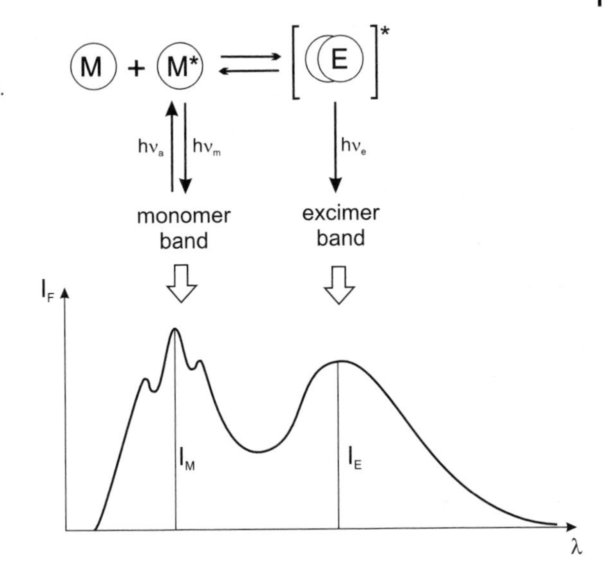

Fig. 4.6. Excimer formation, with the corresponding monomer and excimer bands.

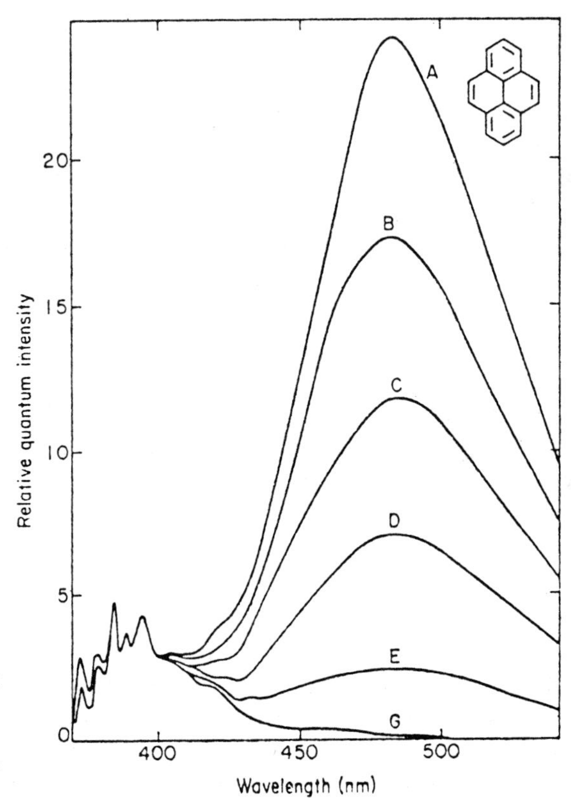

Fig. 4.7. Fluorescence spectra of pyrene at various concentrations in cyclohexane. A: 10^{-2} mol L^{-1}; B: 7.75×10^{-3} mol L^{-1}; C: 5.5×10^{-3} mol L^{-1}; D: 3.25×10^{-3} mol L^{-1}; E: 10^{-3} mol L^{-1}; G: 10^{-4} mol L^{-1} (from Birks and Christophorou, (1963) Spectrochim. Acta, **19**, 401).

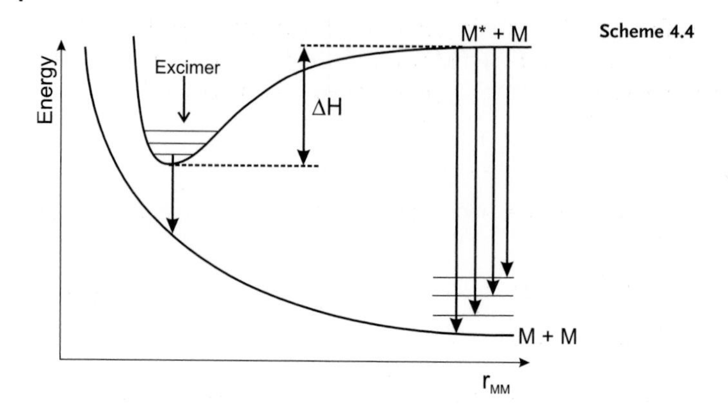

Scheme 4.4

These features can be explained on the basis of energy surfaces, as shown in Scheme 4.4. The lower monotonous curve represents the repulsive energy between the two molecules in the ground state. The upper curve, which is relative to two molecules (one of them being in the ground state), exhibits a minimum corresponding to the formation of an excimer in which the two aromatic rings are facing at a distance of ~3–4 Å. For example, this distance is 3.37 Å for pyrene and the experimental value of the stabilization energy ΔH of the excimer is 170 kJ mol^{-1}. In contrast to the monomer band, the excimer band is structureless because the lowest state is dissociative and can thus be considered as a continuum.

The time evolution of the fluorescence intensity of the monomer M and the excimer E following a δ-pulse excitation can be obtained from the differential equations expressing the evolution of the species. These equations are written according to the kinetic in Scheme 4.5 where k_M and k_E are reciprocals of the excited-state lifetimes of the monomer and the excimer, respectively, and k_1 and k_{-1} are the rate constants for the excimer formation and dissociation processes, respectively. Note that this scheme is equivalent to Scheme 4.3 where $(MQ)^* = (MM)^* = E^*$ and in which the formation of products is ignored.

According to Scheme 4.5, the coupled differential equations can be written as

$$\frac{d[M^*]}{dt} = -k_M[M^*] - k_1[M][M^*] + k^*_{-1}[E^*] \tag{4.41}$$

$$\frac{d[E^*]}{dt} = k_1[M][M^*] - (k_D + k_{-1})[E^*] \tag{4.42}$$

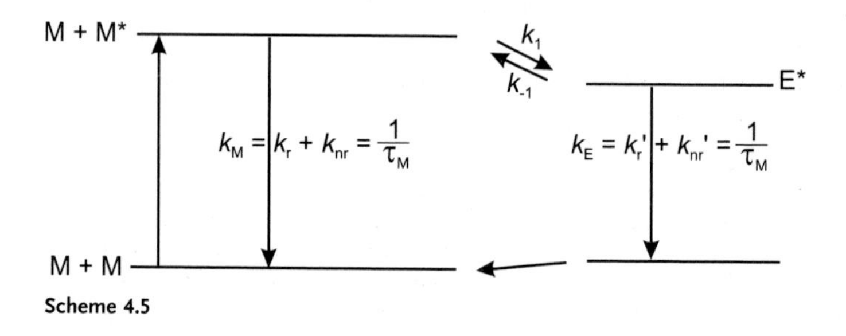

Scheme 4.5

Because formation of excimer E^* is a diffusion-controlled process, Eqs (4.11)–(4.13) apply to the diffusional rate constant k_1 for excimer formation. Under the approximation that k_1 is time-independent, the δ-pulse responses, under the initial conditions (at $t = 0$), $[M^*] = [M^*]_0$ and $[E^*]_0 = 0$, are

$$i_M(t) = k_r[M^*] = \frac{k_r[M^*]_0}{\beta_1 - \beta_2}[(X - \beta_2)e^{-\beta_1 t} + (\beta_1 - X)e^{-\beta_2 t}] \quad (4.43)$$

$$i_E(t) = k_r' k_1[E^*] = \frac{k_r' k_1[M][M^*]_0}{\beta_1 - \beta_2}[e^{-\beta_2 t} - e^{-\beta_1 t}] \quad (4.44)$$

where k_r and k_r' are the radiative rate constants of M^* and E^*, respectively, and β_1 and β_2 are given by

$$\beta_{2,1} = \frac{1}{2}\{X + Y \pm [(Y - X)^2 + 4k_1 k_{-1}[M]]^{1/2}\} \quad (4.45)$$

where $X = k_M + k_1[M] = 1/\tau_M + k_1[M]$ and $Y = k_E + k_{-1} = 1/\tau_E + k_{-1}$.

The decay of monomer emission is thus a sum of two exponentials. In contrast, the time evolution of the excimer emission is a difference of two exponentials, the pre-exponential factors being of opposite signs. The time constants are the same in the expressions of $i_M(t)$ and $i_E(t)$ (β_1 and β_2 are the eigenvalues of the system). The negative term in $i_E(t)$ represents the increase in intensity corresponding to excimer formation; the fluorescence intensity indeed starts from zero because excimers do not absorb light and can only be formed from the monomer (Figure 4.8A).

If the dissociation of the excimer cannot occur during the lifetime of the excited state ($k_{-1} \ll k_E$), we have

$$i_M(t) = k_r[M^*]_0 e^{-Xt} \quad (4.46)$$

$$i_E(t) = \frac{k_r' k_1[M][M^*]_0}{X - k_E}[e^{-k_E t} - e^{-Xt}] \quad (4.47)$$

In this case, the fluorescence decay of the monomer is a single exponential (Figure 4.8B). The relevant decay time ($1/X$) is equal to the rise time of the excimer fluorescence[6)].

The steady-state fluorescence intensities are obtained by integration of Eqs (4.43) and (4.44). The ratio of the fluorescence intensities of the excimer and monomer bands, I_E/I_M (Figure 4.6), is often used to characterize the efficiency of excimer formation. This ratio is given by

6) It should be noted that at high temperatures, a dynamic equilibrium can rapidly be reached in the excited state. The condition $k_1[M]$, $k_{-1} \gg k_M$, k_E can be satisfied above a critical temperature. Then, the fluorescence of monomer and excimer each decay exponentially with a common decay time constant that is an average of the intrinsic lifetimes of the monomer and the excimer (Birks, 1969).

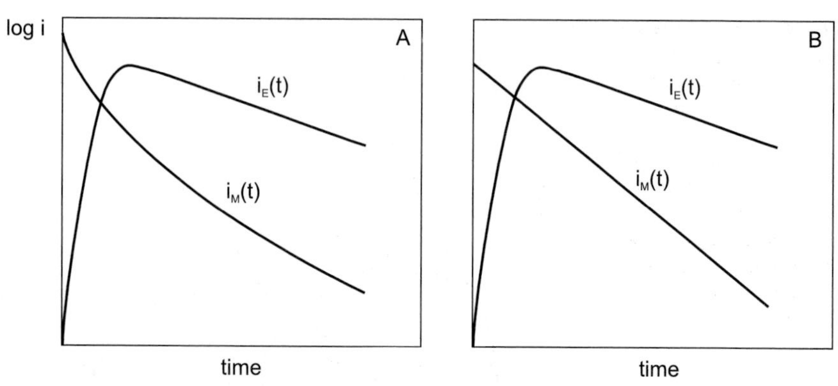

Fig. 4.8. Fluorescence decays of the monomer and the excimer. (a) Dissociation of the excimer within excited-state lifetime. (b) No dissociation of the excimer.

$$\frac{I_E}{I_M} = \frac{k_r'}{k_r} \frac{k_1[M]}{k_{-1} + 1/\tau_E} \tag{4.48}$$

Under conditions in which the dissociation rate of the excimer is slow with respect to de-excitation, Eq. (4.48) is reduced to

$$\frac{I_E}{I_M} = \frac{k_r'}{k_r} \tau_E k_1[M] \tag{4.49}$$

These equations show that the ratio I_E/I_M is proportional to the rate constant k_1 for excimer formation. Assuming that the Stokes–Einstein relation (Eq. 4.12) is valid, k_1 is proportional to the ratio T/η, η being the viscosity of the medium. Application to the estimation of the fluidity of a medium will be discussed in Chapter 8.

When the two monomers are linked by a short flexible chain, intramolecular excimers can be formed. This process is still diffusion-controlled, but in contrast to the preceding case, it is not translational; it requires a close approach between the two molecules via internal rotations during the excited-state lifetime. Equations (4.44), (4.45), (4.47) to (4.49) are still valid after replacing $k_1[M]$ by k_1 because intramolecular excimer formation is independent of the total concentration. Estimation of the local fluidity of a medium can be achieved by means of probes capable of forming intramolecular excimers (see Chapter 8).

In some cases, a monomer in the ground state may already be close to another monomer (e.g. in polymers with pendant fluorophores or on solid surfaces with adsorbed or covalently linked fluorophores), so that the displacement and the rotation required to attain the favorable excimer conformation occurs very quickly. These excimers are called 'excimer-like' or 'preformed excimers', in contrast to the normal case of 'true excimers'. The rise time corresponding to excimer formation may not be detected with instruments whose time resolution is of a few picoseconds.

A popular method for the determination of micellar aggregation numbers is based on self-quenching of pyrene by excimer formation within micelles (see Box 4.2).

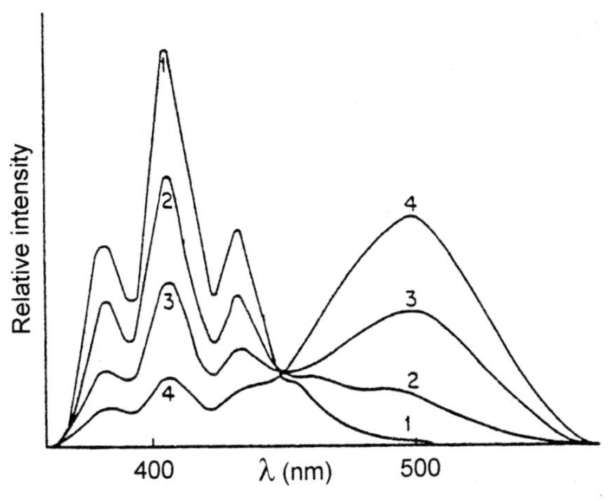

Fig. 4.9. Fluorescence spectra of anthracene (3×10^{-4} mol L^{-1}) in the presence of diethylaniline at various concentrations in toluene. 1: 0 mol L^{-1}; 2: 5×10^{-3} mol L^{-1}; 3: 2.5×10^{-2} mol L^{-1}; 4: 0.10 mol L^{-1} (from Weller, A. (1967), in Claesson S. (Ed.), *Fast Reactions and Primary Processes in Chemical Kinetics*, John Wiley & Sons, New York).

4.4.2
Exciplexes

A well-known example of an exciplex is the excited-state complex of anthracene and *N,N*-diethylaniline resulting from the transfer of an electron from an amine molecule to an excited anthracene molecule. In nonpolar solvents such as hexane, the quenching is accompanied by the appearance of a broad structureless emission band of the exciplex at higher wavelengths than anthracene (Figure 4.9). The kinetic scheme is somewhat similar to that of excimer formation.

When the solvent polarity increases, the exciplex band is red-shifted. The intensity of this band decreases as a result of the competition between de-excitation and dissociation of the exciplex.

It should be noted that de-excitation of exciplexes can lead not only to fluorescence emission but also to ion pairs and subsequently 'free' solvated ions. The latter process is favored in polar media. Exciplexes can be considered in some cases to be intermediate species in electron transfer from a donor to an acceptor (see Section 4.3).

4.5
Photoinduced proton transfer

This section will only cover reactions in aqueous solutions. Water molecules acting as either a proton acceptor or a proton donor will thus be in close contact with an acid or a base undergoing excited-state deprotonation or protonation, respectively. Therefore, these processes will not be diffusion-controlled (Case A in Section 4.2.1).

The acidic or basic properties of a molecule that absorbs light are not the same in the ground state and in the excited state. The redistribution of the electronic density upon excitation may be one of the possible causes of this observation. The most interesting cases are those where acids and bases are stronger in the excited state than in the ground state, because in these cases, excitation may trigger a photo-induced proton transfer. Then, the acidic character of a proton donor group (e.g. OH substituent of an aromatic ring) can be enhanced upon excitation so that the pK^* of this group in the excited state is much lower than the pK in the ground state (Table 4.4). In the same way, the pK^* of a proton acceptor group (e.g. heterocyclic nitrogen atom) in the excited state is much higher than in the ground state (pK) (Table 4.4).

The occurrence of excited-state proton transfer during the lifetime of the excited state depends on the relative rates of de-excitation and proton transfer. The general equations will be presented first, but only for the most extensively studied case where the excited-state process is proton ejection ($pK^* < pK$); the proton donor is thus an acid, AH^*, and the proton acceptor is a water molecule. Methods for the determination of pK^* are then described and finally, the various cases of pH dependence of the absorption and fluorescence spectra are examined.

4.5.1
General equations

Let us consider the possible events following excitation of an acid AH that is stronger in the excited state than in the ground state ($pK^* < pK$). In the simplest case, where there is no geminate proton recombination, the processes are presented in Scheme 4.6, where τ_0 and τ_0' are the excited-state lifetimes of the acidic (AH^*) and basic (A^{-*}) forms, respectively, and k_1 and k_{-1} are the rate constants for deprotonation and reprotonation, respectively. k_1 is a pseudo-first order rate constant, whereas k_{-1} is a second-order rate constant. The excited-state equilibrium constant is $K^* = k_1/k_{-1}$.[7]

The question arises as to whether or not the back-reaction can take place during the excited-state lifetime of A^{-*}. Because this reaction is diffusion-controlled ($k_{-1} \approx 5 \times 10^{10}$ L mol^{-1} s^{-1}), its rate is pH-dependent. If pH ≤ 2, the back-reaction must be taken into account because at this pH, $k_{-1}[H_3O^+] \gtrsim 5 \times 10^8$ s^{-1}. The reciprocal of this value is indeed of the order of the excited-state lifetime of most organic bases.

The time evolution of the fluorescence intensity of the acidic form AH^* and the basic form A^{-*} following δ-pulse excitation can be obtained from the differential equations expressing the evolution of the species. These equations are written according to Scheme 4.6:

$$\frac{d[AH^*]}{dt} = -(k_1 + 1/\tau_0)[AH^*] + k_{-1}[A^{-*}][H_3O^+] \tag{4.50}$$

$$\frac{d[A^{-*}]}{dt} = k_1[AH^*] - (k_{-1}[H_3O^+] + 1/\tau_0')[A^{-*}] \tag{4.51}$$

7) k_1 and k_{-1} are normalized so that K^* is dimensionless.

Tab. 4.4

		COMPOUND	FORMULA	*pK*	*pK**
EXCITED-STATE DEPROTONATION	$ArOH \xrightarrow{-H^+} ArO^-$	phenol	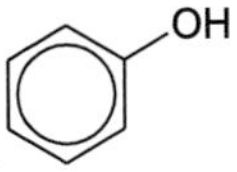	10.6	3.6
		2-naphthol	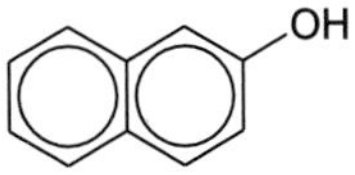	9.3	2.8
		2-naphthol-6-sulfonate	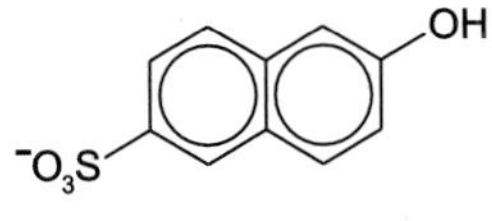	9.12	1.66
		2-naphthol-6,8-disulfonate	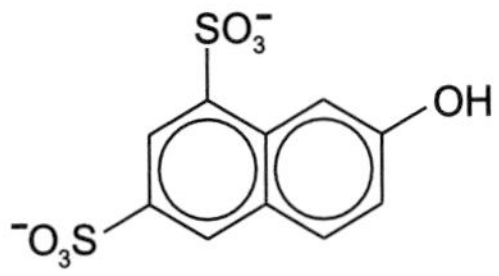	9.3	<1
		8-hydroxypyrene-1,3,6-trisulfonate (pyranine)	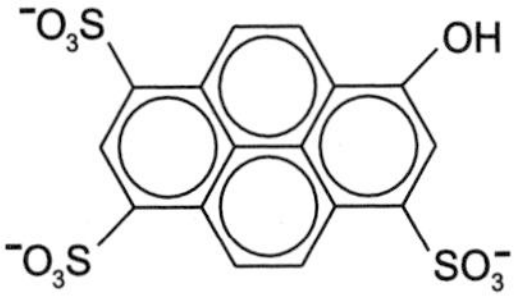	7.7	1.3
	$ArNH_2 \xrightarrow{-H^+} ArNH^-$	2-naphthylamine	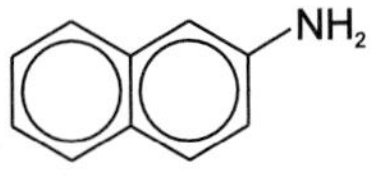	7.1	12.2
EXCITED-STATE PROTONATION	$ArCO_2^- \xrightarrow{+H^+} ArCO_2H$	anthracene-9-carboxylate	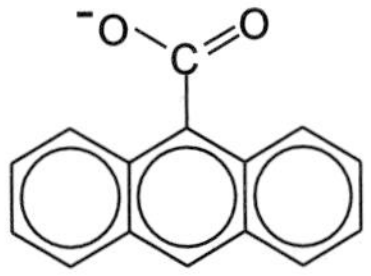	3.7	6.9
	$ArN \xrightarrow{+H^+} ArNH^+$	acridine	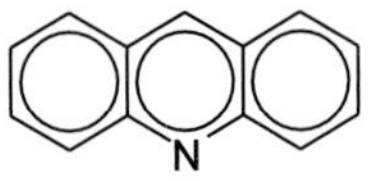	5.5	10.6
		6-methoxyquinoline	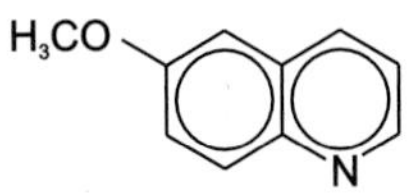	5.2	11.8

Scheme 4.6

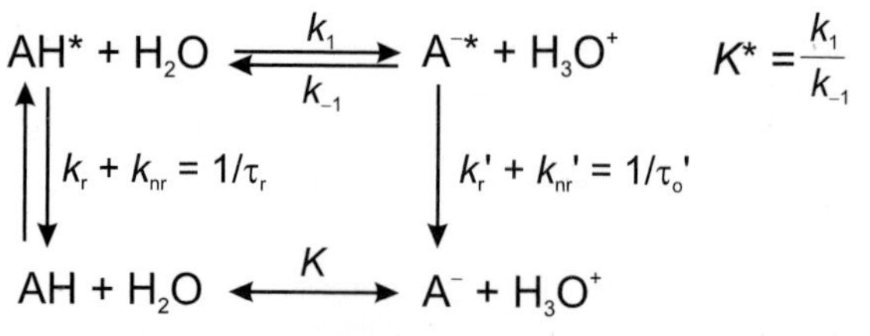

$$\begin{array}{ccc} AH^* + H_2O & \underset{k_{-1}}{\overset{k_1}{\rightleftharpoons}} & A^{-*} + H_3O^+ \\ \uparrow\downarrow \; k_r + k_{nr} = 1/\tau_r & & \downarrow \; k_r' + k_{nr}' = 1/\tau_o' \\ AH + H_2O & \overset{K}{\longleftrightarrow} & A^- + H_3O^+ \end{array} \qquad K^* = \frac{k_1}{k_{-1}}$$

When AH is selectively excited, the δ-pulse responses of the fluorescence intensities, under the initial conditions $[AH^*] = [AH^*]_0$ and $[A^{-*}]_0 = 0$ (at $t = 0$), are

$$i_{AH^*}(t) = k_r[AH^*] = \frac{k_r[AH^*]_0}{\beta_1 - \beta_2}[(X - \beta_2)e^{-\beta_1 t} + (\beta_1 - X)e^{-\beta_2 t}] \tag{4.52}$$

$$i_{A^{-*}} = k_r'[A^{-*}] = \frac{k_r' k_1 [AH^*]_0}{\beta_1 - \beta_2}[e^{-\beta_2 t} - e^{-\beta_1 t}] \tag{4.53}$$

where k_r and k_r' are the radiative rate constants of AH^* and A^{-*}, respectively, and β_1 and β_2 are given by

$$\beta_{2,1} = \frac{1}{2}\{X + Y \pm [(Y - X)^2 + 4k_1^* k_{-1}^* [H_3O^+]]^{1/2}\} \tag{4.54}$$

where $X = k_1^* + 1/\tau_0$ and $Y = k_{-1}^*[H_3O^+] + 1/\tau_0'$.

Under continuous illumination, the steady-state intensities can be easily calculated by considering a light of constant intensity as an infinite sum of infinitely short light pulses. The steady-state intensities are thus simply obtained by integration of Eqs (4.52) and (4.53).

In practice, a multiplication factor C must be introduced to take into account the experimental conditions (total concentration, choice of excitation and emission wavelengths, bandpasses for absorption and emission intensity of the incident light, sensitivity of the instrument).

$$I_{AH^*} = C\Phi_0 \frac{1 + k_{-1}\tau_0'[H_3O^+]}{1 + k_1\tau_0 + k_{-1}\tau_0'[H_3O^+]} \tag{4.55}$$

$$I_{A^{-*}} = C\Phi_0' \frac{k_1\tau_0}{1 + k_1\tau_0 + k_{-1}\tau_0'[H_3O^+]} \tag{4.56}$$

where Φ_0 and Φ_0' are the fluorescence quantum yields of AH and A^-, respectively, in the absence of the excited-state reaction ($\Phi_0 = k_r\tau_0$; $\Phi_0' = k_r'\tau_0'$).

Under pH conditions where the back reaction is too slow to take place during the excited-state lifetime ($k_{-1}[H_3O^+] \ll 1/\tau_0'$), the kinetic scheme is simplified and leads to the following equations:

$$i_{AH^*}(t) = [AH^*]_0 k_r e^{-Xt} \tag{4.57}$$

$$i_{A^{-*}}(t) = \frac{k_r' k_1 [AH^*]_0}{X - Y}[e^{-Yt} - e^{-Xt}] \tag{4.58}$$

where $X = k_1^* + 1/\tau_0$ and $Y = 1/\tau_0'$. Therefore, $i_{AH^*}(t)$ becomes a single exponential and $i_{A^{-*}}(t)$ is still a difference of two exponentials with a rise time $(1/X)$ (equal to the decay time of AH^*), and a decay time $(1/Y)$ (equal to the lifetime of the basic form).

The steady-state intensities reduce to

$$I_{HA^*} = C\Phi_0 \frac{1}{1 + k_1\tau_0} \tag{4.59}$$

$$I_{A^{-*}} = C\Phi_0' \frac{k_1\tau_0}{1 + k_1\tau_0} \tag{4.60}$$

4.5.2 Determination of the excited-state pK*

4.5.2.1 Prediction by means of the Förster cycle

pK* can be theoretically predicted by means of the Förster cycle (Scheme 4.7) in conjunction with spectroscopic measurements.

According to this cycle, we have

$$N_a h\nu_{AH} + \Delta H^{0*} = N_a h\nu_{A^-} + \Delta H^0 \tag{4.61}$$

where ΔH^0 and ΔH^{0*} are the standard molar ionization enthalpies of AH and AH^*, respectively, $h\nu_{AH}$ and $h\nu_{A^-}$ are the energy differences (corresponding to the 0–0 transitions) between the excited state and the ground state of AH and A^-, respectively, and N_a is Avogadro's number. This equation can be rewritten as

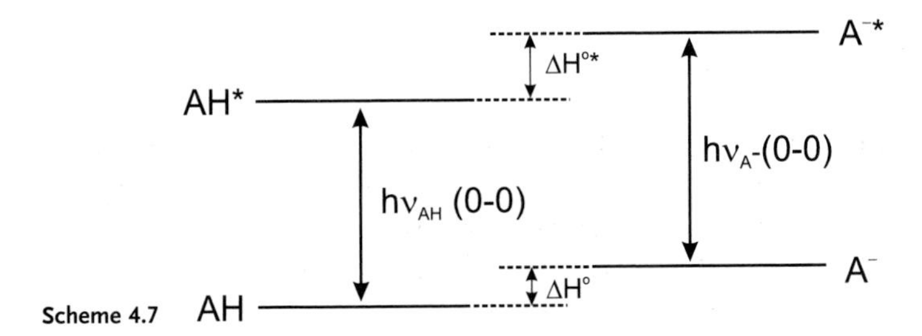

Scheme 4.7

$$\Delta H^{0*} - \Delta H^0 = N_a(h\nu_{A^-} - h\nu_{AH}) \tag{4.62}$$

Assuming that the ionization entropies ΔS^0 and ΔS^{0*} of AH and AH*[8], respectively, are equal, the difference in enthalpies $\Delta H^{0*} - \Delta H^0$ can be replaced by the difference in standard free energies $\Delta G^{0*} - \Delta G^0$, each of them being given by

$$\Delta G^0 = -RT \ln K = 2.3RT\ \mathrm{pK} \tag{4.63}$$

and

$$\Delta G^{0*} = -RT \ln K^* = 2.3RT\ \mathrm{pK}^* \tag{4.64}$$

where R is the gas constant and T is the absolute temperature. Using Eqs (4.63) and (4.64), and converting the frequencies into wavenumbers (in cm^{-1}), Eq. (4.62) becomes (at 298 K)

$$\boxed{\mathrm{pK}^* - \mathrm{pK} = 2.1 \times 10^{-3}(\bar{\nu}_{A^-} - \bar{\nu}_{AH})} \tag{4.65}$$

This equation shows that if the emission band of the basic form is located at higher wavelengths than that of the acidic form, the excited-state pK* is lower than the ground-state pK, and AH* is a stronger acid than AH. Conversely, if the emission band of the basic form is blue-shifted with respect to the emission band of the acidic form, the acid in the excited state is weaker than in the ground state.

In practice, the wavenumbers $\bar{\nu}_{AH}$ and $\bar{\nu}_{A^-}$ corresponding to the 0–0 transitions of AH and A^- are often difficult to determine accurately. They are usually estimated by means of the average between the wavenumbers corresponding to the maxima of absorption and emission:

$$\bar{\nu}_{00} = \frac{\bar{\nu}_{abs}^{max} + \bar{\nu}_{em}^{max}}{2} \tag{4.66}$$

However, the best approximation is found by using the intersection point of the mutually normalized absorption and emission spectra (as was the practice in Förster's laboratory). In the case of large Stokes shift, it may be difficult to determine an intersection point; it is then preferable to take the average of the wavenumbers corresponding to half-heights of the absorption and emission bands, which is a better approximation than the average of the wavenumbers corresponding to the maxima.

8) The approximation $\Delta S \approx \Delta S^*$ is reasonable in most cases because the major contribution to entropy variation is the change in the number of molecules and in the number of charges (and of the corresponding solvation). Both these contributions are the same in the ground and excited states. The changes in the internal degrees of freedom can be shown to be negligible. With respect to changes in dipole moments upon excitation and the consequent solvation effects, they are often similar for the acidic and the basic forms. However, there are noticeable exceptions (e.g. aromatic amines and their conjugates, arylammonium ions).

The determination of the energies of the 0–0 transitions is the major cause of inaccuracy in the estimation of pK*. In fact, a deviation of 4 nm, e.g. around 400 nm, corresponds to a deviation of one unit on the value of pK*.

The Förster cycle method is quite simple, which explains why it has been extensively used. One of the important features of this cycle is that it can be used even in cases where the equilibrium is not established within the excited-state lifetime. However, use of the Förster cycle is difficult or questionable when (i) two absorption bands overlap; (ii) the electronic levels invert during the excited-state lifetime (usually in a solvent-assisted relaxation process); (iii) the excited acidic and basic forms are of different orbital origins (electronic configuration or state symmetry); and (iv) the changes in dipole moment upon excitation are different for the acidic and basic forms.

4.5.2.2 **Steady-state measurements**

The value of the excited-state pK* can be determined by fluorometric titration, but only when the equilibrium is established in the excited state.

In the presence of photoinduced proton transfer, the steady-state fluorescence intensities are given by Eqs (4.55) and (4.56). In the absence of deprotoration (i.e. in a very acidic solution such that $k_{-1}[H_3O^+] \gg 1/\tau_0'$), when the experimental conditions (concentrations, excitation and observation wavelengths, sensitivity of the instrument) are kept strictly identical, the fluorescence intensities is $(I_{AH^*})_0 = C\Phi_0$. Rewriting Eqs (4.55) as $I_{AH^*} = C\Phi$, the following ratio is obtained

$$\frac{I_{AH^*}}{(I_{AH^*})_0} = \frac{\Phi}{\Phi_0} = \frac{1 + k_{-1}\tau_0'[H_3O^+]}{1 + k_1\tau_0 + k_{-1}\tau_0'[H_3O^+]} \tag{4.67}$$

In the same way, in a sufficiently basic medium, $(I_{A^{-*}})_0 = C'\Phi_0'$, and rewriting eq. 4.56 as $I_{A^{-*}} = C'\Phi_0'$, we obtain

$$\frac{I_{A^{-*}}}{(I_{A^{-*}})_0} = \frac{\Phi'}{\Phi_0'} = \frac{k_1\tau_0}{1 + k_1\tau_0 + k_{-1}\tau_0'[H_3O^+]} \tag{4.68}$$

The ratio of these two equations yields

$$\frac{\Phi/\Phi_0}{\Phi_0'/\Phi_0'} = \frac{1}{k_1\tau_0} + \frac{k_{-1}\tau_0'}{k_1\tau_0}[H_3O^+] \tag{4.69}$$

Therefore, if the excited-state lifetimes τ_0 and τ_0' are known, the plot of $(\Phi/\Phi_0)/(\Phi'/\Phi_0')$ versus $[H_3O^+]$ yields the rate constants k_1 and k_{-1}. However, it should be emphasized that corrections have to be made: (i) the proton concentration must be replaced by the proton activity; (ii) the rate constant k_{-1} must be multiplied by a correction factor involving the ionic strength (if the reaction takes place between charged particles), because of the screening effect of the ionic atmosphere on the charged reactive species.

4.5.2.3 Time-resolved experiments

The most reliable method for the determination of k_1 and k_{-1} is based on time-resolved experiments. Either pulse fluorometry or phase fluorometry can be used (see Chapter 6). They provide the values of the decay times from which the rate constants k_1 and k_{-1} are determined from Eqs (4.52) to (4.53) and the ratio k_1/k_{-1} yields K^*.

In some cases, a long tail can be detected in the decay. It has been assigned to geminate recombination according to the following kinetic scheme:

$$\mathrm{AH^* + H_2O} \underset{k_{-1}(t)}{\rightleftharpoons} \mathrm{A^{-*} \ldots H_3O^+} \longrightarrow \mathrm{A^{-*} + H_3O^+}$$

The ions recombine before their mutual distance is greater than the radius of the Coulomb cage. This is in particular the case of pyranine (see formula in Table 4.4) whose acidic form bears 3 negative charges and the basic form 4 negative charges, which results in a large Coulomb cage.

The rate constant for recombination k_{-1} is time-dependent and can be approximated to $t^{-3/2}$ at long times. The geminate recombination explains the residual fluorescence intensity of AH^* despite the fact that $pH \gg pK^*$ (see Section 4.5.3). In restricted media, the tail is even longer because of the higher probability of re-capturing a photoejected proton by geminate recombination (see Box 4.3).

4.5.3
pH dependence of absorption and emission spectra

As a result of the acido–basic properties in the ground and excited states, absorption and fluorescence spectra are pH-dependent. Let us recall that after proton ejection in the excited state, proton back-recombination can occur or not, depending of the pH (see section 4.5.1). For *pH* values greater than ~2, this back reaction does take place. No equilibrium is reached in the excited state. Furthermore, distinciton should be made according to the value of pK^*.

- If pK^* is greater than ~2, a plateau is observed for the relative fluorescence quantum yield of the acidic form and the basic form for pH ranging from pK^* to pK (Figure 4.10A) because of the absence of diffusional recombination. In fact, Eqs (4.59) and (4.60) which are relevant to this case show that I_{HA^*} and $I_{A^{-*}}$ are constants. A typical example is 2-naphthol ($pK = 9.3$, $pK^* = 2.8$).
- If pK^* is less than ~2, the acid is very strong in the excited state and, in general, k_1 is much larger than the reciprocal of the excited-state lifetime, so that the fluorescence of the acidic form is not observed at $pH > pK^* + 2$ but only for lower pH values (Figure 4.10B). Eqs (4.55) and (4.56) expressing the pH dependence of I_{HA^*} and $I_{A^{-*}}$ account for the shape of the curves. However, this is not strictly valid when geminate recombination occurs. A weak emission of the acidic form can then be observed. This is the case for pyranine ($pK = 7.7$, $pK^* = 1.3$) (Figure 4.11).

Box 4.3 Probing the acido–basic properties of water in restricted media or in the vicinity of an interface

The acido–basic properties of water molecules are greatly affected in restricted media such as the active sites of enzymes, reverse micelles, etc. The ability of water to accept or yield a proton is indeed related to its H-bonded structure which is, in a confined environment, different from that of bulk water. Water acidity is then best described by the concept of proton-transfer efficiency – characterized by the rate constants of deprotonation and reprotonation of solutes – instead of the classical concept of pH. Such rate constants can be determined by means of fluorescent acidic or basic probes.

The aqueous cores of reverse micelles are of particular interest because of their analogy with the water pockets in bioaggregates and the active sites of enzymes. Moreover, enzymes solubilized in reverse micelles can exhibit an enhanced catalytic efficiency. Figure B4.3.1 shows a reverse micelle of bis(2-ethylhexyl)sulfosuccinate (AOT) in heptane with three naphthalenic fluorescent probes whose excited-state pK* values are much lower than the ground-state pK (see Table 4.4): 2-naphthol (NOH), sodium 2-naphthol sulfonate (NSOH), potassium 2-naphthol-6,8-disulfonate (NSOH). The spectra and the rate constants for deprotonation and back-recombination (determined by time-resolved experiments) provide information on the location of the probes and the corresponding ability of their microenvironment to accept a proton[a]. (i) NDSOH is located around the center of the water pool, and at water contents $w = [H_2O]/[AOT] >$

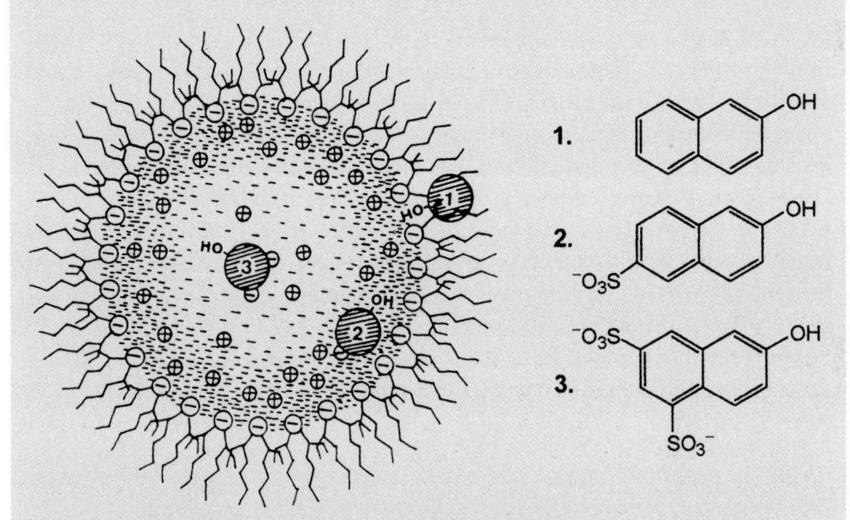

Fig. B4.3.1. Schematic illustration of the average residence sites of the probes NOH (1), NSOH (2), NDSOH (3) in AOT reverse micelles. Length of the surfactant: 11 Å. Diameter of the water pool: 18 Å at $w = 3$, 36 Å at $w = 9$. Largest dimension of the naphthol derivatives ≈ 9 Å (adapted from Bardez et al.[a]).

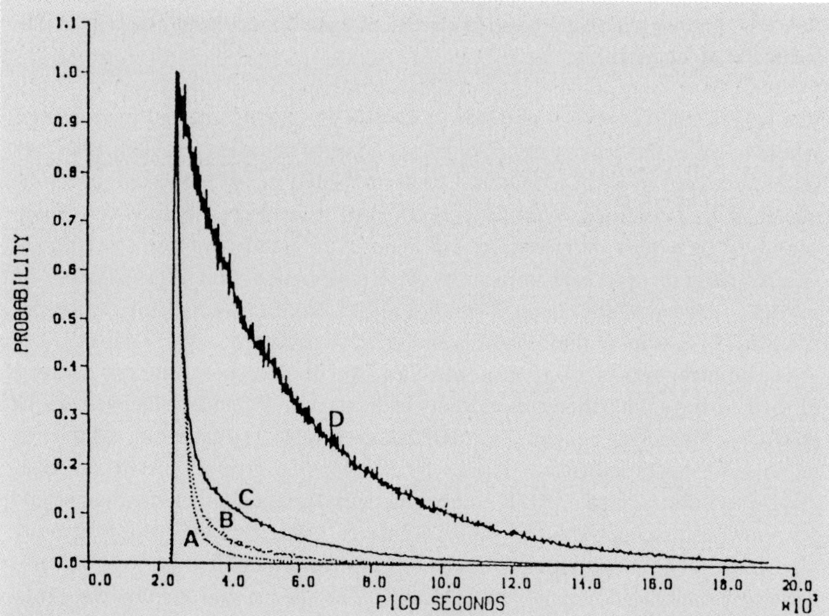

Fig. B4.3.2. Fluorescence decay curves for pyranine in various aqueous environments. See text for the meaning of A, B, C and D (reproduced with permission from Gutman et al.[b]).

10, its protolysis behavior is identical to that of bulk water. (ii) NSOH resides in the vicinity of the interface and an amount of water of $w \approx 40$ is required to observe the same deprotonation rate as in bulk water, whereas the rate of back-recombination is still much faster. (iii) NOH is located at the interface and does not undergo deprotonation in the excited state whatever the water content. Efficiency and kinetics of proton transfer are thus strongly dependent on location. The protolytic reactivity of water is related to its H-bonded structure, which changes as a function of the distance with respect to the interface; in the vicinity of the interface, water molecules are in fact involved in the hydration of the polar heads and sodium ions.

Other restricted media have been probed by measuring the fluorescence decay of pyranine[b], as shown in Figure B4.3.2:

- Curve A: pyranine in water at pH 5.5;
- Curve B: pyranine trapped in the thin water layer (30 Å thickness) of multilamellar vesicles made of dipalmitoylphosphatidyl choline (pH 5.5);
- Curve C: pyranine enclosed in the anion specific pore (18 × 27 Å) of the PhoE protein;
- Curve D: pyranine in the heme binding site of apomyoglobin, a site containing 30 water molecules or less.

It can be seen that the smaller the size of the restricted medium, the longer the tail of the fluorescence decay because of the higher probability of recapturing a photoejected proton by geminate recombination. Thus, the fluorescence decay reports the fate of a proton whose life depends on its microenvironment.

a) Bardez E., Monnier E. and Valeur B. (1985) *J. Phys. Chem.* **89**, 5031–6.
b) Gutman M., Shimoni E. and Tsfadia Y. (1992) Electron and Proton Transfer in Chemistry and Biology, in: A. Müller et al. (Eds), *Studies in Physical and Theoretical Chemistry*, Vol. 78, Elsevier, Amsterdam, pp. 273–85.

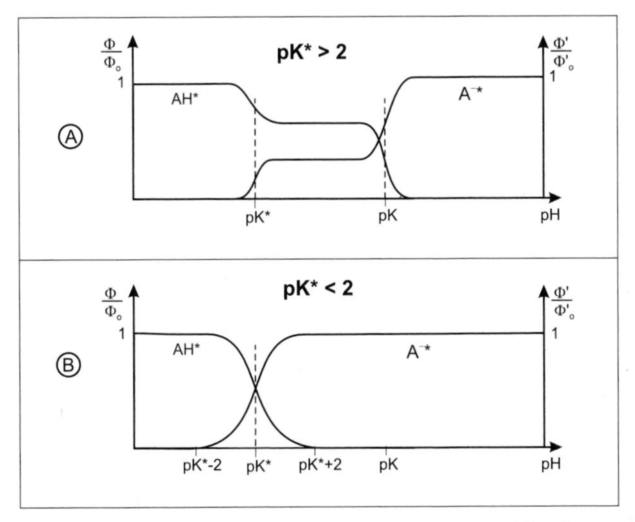

Fig. 4.10. Variations in relative fluorescence quantum yields of acidic and basic forms versus pH for various cases.

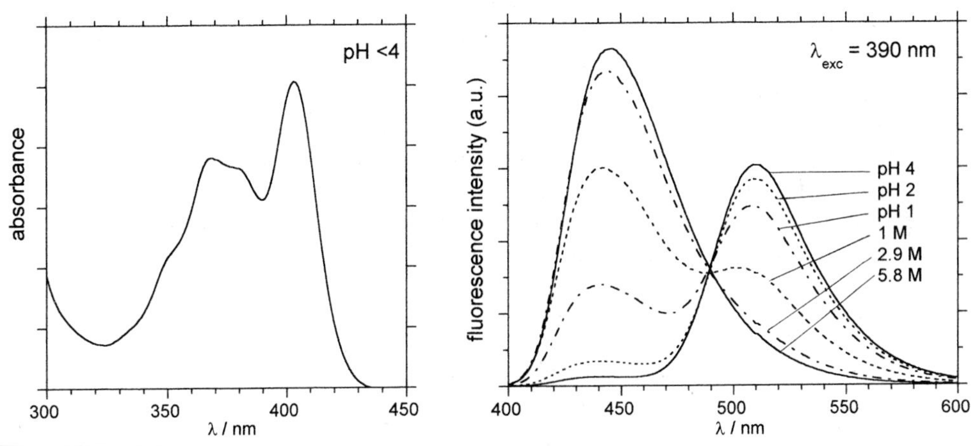

Fig. 4.11. pH dependence of emission spectra of pyranine in very acidic media. The residual fluorescence of the acidic form at pH $>$ 3 is due to geminate recombination.

4.6 Excitation energy transfer

Energy transfer from an excited molecule (donor) to another that is chemically different (acceptor) according to

$$D^* + A \rightarrow D + A^*$$

is called *heterotransfer.* This process is possible provided that the emission spectrum of the donor partially overlaps the absorption spectrum of the acceptor.

If the donor and acceptor are identical, we have *homotransfer*:

$$D^* + D \rightarrow D + D^*$$

When the process can repeat itself so that the excitation migrates over several molecules, it is called *excitation transport* or *energy migration.*

4.6.1 Distinction between radiative and non-radiative transfer

It is important to distinguish between radiative and non-radiative transfer. Radiative transfer corresponds to the absorption by a molecule A (or D) of a photon emitted by a molecule D and is observed when the average distance between D and A (or D) is larger than the wavelength. Such a transfer does not require any interaction between the partners, but it depends on the spectral overlap and on the concentration. In contrast, non-radiative transfer occurs without emission of photons at distances less than the wavelength and results from short- or long-range interactions between molecules. For instance, non-radiative transfer by dipole–dipole interaction is possible at distances up to 80–100 Å. Consequently, such a transfer provides a tool for determining distances of a few tens of Angströms between chromophores (see Chapter 9). It is worth noting that the transfer of excitation energy plays a major role in photosynthetic systems.

Radiative and non-radiative transfers have different effects on the characteristics of fluorescence emission from the donor, which allows us to make a distinction between these two types of transfer. Tables 4.5 and 4.6 summarize the effects of heterotransfer and homotransfer, respectively.

The factors governing the efficiency of radiative and non-radiative transfers are not the same (apart from the spectral overlap, of course, which is required for both processes) (see Table 4.7).

4.6.2 Radiative energy transfer

Radiative transfer is a two-step process: a photon emitted by a donor D is absorbed by an acceptor that is chemically different (A) or identical (D)

Tab. 4.5. Effect of energy transfer on the fluorescence characteristics of the donor in the case of a heterotransfer ($D^* + A \rightarrow D + A^*$)

Characteristics of the donor emission	*Radiative transfer*	*Non-radiative transfer*
Fluorescence spectrum	Modified in the region of spectral overlap	Unchanged
Steady-state fluorescence intensity	Decreased in the region of spectral overlap	Decreased by the same factor whatever λ_{em}
Fluorescence decay	Unchanged	Shortened

Tab. 4.6. Effect of energy transfer on the fluorescence characteristics of the donor in the case of a homotransfer ($D^* + D \rightarrow D + D^*$)

Characteristics of the emission	*Radiative transfer*	*Non-radiative transfer*
Fluorescence spectrum	Modified in the region of spectral overlap	Unchanged
Steady-state fluorescence intensity	Decreased in the region of spectral overlap	Decreased by the same factor whatever λ_{em}
Fluorescence decay	Slower	Unchanged
Steady-state emission anisotropy*	Decreased	Strongly decreased
Decay of emission anisotropy*	Faster	Much faster

* see Chapter 5.

Tab. 4.7. Factors governing the efficiency of radiative and non-radiative transfers

	Spectral overlap	λ_{exc} *and* λ_{em}	Φ_D	r_{DA} *or* r_{DD}	*Viscosity*	*Geometry*
Radiative transfer	yes	yes	no	no	no	yes
Non-radiative transfer	yes	no	yes	yes	yes	no

Φ_D: fluorescence quantum yield of the donor; r_{DA}, r_{DD}: intermolecular distances.

$$D^* \rightarrow D + h\nu$$

$$h\nu + A \rightarrow A^* \quad \text{or} \quad h\nu + D \rightarrow D^*$$

This process is often called *trivial* transfer because of the simplicity of the phenomenon, but in reality the quantitative description is quite complicated because it depends on the size of the sample and its configuration with respect to excitation and observation.

Radiative transfer results in a decrease of the donor fluorescence intensity in the region of spectral overlap. Such a distortion of the fluorescence spectrum is called the *inner filter effect* (see Chapter 6).

The fraction a of photons emitted by D and absorbed by A is given by

$$a = \frac{1}{\Phi_D^0} \int_0^\infty I_D(\lambda)[1 - 10^{-\varepsilon_A(\lambda)C_A l}]\, d\lambda \tag{4.71}$$

where C_A is the molar concentration of acceptor, Φ_D^0 is the fluorescence quantum yield in the absence of acceptor, l is the thickness of the sample, $I_D(\lambda)$ and $\varepsilon_A(\lambda)$ are the donor fluorescence intensity and molar absorption coefficient of the acceptor, respectively, with the normalization condition $\int_0^\infty I_D(\lambda)\, d\lambda = 1$. If the optical density is not too large, a can be approximated by

$$a = \frac{2.3}{\Phi_D^0} C_A l \int_0^\infty I_D(\lambda)\varepsilon_A(\lambda)\, d\lambda \tag{4.72}$$

where the integral represents the overlap between the donor fluorescence spectrum and the acceptor absorption spectrum.

In the case of radiative transfer between identical molecules, the fluorescence decays more slowly as a result of successive re-absorptions and re-emissions. A simple kinetic model has been proposed by Birks (1970). It is based on the assumption of a unique value for the average probability $\bar{\alpha}$ that an emitted photon is absorbed, i.e. without distinction between the generations of photons (a photon of generation n is emitted after n successive re-absorptions). This model leads to the following expressions for the effective lifetime and the macroscopic fluorescence quantum yield:

$$\tau = \frac{\tau_0}{1 - \bar{\alpha}\Phi_0} \tag{4.73}$$

$$\Phi = \frac{\Phi_0 - \bar{\alpha}\Phi_0}{1 - \bar{\alpha}\Phi_0} \tag{4.74}$$

where τ_0 and Φ_0 are the lifetime and fluorescence quantum yield, respectively, in the absence of transfer (i.e. as measured at high dilution). This model provides a simple and fast way of estimating the magnitude of radiative transfer under normal conditions of observation of fluorescence. However, according to this model, the fluorescence decay is expected to be a single exponential and does not depend on wavelength, which is not true in practice. It cannot be assumed that photons of different generations and different wavelengths have the same probability of being absorbed.

Stochastic approaches and Monte Carlo simulations offer a better description of radiative transfer, as shown by Berberan-Santos and coworkers (1999).

4.6.3
Non-radiative energy transfer

Non-radiative transfer of excitation energy requires some interaction between a donor molecule and an acceptor molecule, and it can occur if the emission spectrum of the donor overlaps the absorption spectrum of the acceptor, so that several vibronic transitions in the donor have practically the same energy as the corresponding transitions in the acceptor. Such transitions are coupled (see Figure 4.12), i.e. are in *resonance*. The term *resonance energy transfer* (RET) is often used. In some papers, the acronym FRET is used, denoting *fluorescence resonance energy transfer*, but this expression is incorrect because it is not the fluorescence that is transferred but the electronic energy of the donor. Therefore, it is recommended that either EET (*excitation energy transfer* or *electronic energy transfer*) or RET (*resonance energy transfer*) are used.

Energy transfer can result from different interaction mechanisms. The interactions may be Coulombic and/or due to intermolecular orbital overlap. The Coulombic interactions consist of long-range dipole–dipole interactions (Förster's

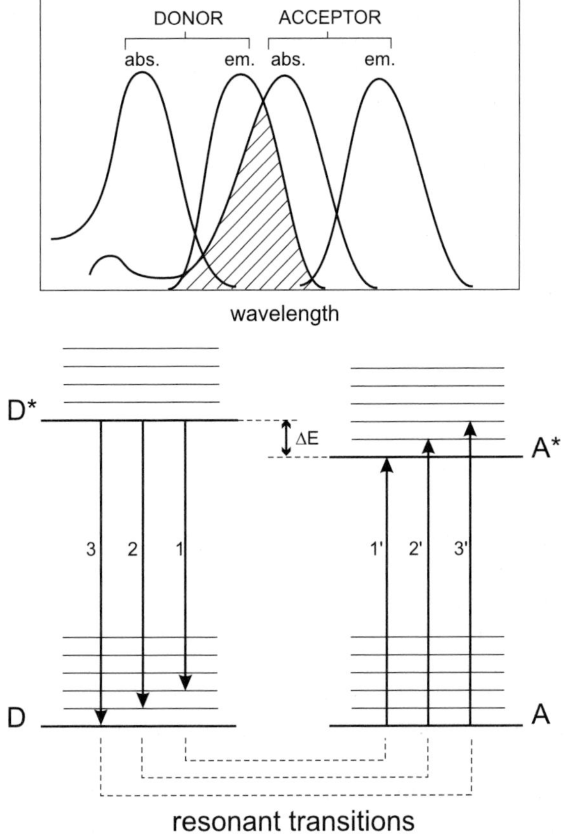

Fig. 4.12. Energy level scheme of donor and acceptor molecules showing the coupled transitions in the case where vibrational relaxation is faster than energy transfer (very weak coupling) and illustration of the integral overlap between the emission spectrum of the donor and the absorption of the acceptor.

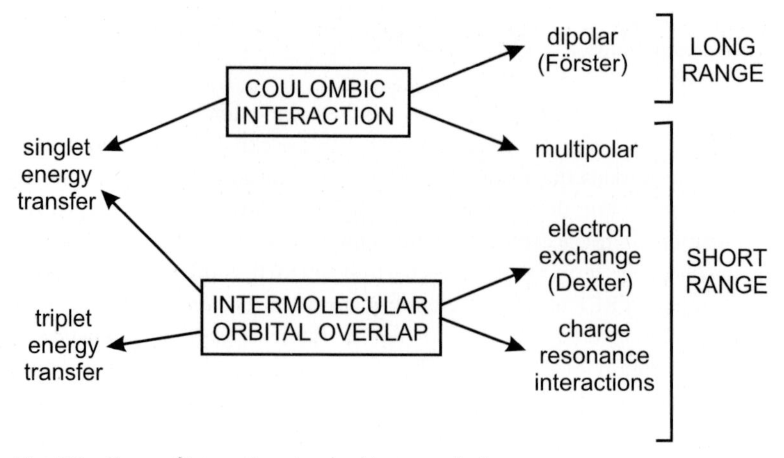

Fig. 4.13. Types of interactions involved in non-radiative transfer mechanisms.

mechanism) and short-range multi-polar interactions. The interactions due to intermolecular orbital overlap, which include electron exchange (Dexter's mechanism) and charge resonance interactions, are of course only short range (Figure 4.13). It should be noted that for singlet–singlet energy transfer ($^1D^* + {}^1A \rightarrow {}^1D + {}^1A^*$), all types of interactions are involved, whereas triplet–triplet energy transfer ($^3D^* + {}^1A \rightarrow {}^1D + {}^3A^*$) is due only to orbital overlap. The reason will be given later.

Box 4.4 shows how the total interaction energy can be expressed as a sum of two terms: a *Coulombic term* U_c and an *exchange term* U_{ex}. The Coulombic term corresponds to the energy transfer process in which the initially excited electron on the donor D returns to the ground state orbital on D, while simultaneously an electron on the acceptor A is promoted to the excited state (Figure 4.14A). The exchange term (which has a quantum-mechanical origin) corresponds to an energy transfer process associated with an exchange of two electrons between D and A (Figure 4.14B).

For allowed transitions on D and A the Coulombic interaction is predominant, even at short distances. For forbidden transitions on D and A (e.g. in the case of transfer between triplet states ($^3D^* + {}^1A \rightarrow {}^1D + {}^3A^*$), in which the transitions $T_1 \rightarrow S_0$ in D* and $S_0 \rightarrow T_1$ in A are forbidden), the Coulombic interaction is negligible and the exchange mechanism is found, but is operative only at short distances (< 10 Å) because it requires overlap of the molecular orbitals. In contrast, the Coulombic mechanism can still be effective at large distances (up to 80–100 Å).

The magnitude of the interaction is even more important than its nature, and it is thus convenient to make a distinction, as proposed by Förster, between three main classes of coupling (strong, weak and very weak), depending on the relative values of the interaction energy (U), the electronic energy difference between D* and A* (ΔE), the absorption bandwidth (Δw) and the vibronic bandwidth ($\Delta\varepsilon$) (Figure 4.15).

Box 4.4 Energy of interaction between a donor molecule and an acceptor molecule

Considering that only two electrons are involved in a transition, one on D and one on A, the properly antisymmetrized wavefunctions for the initial excited state Ψ_i (D excited but not A) and the final excited state Ψ_f (A excited but not D) can be written as

$$\Psi_i = \frac{1}{\sqrt{2}}(\Psi_{D^*}(1)\Psi_A(2) - \Psi_{D^*}(2)\Psi_A(1))$$
$$\Psi_f = \frac{1}{\sqrt{2}}(\Psi_D(1)\Psi_{A^*}(2) - \Psi_D(2)\Psi_{A^*}(1)) \quad \text{(B4.4.1)}$$

where the numbers 1 and 2 refer to the two electrons involved.

The interaction matrix element describing the coupling between the initial and final states is given by

$$U = \langle \Psi_i | V | \Psi_f \rangle \quad \text{(B4.4.2)}$$

where V is the perturbation part of the total Hamiltonian $\hat{H} = \hat{H}_{Donor} + \hat{H}_{Acceptor} + V$.

U can be written as a sum of two terms

$$U = \langle \Psi_{D^*}(1)\Psi_A(2) | V | \Psi_D(1)\Psi_{A^*}(2) \rangle - \langle \Psi_{D^*}(1)\Psi_A(2) | V | \Psi_D(2)\Psi_{A^*}(1) \rangle \quad \text{(B4.4.3)}$$

In the first term, U_c, usually called the Coulombic term, the initially excited electron on D returns to the ground state orbital while an electron on A is simultaneously promoted to the excited state. In the second term, called the exchange term, U_{ex}, there is an exchange of two electrons on D and A. The exchange interaction is a quantum-mechanical effect arising from the symmetry properties of the wavefunctions with respect to exchange of spin and space coordinates of two electrons.

The Coulombic term can be expanded into a sum of terms (multipole–multipole series), but it is generally approximated by the first predominant term representing the dipole–dipole interaction between the transition dipole moments $\mathbf{M}_D$ and $\mathbf{M}_A$ of the transitions $D \rightarrow D^*$ and $A \rightarrow A^*$ (the squares of the transition dipole moments are proportional to the oscillator strengths of these transitions):

$$U_{dd} = \frac{\mathbf{M}_D \cdot \mathbf{M}_A}{r^3} - 3\frac{(\mathbf{M}_A \cdot \mathbf{r})(\mathbf{M}_D \cdot \mathbf{r})}{r^5} \quad \text{(B4.4.4)}$$

where r is the donor–acceptor separation. This expression can be rewritten as:

$$U_{dd} = 5.04 \frac{|\mathbf{M}_D| \, |\mathbf{M}_A|}{r^3} (\cos \theta_{DA} - 3 \cos \theta_D \cos \theta_A) \qquad \text{(B4.4.5)}$$

in which U_{dd} is expressed in cm^{-1}, the transition moments in Debye (1 Debye unit = 3.33×10^{-30} Coulomb meter), and r in nanometers. θ_{DA} is the angle between the two transition moments and θ_D and θ_A are the angles between each transition moment and the vector connecting them.

The magnitude of this term can be large even at long distances (up to 80–100 Å) if the two transitions on D and A are allowed. The excitation energy is transferred through space.

The dipole approximation is valid only for point dipoles, i.e. when the donor–acceptor separation is much larger than the molecular dimensions. At short distances or when the dipole moments are large, it should be replaced by a monopole–monopole expansion. Higher multipole terms should also be included in the calculations.

The exchange term represents the electrostatic interaction between the charge clouds. The transfer in fact occurs via overlap of the electron clouds and requires physical contact between D and A. The interaction is short range because the electron density falls off approximately exponentially outside the boundaries of the molecules. For two electrons separated by a distance r_{12} in the pair D–A, the space part of the exchange interaction can be written as

$$U_{ex} = \langle \Phi_{D^*}(1)\Phi_A(2) \left| \frac{e^2}{r_{12}} \right| \Phi_D(2)\Phi_{A^*}(1) \rangle \qquad \text{(B4.4.6)}$$

where Φ_A and Φ_D are the contributions of the spatial wavefunction to the total wavefunctions Ψ_A and Ψ_D that include the spin functions. The spin selection rules (Wigner's rules) for allowed energy transfer are obtained by integration over the spin coordinates.

The transfer rate k_T is given by Fermi's Golden Rule:

$$k_T = \frac{2\pi}{\hbar} U^2 \rho \qquad \text{(B4.4.7)}$$

where ρ is a measure of the density of the interacting initial and final states, as determined by the Franck–Condon factors, and is related to the overlap integral between the emission spectrum of the donor and the absorption spectrum of the acceptor.

By substituting Eq. (B4.4.5) into Eq. (B4.4.7), we obtain the Förster rate constant k_T^{dd} (Eq. 4.78 in the text) for energy transfer in the case of long-range dipole–dipole interaction, and substitution of Eq. (B4.4.6) into Eq. (B4.4.7) leads to the Dexter rate constant k_T^{ex} (Eq. 4.85 in the text) for the short-range exchange interaction.

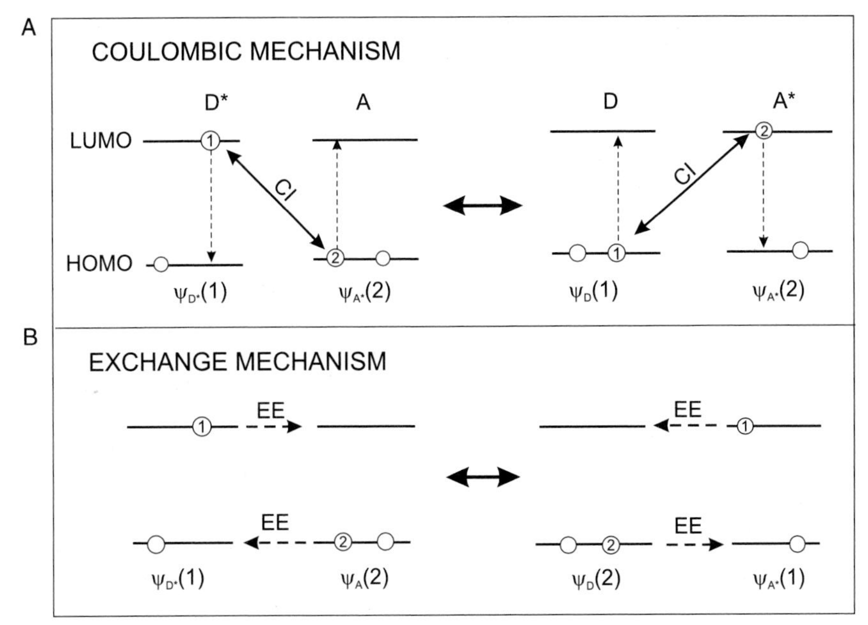

Fig. 4.14. Schematic representation of the (A) Coulombic and (B) exchange mechanisms of excitation energy transfer. CI: Coulombic interaction; EE: electron exchange.

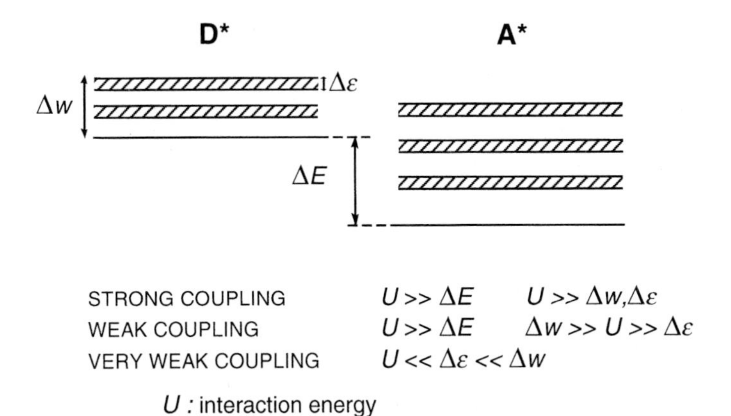

Fig. 4.15. Distinction between strong, weak and very weak coupling.

Strong coupling ($U \gg \Delta E$, $U \gg \Delta w$, $\Delta\varepsilon$) The coupling is called *strong* if the intermolecular interaction is much larger than the interaction between the electronic and nuclear motions within the individual molecules. In this case, the Coulombic term U_c is much larger than the width of the individual transitions $D \rightarrow D^*$ and $A \rightarrow A^*$. Then, all the vibronic subtransitions in both molecules are virtually at resonance with one another.

Strongly coupled systems are characterized by large differences between their absorption spectra and those of their components. For a two-component system, two new absorption bands are observed due to transitions of the in-phase and out-of-phase combinations of the locally excited states. These two transitions are separated in energy by $2|U|$.

In the strong coupling case, the transfer of excitation energy is faster than the nuclear vibrations and the vibrational relaxation ($\sim 10^{-12}$ s). The excitation energy is not localized on one of the molecules but is truly delocalized over the two components (or more in multi-chromophoric systems). The transfer of excitation is a coherent process[9]; the excitation oscillates back and forth between D and A and is never more than instantaneously localized on either molecule. Such a delocalization is described in the frame of the *exciton* theory[10].

A rate of transfer can be defined as the reciprocal of the time required for the excitation, initially localized on D, to reach a maximum density on A. The rate constant is

$$k_T \approx \frac{4|U|}{h} \tag{4.75}$$

where h is Planck's constant. When U is approximated by a dipole–dipole interaction, the distance dependence of U, and consequently of k_T, is r^{-3} (see Eq. B4.4.5 in Box 4.4). It is important to note that, in contrast, the dependence is r^{-6} in the case of very weak coupling (see below).

Weak coupling ($U \gg \Delta E$, $\Delta w \gg U \gg \Delta\varepsilon$) The interaction energy is much lower than the absorption bandwidth but larger than the width of an isolated vibronic level. The electronic excitation in this case is more localized than under strong coupling. Nevertheless, the *vibronic* excitation is still to be considered as delocalized so that the system can be described in terms of stationary *vibronic* exciton states.

Weak coupling leads to minor alterations of the absorption spectrum (hypochromism or hyperchromism, Davidov splitting of certain vibronic bands).

The transfer rate is fast compared to vibrational relaxation but slower than nuclear motions, in contrast to the strong coupling case. It can be approximated as

$$k_T \approx \frac{4|U|S_{vw}^2}{h} \tag{4.76}$$

where S_{vw} is the vibrational overlap integral of the intramolecular transition $v \leftrightarrow w$. This is the transfer rate between an excited molecule with the vibrational quantum number v and an unexcited one with the quantum number w. Because $S_{vw} < 1$, the

9) The relationship between the phases of the locally excited states $\Psi_{D^*}\Psi_A$ and $\Psi_D\Psi_{A^*}$ is fixed.

10) This theory was first developed by Frenkel and further by Davydov and others. A molecular exciton is defined as a 'particle' of excitation that travels through a (supra)molecular structure without electron migration.

transfer rate is slower than in the case of strong coupling. The term US_{vw}^2 represents the interaction energy between the involved vibronic transitions.

Very weak coupling ($U \ll \Delta\varepsilon \ll \Delta w$) The interaction energy is much lower than the vibronic bandwidth. Because the coupling is less than in the preceding cases, the condition for resonance becomes more and more stringent. However, the vibronic bandwidths are generally broadened by thermal and solvent effects (the absorption bands are in fact broad in fluid solutions) so that energy transfer in the very weak coupling limit turns out to be very general.

In this case, there is little or no alteration of the absorption spectra. The vibrational relaxation occurs before the transfer takes place. The transfer rate is given by

$$k_{\mathrm{T}} \approx \frac{4\pi^2 (US_{vw}^2)^2}{h\Delta\varepsilon} \tag{4.77}$$

This expression shows that the transfer rate depends on the square of the interaction energy[11]. For dipole–dipole interaction, the distance dependence is thus r^{-6} instead of r^{-3} for the preceding cases.

Figure 4.16 shows the rates of energy transfer in the three coupling cases. There is some ambiguity in defining a transfer rate in the cases of strong and weak couplings because of more or less delocalized excitation. It is only in the case of very weak coupling that a transfer rate can be defined unequivocally.

Förster's formulation of long-range dipole–dipole transfer (very weak coupling) Förster derived the following expression for the transfer rate constant from classical considerations as well as on quantum-mechanical grounds:

$$\boxed{k_{\mathrm{T}}^{\mathrm{dd}} = k_{\mathrm{D}} \left[\frac{R_0}{r}\right]^6 = \frac{1}{\tau_{\mathrm{D}}^0} \left[\frac{R_0}{r}\right]^6} \tag{4.78}$$

where k_{D} is the emission rate constant of the donor and τ_{D}^0 its lifetime in the absence of transfer, r is the distance between the donor and the acceptor (which is assumed to remain unchanged during the lifetime of the donor), and R_0 is the critical distance or Förster radius, i.e. the distance at which transfer and spontaneous decay of the excited donor are equally probable ($k_{\mathrm{T}} = k_{\mathrm{D}}$). The characteristic inverse sixth power dependence on distance should again be noted.

R_0, which can be determined from spectroscopic data, is given by

$$\boxed{R_0^6 = \frac{9000(\ln 10)\kappa^2 \Phi_{\mathrm{D}}^0}{128\pi^5 N_A n^4} \int_0^\infty I_{\mathrm{D}}(\lambda)\varepsilon_{\mathrm{A}}(\lambda)\lambda^4 \,\mathrm{d}\lambda} \tag{4.79}$$

11) This expression was obtained by Förster and reformulated later on as Fermi's Golden Rule (Eq. B4.4.7 in Box 4.4).

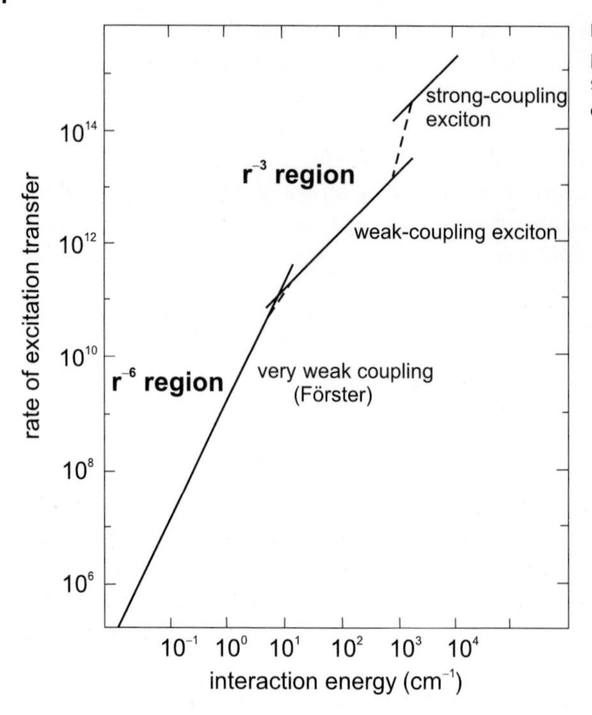

Fig. 4.16. Transfer rates predicted by Förster for strong, weak and very weak coupling.

where κ^2 is the orientational factor, Φ_D^0 is the fluorescence quantum yield of the donor in the absence of transfer, n is the average refractive index of the medium in the wavelength range where spectral overlap is significant, $I_D(\lambda)$ is the fluorescence spectrum of the donor normalized so that $\int_0^\infty I_D(\lambda)\,d\lambda = 1$, and $\varepsilon_A(\lambda)$ is the molar absorption coefficient of the acceptor.[12] Hence, for R_0 in Å, λ in nm, $\varepsilon_A(\lambda)$ in $M^{-1}\ cm^{-1}$ (overlap integral in units of $M^{-1}\ cm^{-1}\ nm^4$), we obtain:

$$R_0 = 0.2108\left[\kappa^2 \Phi_D n^{-4} \int_0^\infty I_D(\lambda)\varepsilon_A(\lambda)\lambda^4\,d\lambda\right]^{1/6} \tag{4.80}$$

R_0 is generally in the range of 15–60 Å (see Chapter 9, Table 9.1).

The orientational factor κ^2 is given by

$$\kappa^2 = \cos\theta_{DA} - 3\cos\theta_D\cos\theta_A = \sin\theta_D\sin\theta_A\cos\varphi - 2\cos\theta_D\cos\theta_A \tag{4.81}$$

12) In many books the overlap integral is written in wavenumbers instead of wavelengths. According to the remark made for quantum yields (see Chapter 3, Box 3.3), $I_D(\lambda)d\lambda = I_D(\bar{\nu})d\bar{\nu}$ and $\varepsilon_A(\lambda) = \varepsilon_A(\bar{\nu})$; therefore, the two expressions are equivalent:

$$\int_0^\infty I_D(\lambda)\varepsilon_A(\lambda)\lambda^4\,d\lambda = \int_0^\infty \frac{I_D(\bar{\nu})\varepsilon_A(\bar{\nu})}{(\bar{\nu})^4}\,d\bar{\nu}$$

If grating monochromators are used, the bandpass $\Delta\lambda$ is constant and the scale is linear in wavelength. The expression in wavelength must then be used for the integral calculation.

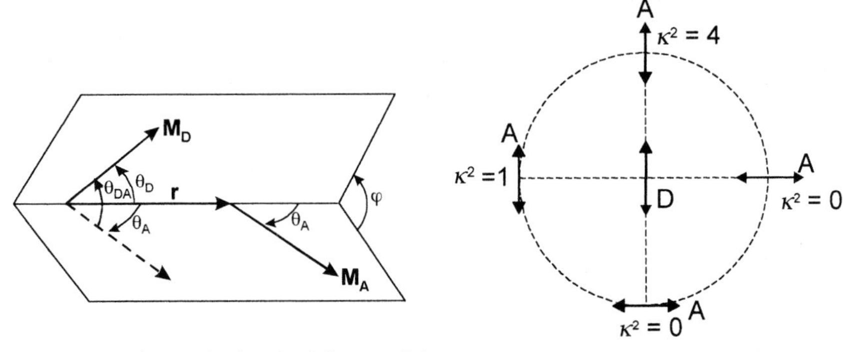

Fig. 4.17. Angles involved in the definition of the orientation factor κ^2 (left) and examples of values of κ^2 (right).

where θ_{DA} is the angle between the donor and acceptor transition moments, and θ_D and θ_A are the angles between these, respectively, and the separation vector; φ is the angle between the projections of the transition moments on a plane perpendicular to the line through the centers (Fig. 4.17). κ^2 can in principle take values from 0 (perpendicular transition moments) to 4 (collinear transition moments). When the transition moments are parallel, $\kappa^2 = 1$.

When the molecules are free to rotate at a rate that is much faster than the de-excitation rate of the donor (isotropic dynamic averaging), the average value of κ^2 is $2/3$. In a rigid medium, the square of the average of κ is 0.476 for an ensemble of acceptors that are statistically randomly distributed about the donor with respect to both distance and orientation (this case is often called the static isotropic average).

The transfer efficiency is defined as

$$\Phi_T = \frac{k_T^{dd}}{k_D + k_T^{dd}} = \frac{k_T^{dd}}{1/\tau_D^0 + k_T^{dd}} \tag{4.82}$$

Using Eq. (4.78), the transfer efficiency can be related to the ratio r/R_0:

$$\Phi_T = \frac{1}{1 + (r/R_0)^6} \tag{4.83}$$

Note that the transfer efficiency is 50% when the donor–acceptor distance is equal to the Förster critical radius. Equation (4.83) shows that the distance between a donor and an acceptor can be determined by measuring the efficiency of transfer, provided that r is not too different from R_0 (which is evaluated by means of Eq. 4.80).

The efficiency of transfer can also be written in the following form:

$$\Phi_T = 1 - \frac{\tau_D}{\tau_D^0} \tag{4.84}$$

where τ_D^0 and τ_D are the donor excited-state lifetimes in the absence and presence of acceptor, respectively.

The use of Förster non-radiative energy transfer for measuring distances at a supramolecular level (*spectroscopic ruler*) will be discussed in detail in Chapter 9.

Dexter's formulation of exchange energy transfer (very weak coupling) In contrast to the inverse sixth power dependence on distance for the dipole–dipole mechanism, an exponential dependence is to be expected from the exchange mechanism. The rate constant for transfer can be written as

$$k_T^{ex} = \frac{2\pi}{h} KJ' \exp(-2r/L) \tag{4.85}$$

where J' is the integral overlap[13)]

$$J' = \int_0^\infty I_D(\lambda)\varepsilon_A(\lambda)\,d\lambda \tag{4.86}$$

with the normalization condition

$$\int_0^\infty I_D(\lambda)\,d\lambda = \int_\infty^0 \varepsilon_A(\lambda)\,d\lambda = 1 \tag{4.87}$$

where L is the average Bohr radius. Because K (Eq. 4.85) is a constant that is not related to any spectroscopic data, it is difficult to characterize the exchange mechanism experimentally.

Selection rules

1. Dipole–dipole mechanism:

- Equation (4.79) shows that R_0, and consequently the transfer rate, is independent of the donor oscillator strength but depends on the acceptor oscillator strength and on the spectral overlap. Therefore, provided that the acceptor transition is allowed (spin conservation) and its absorption spectrum overlaps the donor fluorescence spectrum, the following types of energy transfer are possible:

$^1D^* + {}^1A \rightarrow {}^1D + {}^1A^*$ (singlet–singlet energy transfer) (e.g. anthracene + perylene, or pyrene + perylene).

$^1D^* + {}^3A^* \rightarrow {}^1D + {}^3A^{**}$ (higher triplet). This type of transfer requires overlap of the fluorescence spectrum of D and the T–T absorption spectrum of A (e.g. quenching of perylene by phenanthrene in its lowest triplet state).

$^3D^* + {}^1A \rightarrow {}^1D + {}^1A^*$ (triplet–singlet energy transfer). This type of transfer leads to phosphorescence quenching of the donor (e.g. phenanthrene (T) + rhodamine B (S)).

13) Note that $\int_0^\infty I_D(\lambda)\varepsilon_A(\lambda)\,d\lambda = \int_0^\infty I_D(\bar{\nu})\varepsilon_A(\bar{\nu})\,d\bar{\nu}$.

$^3D^* + {}^3A^* \rightarrow {}^1D + {}^3A^{**}$ (higher triplet). This type of transfer requires overlap of the phosphorescence spectrum of D^* and the T–T absorption spectrum of A.

2. Exchange mechanism:

- Because the energy rate does not imply the transition moments in the exchange mechanism, triplet–triplet energy transfer is possible:

$^3D^* + {}^1A \rightarrow {}^1D + {}^3A^*$ (e.g. naphthalene + benzophenone)

It is worth mentioning that triplet–triplet energy transfer can be used to populate the triplet state of molecules in which intersystem crossing is unlikely (e.g. the triplet state of thymine in frozen solutions can be populated by energy transfer from acetophenone).

- Triplet–triplet annihilation can also occur by an exchange mechanism (according to Wigner's rules):

$^3D^* + {}^3A \rightarrow {}^1D + {}^3A^*$ or $^1A^*$ or $^5A^*$

When D and A are identical ($^3D^* + {}^3D \rightarrow {}^1D + {}^3D^*$), triplet–triplet annihilation leads to a delayed fluorescence, called P-type delayed fluorescence because it was first observed with pyrene. Part of the energy resulting from annihilation allows one of the two partners to return to the singlet state from which fluorescence is emitted, but with a delay after staying in the triplet state. In crystals or polymers, the annihilation process often takes place in the vicinity of defects or impurities that act as energy traps.

- Singlet–singlet transfer is of course allowed as well.

Further details of non-radiative energy transfer will be presented in Chapter 9, together with various applications.

4.7 Bibliography

Andrews D. L. and Demidov A. A. (Eds) **(1999)** *Resonance Energy Transfer*, Wiley, Chichester.

Berberan-Santos M. N., Pereira E. J. N. and Martinho J. M. G. **(1999)** Dynamics of Radiative Transport, in: Andrews D. L. and Demidov A. A. (Eds), *Resonance Energy Transfer*, Wiley, Chichester, pp. 108–150.

Birks J. B. **(1970)** Photophysics of Aromatic Molecules, Wiley, London.

Birks J. B. (Ed.) **(1973)** *Organic Molecular Photophysics*, Wiley, London.

Cheung H. C. **(1991)** Resonance Energy Transfer, in: Lakowicz J. R. (Ed.), *Topics in Fluorescence Spectroscopy, Vol. 2, Principles*, Plenum Press, New York, pp. 127–176.

Clegg R. M. **(1996)** Fluorescence Resonance Energy Transfer, in: Wang X. F. and Herman B. (Eds), *Fluorescence Imaging Spectroscopy and Microscopy*, Wiley, New York, pp. 179–252.

Dexter D. L. **(1953)** A Theory of Sensitized Luminescence in Solids, *J. Chem. Phys.* 21, 836–850.

Eftink M. R. **(1991a)** Fluorescence

Quenching: Theory and Applications, in: Lakowicz J. R. (Ed.), *Topics in Fluorescence Spectroscopy, Vol. 2, Principles*, Plenum Press, New York, pp. 53–126.

Eftink M. R. **(1991b)** Fluorescence Quenching Reactions. Probing Biological Macromolecular Structures, in: Dwey T. G. (Ed.), *Biophysical and Biochemical Aspects of Fluorescence Spectroscopy*, Plenum Press, New York, pp. 1–41.

Eftink M. R. and Ghiron C. A. **(1981)** Fluorescence Quenching Studies with Proteins, *Anal. Biochem.* 114, 199–227.

Förster Th. **(1959)** Transfer Mechanisms of Electronic Excitation, *Disc. Far. Soc.*, 27, 7–17.

Förster Th. **(1960)** Transfer Mechanism of Electronic Excitation Energy, *Radiation Res. Supp.*, 2, 326–339.

Förster Th. **(1965)** Delocalized Excitation and Excitation Transfer, in: Sinanoglu O. (Ed.), *Modern Quantum Chemistry*, Vol. 3, Academic Press, New York. pp. 93–137.

Fox M. A. and Chanon M. (Eds) **(1988)** *Photoinduced Electron Transfer*, Elsevier, Amsterdam.

Kasha M. **(1991)** Energy, Charge Transfer, and Proton Transfer in Molecular Composite Systems, in: Glass W. A. and Varma M. N. (Eds), *Physical and Chemical Mechanisms in Molecular Radiation Biology*, Plenum Press, New York, pp. 231–255.

Turro N. J. **(1978)** *Modern Molecular Photochemistry*, Benjamin/Cummings, Menlo Park, CA.

Van der Meer B. W., Coker G. III and Chen S.-Y. S. **(1994)** *Resonance Energy Transfer. Theory and Data*, VCH, New York.

Weller A. **(1961)** Fast Reactions of Excited Molecules, *Prog. React. Kinetics* 1, 189–214.

5
Fluorescence polarization. Emission anisotropy

Polarisation de la lumière: Disposition des parties composant un rayon lumineux de sorte qu'elles se comportent toutes de la même façon.

[*Polarization of light: Arrangement of the parts which make up a light ray so that they all act in the same way.*]

Encyclopédie Méthodique. Physique
Monge, Cassini, Bertholon et al., 1822

Light is an electromagnetic wave consisting of an electric field **E** and a magnetic field **B** perpendicular both to each other and to the direction of propagation, and oscillating in phase. For natural light, these fields have no preferential orientation, but for linearly polarized light, the electric field oscillates along a given direction; the intermediate case corresponds to partially polarized light (Figure 5.1).

Most chromophores absorb light along a preferred direction[1] (see Chapter 2 for the definition of absorption transition moment, and for examples of transition moments of some fluorophores, see Figure 2.3), depending on the electronic state. In contrast, the emission transition moment is the same whatever the excited state reached by the molecule upon excitation, because of internal conversion towards the first singlet state (Figure 5.2).

If the incident light is linearly polarized, the probability of excitation of a chromophore is proportional to the square of the scalar product $\mathbf{M_A}.\mathbf{E}$, i.e. $\cos^2\theta_A$, θ_A being the angle between the electric vector **E** of the incident light and the absorption transition moment $\mathbf{M_A}$ (Figure 5.2). This probability is maximum when **E** is parallel to $\mathbf{M_A}$ of the molecule; it is zero when the electric vector is perpendicular.

1) The absorption transition moment is not in a single direction for some molecules whose symmetry is D_{6h} (benzene), D_{3h} (triphenylene) or $I_h(C_{60})$.

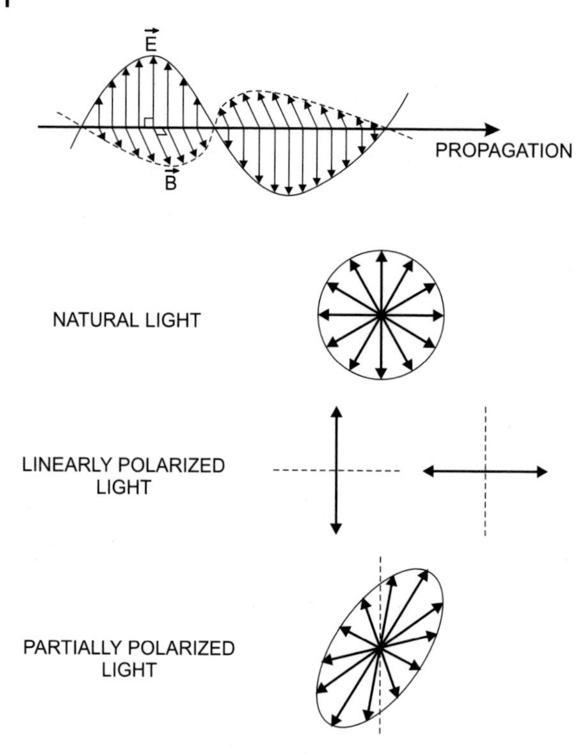

Fig. 5.1. Natural light and linearly polarized light.

Thus, when a population of fluorophores is illuminated by a linearly polarized incident light, those whose transition moments are oriented in a direction close to that of the electric vector of the incident beam are preferentially excited. This is called *photoselection*. Because the distribution of excited fluorophores is anisotropic, the emitted fluorescence is also anisotropic. Any change in direction of the transition moment during the lifetime of the excited state will cause this anisotropy to decrease, i.e. will induce a partial (or total) depolarization of fluorescence.

The causes of fluorescence depolarization are:

- non-parallel absorption and emission transition moments
- torsional vibrations
- Brownian motion
- transfer of the excitation energy to another molecule with different orientation.

Fluorescence polarization measurements can thus provide useful information on molecular mobility, size, shape and flexibility of molecules, fluidity of a medium, and order parameters (e.g. in a lipid bilayer)[2)].

2) Circular polarized luminescence (CPL) is not covered in this book because the field of application of this phenomenon is limited to chiral systems that emit different amounts of left and right circularly polarized light. Nevertheless, it is worth mentioning that valuable information can be obtained by CPL spectroscopy on the electronic and molecular structure in the excited state of chiral organic molecules, inorganic complexes and biomacromolecules (for a review, see Riehl J. P. and Richardson F. S. (1986) *Chem. Rev.* **86**, 1).

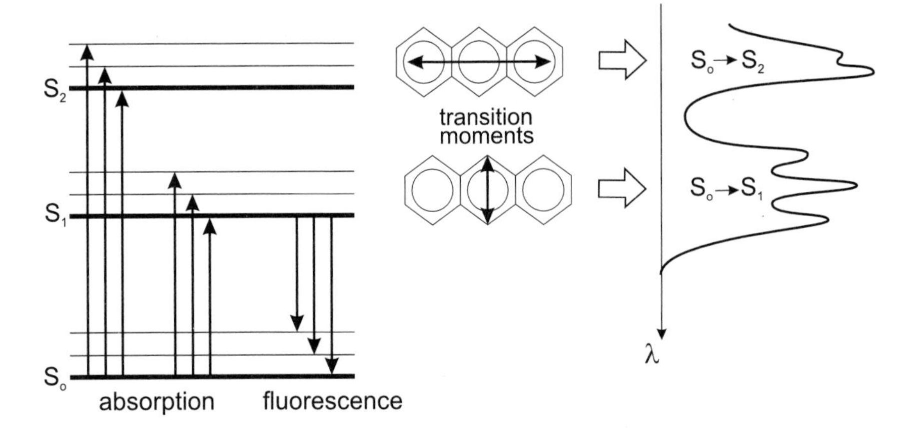

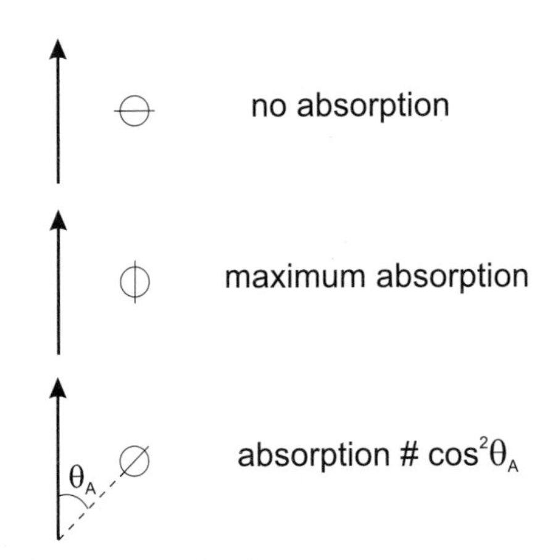

Fig. 5.2. Transition moments (e.g. anthracene) and photoselection.

5.1
Characterization of the polarization state of fluorescence (polarization ratio, emission anisotropy)

Because there is no phase relation between the light emitted by different molecules, fluorescence can be considered as the result of three independent sources of light polarized along three perpendicular axis Ox, Oy, Oz without any phase relation between them. I_x, I_y, I_z are the intensities of these sources, and the total intensity is $I = I_x + I_y + I_z$. The values of the intensity components depend on the polarization of the incident light and on the depolarization processes. Application of the Curie symmetry principle (an effect cannot be more dissymmetric than the

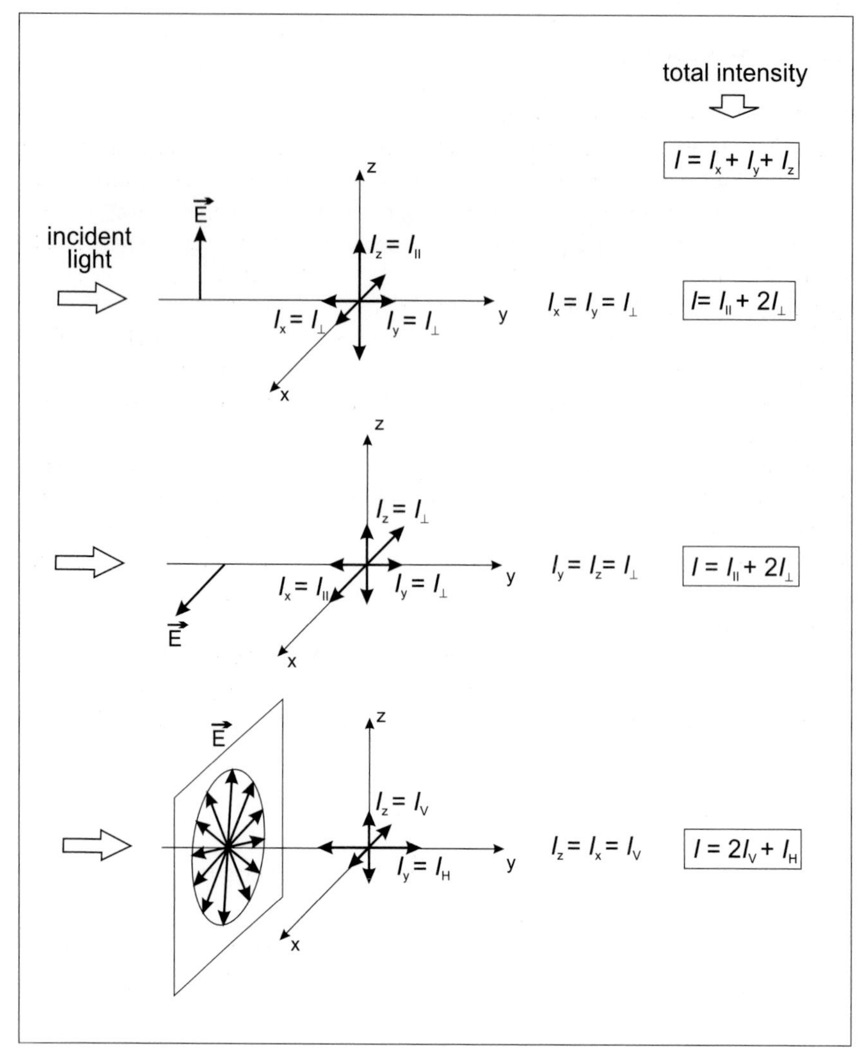

Fig. 5.3. Relations between the fluorescence intensity components resulting from the Curie symmetry principle. The fluorescent sample is placed at the origin of the system of coordinates.

cause from which it results) leads to relations between intensity components, as shown in Figure 5.3. It should be noted that the symmetry principle is strictly valid only for a point source of light, which cannot be rigorously achieved in practice. Moreover, only homogeneous diluted solutions will be considered in this chapter, so that artefacts due for instance to inner filter effects are avoided.

The various cases of excitation are now examined (Figure 5.3).

5.1.1
Excitation by polarized light

5.1.1.1 Vertically polarized excitation

When the incident light is vertically polarized, the vertical axis Oz is an axis of symmetry for the emission of fluorescence according to the Curie principle, i.e. $I_x = I_y$. The fluorescence observed in the direction of this axis is thus unpolarized.

The components of the fluorescence intensity that are parallel and perpendicular to the electric vector of the incident beam are usually denoted as $I_{\parallel}$ and $I_{\perp}$, respectively. For vertically polarized incident light, $I_z = I_{\parallel}$ and $I_x = I_y = I_{\perp}$.

The intensity component I_z corresponding to oscillations of the electric field along the Oz axis cannot be detected by the eye or by a detector placed along this axis. The fluorescence intensity observed in the direction of this axis is thus $I_x + I_y = 2I_{\perp}$.

On the contrary, the Ox and Oy axes are not axes of symmetry for the emission of fluorescence. When fluorescence is observed through a polarizer along the Ox axis (or the Oy axis), the intensity measured is $I_z = I_{\parallel}$ for the vertical position of the polarizer, and $I_y = I_{\perp}$ (or $I_x = I_{\perp}$) for the horizontal position. Without polarizer, the measured intensity is $I_z + I_y = I_{\parallel} + I_{\perp}$ in the Ox direction and $I_z + I_x = I_{\parallel} + I_{\perp}$ in the Oy direction.

In most cases, fluorescence is observed in a horizontal plane at 90° to the propagation direction of the incident beam, i.e. in direction Ox (Figure 5.4). The fluorescence intensity components $I_{\parallel}$ and $I_{\perp}$ are measured by a photomultiplier in

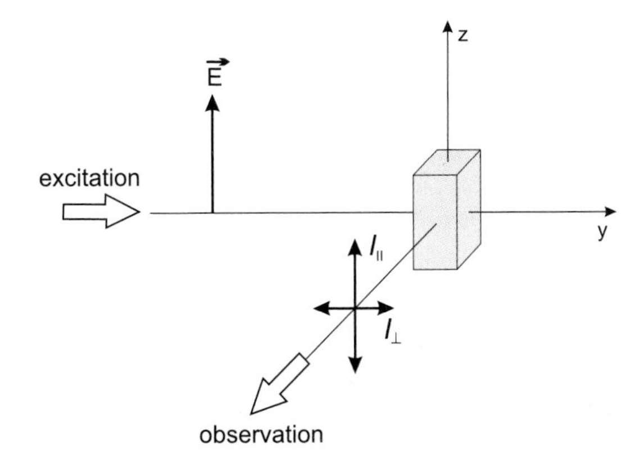

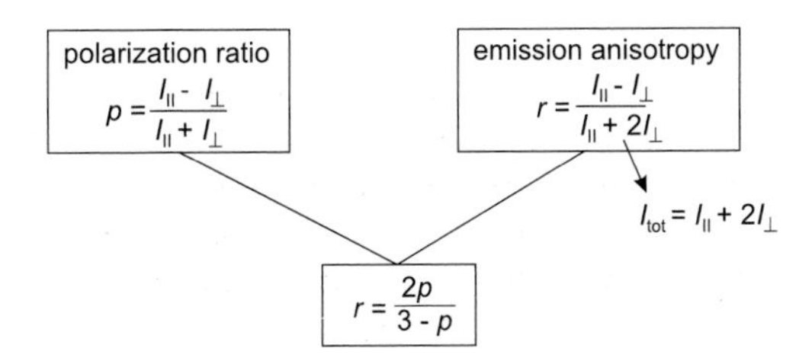

Fig. 5.4. Usual configuration for measuring fluorescence polarization.

front of which a polarizer is rotated. The total fluorescence intensity is $I = I_x + I_y + I_z = I_\parallel + 2I_\perp$ [3].

The polarization state of fluorescence is characterized either by

- the polarization ratio p:

$$p = \frac{I_\parallel - I_\perp}{I_\parallel + I_\perp} \tag{5.1}$$

- or the emission anisotropy r:

$$r = \frac{I_\parallel - I_\perp}{I_\parallel + 2I_\perp} \tag{5.2}$$

In the expression of the polarization ratio, the denominator represents the fluorescence intensity in the direction of observation, whereas in the formula giving the emission anisotropy, the denominator represents the total fluorescence intensity. In a few situations (e.g. the study of radiative transfer) the polarization ratio is to be preferred, but in most cases, the use of emission anisotropy leads to simpler relations (see below).

The relationship between r and p follows from Eqs (5.1) and (5.2):

$$r = \frac{2p}{3 - p} \tag{5.3}$$

5.1.1.2 **Horizontally polarized excitation**

When the incident light is horizontally polarized, the horizontal Ox axis is an axis of symmetry for the fluorescence intensity: $I_y = I_z$. The fluorescence observed in the direction of this axis (i.e. at 90° in a horizontal plane) should thus be unpolarized (Figure 5.3). This configuration is of practical interest in checking the possible residual polarization due to imperfect optical tuning. When a monochromator is used for observation, the polarization observed is due to the dependence of its transmission efficiency on the polarization of light. Then, measurement of the polarization with a horizontally polarized incident beam permits correction to get the true emission anisotropy (see Section 6.1.6).

5.1.2 Excitation by natural light

When the sample is excited by natural light (i.e. unpolarized), the light can be decomposed into two perpendicular components, whose effects on the excitation of a

3) Note that in none of the directions Ox, Oy, Oz, is the observed fluorescence intensity proportional to the total fluorescence intensity. They are respectively $I_\parallel + I_\perp$, $I_\parallel + I_\perp$, and $2I_\perp$. It will be shown in Chapter 6 (see Appendix) how a signal proportional to the total fluorescence intensity can be measured by using excitation and/or emission polarizers at appropriate angles.

population of fluorophores are additive. Upon observation at 90° in a horizontal plane, the incident vertical component has the same effect as previously described, whereas the incident horizontal component leads to unpolarized fluorescence emission in the direction of observation Ox, which is an axis of symmetry: $I_z = I_x$ (Figure 5.3).

The components I_V and I_H, vertically and horizontally polarized respectively, are such that $I_z = I_V = I_x$, $I_y = I_H$ (Figure 5.3). The total fluorescence intensity is then $2I_V + I_H$. The polarization ratio and the emission anisotropy are given by

$$p_n = \frac{I_V - I_H}{I_V + I_H} \qquad r_n = \frac{I_V - I_H}{2I_V + I_H} \tag{5.4}$$

where the subscript n refers to natural exciting light. These two quantities are linked by the following relation:

$$r_n = \frac{2p_n}{3 + p_n} \tag{5.5}$$

It is easy to show that $r_n = r/2$. Therefore, the emission anisotropy observed upon excitation by natural light is half that upon excitation by vertically polarized light. In view of the difficulty of producing perfectly natural light (i.e. totally unpolarized), vertically polarized light is always used in practice. Consequently, only excitation by polarized light will be considered in the rest of this chapter.

5.2 Instantaneous and steady-state anisotropy

5.2.1 Instantaneous anisotropy

Following an infinitely short pulse of light, the total fluorescence intensity at time t is $I(t) = I_{\parallel}(t) + 2I_{\perp}(t)$, and the instantaneous emission anisotropy at that time is

$$\boxed{r(t) = \frac{I_{\parallel}(t) - I_{\perp}(t)}{I_{\parallel}(t) + 2I_{\perp}(t)} = \frac{I_{\parallel}(t) - I_{\perp}(t)}{I(t)}} \tag{5.6}$$

Each polarized component evolves according to

$$I_{\parallel}(t) = \frac{I(t)}{3}[1 + 2r(t)] \tag{5.7}$$

$$I_{\perp}(t) = \frac{I(t)}{3}[1 - r(t)] \tag{5.8}$$

After recording $I_{\parallel}(t)$ and $I_{\perp}(t)$, the emission anisotropy can be calculated by means of Eq. (5.6), provided that the light pulse is very short with respect to the fluorescence decay. Otherwise, we should take into account the fact that the measured polarized components are the convolution products of the δ-pulse responses (5.7) and (5.8) by the instrument response (see Chapter 6).

5.2.2
Steady-state anisotropy

On continuous illumination (i.e. when the incident light intensity is constant), the measured anisotropy is called *steady-state anisotropy* $\bar{r}$. Using the general definition of an averaged quantity, with the total normalized fluorescence intensity as the probability law, we obtain

$$\bar{r} = \frac{\int_0^\infty r(t)I(t)\,\mathrm{d}t}{\int_0^\infty I(t)\,\mathrm{d}t} \tag{5.9}$$

In the case of a single exponential decay with time constant τ (excited-state lifetime), the steady-state anisotropy is given by

$$\bar{r} = \frac{1}{\tau}\int_0^\infty r(t)\exp(-t/\tau)\,\mathrm{d}t \tag{5.10}$$

5.3
Additivity law of anisotropy

When the sample contains a mixture of fluorophores, each has its own emission anisotropy r_i:

$$r_i = \frac{I_\parallel^i - I_\perp^i}{I_\parallel^i + 2I_\perp^i} = \frac{I_\parallel^i - I_\perp^i}{I_i} \tag{5.11}$$

and each contributes to the total fluorescence intensity with a fraction $f_i = I_i/I\left(\sum_i f_i = 1\right)$.

From a practical point of view, we measure the components $I_\parallel$ and $I_\perp$, which are now the sum of all individual components:

$$I_\parallel = \sum_i I_\parallel^i \quad \text{and} \quad I_\perp = \sum_i I_\perp^i \tag{5.12}$$

and usually the same definition of the total emission anisotropy is kept because, in practice, we measure the components $I_\parallel$ and $I_\perp$

$$r = \frac{I_\parallel - I_\perp}{I_\parallel + 2I_\perp} = \frac{\sum_i I_\parallel^i - \sum_i I_\perp^i}{I} = \sum_i \frac{I_\parallel^i - I_\perp^i}{I_i} \times \frac{I_i}{I} = \sum_i f_i r_i \tag{5.13}$$

The important consequence of this is that the total emission anisotropy is the weighted sum of the individual anisotropies[4]:

$$r = \sum_i f_i r_i \tag{5.14}$$

This relationship applies to both steady-state and time-resolved experiments. In the latter case, if each species i exhibits a single exponential fluorescence decay with lifetime τ_i, the fractional intensity of this species at time t is

$$f_i(t) = \frac{a_i \exp(-t/\tau_i)}{I(t)} \tag{5.15}$$

where

$$I(t) = \sum_i a_i \exp(-t/\tau_i) \tag{5.16}$$

Hence

$$r(t) = \sum_i \frac{a_i \exp(-t/\tau_i)}{I(t)} r_i(t) \tag{5.17}$$

This equation shows that, at time t, each anisotropy term is weighted by a factor that depends on the relative contribution to the total fluorescence intensity at that time. This is surprising at first sight, but simply results from the definition used for the emission anisotropy, which is based on the practical measurement of the overall $I_\parallel$ and $I_\perp$ components. A noticeable consequence is that the emission anisotropy of a mixture may not decay monotonously, depending of the values of r_i and τ_i for each species. Thus, $r(t)$ should be viewed as an 'apparent' or a 'technical' anisotropy because it does not reflect the overall orientation relaxation after photoselection, as in the case of a single population of fluorophores.

It should be noted that Eqs (5.7) and (5.8), where $I(t)$ would be the total intensity $\sum I_i(t)$, and $r(t)$ the sum $\sum r_i(t)$, are not valid[5].

Equations (5.14) to (5.17) also apply to the case of a single fluorescent species residing in different microenvironments where the excited-state lifetimes are τ_i.

4) The additivity law can also be expressed with the polarization ratio:

$$\left(\frac{1}{p} - \frac{1}{3}\right)^{-1} = \sum_i f_i \left(\frac{1}{p_i} - \frac{1}{3}\right)^{-1}$$

5) In fact, let us consider as an example a mixture of two fluorophores. The overall $I_\parallel(t)$ and $I_\perp(t)$ components are given by

$$I_\parallel(t) = I_1(t)[1 + 2r_1(t)] + I_2(t)[1 + 2r_2(t)]$$
$$I_\perp(t) = I_1(t)[1 - r_1(t)] + I_2(t)[1 - r_2(t)]$$

It is obvious that these relations cannot be put in the form of Eqs (5.7) and (5.8) where $I(t)$ would be the total intensity $I_1(t) + I_2(t)$ and $r(t)$ the sum $r_1(t) + r_2(t)$.

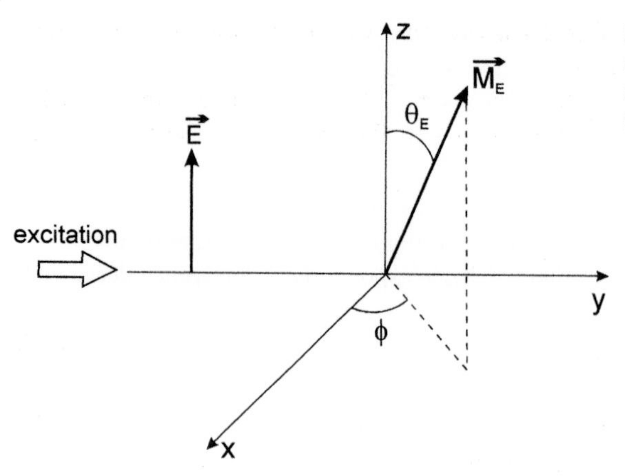

Fig. 5.5. System of coordinates for characterizing the orientation of the emission transition moments.

5.4 Relation between emission anisotropy and angular distribution of the emission transition moments

Let us consider a population of N molecules randomly oriented and excited at time 0 by an infinitely short pulse of light polarized along Oz. At time t, the emission transition moments $\mathbf{M_E}$ of the excited molecules have a certain angular distribution. The orientation of these transition moments is characterized by θ_E, the angle with respect to the Oz axis, and by ϕ (azimuth), the angle with respect to the Oz axis (Figure 5.5). The final expression of emission anisotropy should be independent of ϕ because Oz is an axis of symmetry.

For a particular molecule i, the components of the emission transition moments along the three axes Ox, Oy, Oz are $M_E\alpha_i(t)$, $M_E\beta_i(t)$ and $M_E\gamma_i(t)$, where $\alpha_i(t)$, $\beta_i(t)$ and $\gamma_i(t)$ are the cosines of the angles formed by the emission transition moment with the three axes (such that $\alpha_i^2\beta_i^2\gamma_i^2 = 1$) and M_E is the modulus of the vector transition moment.

The total fluorescence intensity at time t is obtained by summing over all molecules emitting at that time. Because there is no phase relation between the elementary emissions, the contributions of each molecule to the intensity components along Ox, Oy and Oz are proportional to the square of its transition moment components along each axis. Summation over all molecules leads to the following expressions for the fluorescence intensity components:

$$I_x(t) = KM_E^2 \sum_{i=1}^{N} \alpha_i^2(t) = KM_E^2 N\overline{\alpha^2(t)}$$

$$I_y(t) = KM_E^2 \sum_{i=1}^{N} \beta_i^2(t) = KM_E^2 N\overline{\beta^2(t)} \tag{5.18}$$

$$I_z(t) = KM_E^2 \sum_{i=1}^{N} \gamma_i^2(t) = KM_E^2 N\overline{\gamma^2(t)}$$

where K is a proportionality factor. The bars characterize an ensemble average over the N emitting molecules at time t.

Because of the axial symmetry around Oz, $\overline{\alpha^2(t)} = \overline{\beta^2(t)}$, and because $\overline{\alpha^2(t)} + \overline{\beta^2(t)} + \overline{\gamma^2(t)} = 1$, we have $\overline{\gamma^2(t)} = 1 - 2\overline{\alpha^2(t)}$. The emission anisotropy is then given by

$$r(t) = \frac{I_{\parallel}(t) - I_{\perp}(t)}{I(t)} = \frac{I_z(t) - I_y(t)}{I_x(t) + I_y(t) + I_z(t)}$$

$$= \overline{\gamma^2(t)} - \overline{\alpha^2(t)} = \frac{3\overline{\gamma^2(t)} - 1}{2} \tag{5.19}$$

Finally, because $\gamma = \cos\theta_E$, the relation between the emission anisotropy and the angular distribution of the emission transition moments can be written as

$$\boxed{r(t) = \frac{3\,\overline{\cos^2\theta_E(t)} - 1}{2}} \tag{5.20}$$

5.5 Case of motionless molecules with random orientation

5.5.1 Parallel absorption and emission transition moments

When the absorption and emission transition moments are parallel, $\theta_A = \theta_E$, the common value being denoted θ; hence $\overline{\cos^2\theta_A} = \overline{\cos^2\theta_E} = \overline{\cos^2\theta}$. Before excitation, the number of molecules whose transition moment is oriented within angles θ and $\theta + d\theta$, and ϕ and $\phi + d\phi$ is proportional to an elementary surface on a sphere whose radius is unity, i.e. $2\pi \sin\theta\, d\theta\, d\phi$ (Figure 5.6).

Taking into account the excitation probability, i.e. $\cos^2\theta$, the number of excited molecules whose transition moment is oriented within angles θ and $\theta + d\theta$, and ϕ and $\phi + d\phi$, is proportional to $\cos^2\theta \sin\theta\, d\theta\, d\phi$. The fraction of molecules oriented in this direction is

$$W(\theta, \phi)\, d\theta\, d\phi = \frac{\cos^2\theta \sin\theta\, d\theta\, d\phi}{\int_0^{2\pi} d\phi \int_0^{\pi} \cos^2\theta \sin\theta\, d\theta} \tag{5.21}$$

The denominator, which is proportional to the total number of excited molecules, can be calculated by setting $x = \cos\theta$, hence $dx = -\sin\theta\, d\theta$, and its value is $4\pi/3$. Equation (5.21) then becomes

$$W(\theta, \phi)\, d\theta\, d\phi = \frac{3}{4\pi} \cos^2\theta \sin\theta\, d\theta\, d\phi \tag{5.22}$$

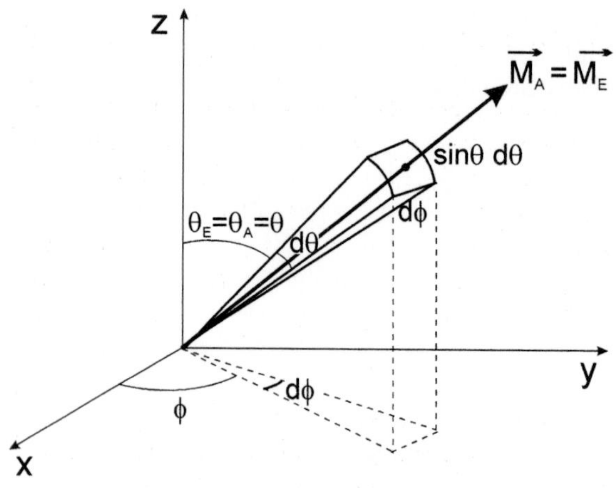

Fig. 5.6. The fraction of molecules whose absorption and emission transition moments are parallel and oriented in a direction within the elementary solid angle. This direction is defined by angles θ and ϕ.

It is then possible to calculate the average of $\cos^2\theta$ over all excited molecules

$$\begin{aligned}\overline{\cos^2\theta} &= \int_0^{2\pi} d\phi \int_0^{\pi} \cos^2\theta W(\theta,\phi)\, d\theta \\ &= \frac{3}{4\pi}\int_0^{2\pi} d\phi \int_0^{\pi} \cos^4\theta \sin\theta\, d\theta \\ &= 3/5\end{aligned} \tag{5.23}$$

Using Eq. (5.20), the emission anisotropy can thus be written as

$$r_0 = \frac{3\,\overline{\cos^2\theta} - 1}{2} = \frac{2}{5} = 0.4 \tag{5.24}$$

r_0 is called the *fundamental anisotropy*, i.e. the theoretical anisotropy in the absence of any motion. In practice, rotational motions can be hindered in a rigid medium. The experimental value, called the *limiting anisotropy*, turns out to be always slightly smaller than the theoretical value. When the absorption and emission transition moments are parallel, i.e. when the molecules are excited to the first singlet state, the theoretical value of r_0 is 0.4[6)], but the experimental value usually ranges from 0.32 to 0.39. The reasons for these differences are discussed in Box 5.1.

6) The corresponding value of the polarization ratio is $p_0 = 0.5$.

Box 5.1 Fundamental and limiting anisotropies

The difference between the theoretical value of the emission anisotropy in the absence of motions (*fundamental anisotropy*) and the experimental value (*limiting anisotropy*) deserves particular attention. The limiting anisotropy can be determined either by steady-state measurements in a rigid medium (in order to avoid the effects of Brownian motion), or time-resolved measurements by taking the value of the emission anisotropy at time zero, because the instantaneous anisotropy can be written in the following form:

$$r(t) = r_0 f(t)$$

where $f(t)$ characterizes the dynamics of rotational motions (orientation autocorrelation function) (see Section 5.6, Eq. 5.32) and whose value is 1 at time zero.

It should first be noted that the measurement of emission anisotropy is difficult, and instrumental artefacts such as large cone angles of the incident and/or observation beams, imperfect or misaligned polarizers, re-absorption of fluorescence, optical rotation, birefringence, etc., might be partly responsible for the difference between the fundamental and limiting anisotropies.

If there is no significant instrumental artefact, any observed difference must be related to the fluorophore itself. There may be several causes. Firstly, a trivial effect can be the overlap of low-lying absorption bands, resulting in an anomalous excitation polarization spectrum. Secondly, the oscillator associated with absorption may not be strictly linear, but may be partly two-dimensional or three-dimensional. However, it is generally considered that the difference between the fundamental and limiting anisotropies is mainly due to *torsional vibrations* of the fluorophores about their equilibrium orientation, as first suggested by Jablonski[a]. A temperature dependence of the limiting anisotropy is then expected, and was indeed reported[b,c]: at low temperature (i.e. when the medium is frozen), r_0 may be considered as a constant quantity whereas at high temperature it decreases linearly with the temperature.

Fast librational motions of the fluorophore within the solvation shell should also be considered[d]. The estimated characteristic time for perylene in paraffin is about 1 ps, which is not detectable by time-resolved anisotropy decay measurement. An 'apparent' value of the emission anisotropy is thus measured, which is smaller than in the absence of libration. Such an explanation is consistent with the fact that fluorescein bound to a large molecule (e.g. polyacrylamide or monoglucoronide) exhibits a larger limiting anisotropy than free fluorescein in aqueous glycerolic solutions. However, the absorption and fluorescence spectra are different for free and bound fluorescein; the question then arises as to whether r_0 could be an intrinsic property of the fluorophore.

In this respect, Johansson[e] reported that r_0 is the same (within experimental accuracy) for fluorophores belonging to the same family, e.g. perylene and perylenyl compounds (0.369 ± 0.002), or xanthene derivatives such as rhodamine

B, rhodamine 6G, rhodamine 101 and fluorescein (0.373 ± 0.002). Therefore, r_0 appears to be an intrinsic property of a fluorophore related to its structure and to the shape of its absorption and fluorescence spectra. A possible explanation for a limiting anisotropy being less than the fundamental anisotropy could then be a change of the molecular geometry between the ground and excited states. In fact, the aromatic plane of perylene was suggested to be slightly twisted in the ground state, and evidence of butterfly-type intramolecular folding in xanthene dyes has been reported.

a) Jablonski A. (1950) *Acta Phys. Pol.* **10**, 193.
b) Kawski A., Kubicki A. and Weyna I. (1985) *Z. Naturforsch.* **40a**, 559.
c) Veissier V., Viovy J. L. and Monnerie L. (1989) *J. Phys. Chem.* **93**, 1709.
d) Zinsli P. E. (1977) *Chem. Phys.* **93**, 1989.
e) Johansson L. B. (1990) *J. Chem. Soc. Faraday Trans.* **86**, 2103.

5.5.2
Non-parallel absorption and emission transition moments

This situation occurs when excitation brings the fluorophores to an excited state other than the first singlet state from which fluorescence is emitted. Let α be the angle between the absorption and emission transition moments. The aim is to calculate $\cos\theta_E$ and then to deduce r by means of Eq. (5.20).

According to the classical formula of spherical trigonometry, $\cos\theta_E$ can be written as

$$\cos\theta_E = \cos\theta_A \cos\alpha + \cos\psi \sin\theta_A \sin\alpha \tag{5.25}$$

where ψ denotes the angle between the planes $(\mathrm{Oz}, \mathbf{M_A})$ and $(\mathrm{Oz}, \mathbf{M_E})$ (Figure 5.7).

By taking the square of the two sides of Eq. (5.25) and taking into account the fact that all values of ψ are equiprobable ($\overline{\cos\psi} = 0; \overline{\cos^2\psi} = 1/2$), we obtain

$$\begin{aligned}\overline{\cos^2\theta_E} &= \cos^2\alpha\,\overline{\cos^2\theta_A} + \frac{1}{2}\sin^2\alpha\,\overline{\sin^2\theta_A}\\ &= \cos^2\alpha\,\overline{\cos^2\theta_A} + \frac{1}{2}(1-\cos^2\alpha)(1-\overline{\cos^2\theta_A})\\ &= \frac{3}{2}\cos^2\alpha\,\overline{\cos^2\theta_A} - \frac{1}{2}\overline{\cos^2\theta_A} - \frac{1}{2}\cos^2\alpha + \frac{1}{2}\end{aligned} \tag{5.26}$$

Hence

$$r_0 = \frac{3\,\overline{\cos^2\theta_E} - 1}{2} = \frac{3\,\overline{\cos^2\theta_A} - 1}{2} \times \frac{3\,\overline{\cos^2\alpha} - 1}{2} \tag{5.27}$$

Because $\overline{\cos^2\theta_A} = 3/5$, the emission anisotropy is given by

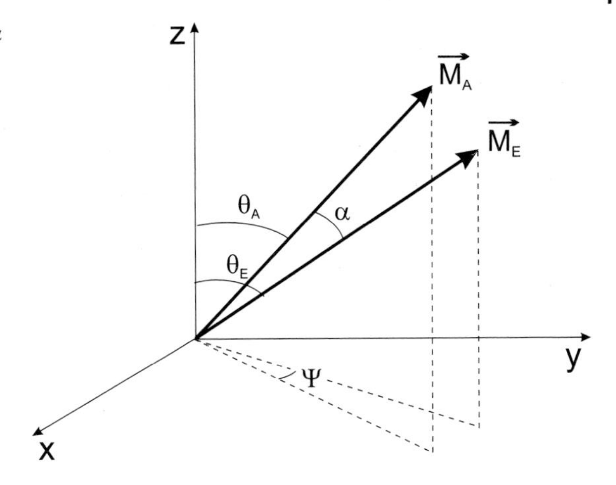

Fig. 5.7. Definition of angles α and ψ when the absorption and emission transition moments are not parallel.

$$r_0 = \frac{2}{5}\,\frac{3\,\overline{\cos^2\alpha} - 1}{2} \tag{5.28}$$

Consequently, the theoretical values of r_0 range from $2/5$ ($= 0.4$) for $\alpha = 0$ (parallel transition moments) and $-1/5$ ($= -0.2$) for $\alpha = 90°$ (perpendicular transition moments)[7]:

$$-0.2 \le r_0 \le 0.4 \tag{5.29}$$

A value close to -0.2 is indeed observed in the case of some aromatic molecules excited to the second singlet state whose transition moment is perpendicular to that of the first singlet state from which fluorescence is emitted (e.g. perylene). r_0 varies with the excitation wavelength for a given observation wavelength and these variations represent the *excitation polarization spectrum*, which allows us to distinguish between electronic transitions. Negative values generally correspond to $S_0 \rightarrow S_2$ transitions. As an illustration, Figure 5.8 shows the excitation polarization spectrum of perylene.

The case of indole and tryptophan is peculiar because the low-lying absorption bands overlap. Box 5.2 shows how the indole absorption spectrum can be resolved into two bands from the combined measurement of the excitation spectrum and the excitation polarization spectrum.

Particular cases should also be mentioned:

- For planar aromatic molecules with a symmetry axis of order 3 or higher (e.g. triphenylene: D_{3h}), r_0 cannot exceed 0.1.
- The fluorescence emitted by fullerene (C_{60}) is intrinsically totally depolarized because of its almost perfect spherical symmetry (I_h).

7) For the polarization ratio: $-1/3 \le p_0 \le 1/2$

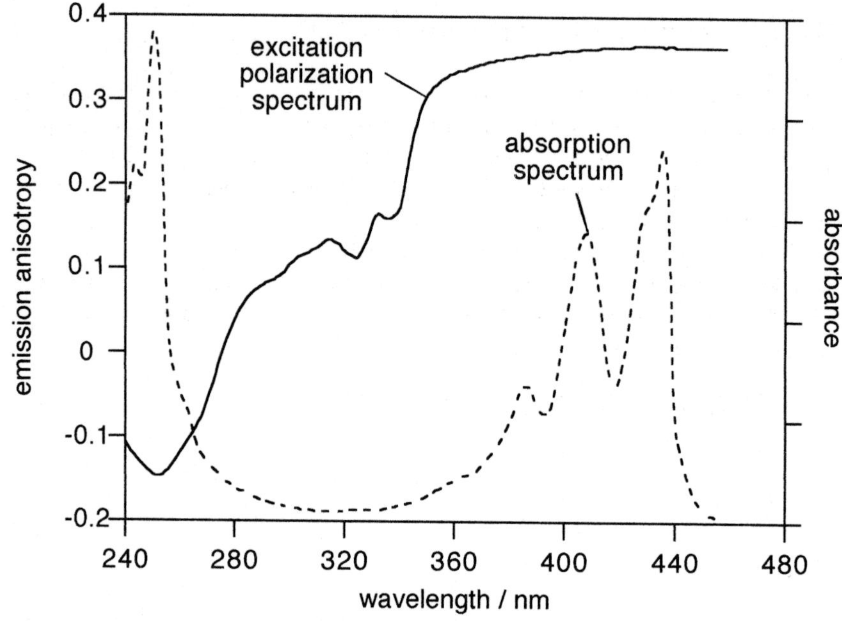

Fig. 5.8. Excitation polarization spectrum of perylene in propane-1,2-diol at −60 °C.

5.6
Effect of rotational Brownian motion

If excited molecules can rotate during the excited-state lifetime, the emitted fluorescence is partially (or totally) depolarized (Figure 5.9). The preferred orientation of emitting molecules resulting from photoselection at time zero is indeed gradually affected as a function of time by the rotational Brownian motions. From the extent of fluorescence depolarization, we can obtain information on the molecular motions, which depend on the size and the shape of molecules, and on the fluidity of their microenvironment.

Quantitative information can be obtained only if the time-scale of rotational motions is of the order of the excited-state lifetime τ. In fact, if the motions are slow with respect to $\tau (r \approx r_0)$ or rapid $(r \approx 0)$, no information on motions can be obtained from emission anisotropy measurements because these motions occur out of the experimental time window.

A distinction should be made between *free rotation* and *hindered rotation*. In the case of free rotation, after a δ-pulse excitation the emission anisotropy decays from r_0 to 0 because the rotational motions of the molecules lead to a random orientation at long times. In the case of hindered rotations, the molecules cannot become randomly oriented at long times, and the emission anisotropy does not decay to zero but to a steady value, r (Figure 5.10). These two cases of free and hindered rotations will now be discussed.

Box 5.2 Resolution of the absorption spectrum of indole[a)]

The long wavelength absorption band of indole consists of two electronic transitions ^1L_a and ^1L_b, whose transition moments are almost perpendicular (more precisely, they are oriented at $-38°$ and $56°$ to the long molecular axis, respectively). Figure B5.2.1 shows the excitation spectrum and the excitation polarization spectrum in propylene glycol at -58 °C.

The emission anisotropy $r_0(\lambda)$ at a wavelength of excitation λ results from the addition of contributions from the ^1L_a and ^1L_b excited states with fractional contributions $f_a(\lambda)$ and $f_b(\lambda)$, respectively. According to the additivity law of emission anisotropies, $r_0(\lambda)$ is given by

$$r_0(\lambda) = f_a(\lambda) r_{0a} + f_b(\lambda) r_{0b}$$

with

$$f_a(\lambda) + f_b(\lambda) = 1$$

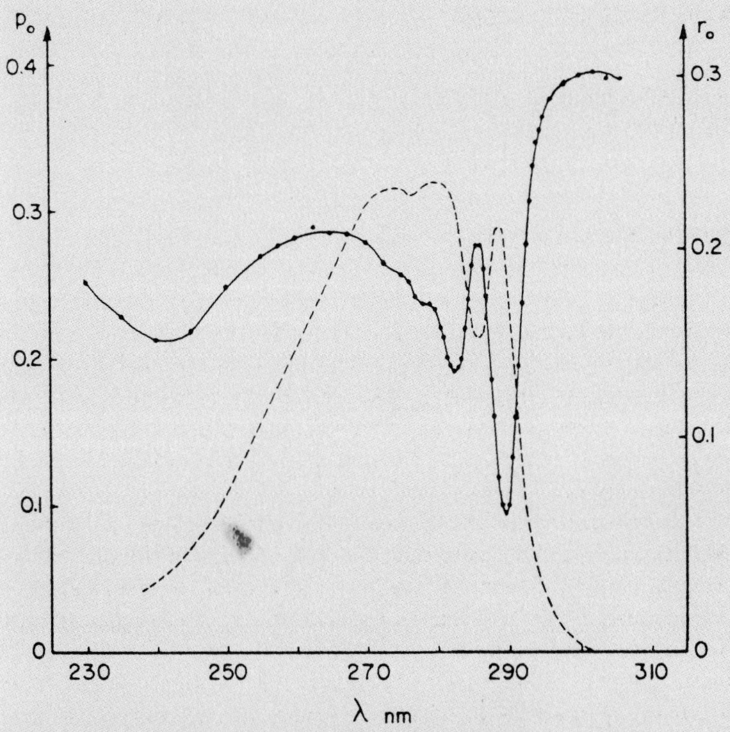

Fig. B5.2.1. Corrected excitation spectrum (broken line) and excitation polarization spectrum of indole in propylene glycol at -58 °C. The fluorescence is observed through a cut-off filter (Corning 7–39 filter) (reproduced with permission from Valeur and Weber[a)]).

where r_{0a} and r_{0b} are the limiting anisotropies corresponding to the ^1L_a and ^1L_b states when excited independently. In the long-wavelength region of the spectrum (305–310 nm), the ^1L_a level is exclusively excited and the limiting anisotropy for this state is thus $r_{0a} = 0.3$ (instead of 0.4, for the reasons given in Box 5.1). Then, using Eq. (5.28), in which 0.4 (2/5) is replaced by 0.3 and α is equal to 90°, we obtain

$$r_{0b} = 0.3\frac{3\,\overline{\cos^2\alpha} - 1}{2} = 0.15$$

The fractional contributions of the ^1L_a and ^1L_b excited states to the emission anisotropies are given by

$$f_a(\lambda) = \frac{r_0(\lambda) - r_{0b}}{r_{0a} - r_{0b}} = \frac{r_0(\lambda) + 0.15}{0.45}$$

$$f_b(\lambda) = \frac{r_{0a} - r_0(\lambda)}{r_{0a} - r_{0b}} = \frac{0.3 - r_0(\lambda)}{0.45}$$

Finally, the contributions $I_a(\lambda)$ and $I_b(\lambda)$ of the ^1L_a and ^1L_b bands to the excitation spectrum $I(\lambda)$ are

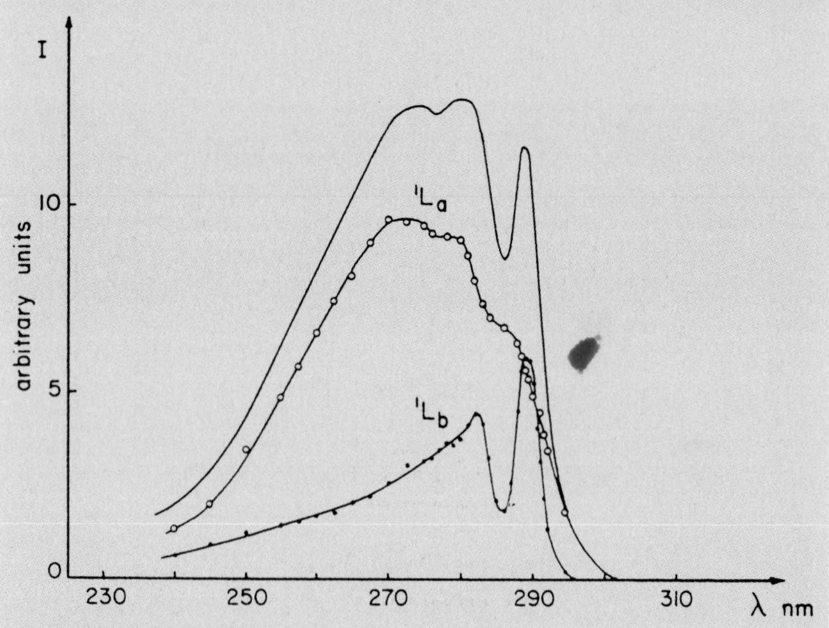

Fig. B5.2.2. Resolution of the excitation spectrum of indole (reproduced with permission from Valeur and Weber[a]).

$I_a(\lambda) = f_a(\lambda)I(\lambda)$

$I_b(\lambda) = f_b(\lambda)I(\lambda)$

Figure B5.2.2 shows the resolution of the excitation spectrum. The ^1L_b band lies below the broader ^1L_a band and exhibits a vibrational structure.

a) Valeur B. and Weber G. (1977) *Photochem. Photobiol.* **25**, 441.

5.6.1
Free rotations

The Brownian rotation of the emission transition moment is characterized by the angle $\omega(t)$ through which the molecule rotates between time zero (δ-pulse excitation) and time t, as shown in Figure 5.11.

Using the same method that led to Eq. (5.27), it is easy to establish the rule of multiplication of depolarization factors: when several processes inducing successive rotations of the transition moments (each being characterized by $\overline{\cos^2 \zeta_i}$) are independent random relative azimuths, the emission anisotropy is the product of the depolarization factors $(3\,\overline{\cos^2 \zeta_i} - 1)/2$:

$$r(t) = \frac{3\,\overline{\cos^2 \theta_E(t)} - 1}{2} = \prod_i \frac{3\,\overline{\cos^2 \zeta_i} - 1}{2} \tag{5.30}$$

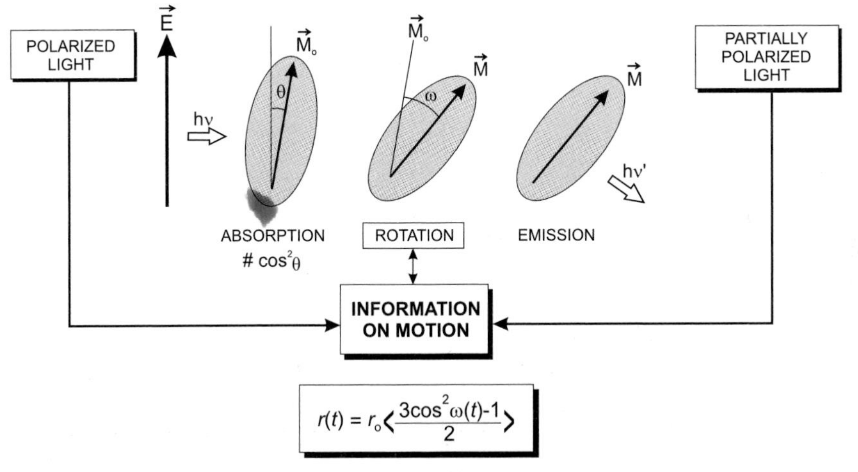

Fig. 5.9. Rotational motions inducing depolarization of fluorescence. The absorption and emission transition moments are assumed to be parallel.

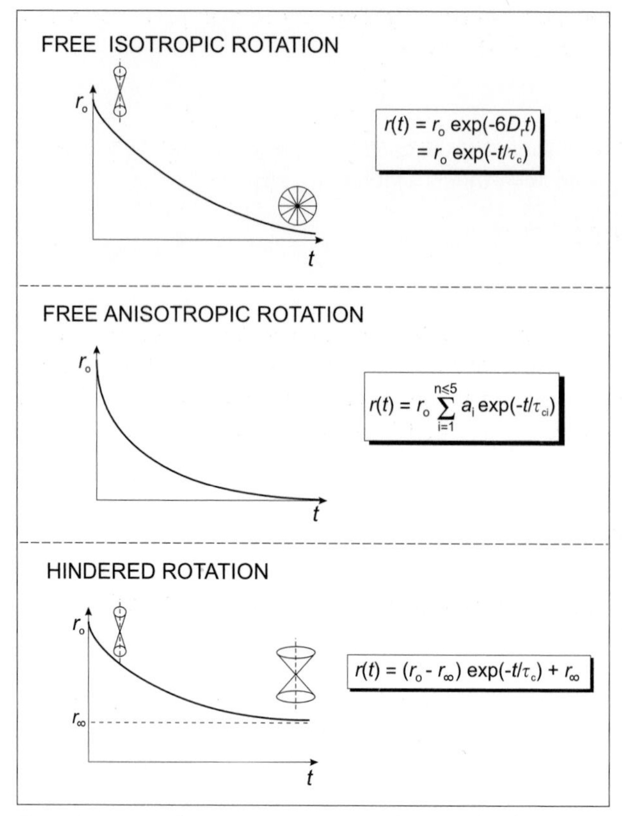

Fig. 5.10. Decay of emission anisotropy in the case of free and hindered rotations.

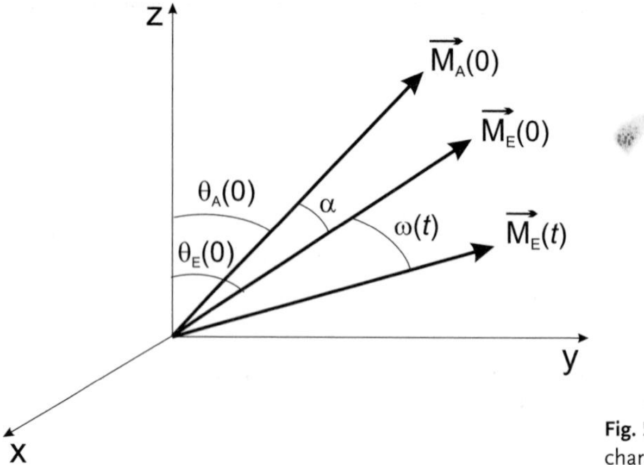

Fig. 5.11. Brownian rotation characterized by $\omega(t)$.

The effect of Brownian rotation is thus simply expressed by multiplying the second member of Eq. (5.27) by $(3\,\overline{\cos^2 \omega(t)} - 1)/2$:

$$r(t) = \frac{3\,\overline{\cos^2 \theta_E(t)} - 1}{2} = \frac{3\,\overline{\cos^2 \theta_A(0)} - 1}{2} \times \frac{3\,\overline{\cos^2 \alpha} - 1}{2} \times \frac{3\,\overline{\cos^2 \omega(t)} - 1}{2} \quad (5.31)$$

$$\boxed{r(t) = r_0 \frac{3\,\overline{\cos^2 \omega(t)} - 1}{2}} \quad (5.32)$$

The quantity $(3\,\overline{\cos^2 \omega(t)} - 1)/2$ is the *orientation autocorrelation function*: it represents the probability that a molecule having a certain orientation at time zero is oriented at ω with respect to its initial orientation. The quantity $(3x - 1)/2$ is the Legendre polynomial of order 2, $P_2(x)$, and Eq. (5.32) is sometimes written in the following form

$$r(t) = r_0 \langle P_2[\cos \omega(t)] \rangle \quad (5.33)$$

The angled brackets $\langle\,\rangle$ indicate an average over all excited molecules[8].

Isotropic rotations

Let us consider first the case of spherical molecules. Their rotations are isotropic and the average of $\cos^2 \omega(t)$ can be calculated by the integral:

$$\overline{\cos^2 \omega(t)} = \int_0^{\infty} \cos^2 \omega W(\omega, t) \sin \omega \, d\omega \quad (5.34)$$

where $W(\omega, t)$ represents the orientation distribution function expressing the probability that a molecule has an orientation ω at time t, taking into account that $\omega = 0$ at time zero. This function must fulfill the following conditions and equations:

- initial condition: $W(0, 0) = 1$
- normalization condition

$$2\pi \int_0^{\pi} W(\omega, t) \sin \omega \, d\omega = 1 \quad (5.35)$$

8) It is interesting to note that in dielectric relaxation experiments, the orientation autocorrelation function that is involved in the theory is

$$\langle P_1[\cos \omega(t)] \rangle = \overline{\cos \omega(t)}$$

- Brownian diffusion equation for a spherical particle

$$\frac{\partial W(\omega, t)}{\partial t} = D_r \nabla^2 W(\omega, t) \tag{5.36}$$

where D_r is the rotation diffusion coefficient. In spherical coordinates, Eq. (5.36) can be rewritten as

$$\frac{\partial W}{\partial t} = D_r \frac{1}{\sin \omega} \frac{\partial}{\partial \omega} \left(\sin \omega \frac{\partial W}{\partial \omega} \right) \tag{5.37}$$

By setting $\overline{\cos^2 \omega(t)} = u$, the above equations lead to

$$\frac{du}{dt} = 2D_r - 6D_r u \tag{5.38}$$

Taking into account the initial condition $\overline{\cos^2 \omega(0)} = 1$, the solution of this equation is

$$u = \frac{1}{3}[1 + 2 \exp(-6D_r t)] \tag{5.39}$$

The autocorrelation function $(3u - 1)/2$ is then a single exponential

$$\langle P_2[\cos \omega(t)] \rangle = \exp(-6D_r t) \tag{5.40}$$

Hence

$$\boxed{r(t) = r_0 \exp(-6D_r t)} \tag{5.41}$$

D_r can be determined by time-resolved fluorescence polarization measurements, either by pulse fluorometry from the recorded decays of the polarized components $I_{\parallel}$ and $I_{\perp}$, or by phase fluorometry from the variations in the phase shift between $I_{\parallel}$ and $I_{\perp}$ as a function of frequency (see Chapter 6). If the excited-state lifetime is unique and determined separately, steady-state anisotropy measurements allow us to determine D_r from the following equation, which results from Eqs (5.10) and (5.41):

$$\boxed{\frac{1}{\bar{r}} = \frac{1}{r_0}(1 + 6D_r \tau)} \tag{5.42}$$

This relationship is called Perrin's equation, because it was established for the first time by Francis Perrin; it was written at that time with polarization ratios

$$\left(\frac{1}{p} - \frac{1}{3}\right) = \left(\frac{1}{p_0} - \frac{1}{3}\right)(1 + 6D_r \tau)$$

Once D_r is determined by fluorescence polarization measurements, the Stokes–Einstein relation can be used:

$$D_r = \frac{RT}{6V\eta} \tag{5.43}$$

where V is the hydrodynamic molecular volume of the fluorophore, η is the viscosity of the medium, T is the absolute temperature and R is the gas constant. However, it should be emphasized that the validity of the Stokes–Einstein relation is questionable on a molecular scale (see Chapter 8). In particular, it is not valid to assign a numerical value to the viscosity of a microenvironment from fluorescence polarization measurements in conjunction with the Stokes–Einstein relation.

Equations (5.41) and (5.42) are often written with the rotational correlation time $\tau_c = 1/(6D_r)$

$$r(t) = r_0 \exp(-t/\tau_c) \tag{5.44}$$

$$\frac{1}{\bar{r}} = \frac{1}{r_0}\left(1 + \frac{\tau}{\tau_c}\right) \tag{5.45}$$

We have considered spherical molecules so far, but it should be noted that isotropic rotations can also be observed in the case of molecules with cylindrical symmetry and whose absorption and emission transition moments are parallel and oriented along the symmetry axis. In fact, any rotation around this axis has no effect on the fluorescence polarization. Only rotations perpendicular to this axis have an effect. A typical example is diphenylhexatriene whose transition moment is very close to the molecular axis (see Chapter 8).

Anisotropic rotations

In most cases, fluorescent molecules undergo anisotropic rotations because of their asymmetry. A totally asymmetric rotor has three different rotational diffusion coefficients, and in cases where the absorption and emission transition moments are not directed along one of the principal diffusion axes, the decay of $r(t)$ is a sum of five exponentials (see Box 5.3).

When the instantaneous emission anisotropy $r(t)$ is a sum of exponentials

$$r(t) = r_0 \sum_i a_i \exp(-t/\tau_{ci}) \tag{5.46}$$

the steady-state anisotropy is given by

$$\bar{r} = r_0 \sum_i \frac{a_i}{1 + \tau/\tau_{ci}} \tag{5.47}$$

Steady-state anisotropy measurements are then insufficient for fully characterizing rotational motions and time-resolved experiments are required.

Box 5.3 Emission anisotropy of totally asymmetric rotors and ellipsoids

A totally asymmetric rotor has three different rotational diffusion coefficients D_1, D_2 and D_3, around the three principal diffusion axes of the molecule (which are different from the laboratory axes x, y and z). In cases where the absorption and emission transition moments are not directed along one of the principal axes, the decay of $r(t)$ is a sum of five exponential terms. There was originally some controversy about the expression of $r(t)$[a–d]. The correct solution was derived by Belford et al.[d]

$$r(t) = \frac{6}{5}\sum_{i=1}^{3} C_i \exp(-t/\tau_i) + [(F+G)/4]\exp(-6D-2\Delta)t$$
$$+ [(F-G)/4]\exp(-6D+2\Delta)t$$

where D, the mean rotational diffusion coefficient, Δ, C_i and F are given by

$$D = (D_1 + D_2 + D_3)/3$$
$$\Delta = (D_1^2 + D_2^2 + D_3^2 - D_1D_2 - D_1D_3 - D_2D_3)^{1/2}$$
$$C_i = \alpha_j\alpha_k\varepsilon_j\varepsilon_k \quad (i, j, k = 123, 231 \text{ or } 321)$$

where α_1, α_2 and α_3 are the cosines of the angles formed by the absorption transition moments with the three principal axes, and ε_1, ε_2 and ε_3 are the corresponding direction cosines of the emission transition moments.

$$\tau_i = 1/(3D + 3D_i)$$
$$F = \sum_{i=1}^{3} \alpha_i^2\varepsilon_i^2 - 1/3$$
$$G\Delta = \sum_{i=1}^{3} D_i(\alpha_i^2\varepsilon_i^2 + \alpha_j^2\varepsilon_k^2 + \alpha_k^2\varepsilon_j^2) - D \quad i \neq j \neq k$$

For a particle with spherical symmetry ($D_1 = D_2 = D_3 = D$), the above expression reduces to Eq. (5.41).

In the particular case of prolate and oblate ellipsoids, the number of exponentials is reduced to three because two of the three axes are equivalent. The rotation diffusion coefficients around the axis of symmetry and the equatorial axis are denoted D_1 and D_2, respectively. The emission anisotropy can then be written as

$$r(t) = a_1 \exp[-6D_2t] + a_2 \exp[-(D_1 + 5D_2)t] + a_3 \exp[-(4D_1 + 2D_2)t]$$

where the pre-exponential factors depend on the orientation of the transition moments with respect to the ellipsoid axes as follows:

$$a_1 = 0.1(3\cos^2\theta_1 - 1)(3\cos^2\theta_2 - 1)$$

$$a_2 = 0.3\sin 2\theta_1 \sin 2\theta_2 \cos\phi$$

$$a_3 = 0.3\sin^2\theta_1 \sin^2\theta_2(\cos^2\phi - \sin^2\phi)$$

where θ_1 and θ_2 are the angles formed by the absorption and emission transitions moments, respectively, with the axis of symmetry of the ellipsoid, and ϕ is the angle formed by the projections of the two moments in the plane perpendicular to the axis of symmetry.

When the absorption and emission transition moments are collinear ($\theta_1 = \theta_2 = \theta$ and $\phi = 0$), the pre-exponential factors become

$$a_1 = 0.1(3\cos^2\theta - 1)^2$$

$$a_2 = 0.3\sin^2 2\theta$$

$$a_3 = 0.3\sin^4\theta$$

In the case of an ellipsoid of revolution for which the absorption or emission transition moment is parallel to the axis of symmetry (θ_1 or $\theta_2 = 0$), the anisotropy decay is a single exponential

$$r(t) = 0.4\exp(-6D_2 t)$$

A more complex expression is obtained when the absorption and emission transition moments lie in a plane perpendicular to the axis of symmetry

$$r(t) = 0.1\exp(-6D_2 t) + 0.3(2\cos^2\alpha - 1)\exp[-(4D_1 + 2D_2)t]$$

where α is the angle between the absorption and emission transition moments. In the particular case where this angle is 45°, this expression reduces to the first term.

In principle, the shape parameters of asymmetric rotors can be estimated from time-resolved anisotropy decay measurements, but in practice it is difficult to obtain accurate anisotropy decay curves over much more than one decade, which is often insufficient to determine more than two rotational correlation times.

a) Tao T. (1969) *Biopolymers* **8**, 609.
b) Chuang, T. J. and Eisenthal K. B. (1972) *J. Chem. Phys.* **57**, 5094.
c) Ehrenberg M. and Rigler R. (1972) *Chem. Phys. Lett.* **14**, 539.
d) Belford G. G., Belford R. L. and Weber G. (1972) *Proc. Natl. Acad. Sci. USA*, **69**, 1392.

5.6.2
Hindered rotations

Special attention should be paid to anisotropic media such as lipid bilayers and liquid crystals. Let us consider first the *'wobble-in-cone' model* (Kinosita et al., 1977; Lipari and Szabo, 1980) in which the rotations of a rod-like probe (with the direction of its absorption and emission transition moments coinciding with the long molecular axis) are restricted within a cone. The rotational motions are described by the rotational diffusion coefficient D_w around an axis perpendicular to the long molecular axis (the rotations around this axis having no effect on the emission anisotropy) and an order parameter (half-angle of the cone θ_c) reflecting the degree of orientational constraint due to the surrounding paraffinic chains (Figure 5.12). θ_c can be determined from the ratio r/r_0:

$$\frac{r_\infty}{r_0} = \left[\frac{1}{2}\cos\theta_c(1+\cos\theta_c)\right]^2 \tag{5.48}$$

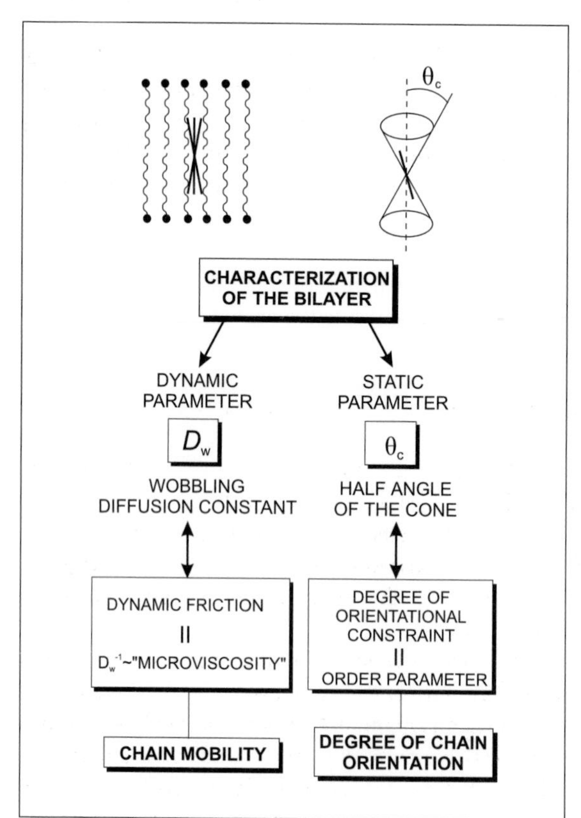

Fig. 5.12. 'Wobble-in-cone' model for the characterization of bilayers. The absorption and emission moments are assumed to coincide with the long molecular axis.

An approximate expression of the anisotropy decay is

$$r(t) = (r_0 - r_\infty)\exp(-t/\tau_c) + r_\infty \tag{5.49}$$

where τ_c is the effective relaxation time of $r(t)$, i.e. the time in which the initially photoselected distribution of orientations approaches the stationary distribution. This time is related to D_w and θ_c by:

$$\tau_c D_w\left(1 - \frac{r_\infty}{r_0}\right) = -x_0^2(1+x_0)^2\{\ln[(1+x_0)/2] + (1+x_0)/2\}/[2(1+x_0)] + (1+x_0)(6+8x_0 - x_0^2 - 12x_0^3 - 7x_0^4)/24 \tag{5.50}$$

where $x_0 = \cos\theta_c$.

In practice, the parameters r_0, r_∞ and τ_c are obtained from the best fit of $I_\parallel(t)$ and $I_\perp(t)$ given by Eqs (5.7) and (5.8), in which $r(t)$ has the form of Eq. (5.49). Then, θ_c is evaluated from r_∞/r_0, and Eq. (5.50) allows us to calculate the wobbling diffusion constant D_w.

If data analysis with a single exponential decay is not satisfactory, a double exponential can be used, but such a decay must be considered as a purely mathematical model.

It is interesting to note that $\tau_c(1 - r_\infty/r_0)$ is exactly the area A under $[r(t) - r_\infty]/r_0$. Therefore, even if the anisotropy decay is not a single exponential, D_w can be determined by means of Eq. (5.50) in which $\tau_c(1 - r_\infty/r_0)$ is replaced by the measured area A. An example of application of the wobble-in-cone model to the study of vesicles and membranes is given in Chapter 8 (Box 8.3). More general theories have also been developed (see Box 5.4).

5.7 Applications

The various fields concerned with the applications of fluorescence polarization are listed in Table 5.1.

The fluorescence polarization technique is a very powerful tool for studying the fluidity and orientational order of organized assemblies (see Chapter 8): aqueous micelles, reverse micelles and microemulsions, lipid bilayers, synthetic non-ionic vesicles, liquid crystals. This technique is also very useful for probing the segmental mobility of polymers and antibody molecules. Information on the orientation of chains in solid polymers can also be obtained.

Fluorescence polarization is also well suited to equilibrium binding studies when the free and bound species involved in the equilibrium have different rotational rates (Scheme 5.1). Most molecular interactions can be analyzed by this method. It should be emphasized that, in contrast to other methods using tracers, fluorescence polarization provides a direct measurement of the ratio of bound and free tracer without prior physical separation of these species. Moreover, measure-

Box 5.4 General model for hindered rotations[a–f]

The wobble-in-cone model has been generalized by introducing three autocorrelation functions $G_0(t)$, $G_1(t)$ and $G_2(t)$ into the expression for $r(t)$

$$r(t) = r_0[G_0(t) + 2G_1(t) + 2G_2(t)]$$

The values of these autocorrelation functions at times $t = 0$ and $t = \infty$ are related to the two order parameters $\langle P_2 \rangle$ and $\langle P_4 \rangle$, which are orientational averages of the second- and fourth-rank Legendre polynomial $P_2(\cos\beta)$ and $P_4(\cos\beta)$, respectively, relative to the orientation β of the probe axis with respect to the normal to the local bilayer surface or with respect to the liquid crystal direction. The order parameters are defined as

$$\langle P_2 \rangle = \langle 3\cos^2\beta - 1 \rangle / 2$$

$$\langle P_4 \rangle = \langle 35\cos^4\beta - 30\cos^2\beta + 3 \rangle / 8$$

and the autocorrelation functions are given by

$$G_0(t) = \frac{1}{5} + \frac{2}{7}\langle P_2 \rangle + \frac{18}{35}\langle P_4 \rangle$$

$$G_1(t) = \frac{1}{5} + \frac{1}{7}\langle P_2 \rangle - \frac{12}{35}\langle P_4 \rangle$$

$$G_2(t) = 4\frac{1}{5} - \frac{2}{7}\langle P_2 \rangle + \frac{3}{35}\langle P_4 \rangle$$

$$G_0(\infty) = \langle P_2 \rangle^2$$

$$G_1(\infty) = G_2(\infty) = 0$$

It is assumed that the probe molecules undergo Brownian rotational motions with an angle-dependent ordering potential $U(\beta)$

$$U(\beta) = kT[\lambda_2 P_2(\cos\beta) + \lambda_4 P_4(\cos\beta)]$$

For a rod-like probe with its absorption transition moment direction coinciding with the long molecular axis, the rotational motion in this potential well is described by the diffusion coefficient $D_\perp$. The decay of the autocorrelation functions is then shown to be an infinite sum of exponential terms:

$$G_k(t) = \sum_{m=0}^{\infty} b_{km} \exp(-a_{km} D_\perp t) \quad k = 0, 1, 2$$

The coefficients a_{km} and b_{km} are complex functions of the parameters λ_2 and λ_4 that describe the ordering potential. In many practical situations, $G_k(t)$ is essentially mono-exponential:

$$G_k(t) = [G_k(0) - G_k(\infty)] \exp(-a_{k1} D_\perp t) + G_k(\infty)$$

The diffusion constant $D_\perp$ with the underlying 'microviscosity', and the two order parameters $\langle P_2 \rangle$, $\langle P_4 \rangle$ reflecting the degree of orientational constraint have been successfully determined from the fluorescence anisotropy decay in vesicles and liquid crystals.

a) Van der Meer W., Kooyman R. P. H. and Levine Y. K. (1982) *Chem. Phys.* **66**, 39.
b) Van der Meer W., Pottel H., Herreman W., Ameloot M., Hendrickx H. and Schröder H. (1984) *Biophys. J.*, **46**, 515.
c) Zannoni C., Arcioni A. and Cavatorta P. (1983) *Chem. Phys. Lipids* **32**, 179.
d) Szabo A. (1984) *J. Chem. Phys.* **81**, 150.
e) Fisz J. J. (1985) *Chem. Phys.* **99**, 177; ibid. (1989) **132**, 303; ibid. (1989) **132**, 315.
f) Pottel H., Herreman W., Van der Meer B. W. and Ameloot M. (1986) *Chem. Phys.* **102**, 37.

ments are carried out in real time, thus giving information on the kinetics of association and dissociation reactions.

Immunoassays based on fluorescence polarization have become very popular because, in contrast to radioimmunoassays, they require no steps to separate free and bound tracer. In a fluoroimmunoassay, the fluorescently labeled antigen is

Tab. 5.1. Fields of application of fluorescence polarization

Field	*Information*
Spectroscopy	Excitation polarization spectra: distinction between excited states
Polymers	Chain dynamics
	Local viscosity in polymer environments
	Molecular orientation in solid polymers
	Migration of excitation energy along polymer chains
Micellar systems	Internal 'microviscosity' of micelles
	Fluidity and order parameters (e.g. bilayers of vesicles)
Biological membranes	Fluidity and order parameters
	Determination of the phase transition temperature
	Effect of additives (e.g. cholesterol)
Molecular biology	Proteins (size, denaturation, protein–protein interactions, etc.)
	DNA–protein interactions
	Nucleic acids (flexibility)
Immunology	Antigen–antibody reactions
	Immunoassays
Artificial and natural antennae	Migration of excitation energy

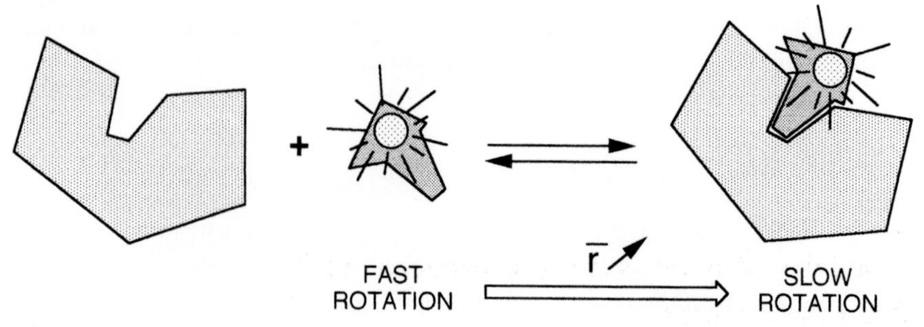

Scheme 5.1

bound to an antibody such that most of the antigen is bound, thereby maximizing the value of emission anisotropy. Upon addition of unlabeled antigen, bound labeled antigen will be displaced from the antibody and the emission anisotropy will decrease accordingly.

Finally, energy migration, i.e. excitation energy hopping, in polymers, artificial antenna systems, photosynthetic units, etc. can be investigated by fluorescence polarization (see Chapter 9).

5.8 Bibliography

Kinosita K., Kawato S. and Ikegami A. **(1977)** A Theory of Fluorescence Polarization Decay in Membranes, *Biophys. J.* 20, 289–305.

Lipari G. and Szabo A. **(1980)** Effect of Vibrational Motion on Fluorescence Depolarization and Nuclear Magnetic Resonance Relaxation in Macromolecules and Membranes, *Biophys. J.* 30, 489–506.

Steiner R. F. **(1991)** Fluorescence Anisotropy: Theory and Applications, in: Lakowicz J. R. (Ed.), *Topics in Fluorescence Spectroscopy, Vol. 2, Principles*, Plenum Press, New York, pp. 127–176.

Thulstrup E. W. and Michl J. **(1989)** *Elementary Polarization Spectroscopy*, VCH, New York.

Valeur B. **(1993)** Fluorescent Probes for Evaluation of Local Physical, Structural Parameters, in: Schulman S. G. (Ed.), *Molecular Luminescence Spectroscopy. Methods and Applications*, Part 3, Wiley-Interscience, New York, pp. 25–84.

Weber G. (1953) Rotational Brownian Motions and Polarization of the Fluorescence of Solutions, *Adv. Protein Chem.* 8, 415–459.

6
Principles of steady-state and time-resolved fluorometric techniques

> Experiments never
> deceive. It is our
> judgement that deceives
> itself because it expects
> results which experiments
> do not give.
>
> Leonardo da Vinci,
> 15th century

The aim of this chapter is to give the reader a good understanding of the instrumentation necessary to carry out reliable measurements. Only the principles will be described and the reader is referred to more specialized books or reviews for further details (see Bibliography at the end of the chapter).

6.1
Steady-state spectrofluorometry

Fluorescence measurements are indeed more difficult than they appear at first sight, for several reasons:

- wavelength dependence of (i) the light source intensity, (ii) the transmission efficiency of the monochromators, (iii) the detector sensitivity. Proper correction of the emission and excitation spectra is not simple;
- effect of polarization of the incident light and fluorescence emission (the transmission efficiency of a monochromator depends on the polarization of light);
- possible contamination of the signal by scattered light (Rayleigh or Raman), especially from turbid samples or by background fluorescence of the medium (fluorescent impurities of the solvent, self-fluorescence of biological samples, etc.).

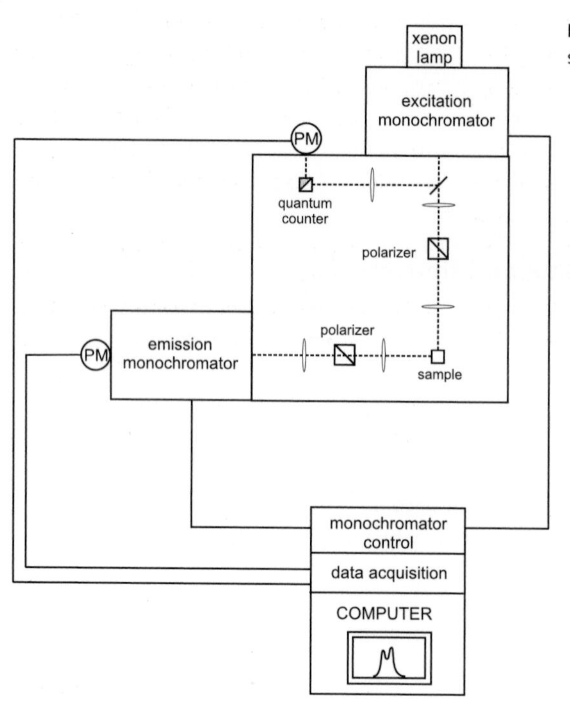

Fig. 6.1. A conventional spectrofluorometer.

6.1.1
Operating principles of a spectrofluorometer

Figure 6.1 shows the components of a conventional spectrofluorometer. The light source is generally a high-pressure xenon arc lamp, which offers the advantage of continuous emission from ~250 nm to the infrared. A monochromator is used to select the excitation wavelength. Fluorescence is collected at right angles with respect to the incident beam and detected through a monochromator by a photomultiplier. Automatic scanning of wavelengths is achieved by the motorized monochromators, which are controlled by the electronic devices and the computer (in which the data are stored).

The optical module contains various parts: a sample holder, shutters, polarizers if necessary, and a beam splitter consisting of a quartz plate reflecting a few per cent of the exciting light towards a quantum counter or a photodiode. A quantum counter usually consists of a triangular cuvette containing a concentrated solution of a dye whose fluorescence quantum yield is independent of the excitation wavelength (e.g. rhodamine 101 in ethylene glycol, concentration $\approx 3\ gL^{-1}$)[1]. At such a

1) Rhodamine B is most frequently used as a quantum counter, but the advantage of rhodamine 101 is that its fluorescence quantum yield is almost independent of temperature, unlike rhodamine B.

high concentration, the solution absorbs all incident photons and the light collected at right angles (see Figure 6.1) by a reference photomultiplier is proportional to the photon flux of the incident light over the excitation wavelength range corresponding to the absorption range of the dye (220–600 nm for rhodamine 101). However, a quantum counter is limited in wavelength to the absorption range of the dye. Therefore, a photodiode whose wavelength response is as flat as possible may be preferred to a quantum counter.

A reference channel (quantum counter or photodiode) has two advantages: (i) it compensates for the time fluctuations of the lamp via a ratiometric measurement (ratio of the output signals of the photomultiplier detecting the fluorescence of the sample to the output signal of the reference detector); (ii) it permits correction of excitation spectra (see below).

Emission and excitation spectra were defined in Chapter 3 using the following expression for the measured fluorescence intensity

$$I_F(\lambda_E, \lambda_F) = kF(\lambda_F)I_0(\lambda_E)\{1 - \exp[-2.3\varepsilon(\lambda_E)lc]\}$$

or

$$I_F(\lambda_E, \lambda_F) = kF(\lambda_F)I_0(\lambda_E)\{1 - 10^{-A(\lambda_E)}\} \tag{6.1}$$

where λ_E is the excitation wavelength, λ_F is the wavelength at which the fluorescence is observed, I_0 is the intensity of the incident beam on the sample, $\varepsilon(\lambda_E)$ is the molar absorption coefficient at the excitation wavelength, l is the optical path in the sample, c is the molar concentration and $A(\lambda_E)$ is the absorbance at the excitation wavelength. $F(\lambda_F)$ is the variation in fluorescence intensity per unit wavelength with the normalization condition $\int_0^\infty F(\lambda_F)\,d\lambda_F = \Phi_F$ (see Chapter 3). k is an instrumental factor that depends on many parameters: geometry of observation (solid angle through which the light is collected), transmission efficiency of the monochromators, width of the monochromator slits, high voltage of the photomultiplier, gain of the electronic devices, etc. Therefore, the numerical value of the measured fluorescence intensity has no meaning and the fluorescence spectrum is usually plotted in arbitrary units[2)]; sometimes it is normalized so that $\int_0^\infty I_F\,d\lambda_F = \Phi_F$.

It is important to recall that I_F is proportional to the concentration only in dilute solutions [$A(\lambda_E) < 0.05$); Eq. (6.1) is then reduced to

$$\begin{aligned} I_F(\lambda_E, \lambda_F) &= 2.3kF(\lambda_F)I_0(\lambda_E)A(\lambda_E) \\ &= 2.3kF(\lambda_F)I_0(\lambda_E)\varepsilon(\lambda_E)lc \end{aligned} \tag{6.2}$$

2) In contrast, a spectrophotometer measures an absorbance, a quantity that is expressed on an absolute scale.

The emission spectrum corresponds to the variations of I_F as a function of λ_F, the excitation wavelength λ_E being fixed, whereas the excitation spectrum reflects the variations of I_F as a function of λ_E, the observation wavelength λ_F being fixed. Note that the spectra are recorded as a function of wavelength and not wavenumber because the monochromators of spectrofluorometers are presently all equipped with gratings (no longer with prisms), so that for a given width of the input and output slits, the monochromators operate at a constant bandpass expressed in wavelength.

As outlined above, the spectra are distorted by the wavelength dependence of several components of the instrument. Correction of spectra is of major importance for quantitative measurements (determination of quantum yields and calculation of overlap integrals), for comparison of excitation and absorption spectra, and for comparison of fluorescence data obtained under different experimental conditions.

6.1.2
Correction of excitation spectra

According to Eqs (6.1) and (6.2), in which the emission wavelength λ_F is constant, the excitation spectrum is distorted by the variations $I_0(\lambda_E)$ of the intensity of the exciting light; these variations are due to the wavelength dependence of the lamp intensity and of the transmission efficiency of the excitation monochromator. Because the quantum counter circumvents the wavelength dependence of the sensitivity of the reference photomultiplier, the ratio of the fluorescence signal from the sample to that from the quantum counter or photodiode, as a function of the excitation wavelength, provides in principle corrected excitation spectra.

However, such correction procedures may be insufficient when very accurate measurements are needed (for instance when information is expected from the comparison of the absorption and excitation spectra). In fact, the optical geometry of the reference channel is not identical to that of the main channel, and the wavelength dependence of optical parts (e.g. focal length of lenses) may introduce some distortion into the excitation spectrum. It is then recommended to use correction factors obtained by using a fluorescent compound absorbing in the same wavelength range as that of the sample to be studied, and whose absorption spectrum is identical to its excitation spectrum. The ratio of the measured excitation spectrum of this reference compound – as described above using the quantum counter – to the absorption spectrum provides the correction factors that can be stored in the computer.

Spectrofluorometers equipped with a photodiode instead of a quantum counter provide excitation spectra that should be further corrected because, in addition to the reasons explained above, the wavelength response of the photodiode is not strictly flat over the whole wavelength range available.

It should be noted that most commercially available instruments are delivered with a file containing the correction factors.

6.1.3
Correction of emission spectra

The emission spectrum is distorted by the wavelength dependence of the emission monochromator efficiency and the photomultiplier response. In general, the correction factors are measured by the manufacturer using a calibrated tungsten lamp. A reflector (e.g. made of freshly prepared magnesium oxide) is introduced into the sample holder and set at 45°, and is illuminated by the lamp externally positioned at right angles. The spectral response of the detection system is recorded and the correction factors are obtained by dividing this spectral response by the spectral output data provided with the lamp. For wavelengths shorter than about 320 nm, where the intensity of tungsten lamps is too low to get reliable correction factors, a hydrogen or deuterium lamp can be used. It is recommended that the correction factors are recorded with a polarizer placed – with a defined orientation – between the sample and the emission monochromator, because of the polarization dependence of the transmission efficiency of the monochromator (see Section 6.1.5).

Alternatively, a standard fluorescent compound can be used whose corrected emission spectrum has been reported[3]. Comparison of this spectrum with the 'technical' spectrum recorded with the detection system provides the correction factors. The wavelength range must obviously cover the fluorescence spectrum to be corrected. Unfortunately, there is a limited number of reliable standards.

Emission correction factors are provided with most commercially available spectrofluorometers.

6.1.4
Measurement of fluorescence quantum yields

Proper correction of the emission spectrum is a prerequisite for the measurement of the fluorescence quantum yield of a compound. Quantum yields are usually determined by comparison with a fluorescence standard[4], i.e. a compound of known quantum yield that would ideally satisfy the following criteria (Demas, 1982):

3) Numerical values of the corrected fluorescence spectrum of quinine sulfate can be found in: Velapoldi R. A. and Mielenz K. D. (1980) Standard reference materials: A fluorescence SRM: quinine sulfate dihydrate (SRM 936), NBS Spec. Publ. 260–64, Jan. 1980, PB 80132046, (NTIS), Springfield, VA, USA. Samples of quinine sulfate dihydrate can be ordered from the Standard Reference Materials Program, National Institute of Standards and Technology, Gaithersburg, MD 20678-9950, USA.
A collection of corrected excitation and emission spectra can be found in: Miller J. N. (Ed.) (1981) *Standards for Fluorescence Spectrometry*, Chapman and Hall, London. Corrected emission spectra can also be found in Appendix 1 of Lakowicz J. R. (1999) *Principles of Fluorescence Spectroscopy*, Kluwer Academic/Plenum Publishers, New York.

4) Absolute determinations of quantum yield are possible using integrating spheres, by calorimetry and by other methods, but they are difficult and uncommon.

Tab. 6.1. Standards for the determination of fluorescence quantum yields

Range	*Compound*	*Temp. (°C)*	*Solvent*	Φ_F	*Ref.*
270–300 nm	Benzene	20	Cyclohexane	0.05 ± 0.02	1
300–380 nm	Tryptophan	25	H_2O (pH 7.2)	0.14 ± 0.02	2
300–400 nm	Naphthalene	20	Cyclohexane	0.23 ± 0.02	3
315–480 nm	2-Aminopyridine	20	0.1 mol L^{-1} H_2SO_4	0.60 ± 0.05	4
360–480 nm	Anthracene	20	Ethanol	0.27 ± 0.03	1, 5
400–500 nm	9,10-diphenylanthracene	20	Cyclohexane	0.90 ± 0.02	6, 7
400–600 nm	Quinine sulfate dihydrate	20	0.5 mol L^{-1} H_2SO_4	0.546	5, 7
600–650 nm	Rhodamine 101	20	Ethanol	1.0 ± 0.02	8
				0.92 ± 0.02	9
600–650 nm	Cresyl violet	20	Methanol	0.54 ± 0.03	10

1) Dawson W. R. and Windsor M. W. (1968) *J. Phys. Chem.* **72**, 3251.
2) Kirby E. P. and Steiner R. F. (1970) *J. Phys. Chem.* **74**, 4480.
3) Berlman I. B. (1965) *Handbook of Fluorescence Spectra of Aromatic Molecules*, Academic Press, London.
4) Rusakowicz R. and Testa A. C. (1968) *J. Phys. Chem.* **72**, 2680.
5) Melhuish W. H. (1961) *J. Phys. Chem.* **65**, 229.
6) Hamai S. and Hirayama F. (1983) *J. Phys. Chem.* **87**, 83.
7) Meech S. R. and Phillips D. (1983) *J. Photochem.* **23**, 193.
8) Karstens T. and Kobs K. (1980) *J. Phys. Chem.* **84**, 1871.
9) Arden-Jacob J., Marx N. J. and Drexhage K. H. (1997) *J. Fluorescence* **7(Suppl.)**, 91S.
10) Magde D., Brannon J. H., Cramers T. L. and Olmsted J. III (1979) *J. Phys. Chem.* **83**, 696.

- readily available in a highly pure form;
- photochemically stable in solution and in storage in the solid state;
- high fluorescence quantum yield accurately known;
- broad absorption and fluorescence spectra with no sharp spectral features, to avoid bandpass errors;
- fluorescence spectrum and quantum yield independent of excitation wavelength;
- small overlap between absorption and emission spectra to avoid self-absorption errors;
- unpolarized emission.

A list of fluorescence standards is given in Table 6.1.

In order to minimize the effects of possible inaccuracy of the correction factors for the emission spectrum, the standard is preferably chosen to be excitable at the same wavelength as the compound, and with a fluorescence spectrum covering a similar wavelength range.

The fluorescence spectra of dilute solutions ($A < 0.05$) of the compound and the standard must be recorded under exactly the same experimental conditions (slits of the monochromators, high voltage of the photomultipliers, gain of the electronic devices). The temperature of the sample holder must be controlled because the fluorescence quantum yield of most compounds is temperature-dependent. If the solvents used for the standard and the compound are not the same, it is necessary to introduce a correction for the refractive index n[5]. The ratio of the quantum yields is given by

$$\frac{\Phi_F}{\Phi_{FR}} = \frac{n^2}{n_R^2} \frac{\int_0^\infty F(\lambda_F)\, d\lambda_F}{\int_0^\infty F_R(\lambda_F)\, d\lambda_F} \tag{6.3}$$

where the subscript R refers to the reference solution (standard). Using Eq. (6.1), this ratio becomes

$$\frac{\Phi_F}{\Phi_{FR}} = \frac{n^2}{n_R^2} \times \frac{\int_0^\infty I_F(\lambda_E, \lambda_F)\, d\lambda_F}{\int_0^\infty I_{FR}(\lambda_E, \lambda_F)\, d\lambda_F} \times \frac{1 - 10^{-A_R(\lambda_E)}}{1 - 10^{-A(\lambda_E)}} \tag{6.4}$$

where I_F and I_{FR} are the *corrected* fluorescence spectra. The integrals represent the area under the fluorescence spectra. It should be recalled that, from a practical point of view, the integral calculation must be done on a wavelength scale and not a wavenumber scale because all spectrofluorometers are equipped with grating monochromators; the fluorescence spectrum is thus recorded on a linear wavelength scale at constant wavelength bandpass $\Delta\lambda_F$ (which is the integration step) (see Chapter 3, Box 3.3 for further details).

Equation (6.4) is still valid if the compound and the standard are not excited at the same wavelength, provided that the instrument is well corrected for the wavelength dependence of the lamp intensity and the excitation monochromator efficiency. Otherwise, the second term of Eq. (6.4) must be multiplied by the ratio $I_0(\lambda_E)/I_0(\lambda_{ER})$.

It should be noted that the accuracy of the determination of fluorescence quantum yields cannot be better than 5–10%, due to the small additive errors relevant to the absorbances at the excitation wavelength, the correction factors of the detection system and the quantum yield of the standard.

Attention should be paid to possible problems in the measurement of fluorescence quantum yields (some of which are discussed Section 6.1.5): inner filter effects, possible wavelength effects on Φ_F, refractive index corrections, polarization effects, temperature effects, impurity effects, photochemical instability and Raman scattering.

6.1.5
Pitfalls in steady-state fluorescence measurements: inner filter effects and polarization effects

It should again be emphasized that the determination of a 'true' fluorescence spectrum and of a quantum yield requires the use of very dilute solutions or there may be some undesirable effects.

5) This correction is due to the fact that the fluorescence is refracted at the surface separating the solution and air. The fluorescence flux falling on the aperture of the detection system is inversely proportional to the square of the refractive index.

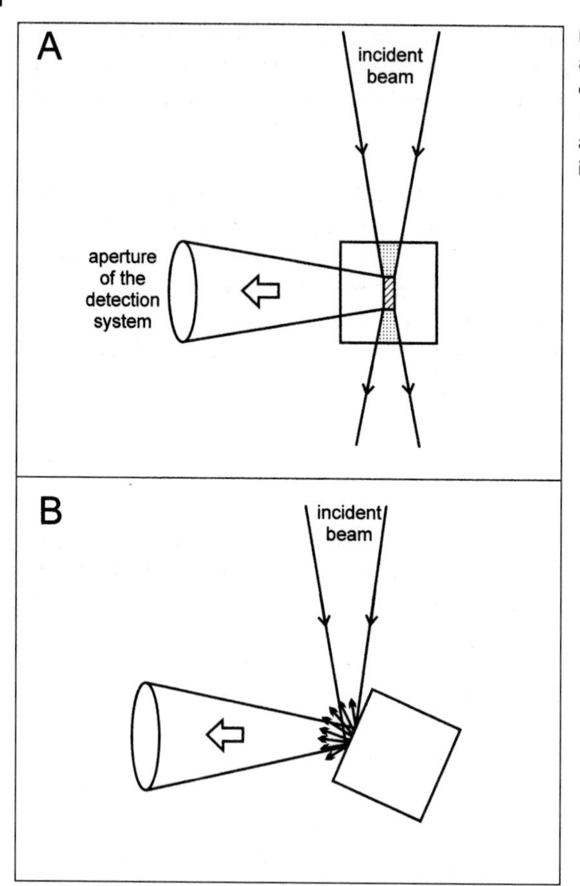

Fig. 6.2. Geometric arrangements for the observation of fluorescence: (a) conventional right-angle arrangement; (b) front-face illumination.

Excitation inner filter effect With the usual right angle observation, the detection system collects the fluorescence emitted only from the central part of the exciting beam (Figure 6.2a). When the concentration of the solution is low enough ($A < 0.1$), the incident light is only slightly attenuated through the cuvette, whereas when the concentration is large, a significant part of the incident light is absorbed before reaching the central part of the cuvette. This leads to a paradox: as the concentration increases, the observed intensity increases, goes through a maximum and decreases.

Consequently, it is important to note that the fluorescence intensity of a compound is proportional to the concentration over only a limited range of absorbances, due to this geometrical reason, in addition to the mathematical reason (as shown by Eq. 6.1).

For analytical applications, when a linear relationship between fluorescence intensity and concentration is desirable, a correction curve must be built up under the same conditions as those that will be used for the actual experiment.

Emission inner filter effect (self-absorption) The fluorescence photons emitted in the region overlapping the absorption spectrum can be absorbed (radiative energy transfer; see Chapter 4, section 4.6.2). This results in a distortion of the shape of the fluorescence spectrum in this region. The larger the spectral overlap, the larger the distortion. This effect depends on the geometry of the sample cell arrangement. For the usual right angle observation, a theoretical correction is difficult.

Inner filter effects due to the presence of other substances When the solution contains other chromophores that absorb light in the same wavelength range as the fluorescent compound under study, the chromophores act as filters at the excitation wavelength and the fluorescence intensity must be multiplied by a correction factor. If the chromophores do not interact with the fluorescent compound, the correction factor is simply the fraction of light absorbed by the compound at the chosen excitation wavelength, so that the corrected fluorescent intensity is given by:

$$I_F^{corr}(\lambda_E, \lambda_F) = I_F(\lambda_E, \lambda_F)\frac{A(\lambda_E)}{A_{tot}(\lambda_E)} \tag{6.5}$$

where $A(\lambda_E)$ is the absorbance of the fluorescent compound $[\varepsilon(\lambda_E)lc]$ and $A_{tot}(\lambda_E)$ is the total absorbance of the solution. If the chromophores absorb at the observation wavelength, the fluorescence of the compound is attenuated. This effect is again called the *emission inner filter effect.*

All these inner filters effects are difficult to correct and it is advisory to work as much as possible with dilute solutions.

When the use of high concentrations is required, front-face illumination (Figure 6.2b) offers the advantage of being much less sensitive to the excitation inner filter effect. The illuminated surface is better oriented at 30° than at 45°, because at 45° the unabsorbed incident light is partially reflected towards the detection system, which may increase the stray light interfering with the fluorescence signal.

Polarization effects The transmission efficiency of a monochromator depends on the polarization of light. This can easily be demonstrated by placing a polarizer between the sample and the emission monochromator: it is observed that the position and shape of the fluorescence spectrum may significantly depend on the orientation of the polarizer. Consequently, the observed fluorescence intensity depends on the polarization of the emitted fluorescence, i.e. on the relative contribution of the vertically and horizontally polarized components. This problem can be circumvented in the following way.

Let I_x, I_y and I_z be the intensity components of the fluorescence, respectively (Figure 6.3). If no polarizer is placed between the sample and the emission monochromator, the light intensity viewed by the monochromator is $I_z + I_y$, which is not proportional to the total fluorescence intensity $(I_x + I_y + I_z)$. Moreover, the transmission efficiency of the monochromator depends on the polarization of the incident light and is thus not the same for I_z and I_y. To get a response proportional to the total fluorescence intensity, independently of the fluorescence polarization, polarizers must be used under 'magic angle' conditions (see appendix, p. 196): a polarizer is introduced between the excitation monochromator and the sample and

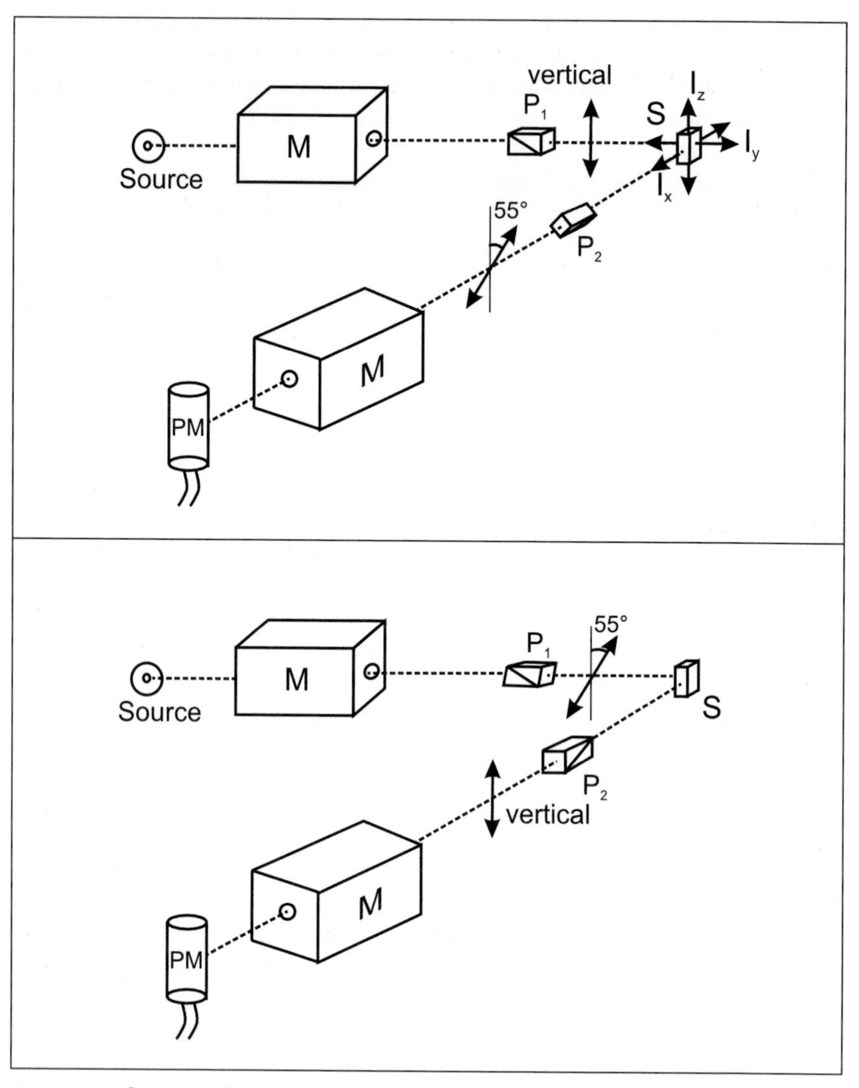

Fig. 6.3. Configuration of the excitation and emission polarizers for observing a signal proportional to the total fluorescence intensity. M: monochromator; S: sample; P_1, P_2: polarizers.

set in the vertical position, and another between the sample and the emission monochromator and set at the magic angle (54.7°). Alternatively, the excitation polarizer can be set at the magic angle, and the emission polarizer in the vertical position (Figure 6.3).

Consequently, when fluorescence is polarized, it is recommended that excitation or emission spectra are recorded with properly oriented polarizers.

As previously outlined, the emission correction factors must be recorded with an emission polarizer in a defined orientation (preferably at the magic angle), and this orientation must be kept unchanged for recording emission or excitation spectra irrespective of whether or not the fluorescence is polarized.

6.1.6 Measurement of steady-state emission anisotropy. Polarization spectra

Let us recall the definition of the emission anisotropy r (see Chapter 5) in a configuration where the exciting light is vertically polarized and the emitted fluorescence is observed at right angles in a horizontal plane (Figure 6.4):

$$r = \frac{I_{\parallel} - I_{\perp}}{I_{\parallel} + 2I_{\perp}} \tag{6.6}$$

where $I_{\parallel}$ and $I_{\perp}$ are the polarized components parallel and perpendicular to the direction of polarization of the incident light, respectively.

r can be determined with a spectrofluorometer equipped with polarizers. It is not sufficient to keep the excitation polarizer vertical and to rotate the emission polarizer because, as already mentioned, the transmission efficiency of a monochromator depends on the polarization of the light. The determination of the emission anisotropy requires four intensity measurements: I_{VV}, I_{VH}, I_{HV} and I_{HH} (V: vertical; H: horizontal; the first subscript corresponds to the orientation of the excitation polarizer and the second subscript to the emission polarizer) (Figure 6.4).

For a horizontally polarized exciting light, the vertical component (I_{HV}) and the horizontal component (I_{HH}) of the fluorescence detected through the emission monochromator are different, although these components are identical before entering the monochromator according to the Curie symmetry principle (see Chapter 5). The G factor[6], $G = I_{HV}/I_{HH}$, measured through the monochromator, varies from 1 and can be used as a correcting factor for the ratio $I_{\parallel}/I_{\perp}$ because it represents the ratio of the sensitivities of the detection system for vertically and horizontally polarized light:

$$\frac{I_{VV}}{I_{VH}} = G\frac{I_{\parallel}}{I_{\perp}} \tag{6.7}$$

Equation (6.6) then becomes

$$r = \frac{I_{VV} - GI_{VH}}{I_{VV} + 2GI_{VH}} \tag{6.8}$$ [7]

6) The letter G comes from 'grating'. In fact, the majority of the polarization effects arise from the polarization dependence of the monochromator grating, but the detector response and the optics can contribute to a lesser extent.

7) For the sake of simplicity, this relation is often written as

$$r = \frac{I_{\parallel} - GI_{\perp}}{I_{\parallel} + 2GI_{\perp}}$$

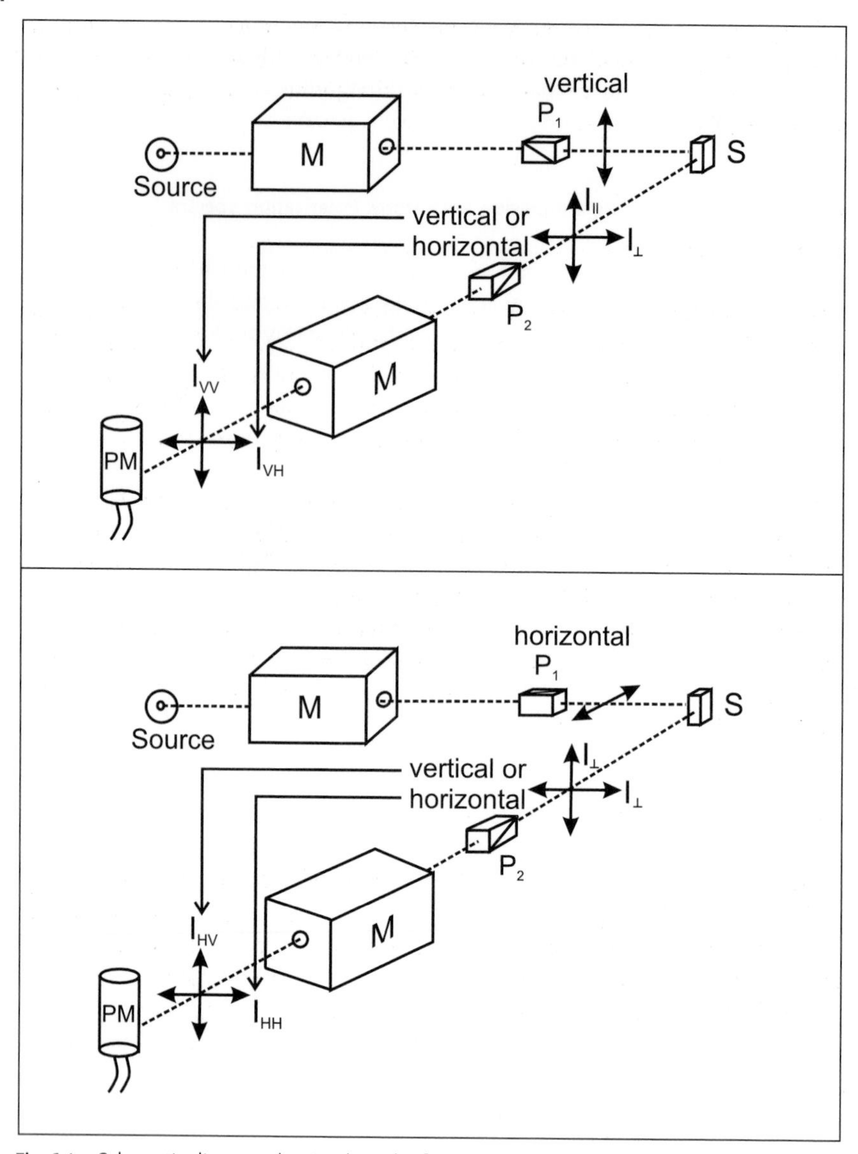

Fig. 6.4. Schematic diagram showing how the four intensity components are measured for determination of the steady-state anisotropy.

The excitation polarization spectrum (which represents the variation of r as a function of the excitation wavelength; see Chapter 5) is obtained by recording the variations of $I_{VV}(\lambda_E)$, $I_{VH}(\lambda_E)$, $I_{HV}(\lambda_E)$, and $I_{HH}(\lambda_E)$ and by calculating $r(\lambda_E)$ by means of Eq. (6.8) with $G(\lambda_E) = I_{VH}(\lambda_E)/I_{HH}(\lambda_E)$. The correction factors for wavelength dependence can be ignored because they cancel out in the ratio.

In some cases, filters are used instead of the emission monochromator. In principle, no G factor is then considered, but in practice, effects may be due to the sensitivity of the photomultipliers to polarization (in particular, photomultipliers with side-on photocathodes).

6.2 Time-resolved fluorometry

Knowledge of the dynamics of excited states is of major importance in understanding photophysical, photochemical and photobiological processes. Two time-resolved techniques, *pulse fluorometry* and *phase-modulation fluorometry*, are commonly used to recover the lifetimes, or more generally the parameters characterizing the *δ-pulse response* of a fluorescent sample (i.e. the response to an infinitely short pulse of light expressed as the Dirac function δ).

Pulse fluorometry uses a short exciting pulse of light and gives the δ-pulse response of the sample, convoluted by the instrument response. Phase-modulation fluorometry uses modulated light at variable frequency and gives the *harmonic response* of the sample, which is the Fourier transform of the δ-pulse response. The first technique works in the *time domain*, and the second in the *frequency domain*. Pulse fluorometry and phase-modulation fluorometry are theoretically equivalent, but the principles of the instruments are different. Each technique will now be presented and then compared.

6.2.1 General principles of pulse and phase-modulation fluorometries

The principles of pulse and phase-modulation fluorometries are illustrated in Figures 6.5 and 6.6. The δ-pulse response $I(t)$ of the fluorescent sample is, in the simplest case, a single exponential whose time constant is the excited-state lifetime, but more often it is a sum of discrete exponentials, or a more complicated function; sometimes the system is characterized by a distribution of decay times. For any excitation function $E(t)$, the response $R(t)$ of the sample is the convolution product of this function by the δ-pulse response:

$$R(t) = E(t) \otimes I(t) = \int_{-\infty}^{t} E(t')I(t-t')\,\mathrm{d}t' \qquad (6.9)^{8)}$$

Pulse fluorometry The sample is excited by a short pulse of light and the fluorescence response is recorded as a function of time. If the duration of the pulse is long

8) The convolution integral appearing in this equation can be easily understood by considering the excitation function as successive Dirac functions at various times t.

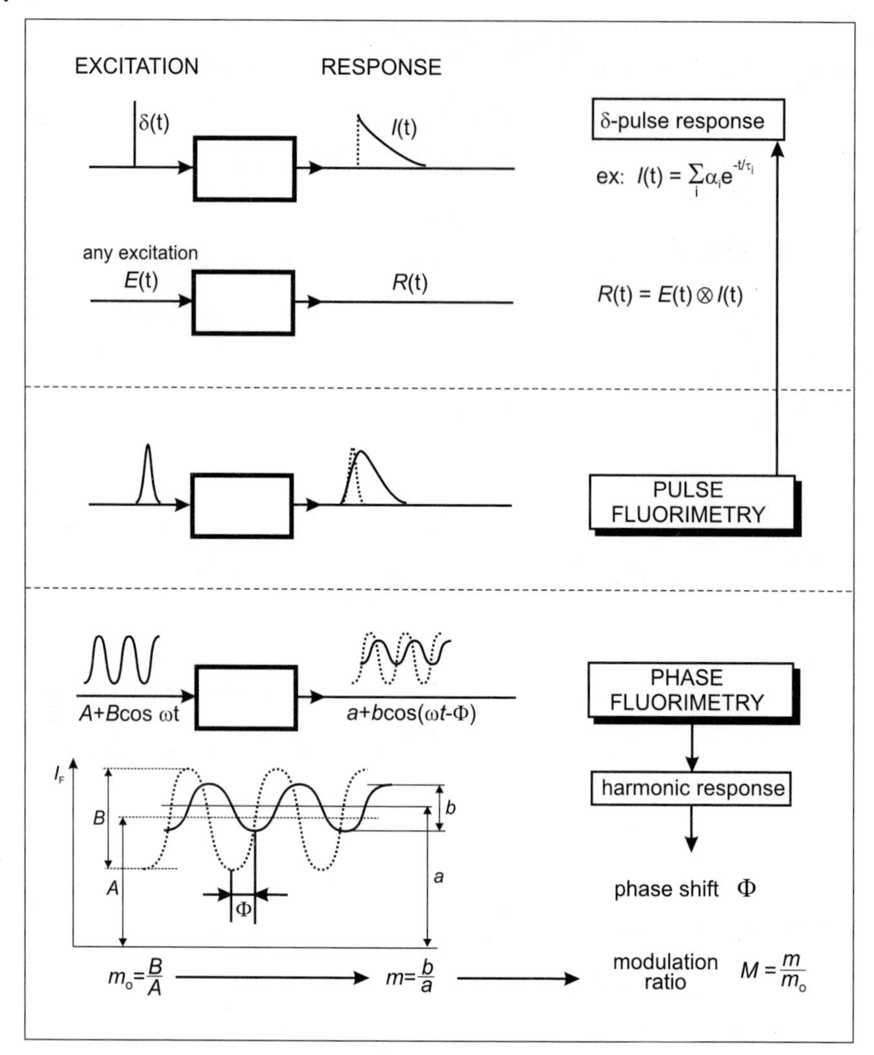

Fig. 6.5. Principles of time-resolved fluorometry.

with respect to the time constants of the fluorescence decay, the fluorescence response is the convolution product given by Eq. (6.9): the fluorescence intensity increases, goes through a maximum and becomes identical to the true δ-pulse response $i(t)$ as soon as the intensity of the light pulse is negligible (Figure 6.6). In this case, data analysis for the determination of the parameters characterizing the δ-pulse response requires a deconvolution of the fluorescence response (see Section 6.2.5).

Phase-modulation fluorometry The sample is excited by a sinusoidally modulated light at high frequency. The fluorescence response, which is the convolution product (Eq. 6.9) of the δ-pulse response by the sinusoidal excitation function, is sinusoidally

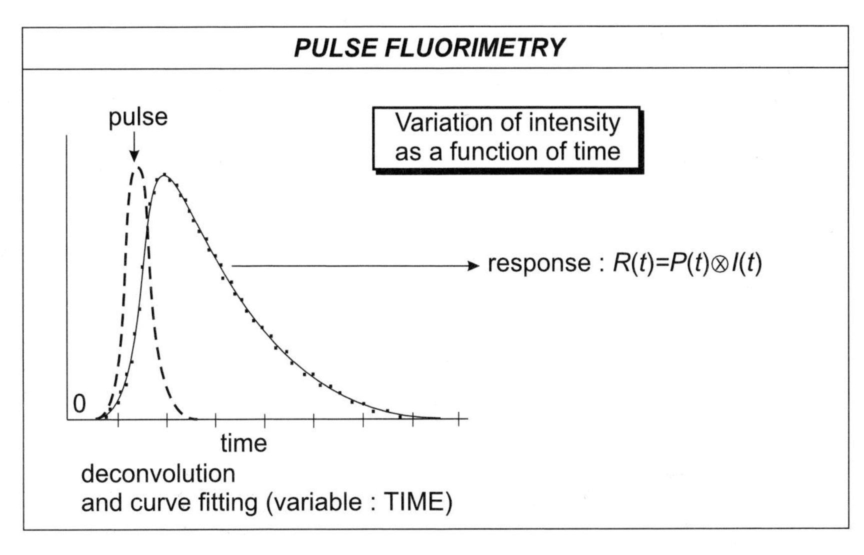

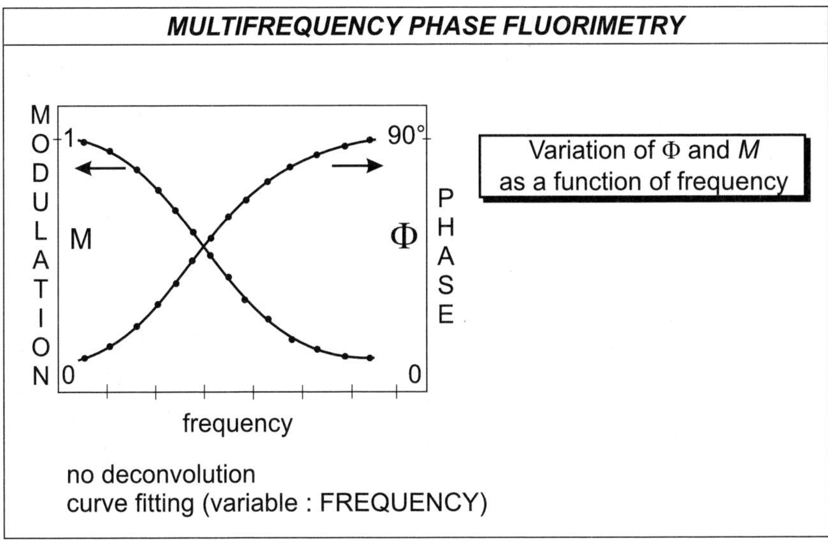

Fig. 6.6. Principles of pulse fluorometry and multi-frequency phase-modulation fluorometry.

modulated at the same frequency but delayed in phase and partially demodulated with respect to the excitation. The phase shift Φ and the modulation ratio M (equal to m/m_0), i.e. the ratio of the modulation depth m (AC/DC ratio) of the fluorescence and the modulation depth of the excitation m_0; see Figure 6.5) characterize the harmonic response of the system. These parameters are measured as a function of the modulation frequency. No deconvolution is necessary because the data are directly analyzed in the frequency domain (Figure 6.6).

Relationship between harmonic response and δ-pulse response It is worth demonstrating that the harmonic response is the Fourier transform of the δ-pulse response. The sinusoidal excitation function can be written as

$$E(t) = E_0[1 + m_0 \exp(j\omega t)] \tag{6.10}$$ [9]

where ω is the angular frequency ($= 2\pi f$). The response of the system is calculated using Eq. (6.9), which gives, after introducing the new variable $u = t - t'$

$$R(t) = E_0\left[\int_0^\infty I(u)\,du + m_0 \exp(j\omega t)\int_0^\infty I(u)\exp(-j\omega u)\,du\right] \tag{6.11}$$

It is convenient for the calculations to use the normalized δ-pulse response $i(t)$ according to

$$\int_0^\infty i(t)\,dt = 1 \tag{6.12}$$

The fluorescence response can then be written:

$$R(t) = E_0[1 + m\exp(j\omega t - \Phi)] \tag{6.13}$$

with

$$m\exp(-j\Phi) = m_0\int_0^\infty i(u)\exp(-j\omega u)\,du \tag{6.14}$$

Using the modulation ratio $M = m/m_0$ and replacing u by t for clarity, we obtain

$$\boxed{M\exp(-j\Phi) = \int_0^\infty i(t)\exp(-j\omega t)\,dt} \tag{6.15}$$

This important expression shows that the harmonic response expressed as $M\exp(-j\Phi)$ is the Fourier transform of the δ-pulse response.

It is convenient to introduce the sine and cosine transforms P and Q of the δ-pulse response:

$$P = \int_0^\infty i(t)\sin(\omega t)\,dt \tag{6.16}$$

$$Q = \int_0^\infty i(t)\cos(\omega t)\,dt \tag{6.17}$$

9) $\exp(j\omega t) = \cos(\omega t) + j\sin(\omega t)$

If the δ-pulse response is not normalized according to Eq. (6.12), then Eqs (6.16) and (6.17) should be replaced by

$$P = \frac{\int_0^\infty I(t)\sin(\omega t)\,dt}{\int_0^\infty I(t)\,dt} \tag{6.18}$$

$$Q = \frac{\int_0^\infty I(t)\cos(\omega t)\,dt}{\int_0^\infty I(t)\,dt} \tag{6.19}$$

Equation (6.15) can be rewritten as

$$M\cos\Phi - jM\sin\Phi = Q - jP \tag{6.20}$$

Therefore,

$$M\sin\Phi = P \tag{6.21}$$

$$M\cos\Phi = Q \tag{6.22}$$

Appropriate combinations of these two equations lead to

$$\Phi = \tan^{-1}\left(\frac{P}{Q}\right) \tag{6.23}$$

$$M = [P^2 + Q^2]^{1/2} \tag{6.24}$$

In practice, the phase shift Φ and the modulation ratio M are measured as a function of ω. Curve fitting of the relevant plots (Figure 6.6) is performed using the theoretical expressions of the sine and cosine Fourier transforms of the δ-pulse response and Eqs (6.23) and (6.24). *In contrast to pulse fluorometry, no deconvolution is required.*

General relations for single exponential and multi-exponential decays For a single exponential decay, the δ-pulse response is

$$I(t) = \alpha\exp(-t/\tau) \tag{6.25}$$

where τ is the decay time and α is the pre-exponential factor or amplitude.
The phase shift and relative modulation are related to the decay time by

$$\tan\Phi = \omega\tau \tag{6.26}$$

$$M = \frac{1}{(1+\omega^2\tau^2)^{1/2}} \tag{6.27}$$

For a multi-exponential decay with n components, the δ-pulse response is

$$I(t) = \sum_{i=1}^{n} \alpha_i \exp(-t/\tau_i) \tag{6.28}$$

Note that the fractional intensity of component i, i.e. the fractional contribution of component i to the total steady-state intensity is

$$f_i = \frac{\int_0^\infty I_i(t)\,\mathrm{d}t}{\int_0^\infty I(t)\,\mathrm{d}t} = \frac{\alpha_i \tau_i}{\sum_{i=1}^{n} \alpha_i \tau_i} \tag{6.29}$$

with, of course, $\sum_{i=1}^{n} f_i = 1$.

Using Eqs (6.18) and (6.19), the sine and cosine Fourier transforms, P and Q, are given by

$$P = \frac{\omega \sum_{i=1}^{n} \frac{\alpha_i \tau_i^2}{1+\omega^2\tau_i^2}}{\sum_{i=1}^{n} \alpha_i \tau_i} = \omega \sum_{i=1}^{n} \frac{f_i \tau_i}{1+\omega^2\tau_i^2} \tag{6.30}$$

$$Q = \frac{\sum_{i=1}^{n} \frac{\alpha_i \tau_i}{1+\omega^2\tau_i^2}}{\sum_{i=1}^{n} \alpha_i \tau_i} = \sum_{i=1}^{n} \frac{f_i}{1+\omega^2\tau_i^2} \tag{6.31}$$

These equations are to be used in conjunction with Eqs (6.23) and (6.24) giving Φ and M.

When the fluorescence decay of a fluorophore is multi-exponential, the natural way of defining an *average decay time* (or *lifetime*) is:

$$\langle\tau\rangle_f = \frac{\int_0^\infty t I(t)\,\mathrm{d}t}{\int_0^\infty I(t)\,\mathrm{d}t} = \frac{\int_0^\infty t \sum_{i=1}^{n} \alpha_i \exp(-t/\tau_i)\,\mathrm{d}t}{\int_0^\infty \sum_{i=1}^{n} \alpha_i \exp(-t/\tau_i)\,\mathrm{d}t}$$

$$\langle\tau\rangle_f = \frac{\sum_{i=1}^{n} \alpha_i \tau_i^2}{\sum_{i=1}^{n} \alpha_i \tau_i} = \sum_{i=1}^{n} f_i \tau_i \tag{6.32}$$

In this definition, each decay time is weighted by the corresponding fractional intensity. This average is called the *intensity-averaged decay time* (or *lifetime*).

Another possibility is to use the amplitudes (pre-exponential factors) as weights:

$$\langle\tau\rangle_a = \frac{\sum_{i=1}^{n} \alpha_i \tau_i}{\sum_{i=1}^{n} \alpha_i} = \sum_{i=1}^{n} a_i \tau_i \tag{6.33}$$

where a_i are the fractional amplitudes

$$a_i = \frac{\alpha_i}{\sum_{i=1}^{n} \alpha_i} \tag{6.34}$$

with, of course, $\sum_{i=1}^{n} a_i = 1$. This average is called the *amplitude-averaged decay time* (or *lifetime*).

The definition used depends on the phenomenon under study. For instance, the intensity-averaged lifetime must be used for the calculation of an average collisional quenching constant, whereas in resonance energy transfer experiments, the amplitude-averaged decay time or lifetime must be used for the calculation of energy transfer efficiency (see Section 9.2.1).

6.2.2 Design of pulse fluorometers

6.2.2.1 Single-photon timing technique

Pulse fluorometry is the most popular technique for the determination of lifetimes (or decay parameters). Most instruments are based on the *time-correlated single-photon counting* (TCSPC) method, better called as *single-photon timing* (SPT). The basic principle relies on the fact that the probability of detecting a single photon at time t after an exciting pulse is proportional to the fluorescence intensity at that time. After timing and recording the single photons following a large number of exciting pulses, the fluorescence intensity decay curve is reconstructed.

Figure 6.7 shows a conventional single-photon counting instrument. The excitation source can be either a flash lamp or a mode-locked laser. An electrical pulse associated with the optical pulse is generated (e.g. by a photodiode or the electronics associated with the excitation source) and routed – through a discriminator – to the start input of the time-to-amplitude converter (TAC). Meanwhile, the sample is excited by the optical pulse and emits fluorescence. The optics are tuned (e.g. by means of a neutral density filter) so that the photomultiplier detects no more than one photon for each exciting pulse. The corresponding electrical pulse is routed – through a discriminator – to the stop input of the TAC. The latter generates an output pulse whose amplitude is directly proportional to the delay time between the start and the stop pulses[10]. The height analysis of this pulse is achieved by an analogue-to-digital converter and a multichannel analyzer (MCA), which increases by one the contents of the memory channel corresponding to the digital value of the pulse. After a large number of excitation and detection events, the histogram of pulse heights represents the fluorescence decay curve. Obviously, the larger the number of events, the better the accuracy of the decay curve. The required accuracy

10) The start pulse initiates charging of a capacitor and the stop pulse stops the charging ramp. The pulse delivered by the TAC is proportional to the final voltage of the capacitor, i.e. proportional to the delay time between the start and the stop pulses.

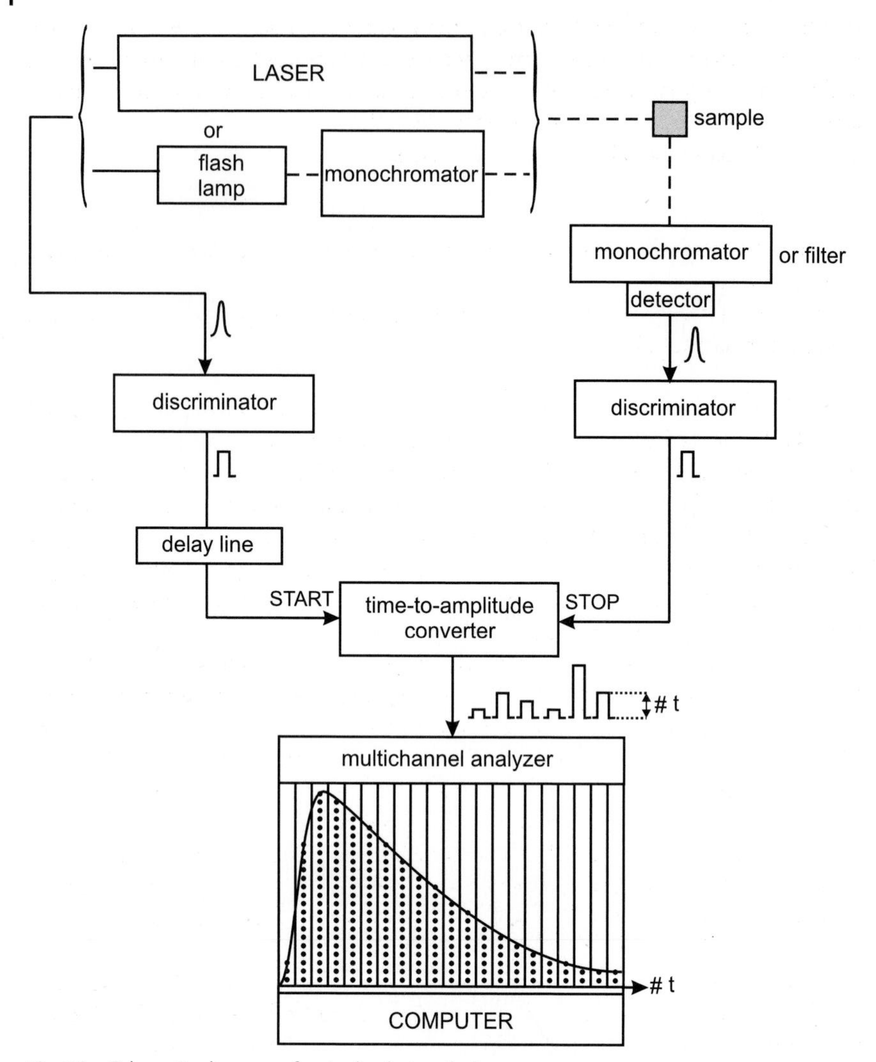

Fig. 6.7. Schematic diagram of a single-photon timing fluorometer.

depends on the complexity of the δ-pulse response of the system; for instance, a high accuracy is necessary for recovering a distribution of decay times.

When deconvolution is required, the time profile of the exciting pulse is recorded under the same conditions by replacing the sample with a scattering solution (Ludox (colloidal silica) or glycogen).

It is important to note that the number of fluorescence pulses must be kept much smaller than the number of exciting pulses ($<$ 0.01–0.05 stops per pulse), so that the probability of detecting two fluorescence pulses per exciting pulse is negligible. Otherwise, the TAC will take into account only the first fluorescence pulse and the counting statistics will be distorted: the decay will appear shorter than it actually is. This effect is called the 'pileup effect'.

The SPT technique offers several advantages:

- high sensitivity;
- outstanding dynamic range and linearity: three or four decades are common and five is possible;
- well-defined statistics (Poisson distribution) allowing proper weighting of each point in data analysis.

The excitation source is of major importance. Flash lamps running in air, or filled with N_2, H_2 or D_2 are not expensive but the excitation wavelengths are restricted to the 200–400 nm range. They deliver nanosecond pulses, so that decay times of a few hundreds of picoseconds can be measured. Furthermore, the repetition rate is not high (10^4–10^5 Hz) and because the number of fluorescence pulses per exciting pulse must be kept below 5%, the collection period may be quite long, depending on the required accuracy (a few tens of minutes to several hours). For long collection periods, lamp drift may be a serious problem.

Lasers as excitation sources are of course much more efficient and versatile, at the penalty of high cost. Mode-locked lasers (e.g. argon ion lasers) associated with a dye laser or Ti:sapphire laser can generate pulses over a broad wavelength range. The pulse widths are in the picosecond range with a high repetition rate. This rate must be limited to a few MHz in order to let the fluorescence of long lifetime samples vanish before a new exciting pulse is generated.

Synchrotron radiation can also be used as an excitation source with the advantage of almost constant intensity versus wavelength over a very broad range, but the pulse width is in general of the order of hundreds of picosecond or not much less. There are only a few sources of this type in the world.

The time resolution of the instrument is governed not only by the pulse width but also by the electronics and the detector. The linear time response of the TAC is most critical for obtaining accurate fluorescence decays. The response is more linear when the time during which the TAC is in operation and unable to respond to another signal (dead time) is minimized. For this reason, it is better to collect the data in the reverse configuration: the fluorescence pulse acts as the start pulse and the corresponding excitation pulse (delayed by an appropriate delay line) as the stop pulse. In this way, only a small fraction of start pulses result in stop pulses and the collection statistics are better.

Microchannel plate photomultipliers are preferred to standard photomultipliers, but they are much more expensive. They exhibit faster time responses (10- to 20-fold faster) and do not show a significant 'color effect' (see below).

With mode-locked lasers and microchannel plate photomultipliers, the instrument response in terms of pulse width is 30–40 ps so that decay times as short as 10–20 ps can be measured.

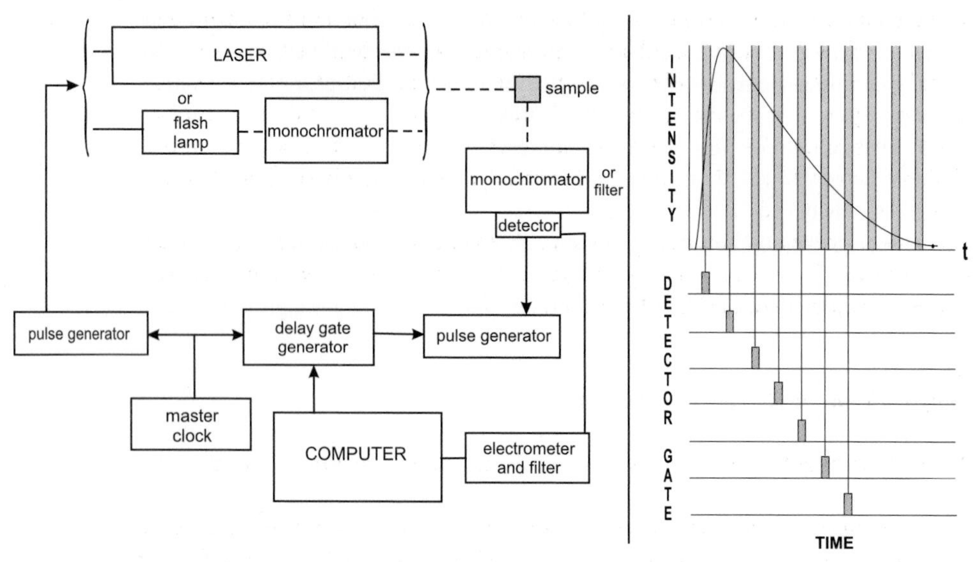

Fig. 6.8. Principle of the stroboscopic technique.

6.2.2.2 **Stroboscopic technique**

This technique is an interesting alternative to the single-photon timing technique (James et al., 1992). The sample is excited by a train of light pulses delivered by a flash lamp. A computer-controlled digital delay unit is used to gate (or 'strobe') the photomultiplier by a voltage pulse (whose width defines a time window) at a time accurately delayed with respect to the light pulse. Synchronization with the flash lamp is achieved by a master clock. By this procedure, only photons emitted from the sample that arrive at the photocathode of the photomultiplier during the gate time will be detected. The fluorescence intensity as a function of time can be constructed by moving the time window after each pulse from before the pulse light to any suitable end time (e.g. ten times the excited-state lifetime). The principle of the stroboscopic technique is shown in Figure 6.8.

The strobe technique offers some advantages. It does not require expensive electronics. High-frequency light sources are unnecessary because the intensity of the signal is directly proportional to the intensity of the light pulse, in contrast to the single-photon timing technique. High-intensity picosecond lasers operating at low repetition rate can be used, with the advantage of lower cost than cavity-dumped, mode-locked picosecond lasers. However, the time resolution of the strobe technique is less than that of the single-photon timing technique, and lifetime measurements with samples of low fluorescence intensity are more difficult.

6.2.2.3 **Other techniques**

Lifetime instruments using a streak camera as a detector provide a better time resolution than those based on the single-photon timing technique. However, streak cameras are quite expensive. In a streak camera, the photoelectrons emitted by the photocathode are accelerated into a deflection field. The beam is then swept linearly in time across a microchannel plate electron amplifier that preserves the spatial information. The amplified electron beam is collected by a video camera–computer system. Temporal information is thus converted into spatial information. The time resolution of streak cameras (few picoseconds or less) is better than that of single-photon timing instruments, but the dynamic range is smaller (2–3 decades instead of 3–5 decades).

The instruments that provide the best time resolution (about 100 femtoseconds) are based on fluorescence up-conversion. This very sophisticated and expensive technique will be described in Chapter 11.

6.2.3
Design of phase-modulation fluorometers

Before describing the instruments, it is worth making two preliminary remarks:

(i) The optimum frequency for decay time measurements using either the phase shift or the modulation ratio is, according to Eqs (6.26) and (6.27), such that $\omega\tau$ is close to 1, i.e. $f \approx 1/(2\pi\tau)$. Therefore, for decay times of 10 ps, 1 ns or 100 ns, the optimum frequencies are about 16 GHz, 160 MHz or 1.6 MHz, respectively. These values give an order of magnitude of the frequency ranges that are in principle to be selected depending on the decay times. Figure 6.9 shows simulation data.

(ii) In the case of a single exponential decay, Eqs (6.26) and (6.27) provide two independent ways of measuring the decay time:

- by phase measurements:

$$\tau_\Phi = \frac{1}{\omega} \tan^{-1} \Phi \tag{6.35}$$

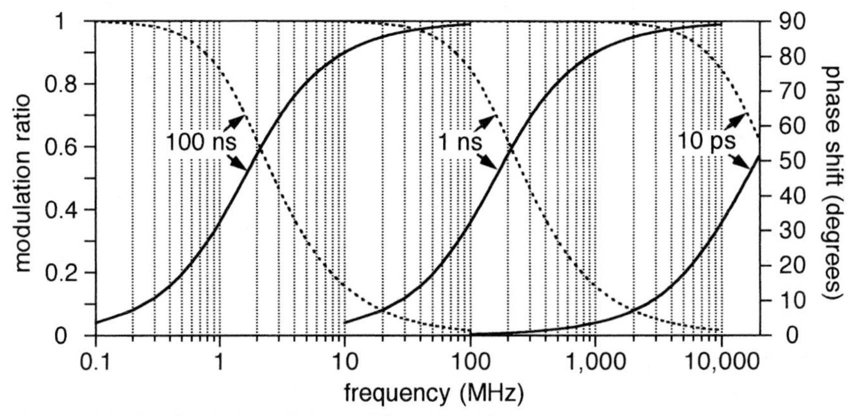

Fig. 6.9. Simulated variations of phase shift and modulation ratio versus frequency using Eqs (6.25) and (6.26).

- by modulation measurements:

$$\tau_M = \frac{1}{\omega}\left(\frac{1}{M^2} - 1\right)^{1/2} \tag{6.36}$$

The values measured in these two ways should of course be identical and independent of the modulation frequency. This provides two criteria to check whether an instrument is correctly tuned by using a lifetime standard whose fluorescence decay is known to be a single exponential.

Note that the measurement of a decay time is fast (a fraction of a second) for a single exponential decay because a single frequency is sufficient. Note also that a significant difference between the values obtained by means of Eqs (6.35) and (6.36) is compelling evidence of non-exponentiality of the fluorescence decay.

Historically, the first instrument for the determination of lifetime was a phase fluorometer (designed by Gaviola in 1926) operating at a single frequency. Progress in instrumentation enabled variable modulation frequency by employing a cw laser (or a lamp) and an electro-optic modulator (0.1–250 MHz), or by using the harmonic content of a pulsed laser source (up to 2 GHz). These two techniques will now be described.

6.2.3.1 Phase fluorometers using a continuous light source and an electro-optic modulator

The light source can be a xenon lamp associated with a monochromator. The optical configuration should be carefully optimized because the electro-optic modulator (usually a Pockel's cell) must work with a parallel light beam. The advantages are the low cost of the system and the wide availability of excitation wavelengths. In terms of light intensity and modulation, it is preferable to use a cw laser, which costs less than mode-locked pulsed lasers.

Figure 6.10 shows a multi-frequency phase-modulation fluorometer. A beam splitter reflects a few per cent of the incident light towards a reference photomultiplier (via a cuvette containing a reference scattering solution or not). The fluorescent sample and a reference solution (containing either a scatter or a reference fluorescent compound) are placed in a rotating turret. The emitted fluorescence or scattered light is detected by a photomultiplier through a monochromator or an optical filter. The Pockel's cell is driven by a frequency synthesizer and the photomultiplier response is modulated by varying the voltage at the second dynode by means of another frequency synthesizer locked in phase with the first one. The two synthesizers provide modulated signals that differ in frequency by a few tens of Hz in order to achieve cross-correlation (heterodyne detection). This procedure is very accurate because the phase and modulation information contained in the signal is transposed to the low-frequency domain where phase shifts and modulation depths can be measured with a much higher accuracy than in the high-frequency domain. For instruments using a cw laser, the standard deviations are currently 0.1–0.2° for the phase shift and 0.002–0.004 for the modulation ratio. When the light source is a xenon lamp, these standard deviations are ~0.5° and 0.005, respectively.

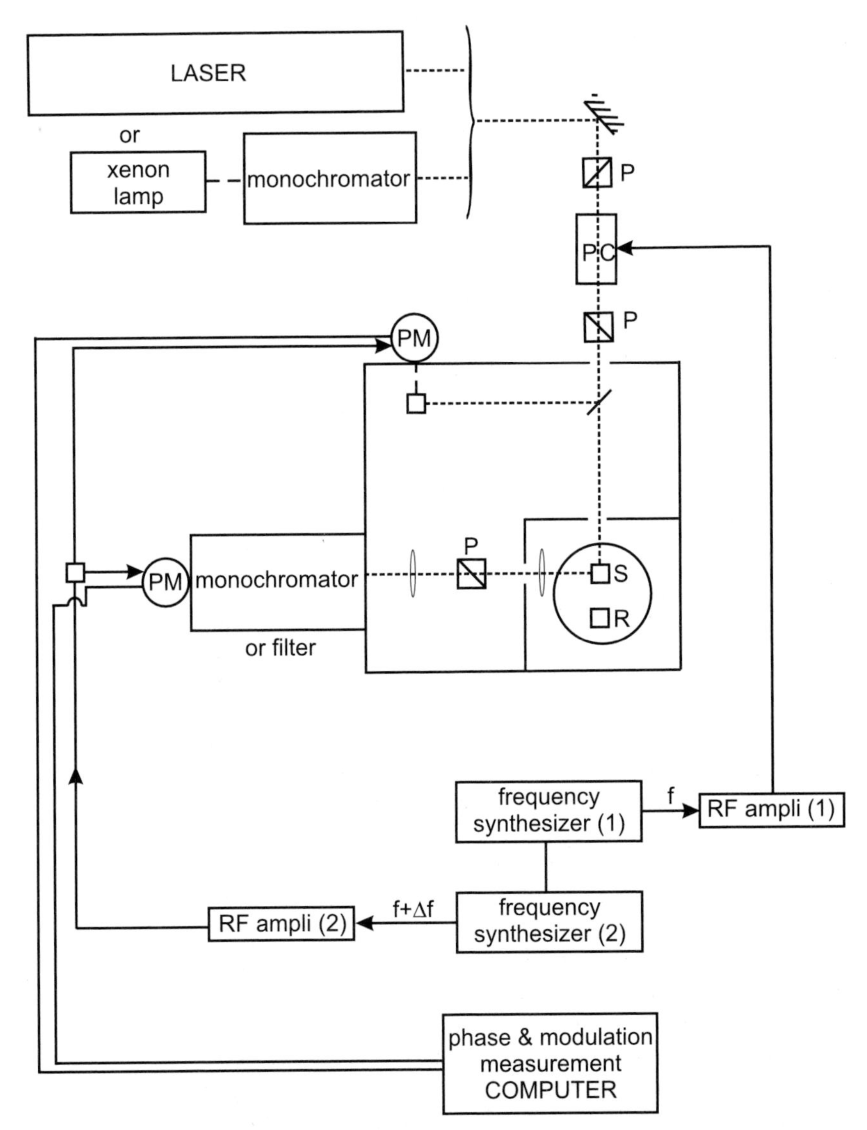

Fig. 6.10. Schematic diagram of a multi-frequency phase-modulation fluorometer. P: polarizers; PC: Pockel's cell; S: sample; R: reference.

Practically, the phase delay ϕ_R and the modulation ratio m_R of the light emitted by the scattering solution (solution of glycogen or suspension of colloidal silica) are measured with respect to the signal detected by the reference photomultiplier. Then, after rotation of the turret, the phase delay ϕ_F and the modulation ratio m_F for the sample fluorescence are measured with respect to the signal detected by the reference photomultiplier. The absolute phase shift and modulation ratio of the sample are then $\Phi = \phi_R - \phi_R$ and $M = m_F/m_R$, respectively.

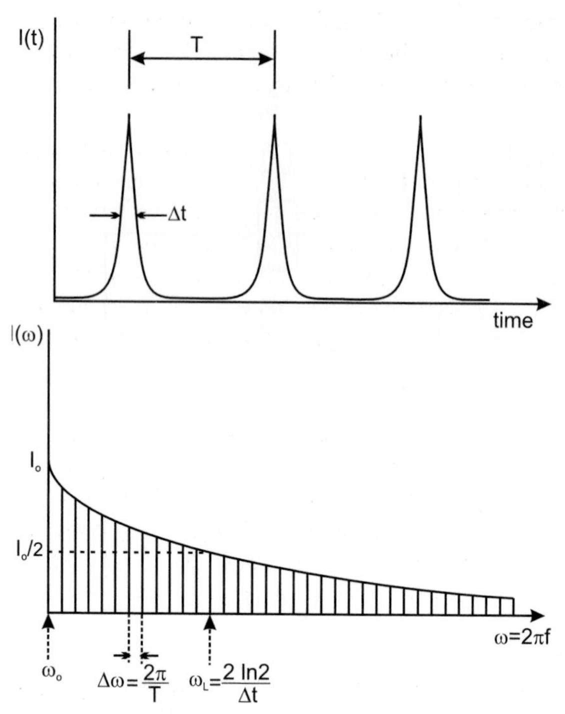

Fig. 6.11. Harmonic content of a pulse train. For a repetition rate of 4 MHz and $\Delta t = 5$ ps, $\omega_L = 277$ GHz, $\Delta\omega = 25.1$ MHz.

6.2.3.2 **Phase fluorometers using the harmonic content of a pulsed laser**

The type of laser source that can be used is exactly the same as for single-photon counting pulse fluorometry (see above). Such a laser system, which delivers pulses in the picosecond range with a repetition rate of a few MHz can be considered as an intrinsically modulated source. The harmonic content of the pulse train – which depends on the width of the pulses (as illustrated in Figure 6.11) – extends to several gigahertz.

For high-frequency measurements, normal photomultipliers are too slow, and microchannel plate photomultipliers are required. However, internal cross-correlation is not possible with the latter and an external mixing circuit must be used.

The time resolution of a phase fluorometer using the harmonic content of a pulsed laser and a microchannel plate photomultiplier is comparable to that of a single-photon counting instrument using the same kind of laser and detector.

6.2.4
Problems with data collection by pulse and phase-modulation fluorometers

6.2.4.1 **Dependence of the instrument response on wavelength. Color effect**

The transit times of photoelectrons in normal photomultipliers depend significantly on the wavelength of the incident photon, whereas this dependence is negligible for microchannel plate detectors. Using the former, this dependence results in a shift between the time profile of the fluorescence and that of the exciting pulse, because they are recorded at different wavelengths. This effect is called the 'color effect'.

An efficient way of overcoming this difficulty is to use a reference fluorophore (instead of a scattering solution) (i) whose fluorescence decay is a single exponential, (ii) which is excitable at the same wavelength as the sample, and (iii) which emits fluorescence at the observation wavelength of the sample. In pulse fluorometry, the deconvolution of the fluorescence response can be carried out against that of the reference fluorophore. In phase-modulation fluorometry, the phase shift and the relative modulation can be measured directly against the reference fluorophore.

6.2.4.2 Polarization effects

Distortion of the fluorescence response measured by the detection system (monochromator + detector) arises when the emitted fluorescence is partially polarized. As explained in the Appendix, a response proportional to the total fluorescence intensity can be observed by using two polarizers: an excitation polarizer in the vertical position, and an emission polarizer set at the magic angle (54.7°) with respect to the vertical, or vice versa (see the configurations in Figure 6.3).

6.2.4.3 Effect of light scattering

It is sometimes difficult to totally remove (by the emission monochromator and appropriate filters) the light scattered by turbid solutions or solid samples. A subtraction algorithm can then be used in the data analysis to remove the light scattering contribution.

6.2.5 Data analysis

6.2.5.1 Pulse fluorometry

Considerable effort has gone into solving the difficult problem of deconvolution and curve fitting to a theoretical decay that is often a sum of exponentials. Many methods have been examined (O'Connor et al., 1979): methods of least squares, moments, Fourier transforms, Laplace transforms, phase-plane plot, modulating functions, and more recently maximum entropy. The most widely used method is based on nonlinear least squares. The basic principle of this method is to minimize a quantity that expresses the mismatch between data and fitted function. This quantity χ^2 is defined as the weighted sum of the squares of the deviations of the experimental response $R(t_i)$ from the calculated ones $R_c(t_i)$:

$$\chi^2 = \sum_{i=1}^{N} \left[\frac{R(t_i) - R_c(t_i)}{\sigma(i)} \right]^2 \tag{6.37}$$

where N is the total number of data points and $\sigma(i)$ is the standard deviation of the ith data point, i.e. the uncertainty expected from statistical considerations (noise). It is convenient to normalize χ^2 so that the value is independent of N and the number p of fitted parameters (e.g. for a double-exponential decay, $p = 3$: two decay times and one pre-exponential factor). Division of χ^2 by the number of degrees of freedom, $\nu = N - p$, leads to the reduced chi square χ_r^2 whose value should be close to 1 for a good fit.

$$\chi_r^2 = \frac{1}{\nu}\sum_{i=1}^{N}\left[\frac{R(t_i) - R_c(t_i)}{\sigma(i)}\right]^2 \tag{6.38}$$

In single-photon counting experiments, the statistics obey a Poisson distribution and the expected deviation $\sigma(i)$ is approximated to $[R(t_i)]^{1/2}$ so that Eq. (6.38) becomes

$$\chi_r^2 = \frac{1}{\nu}\sum_{i=1}^{N}\left[\frac{R(t_i) - R_c(t_i)}{R(t_i)}\right]^2 \tag{6.39}$$

In practice, initial guesses of the fitting parameters (e.g. pre-exponential factors and decay times in the case of a multi-exponential decay) are used to calculate the decay curve; the latter is reconvoluted with the instrument response for comparison with the experimental curve. Then, a minimization algorithm (e.g. Marquardt method) is employed to search the parameters giving the best fit. At each step of the iteration procedure, the calculated decay is reconvoluted with the instrument response. Several softwares are commercially available.

6.2.5.2 Phase-modulation fluorometry

The least-squares method is also widely applied to curve fitting in phase-modulation fluorometry; the main difference with data analysis in pulse fluorometry is that no deconvolution is required: curve fitting is indeed performed in the frequency domain, i.e. directly using the variations of the phase shift Φ and the modulation ratio M as functions of the modulation frequency. Phase data and modulation data can be analyzed separately or simultaneously. In the latter case the reduced chi squared is given by

$$\chi_r^2 = \frac{1}{\nu}\left[\sum_{i=1}^{N}\frac{[\Phi(\omega_i) - \Phi_c(\omega_i)]^2}{\sigma_\Phi(\omega_i)} + \sum_{i=1}^{N}\frac{[M(\omega_i) - M_c(\omega_i)]^2}{\sigma_M(\omega_i)}\right] \tag{6.40}$$

where N is the total number of frequencies. In this case, the number of data points is twice the number of frequencies, so that the number of degrees of freedom is $\nu = 2N - p$.

Data analysis in phase fluorometry requires knowledge of the sine and cosine of the Fourier transforms of the δ-pulse response. This of course is not a problem for the most common case of multi-exponential decays (see above), but in some cases the Fourier transforms may not have analytical expressions, and numerical calculations of the relevant integrals are then necessary.

6.2.5.3 Judging the quality of the fit

Once the best fit parameters are obtained, the quality of the fitted decay function must be judged using statistical and graphical criteria.

In the least-squares method, the first criterion is of course the reduced chi squared χ_r^2 (Eqs 6.39 and 6.40), whose value should be close to 1 for a good fit. Acceptable values are in the 0.8–1.2 range. Lower values indicate that the data set is too small for a meaningful fit and higher values are caused by a significant deviation from the theoretical model (e.g. insufficient number of exponential terms). Systematic errors (arising for instance from radiofrequencies interfering with the detection) can also explain higher values.

The values of χ_r^2 in phase-modulation data analysis deserve further comment. Values of the standard deviations, σ_Φ and σ_M, must be assigned; these values are in principle frequency-dependent. At each frequency, a large number of phase shift and modulation measurements is averaged so that values of the standard deviations are easily calculated. However, it is usually the case that when such values are used in the analysis, abnormally high values of χ_r^2 are obtained, which means that values of the standard deviations are underestimated (because of some systematic errors). Therefore, 'realistic' frequency-independent values must be assigned. The important consequence is that the value of χ_r^2 is not sufficient as a criterion of good fit and that only relative values can be used in accepting or rejecting a theoretical model. For instance, in the case of a multi-exponential decay, when χ_r^2 decreases two-fold or more when an exponential term is added, then addition of this term is justified.

In addition to the value of χ_r^2, graphical tests are useful. The most important of these is the plot of the *weighted residuals*, defined as

$$W(t_i) = \frac{R(t_i) - R_c(t_i)}{\sigma(i)} \tag{6.41}$$

with $\sigma(i) = [R(t_i)]^{1/2}$ for single-photon counting data. If the fit is good, the weighted residuals are randomly distributed around zero.

When the number of data points is large (i.e. in the single-photon timing technique, or in phase fluorometry when using a large number of modulation frequencies), the autocorrelation function of the residuals, defined as

$$C(j) = \frac{\dfrac{1}{N-j}\sum_{i=1}^{N-j} W(t_i)W(t_{i+j})}{\dfrac{1}{N}\sum_{i=1}^{N}[W(t_i)]^2} \tag{6.42}$$

is also a useful graphical test of the quality of the fit. $C(j)$ expresses the correlation between the residual in channels j and $i+j$. A low-frequency periodicity is symptomatic of radiofrequency interferences. Several other tests (e.g. Durbin–Watson parameter) can also be used (Eaton, 1990).

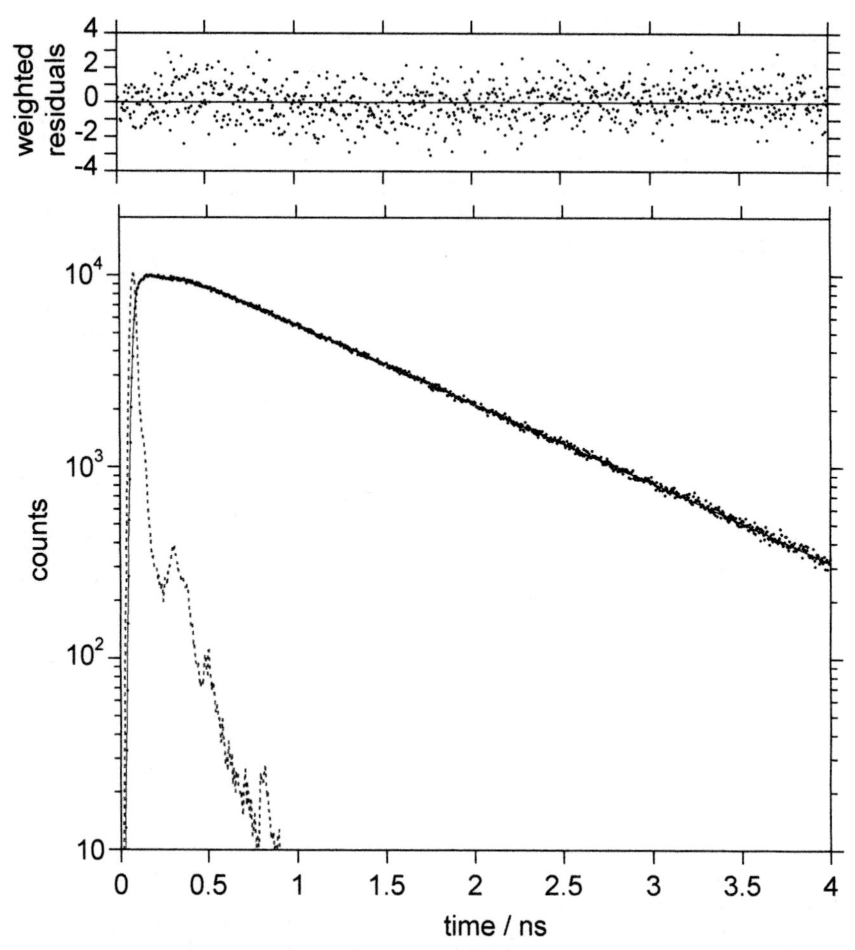

Fig. 6.12. Data obtained by the single-photon timing technique using a mode-locked ion–argon laser that synchronously pumps a cavity-dumped dye laser. Sample: solution of POPOP in cyclohexane (undegassed). Excitation wavelength: 340 nm; observation wavelength: 390 nm. Reference: scattering solution (Ludox). Number of channels: 900; channel width: 4.68 ps. Result: $\tau = 1.05 \pm 0.01$ ns; $\chi_r^2 = 1.055$.

Typical sets of data obtained by pulse fluorometry and phase-modulation fluorometry are shown in Figures 6.12 and 6.13, respectively.

6.2.5.4 **Global analysis**

In a global analysis, related experimental sets of data are simultaneously analyzed. This results in improved accuracy of the recovered parameters. In the typical case of excited-state reactions that involve species having different fluorescence spectra, the data sets can be collected at different emission wavelengths and simultaneously analyzed. For instance, when A $\rightarrow$ B in the excited state (e.g. excimer formation,

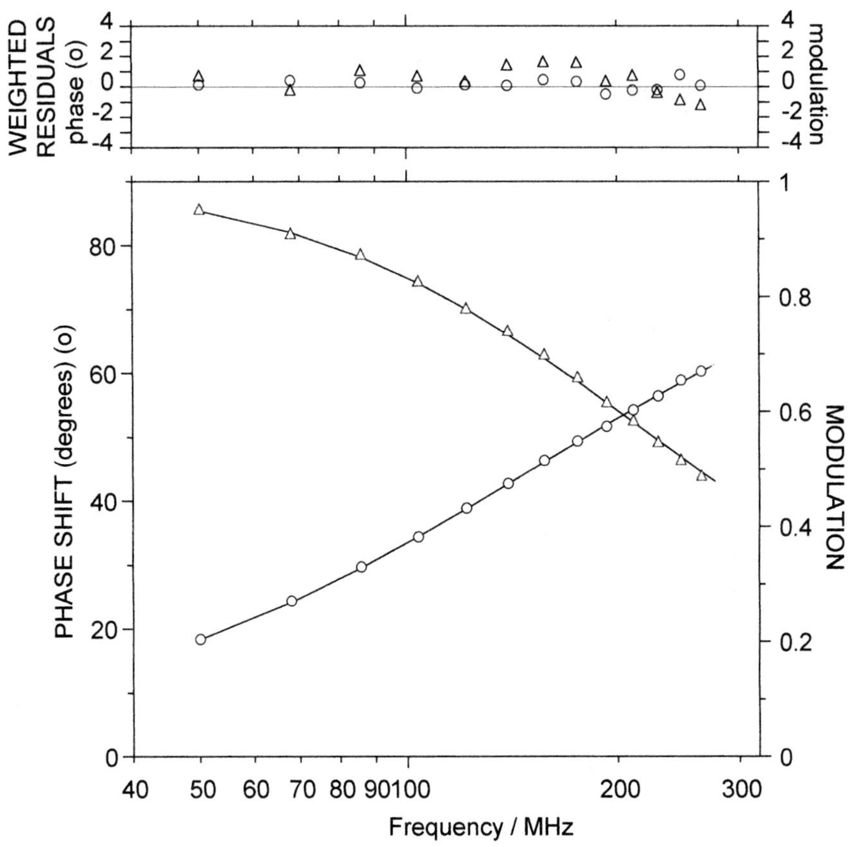

Fig. 6.13. Data obtained by the phase-modulation technique with a Fluorolog tau-3 instrument (Jobin Yvon-Spex) operating with a xenon lamp and a Pockel's cell. Note that because the fluorescence decay is a single exponential, a single appropriate modulation frequency suffices for the lifetime determination. The broad set of frequencies permits control of the proper tuning of the instrument. Sample: solution of POPOP in cyclohexane (undegassed). Excitation wavelength: 340 nm; observation through a cut-off filter at 390 nm (no emission monochromator). Reference: scattering solution (Ludox). Number of frequencies: 13 frequencies from 50 to 280 MHz. Result: $\tau = 1.06 \pm 0.01$ ns; $\chi_r^2 = 0.924$.

proton transfer, etc.), we can fit all the curves recorded at different wavelengths by constraining the decay times to be the same for all sets, but allowing different values of the pre-exponential factors.

6.2.5.5 **Complex fluorescence decays. Lifetime distributions**

When the fit with a single exponential is unsatisfactory, a second exponential is often added. If the quality of the fit is still unsatisfactory, a third exponential term is added. In most cases, a fourth component does not improve the quality of the fit. A satisfactory fit with a sum of two (or three, or four) exponential terms with more or less closely spaced time constants means that the number of fitting parameters is large enough – considering the limited accuracy of the experimental data – but it does not prove that the number of components is two (or three, or four), unless it is already known that the system can be described by a physical model involving distinct exponentials.

In some circumstances, it can be anticipated that continuous lifetime distributions would best account for the observed phenomena. Examples can be found in biological systems such as proteins, micellar systems and vesicles or membranes. If an *a priori* choice of the shape of the distribution (i.e. Gaussian, sum of two Gaussians, Lorentzian, sum of two Lorentzians, etc.) is made, a satisfactory fit of the experimental data will only indicate that the assumed distribution is compatible with the experimental data, but it will not demonstrate that this distribution is the only possible one, and that a sum of a few distinct exponentials should be rejected.

To answer the question as to whether the fluorescence decay consists of a few distinct exponentials or should be interpreted in terms of a continuous distribution, it is advantageous to use an approach without *a priori* assumption of the shape of the distribution. In particular, the maximum entropy method (MEM) is capable of handling both continuous and discrete lifetime distributions in a single analysis of data obtained from pulse fluorometry or phase-modulation fluorometry (Brochon, 1994) (see Box 6.1).

6.2.6 Lifetime standards

The excited-state lifetimes of some compounds that can be used as standards are given in Table 6.2. These values are the averages of data obtained from eight laboratories. There is some deviation from one laboratory to another. For instance, the values reported for POPOP in deoxygenated cyclohexane range from 1.07 to 1.18 ns. The unavoidable small artefacts of any instrument, the different origins of the samples and solvents, the difficulty of preparing identical solutions without any trace of impurity, quencher, etc., can explain the small deviations observed from laboratory to laboratory. However, more important than the lifetime value itself, is the control that the instrument is properly tuned. After selecting one of the standards in Table 6.1 whose characteristics (lifetime, excitation and observation wavelengths) are preferentially not very different from those of the sample to be studied, it is advisable to check that this selected standard indeed exhibits a single exponential decay with a value consistent with that reported in Table 6.2. In the case of short lifetime measurements requiring deconvolution, it should be checked that the reconvoluted curve satisfactorily superimposes the experimental curve, especially in the rise part of the curve.

When normal photomultipliers that exhibit a 'color effect' (see above) are used, the compounds of Table 6.2 that have a short lifetime (e.g. POPOP, PPO) can be used in place of a scattering solution in order to remove this effect (this method is valid for both pulse and phase fluorometries). Such a reference fluorophore must

Box 6.1 The maximum entropy method (MEM)

The maximum entropy method has been successfully applied to pulse fluorometry and phase-modulation fluorometry[a–e]. Let us first consider pulse fluorometry. For a multi-exponential decay with n components whose fractional amplitudes are α_i, the δ-pulse response is

$$I(t) = \sum_{i=1}^{n} \alpha_i \exp(-t/\tau_i)$$

or, for a distribution of decay times,

$$I(t) = \int \alpha(t) \exp(-t/\tau)\, \mathrm{d}\tau$$

The problem is to determine the spectrum of decay times $\alpha(\tau)$ from time-resolved fluorescence data. $\alpha(\tau)$ appears as the inverse Laplace transform of the fluorescence decay deconvolved by the light pulse. However, inverting the Laplace transform is very ill-conditioned. Consequently, the unavoidable small errors in the measurement of the fluorescence decay or light pulse profile can lead to very large changes in the reconstruction of $\alpha(\tau)$, i.e. a large multiplicity of allowable solutions. Among these solutions that fit equally well the experimental decay on the basis of statistical tests (mainly chi-squared values, weighted residuals and autocorrelation functions), some of them can be rejected when they contain unphysical features. The remaining set of solutions is called the *feasible set* (Figure B6.1.1) from which we must choose one member.

The choice of the *preferred solution* within the feasible set can be achieved by maximizing some function $F[\alpha(\tau)]$ of the spectrum that introduces the fewest artefacts into the distribution. It has been proved that only the Shannon–Jaynes entropy S will give the least correlated solution[f]. All other maximization (or regularization) functions introduce correlations into the solution not demanded by the data. The function S is defined as

$$S = \int_0^{\infty} \left[\alpha(\tau) - m(\tau) - \alpha(\tau) \log \frac{\alpha(\tau)}{m(\tau)} \right] \mathrm{d}\tau$$

where m is the model that encodes our prior knowledge about the system. It has been shown that if we have no knowledge of which values of the fractional amplitudes α are expected, they all have equal prior probability, and m is a flat distribution (m_i = constant).

In phase-modulation fluorometry, it is convenient to use the fractional contributions f_i instead of the fractional amplitudes, as shown in Eqs (6.30) and (6.31). The above equation is thus rewritten as

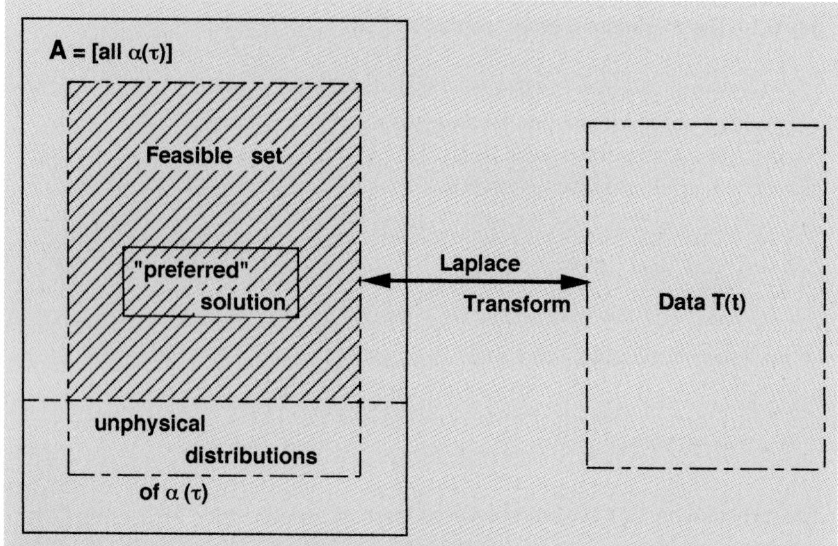

Fig. B6.1.1. The feasible set (hatched) in which MEM will choose a 'preferred' solution (redrawn from Livesey and Brochon[a]).

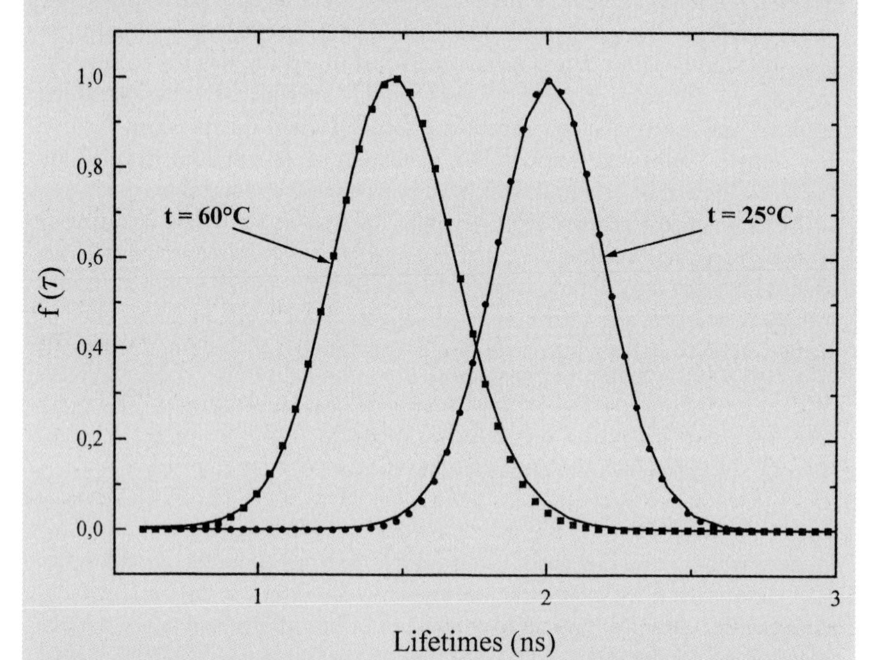

Fig. B6.1.2. Recovered distributions from phase-modulation data for ANS in an aqueous solution containing micelles of sodium dodecylsulfate at 25 and 60 °C. The number of modulation frequencies is 24. Analysis by MEM with 100 image points (log scale from 0.6 to 25 ns) was performed with an additional exponential component (presence of free ANS). Best fits of the recovered distributions by a Gaussian (●, ■) (reprinted with permission from Brochon et al.[g]).

$$S = \int_0^\infty \left[f(\tau) - m(\tau) - f(\tau) \log \frac{f(\tau)}{m(\tau)} \right] d\tau$$

(with $\int m(\tau)\, d\tau = 1$ if no prior knowledge of the distribution).

The commercially available software (Maximum Entropy Data Consultant Ltd, Cambridge, UK) allows reconstruction of the distribution $\alpha(\tau)$ (or $f(\tau)$) which has the maximal entropy S subject to the constraint of the chi-squared value. The quantified version of this software has a full Bayesian approach and includes a precise statement of the accuracy of quantities of interest, i.e. position, surface and broadness of peaks in the distribution. The distributions are recovered by using an automatic stopping criterion for successive iterates, which is based on a Gaussian approximation of the likelihood.

An example of recovered distribution is shown in Figure B6.1.2. It concerns the distribution of lifetimes of 2,6-ANS solubilized in the outer core region of sodium dodecylsulfate micelles[g]. In fact, the microheterogeneity of solubilized sites results in a distribution of lifetimes.

a) Livesey A. K. and Brochon J.-C. (1987) *Biophys. J.* **52**, 517.
b) Brochon J.-C., Livesey A. K., Pouget J. and Valeur B. (1990) *Chem. Phys. Lett.* **174**, 517.
c) Siemarczuk A., Wagner B. D. and Ware W. R. (1990) *J. Phys. Chem.* **94**, 1661.
d) Brochon J.-C. (1994) *Methods in Enzymology*, **240**, 262.
e) Shaver J. M. and McGown L. B. (1996) *Anal. Chem.* **68**, 9; 611.
f) Livesey A. K. and Skilling J. (1985) *Acta Crystallogr. Sect. B. Struct. Crystallogr. Crystallogr. Chem.* **A41**, 113.
g) Brochon J.-C., Pouget J. and Valeur B. (1995) *J. Fluorescence* **5**, 193.

be excitable and observable under the same conditions as those of the sample. However, it should be emphasized that the lifetime value of the reference fluorophore must not be taken from Table 6.2 (because the value may slightly differ from one instrument to another), but be measured on the same instrument and under the same conditions as those of the investigations on new samples to be done using this reference fluorophore.

It is interesting to note that when using two fluorophores (whose fluorescence decays are known to be single exponentials), one as a sample and the other as a reference, it is possible to determine the lifetimes of both the fluorophores without external reference: this can be achieved in data analysis by varying the reference lifetime until a minimum value of χ_r^2 is reached.

6.2.7 Time-dependent anisotropy measurements

6.2.7.1 Pulse fluorometry

The time dependence of emission anisotropy is defined as (see Chapter 5):

$$r(t) = \frac{I_\parallel(t) - I_\perp(t)}{I_\parallel(t) + 2I_\perp(t)} = \frac{I_\parallel(t) - I_\perp(t)}{I(t)} \tag{6.43}$$

Tab. 6.2. Lifetime of various compounds in deoxygenated fluid solutions at 20 °C. Averages of the values measured by eight laboratories by either pulse fluorometry (four laboratories) or phase fluorometry (four laboratories)[a]

Compound[b]	***Solvent***	***Lifetime*** $\bar{\tau}$ **(ns)**[c]	**100** $s/\bar{\tau}$	λ^{ex} **(nm)**	λ^{em} **(nm)**	***d***	***e***
NATA	Water	3.04 ± 0.04	1.2	295–325	325–415	5	4
Anthracene	Methanol	5.1 ± 0.3	6.4	300–330	380–442	6	6
	Cyclohexane	5.3 ± 0.2	3.0	295–325	345–442	5	5
9-Cyanoanthracene	Methanol	16.5 ± 0.5	6.0	295–325	370–442	6	5
	Cyclohexane	12.4 ± 0.5	4.1	295–325	345–380	4	3
Erythrosin B	Water	0.089 ± 0.002	2.5	488, 514, 568	515–575	5	4
	Methanol	0.48 ± 0.02	5.0	488, 514	515–560	5	5
9-Methylcarbazole	Cyclohexane	14.4 ± 0.4	2.5	295–325	360–400	5	4
DPA	Methanol	8.7 ± 0.5	5.9	295–344	370–475	7	7
	Cyclohexane	7.3 ± 0.5	6.2	295–344	345–480	7	6
PPO	Methanol	1.64 ± 0.04	2.4	295–330	345–425	7	7
	Cyclohexane	1.35 ± 0.03	2.5	295–325	345–425	6	6
POPOP	Cyclohexane	1.13 ± 0.05	4.3	295–325	380–450	4	4
Rhodamine B	Water	1.71 ± 0.07	4.1	488–514	515–630	5	4
	Methanol	2.53 ± 0.08	3.1	295, 488, 514	515–630	6	5
Rubrene	Methanol	9.8 ± 0.3	2.6	300, 330, 488, 514	530–590	5	5
SPA	Water	31.2 ± 0.4	1.4	300–330	370–510	5	5
p-Terphenyl	Methanol	1.16 ± 0.08	7.0	284–315	330–380	6	6
	Cyclohexane	0.99 ± 0.03	2.9	295–315	330–390	4	4

a) Data collected by N. Boens and M. Ameloot.
b) Abbreviations used: NATA: N-acetyl-L-tryptophanamide, DPA: 9,10-diphenylanthracene, POPOP: 1,4-bis(5-phenyloxazol-2-yl)benzene, PPO: 2,5-diphenyloxazole, SPA: N-(3-sulfopropyl)acridinium. All solutions are deoxygenated by repetitive freeze–pump–thaw cycles or by bubbling N_2 or Ar through the solutions.
c) The quoted errors are sample standard deviations

$$s = \sqrt{\frac{1}{n-1}\sum_{i=1}^{n}(\tau_i - \bar{\tau})^2}.$$

d) Number of lifetime data measured.
e) Number of lifetime data used in the calculation of the mean lifetime $\bar{\tau}$ and its standard deviation s. The difference between columns d and e gives the number of outliers.

where $I_{\parallel}(t)$ and $I_{\perp}(t)$ are the polarized components parallel and perpendicular to the direction of polarization of the incident light, and $I(t)$ is the total fluorescence intensity.

For a homogeneous population of fluorophores, i.e. for fluorophores characterized by the same decay of $I(t)$ and the same $r(t)$, the polarized components are

$$I_{\parallel}(t) = \frac{I(t)}{3}[1 + 2r(t)] \tag{6.44}$$

$$I_{\perp}(t) = \frac{I(t)}{3}[1 - r(t)] \tag{6.45}$$

In the simplest case of a single exponential decay of the total fluorescence intensity, $I(t) = I_0 \exp(-t/\tau)$, and a single rotational correlation time, i.e. $r(t) = r_0 \exp(-t/\tau_c)$, the polarized components are:

$$I_{\parallel}(t) = \frac{I_0 \exp(-t/\tau)}{3}[1 + 2r_0 \exp(-t/\tau_c)] \tag{6.46}$$

$$I_{\perp}(t) = \frac{I_0 \exp(-t/\tau)}{3}[1 - r_0 \exp(-t/\tau_c)] \tag{6.47}$$

This can easily be extended to multi-exponential decays of $I(t)$ and $r(t)$.

The emission anisotropy can be calculated from the experimental decays of the polarized components by means of Eq. (6.43), but only if the decay times are much larger than the width of the excitation pulse. Otherwise, Eq. (6.43) cannot be used because the responses $R(t)$ $(= E(t) \otimes I(t))$ and $R_{\perp}(t)$ $(= E(t) \otimes I_{\perp}(t))$ must be deconvoluted. Several methods can then be used to recover the parameters:

(i) the polarized components are analyzed by global analysis because the same decay times appear in both components (see for instance Eqs 6.46 and 6.47);
(ii) the curves $R_{\parallel}(t) - R_{\perp}(t)$ and $R_{\parallel}(t) + 2R_{\perp}(t)$ $(= R(t))$ are calculated from the experimental decays, $R_{\parallel}(t)$ and $R_{\perp}(t)$, and then analyzed by a method including deconvolution. The drawback of this method is the poor signal-to-noise ratio of the difference curve;
(iii) the polarized components are deconvoluted separately without *a priori* assumption of the shape of the decays, e.g. using the sum of a large number of exponential terms. The constructed anisotropy from the fitting parameters, now free of convolution effects, can be analyzed by curve fitting procedures. The advantage is that the constructed anisotropy does not contain the parameters of the total fluorescence intensity $I(t)$.

The case of several populations of fluorophores having their own fluorescence decay $I_i(t)$ and time constants characterizing $r_i(t)$ deserves particular attention. In Section 5.3, it was concluded that an 'apparent' or a 'technical' emission anisotropy $r(t)$ can be obtained by considering that the measured polarized components, $I(t)$ and $I_{\perp}(t)$, are the sums of the individual components (i.e. of each population) and by using Eq. (6.43). Hence

$$r(t) = \sum_i f_i(t) r_i(t) \tag{6.48}$$

where

$$f_i(t) = \frac{a_i \exp(-t/\tau_i)}{I(t)} = \frac{a_i \exp(-t/\tau_i)}{\sum_i a_i \exp(-t/\tau_i)} \quad (6.49)$$

6.2.7.2 Phase-modulation fluorometry

Instead of recording separately the decays of the two polarized components, we measure the differential polarized phase angle $\Delta(\omega) = \Phi - \Phi_\perp$ between these two components and the polarized modulation ratio $\Lambda(\omega) = m/m_\perp$. It is interesting to define the frequency-dependent anisotropy as follows:

$$r(\omega) = \frac{\Lambda(\omega) - 1}{\Lambda(\omega) + 2} \quad (6.50)$$

At low frequency, $r(\omega)$ tends towards the steady-state anisotropy, and at high frequency $r(\omega)$ approaches r_0, the emission anisotropy in the absence of rotational motions.

The following relationships are used for data analysis:

$$\Delta(\omega) = \tan^{-1}\left(\frac{Q_\parallel P_\perp - P_\parallel Q_\perp}{P_\parallel P_\perp + Q_\parallel Q_\perp}\right) \quad (6.51)$$

and

$$\Lambda(\omega) = \left(\frac{P_\parallel^2 + Q_\parallel^2}{P_\perp^2 + Q_\perp^2}\right)^{1/2} \quad (6.52)$$

where $P_\parallel$ and $Q_\parallel$ are the sine and cosine transforms of $I(t)$, and $P_\perp$ and $Q_\perp$ are the sine and cosine transforms of $I_\perp(t)$.

6.2.8 Time-resolved fluorescence spectra

Evolution of fluorescence spectra during the lifetime of the excited state can provide interesting information. Such an evolution occurs when a fluorescent compound is excited and then evolves towards a new configuration whose fluorescent decay is different. A typical example is the solvent relaxation around an excited-state compound whose dipole moment is higher in the excited state than in the ground state (see Chapter 7): the relaxation results in a gradual red-shift of the fluorescence spectrum, and information on the polarity of the microenvironment around a fluorophore is thus obtained (e.g. in biological macromolecules).

Figure 6.14 shows how the fluorescence spectra at various times are recovered from the fluorescence decays at several wavelengths. The principle of the measurements is different in pulse and in phase fluorometry.

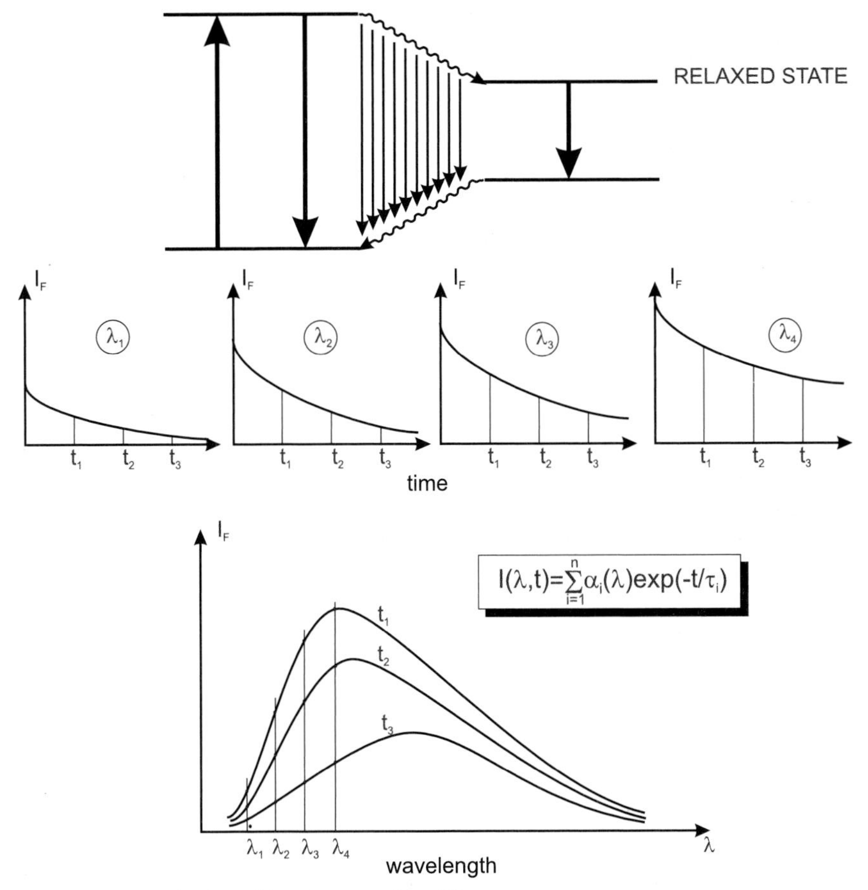

Fig. 6.14. Determination of the time evolution of fluorescence spectra.

In phase fluorometry, the phase (and modulation) data are recorded at a given wavelength and analyzed in terms of a multi-exponential decay (without *a priori* assumption of the shape of the decay). The fitting parameters are then used to calculate the fluorescence intensities at various times $t_1, t_2, t_3, \ldots$ The procedure is repeated for each observation wavelength $\lambda_1, \lambda_2, \lambda_3, \ldots$ It is then easy to reconstruct the spectra at various times.

In pulse fluorometry, we take advantage of the fact that the amplitudes of the output pulses of the TAC are proportional to the times of arrival of the fluorescence photons on the photomultiplier. Selection of a given height of pulse, i.e. of a given time of arrival, is electronically possible (by means of a single-channel analyzer) and allows us to record the fluorescence spectra at a given time t after the excitation pulse. This is repeated for various times. The method described above for phase fluorometry can also be used in pulse fluorometry.

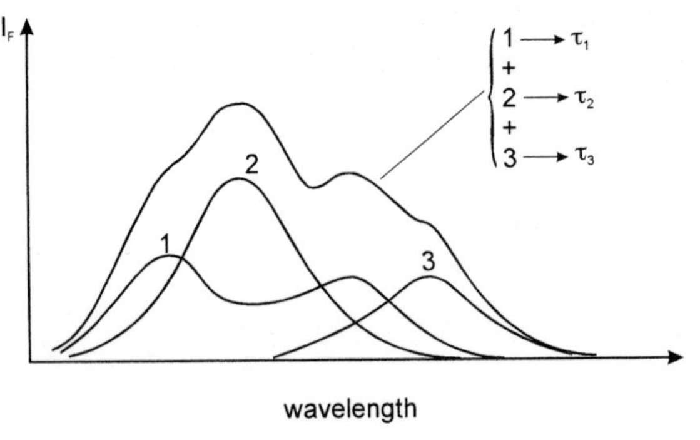

Fig. 6.15. Decomposition of a fluorescence spectrum into its components.

6.2.9
Lifetime-based decomposition of spectra

A fluorescence spectrum may result from overlapping spectra of several fluorescent species (or several forms of a fluorescent species). If each of them is characterized by a single lifetime, it is possible to decompose the overall spectrum into its components.

Let us consider for instance a spectrum consisting of three components whose lifetimes, $\tau_1, \tau_2, \tau_3, \ldots$ have been determined separately (Figure 6.15).

Decomposition of the fluorescence spectrum is possible in pulse fluorometry by analyzing the decay with a three-exponential function at each wavelength

$$I_\lambda(t) = \alpha_{1\lambda} \exp(-t/\tau_1) + \alpha_{2\lambda} \exp(-t/\tau_2) + \alpha_{3\lambda} \exp(-t/\tau_3) \tag{6.53}$$

and by calculating the fractional intensities $f_{1\lambda}$, $f_{2\lambda}$ and $f_{3\lambda}$ as follows

$$f_{i\lambda} = \frac{\alpha_{i\lambda}\tau_i}{\sum_{i=1}^{3} \alpha_{i\lambda}\tau_i} \tag{6.54}$$

The procedure in phase-modulation fluorometry is more straightforward. The sine and cosine Fourier transforms of the δ-pulse response are, according to Eqs (6.30) and (6.31), given by

$$P_\lambda = \frac{f_{1\lambda}\omega\tau_1}{1+\omega^2\tau_1^2} + \frac{f_{2\lambda}\omega\tau_2}{1+\omega^2\tau_2^2} + \frac{f_{3\lambda}\omega\tau_3}{1+\omega^2\tau_3^2} \tag{6.55}$$

$$Q_\lambda = \frac{f_{1\lambda}}{1+\omega^2\tau_1^2} + \frac{f_{2\lambda}}{1+\omega^2\tau_2^2} + \frac{f_{3\lambda}}{1+\omega^2\tau_3^2} \tag{6.56}$$

with

$$1 = f_{1\lambda} + f_{2\lambda} + f_{3\lambda} \tag{6.57}$$

Decomposition in real time is possible by measuring Φ_λ and M_λ as a function of wavelength at a single frequency and by calculating P_λ and Q_λ by means of Eqs (6.21) and (6.22)

$$P_\lambda = M_\lambda \sin \Phi_\lambda \tag{6.58}$$

$$Q_\lambda = M_\lambda \cos \Phi_\lambda \tag{6.59}$$

$f_{1\lambda}, f_{2\lambda}$ and $f_{3\lambda}$ are then solutions of the system of Eqs (6.55) to (6.57).

6.2.10
Comparison between pulse and phase fluorometries

Comparison between the two techniques will be presented from three points of view: *theoretical*, *instrumental* and *methodological*.

(i) Pulse and phase fluorometries are theoretically equivalent: they provide the same kind of information because the harmonic response is the Fourier transform of the δ-pulse response.

(ii) From the instrumental point of view, the latest generations of instruments use both pulsed lasers and microchannel plate detectors. Only the electronics are different. Because the *time resolution* is mainly limited by the time response of the detector, this parameter is the same for both techniques. Moreover, the optical module is identical so the total cost of the instruments is similar.

(iii) The *methodologies* are quite different because they are relevant to time domain and frequency domain. The advantages and drawbacks are as follows:
- Pulse fluorometry permits *visualization of the fluorescence decay*, whereas visual inspection of the variations of the phase shift versus frequency does not allow the brain to visualize the inverse Fourier transform!
- Pulse fluorometry using the single-photon timing technique has an outstanding *sensitivity*: experiments with very low levels of light (e.g. owing to low quantum yields or strong quenching) simply require longer acquisition times (but attention must be paid to the possible drift of the excitation source), whereas in phase fluorometry, the fluorescence intensity must be high enough to get an analog signal whose zero crossing (for phase measurements) and amplitude (for modulation measurements) can be measured with enough accuracy.
- No *deconvolution* is necessary in phase fluorometry, while this operation is often necessary in pulse fluorometry and requires great care in recording the instrument response, especially for very short decay times.
- The *well-defined statistics* in single-photon counting is an advantage for data analysis. In phase fluorometry, the evaluation of the standard deviation of phase shift and modulation ratio may not be easy.
- *Time-resolved emission anisotropy* measurements are more straightforward in pulse fluorometry.

- *Time-resolved spectra* are more easily recorded in pulse fluorometry.
- *Lifetime-based decomposition of spectra* into components is simpler in phase fluorometry.
- The *time of data collection* depends on the complexity of the δ-pulse response. For a single exponential decay, phase fluorometry is more rapid. For complex δ-pulse responses, the time of data collection is about the same for the two techniques: in pulse fluorometry, a large number of photon events is necessary, and in phase fluorometry, a large number of frequencies has to be selected. It should be emphasized that the short acquisition time for phase shift and modulation ratio measurements at a given frequency is a distinct advantage in several situations, especially for lifetime-imaging spectroscopy.

In conclusion, pulse and phase fluorometries each have their own advantages and drawbacks. They appear to be complementary methods and are by no means competitive.

6.3
Appendix: Elimination of polarization effects in the measurement of fluorescence intensity and lifetime

When the fluorescence is polarized, the use of polarizers with appropriate orientations allows detection of a signal proportional to the total fluorescence intensity.

The angles θ and ϕ will characterize the transmission directions of the polarizers introduced in the excitation and emission beam (Figure 6.16). Let us consider first the relations between the components of the fluorescence intensity I_x, I_y, I_z

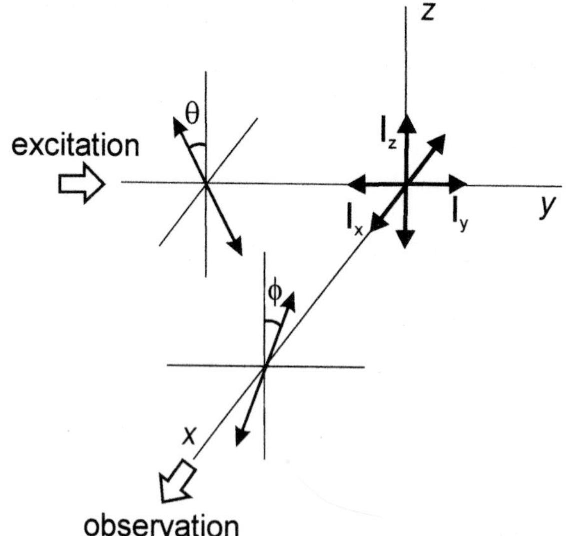

Fig. 6.16. Definition of angles θ and ϕ characterizing the transmission directions of the polarizers introduced in the excitation and emission beam.

depending on angle θ. According to symmetry considerations explained in Chapter 5, we have:

for vertical excitation ($\theta = 0°$):

$$I_z = I_{\|}$$

$$I_x = I_y = I_{\perp}$$

$$I = I_x + I_y + I_z = I_{\|} + 2I_{\perp}$$

for horizontal excitation ($\theta = 90°$):

$$I_x = I_{\|}$$

$$I_y = I_z = I_{\perp}$$

$$I = I_x + I_y + I_z = I_{\|} + 2I_{\perp}$$

Now, for a given angle θ, excitation can be considered as the superimposition of two beams, one vertically polarized with a weight of $\cos^2\theta$, and the other horizontally polarized with a weight of $\sin^2\theta$. Therefore, the intensity components are

$$I_x = I_{\|}\sin^2\theta + I_{\perp}\cos^2\theta$$

$$I_z = I_{\|}\cos^2\theta + I_{\perp}\sin^2\theta$$

$$I_y = I_{\perp}$$

$$I = I_x + I_y + I_z = I_{\|} + 2I_{\perp}$$

The total fluorescence intensity is thus again equal to $I_{\|} + 2I_{\perp}$. This equality is therefore valid whatever the value of θ.

Regarding the intensity observed in the Ox direction without a polarizer, I_x is not detected and thus $I_{obs} = I_y + I_z$. With a polarizer at an angle ϕ with respect to the vertical, the observed intensity becomes

$$I_{obs} = I_y\sin^2\phi + I_z\cos^2\phi$$

Taking into account the θ-dependent expressions for I_y and I_z, we obtain

$$\begin{aligned} I_{obs} &= I_{\perp}\sin^2\phi + (I_{\|}\cos^2\theta + I_{\perp}\sin^2\theta)\cos^2\phi \\ &= I_{\|}\cos^2\theta\cos^2\phi + I_{\perp}(\sin^2\phi + \sin^2\theta\cos^2\phi) \end{aligned}$$

Because we want this observed intensity to be proportional to the total intensity, i.e. $I_{\|} + 2I_{\perp}$, the condition to be fulfilled is:

$$2\cos^2\theta\cos^2\phi = \sin^2\phi + \sin^2\theta\cos^2\phi$$

or

$$\boxed{\cos^2\theta\cos^2\phi = 1/3}$$

The consequences of this relation are:

- if the excitation polarizer is in the vertical position ($\theta = 0$), the emission polarizer must be set at the 'magic angle' $\phi = 54.7°$ ($\cos^2\phi = 1/3$);
- if the emission polarizer is the vertical position ($\phi = 0$), the excitation polarizer must be set at the magic angle $\theta = 54.7°$ ($\cos^2\theta = 1/3$) (see Figure 6.3).

Moreover, it is easy to show that, if the emission is observed without a polarizer, an excitation polarizer must be set at $\theta = 35.3°$ ($\cos^2\theta = 2/3$). This arrangement is suitable when the fluorescence is detected through an optical filter (to reject scattering light) and not through a monochromator, because of the polarization dependence of the transmission efficiency of the latter.

6.4 Bibliography

Beechem J. M., Ameloot M. and Brand L. **(1985)** Global Analysis of Fluorescence Decay Surfaces: Excited-State Reactions, *Chem. Phys. Lett.* 120, 466–472.

Brochon J. C. **(1994)** Maximum Entropy Method of Data Analysis in Time-Resolved Spectroscopy, *Methods in Enzymology*, 240, 262–311.

Cundall R. B. and Dale R. E. (Eds) **(1983)** *Time-Resolved Fluorescence Spectroscopy in Biochemistry and Biology*, Plenum Press, New York.

Demas J. N. **(1982)** Measurement of Photon Yields, in: Mielenz K. D. (Ed.), *Measurement of Photoluminescence*, Academic Press, New York, pp. 195–247.

Demas J. N. **(1983)** *Excited-State Lifetime Measurements*, Academic Press, New York.

Demas J. N. and Crosby G. A. **(1971)** The Measurement of Photoluminescence Quantum Yields. A Review. *J. Phys. Chem.* 75, 991–1024.

Eaton D. F. **(1988)** Reference Materials for Fluorescence Measurements, *Pure Appl. Chem.* 60, 1107–1114.

Eaton D. F. **(1990)** Recommended Methods for Fluorescence Decay Analysis, *Pure Appl. Chem.* 62, 1631–1648.

Fleming G. R. **(1986)** *Chemical Applications of Ultrafast Spectroscopy*, Oxford University Press, New York.

James D. R., Siemiarczuk A. and Ware W. R. **(1992)** Stroboscopic Optical Boxcar Technique for the Determination of Fluorescence Lifetimes, *Rev. Sci. Instrum.* 63, 1710–1716.

Jameson D. M., Gratton E. and Hall R. D. **(1984)** The Measurement and Analysis of Heterogeneous Emissions by Multifrequency Phase and Modulation Fluorometry, *Appl. Spectrosc. Rev.* 20, 55–106.

Knutson J. M., Beechem J. M. and Brand L. **(1983)** Simultaneous Analysis of Multiple Fluorescence Decay Curves: A Global Approach, *Chem. Phys. Lett.* 102, 501–507.

Lakowicz J. R. (Ed.) **(1991)** *Topics in Fluorescence Spectroscopy*, Vol. 1, Techniques, Plenum Press, New York.

Mielenz K. D. (Ed.) **(1982)** *Measurement of Photoluminescence*, Academic Press, New York.

Miller J. N. (Ed.) **(1981)** *Standards for Fluorescence Spectrometry*, Chapman and Hall, London.

O'Connor D. V. and Phillips D. **(1984)** *Time-*

Correlated Single Photon Counting, Academic Press, London.

O'Connor D. V., Ware W. R. and André J. C. **(1979)** Deconvolution of Fluorescence Decay Curves. A Critical Comparison of Techniques, *J. Phys. Chem.* 83, 1333–1343.

Parker C. A. **(1968)** *Photoluminescence of Solutions*, Elsevier, Amsterdam.

Pouget J., Mugnier J. and Valeur B. **(1989)** Correction of Timing Errors in Multifrequency Phase/Modulation Fluorometry. *J. Phys. E: Sci. Instrum.* 22, 855–862.

7
Effect of polarity on fluorescence emission. Polarity probes

Newton (...) supposait que toutes les molécules lumineuses avaient deux pôles analogues à ceux d'un aimant (...). Ainsi, d'après cet homme illustre, toutes les molécules lumineuses jouissaient de la polarité, et c'est cette hypothèse qui a déterminé la dénomination de polarisation appliquée aux phénomènes lumineux.

[Newton (...) assumed that all luminous molecules had two poles analogous to those of a magnet (...). Thus, according to this illustrious man, all luminous molecules possess polarity, and it is this hypothesis which led to the designation of polarization as applied to optical phenomena.]

Encyclopédie Méthodique. Physique
Monge, Cassini, Bertholon et al., 1822

Polarity plays a major role in many physical, chemical, biochemical and biological phenomena. This chapter aims to describe how 'local polarity' can be estimated using a fluorescent probe. But what is polarity? This apparently simple question deserves some attention before describing the methodologies based on polarity-sensitive fluorescent probes.

7.1
What is polarity?

Since the pioneering work of Berthelot and Péan de Saint-Gilles in 1862, it has been well known that solvents strongly influence both reaction rates and the position of chemical equilibria. Such a solvent dependence is also observed for the spectral bands of individual species measured by various spectrometric techniques (UV–visible and infrared spectrophotometries, fluorescence spectroscopy, NMR spectrometry, etc.).

It was mentioned in Chapters 2 and 3 that broadening of the absorption and fluorescence bands results from fluctuations in the structure of the solvation shell around a solute (this effect, called *inhomogeneous broadening*, superimposes homogeneous broadening because of the existence of a continuous set of vibrational sublevels). Moreover, shifts in absorption and emission bands can be induced by a change in solvent nature or composition; these shifts, called *solvatochromic shifts*, are experimental evidence of changes in solvation energy. In other words, when a solute is surrounded by solvent molecules, its ground state and its excited state are more or less stabilized by solute–solvent interactions, depending on the chemical nature of both solute and solvent molecules. Solute–solvent interactions are commonly described in terms of Van der Waals interactions and possible specific interactions like hydrogen bonding. What is the relationship between these interactions and 'polarity'?

To answer this question, let us first consider a neutral molecule that is usually said to be 'polar' if it possesses a dipole moment (the term 'dipolar' would be more appropriate)[1]. In solution, the solute–solvent interactions result not only from the permanent dipole moments of solute or solvent molecules, but also from their polarizabilities. Let us recall that the polarizability α of a spherical molecule is defined by means of the dipole $\boldsymbol{\mu}_i = \alpha\mathbf{E}$ induced by an external electric field $\mathbf{E}$ in its own direction. Figure 7.1 shows the four major dielectric interactions (dipole–dipole, solute dipole–solvent polarizability, solute polarizability–solvent dipole, polarizability–polarizability). Analytical expressions of the corresponding energy terms can be derived within the simple model of spherical-centered dipoles in isotropically polarizable spheres (Suppan, 1990). These four *non-specific* dielectric in-

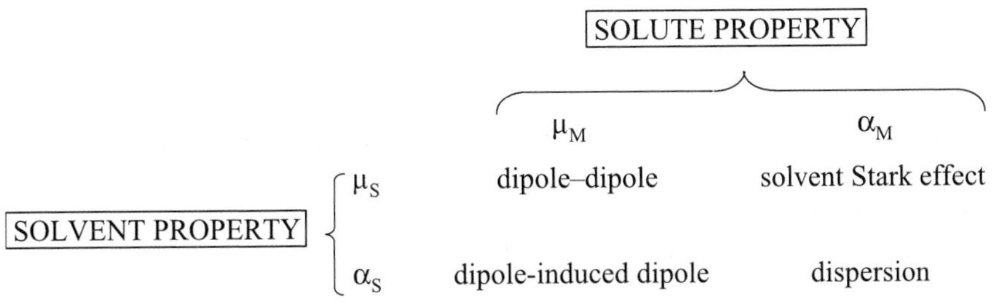

Fig. 7.1. Dielectric solute–solvent interactions resulting from the dipole moments and average polarizabilities (from Suppan, 1990).

1) The dipole moment is defined as $\boldsymbol{\mu} = q\mathbf{d}$ (it consists of two equal charges $+q$ and $-q$ separated by a distance $\mathbf{d}$). If the molecule contains several dipoles (bond dipoles), the total dipole moment is their vector sum.

teractions should be distinguished from *specific* interactions such as hydrogen bonding.

To describe solvatochromic shifts, an additional energy term relative to the solute should be considered. This term is related to the transition dipole moment that results from the migration of electric charges during an electronic transition. Note that this transient dipole has nothing to do with the difference $\mu_e - \mu_g$ between the permanent dipole moment in the excited state and that in the ground state.

After these preliminary remarks, the term 'polarity' appears to be used loosely to express the complex interplay of all types of solute–solvent interactions, i.e. non-specific dielectric solute–solvent interactions and possible specific interactions such as hydrogen bonding. Therefore, polarity cannot be characterized by a single parameter, although the 'polarity' of a solvent (or a microenvironment) is often associated with the static dielectric constant ε (macroscopic quantity) or the dipole moment μ of the solvent molecules (microscopic quantity). Such an oversimplification is unsatisfactory.

In reality, underlying the concept of polarity, solvation aspects with the relevant solvation energy should be considered. As noticed by Suppan (1990), each term of the solvation energy E_{solv} is a product of two factors expressing separately the polar characteristics P of the solute and the polar characteristics Π of the solvent ($E_{solv} = P\Pi$). Suppan has emphasized that P and Π are not simple numbers, but matrices describing the properties of the solute molecule (dipole moment, polarizability, transition moment, hydrogen bonding capability) and the solvent molecule (dielectric constant, refractive index, hydrogen bonding capability). This approach is conceptually interesting but it has not so far permitted quantitative evaluation of polarity.

7.2 Empirical scales of solvent polarity based on solvatochromic shifts

Compounds are called *solvatochromic* when the location of their absorption (and emission) spectra depend on solvent polarity. A bathochromic (red) shift and a hypsochromic (blue) shift with increasing solvent polarity pertain to *positive* and *negative solvatochromism*, respectively. Such shifts of appropriate solvatochromic compounds in solvents of various polarity can be used to construct an empirical polarity scale (Reichardt, 1988; Buncel and Rajagopal, 1990).

7.2.1 Single-parameter approach

Kosower in 1958 was the first to use solvatochromism as a probe of solvent polarity. The relevant *Z*-scale is based on the solvatochromic shift of 4-methoxycarbonyl-1-ethylpyridinium iodide (**1**). Later, Dimroth and Reichardt suggested using betain dyes, whose negative solvatochromism is exceptionally large. In particular, 2,6-

diphenyl-4-(2,4,6-triphenyl-1-pyridino)-phenolate (**2**) and its more lipophilic derivative (**3**) (soluble even in hydrocarbons) are the basis of the $E_T(30)$[2] scale.

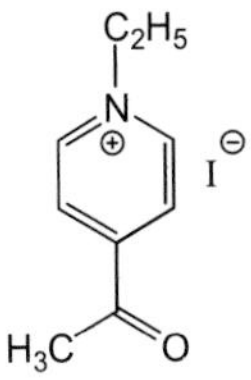

1

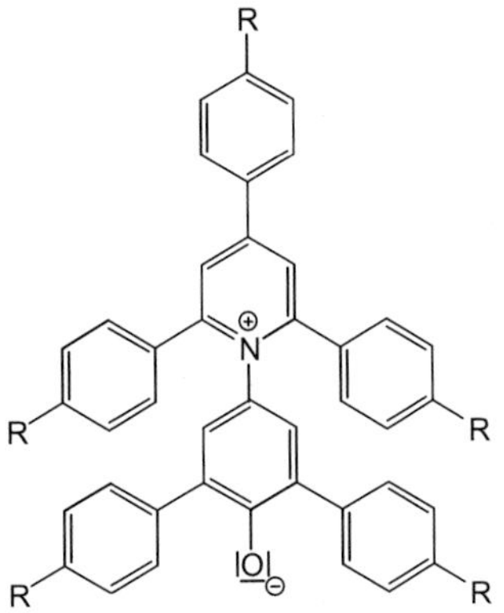

2 : R = H
3 : R = $C(CH_3)_3$

These compounds can be considered as zwitterions in the ground state. Upon excitation, electron transfer occurs from the oxygen atom to the center of the aromatic system. The dipole moment is thus about 15 D in the ground state whereas it is nearly zero in the excited state.

The $E_T(30)$ value for a solvent is simply defined as the transition energy for the longest wavelength absorption band of the dissolved pyridinium-*N*-phenoxide betain dye **2** measured in kcal mol^{-1}[3]. An extensive list of $E_T(30)$ values is available (Reichardt, 1988): they range from 30.9 in *n*-heptane to 63.1 in water.

Attention should be paid to the additional hydrogen bonding effect in protic solvents like alcohols. It has indeed been observed that correlations of solvent-dependent properties (especially positions and intensities of absorption and emission bands) with the $E_T(30)$ scale often follow two distinct lines, one for non-protic solvents and one for protic solvents.

Moreover, in many investigations using probes other than betains, the parameter $E_T(30)$ has still been used to characterize the solvent polarity. However, a polarity scale based on a particular class of molecules applies, in principle, only to probe molecules that resemble these molecules.

The sensitivity of betain dyes to solvent polarity is exceptionally high, but unfortunately they are not fluorescent. Yet polarity-sensitive fluorescent dyes offer

2) Compound **2** was the betain dye no. 30 in the original paper of Dimroth and Reichardt, which explains the notation $E_T(30)$.

3) The transition energy (in kcal mol^{-1}) is given by

$$E_T = hc\bar{\nu}N_a = 2.859 \times 10^{-3}\bar{\nu}$$

where h is Planck's constant, c is the velocity of light, $\bar{\nu}$ is the wavenumber (expressed in cm^{-1}) corresponding to the transition, and N_a is Avogadro's number.

distinct advantages, especially in biological studies. With this in mind, efforts have been focused on the design of such fluorescent probes (see Section 7.5).

It should be emphasized again that there are many parameters underlying the concept of polarity, and therefore the validity of empirical scales of polarity based on a single parameter is questionable.

7.2.2 Multi-parameter approach

A multi-parameter approach is preferable and the π^* scale of Kamlet and Taft (Kamlet et al., 1977) deserves special recognition because it has been successfully applied to the positions or intensities of maximal absorption in IR, NMR, ESR and UV–visible absorption and fluorescence spectra, and many other physical or chemical parameters (reaction rate, equilibrium constant, etc.). Such observables are denoted XYZ and their variations as a function of solvent polarity can be expressed by the generalized equation

$$\mathrm{XYZ} = \mathrm{XYZ}_0 + s\pi^* + a\alpha + b\beta \tag{7.1}$$

where π^* is a measure of the polarity/polarizability effects of the solvent; the α scale is an index of solvent HBD (hydrogen bond donor) acidity and the β scale is an index of solvent HBA (hydrogen bond acceptor) basicity. The coefficients s, a and b describe the sensitivity of a process to each of the individual contributions[4]. The advantage of the Kamlet–Taft treatment is to sort out the quantitative role of properties such as hydrogen bonding. Examples of values of π^*, α and β are given in Table 7.1.

The observable XYZ can be in particular the wavenumber $\bar{\nu}$ of an absorption or emission band and the preceding equation is rewritten as

$$\boxed{\bar{\nu} = \bar{\nu}_0 + s\pi^* + a\alpha + b\beta} \tag{7.2}$$

where $\bar{\nu}$ and $\bar{\nu}_0$ are the wavenumbers of the band maxima in the considered solvent and in the reference solvent (generally cyclohexane), respectively.

It is remarkable that the π^* scale has been established from the averaged spectral behavior of numerous solutes. It offers the distinct advantage of taking into account both non-specific and specific interactions.

A good illustration is provided by 4-amino-7-methylcoumarin (Kamlet et al., 1981), which possesses both hydrogen bond donor and acceptor groups (see

4) Equation (7.1) applies when only non-chlorinated aliphatic solvents or aromatic solvents are considered. When chlorinated and non-chlorinated aliphatic solvents and aromatic solvents need to be considered together, Eq. (7.1) must be rewritten as $\mathrm{XYZ} = \mathrm{XYZ}_0 + s(\pi^* + d\delta) + a\alpha + b\beta$ where δ is a polarizability correction term equal to 0.0 for non-chlorinated aliphatic solvents, 0.5 for polychlorinated aliphatics, and 1.0 for aromatic solvents.

Tab. 7.1. Parameters of the π^* scale of polarity (data from Kamlet et al., 1983)

Solvent	π^*	α	β
Cyclohexane	0.00	0.00	0.00
n-Hexane, *n*-heptane	−0.08	0.00	0.00
Benzene	0.59	0.10	0.00
Toluene	0.54	0.11	0.00
Dioxane	0.55	0.00	0.37
Tetrahydrofuran	0.58	0.00	0.55
Acetone	0.71	0.08	0.48
Carbon tetrachloride	0.28	0.00	0.00
1,2-Dichloroethane	0.81	0.00	0.00
Diethyl ether	0.27	0.00	0.47
Ethyl acetate	0.55	0.00	0.45
Dimethylsulfoxide	1.00	0.00	0.76
N,*N*-Dimethylformamide	0.88	0.00	0.69
Acetonitrile	0.75	0.19	0.31
Ethanol	0.54	0.83	0.77
Methanol	0.60	0.93	0.62
n-Butanol	0.47	0.79	0.88
Trifluoroethanol	0.73	1.51	0.00
Ethylene glycol	0.92	0.90	0.52
Water	1.09	1.17	0.18

Scheme 7.1, relevant to an amphiprotonic solvent). Using Eq. (7.2), the multiple linear regression equation for the fluorescence maximum (expressed in 10^3 cm^{-1}) is

$$\overline{\nu_f^{max}} = 26.71 - 2.02\pi^* - 1.58\alpha - 1.32\beta$$

with a very good correlation coefficient ($r = 0.997$).

It should be noted that 4-amino-7-methyl-coumarin undergoes photoinduced charge transfer (see Section 7.3) from the amino group to the carbonyl group. Therefore, in the excited state, the hydrogen bonds involving the more negatively charged oxygen and the more positively charged amino group are stronger than in

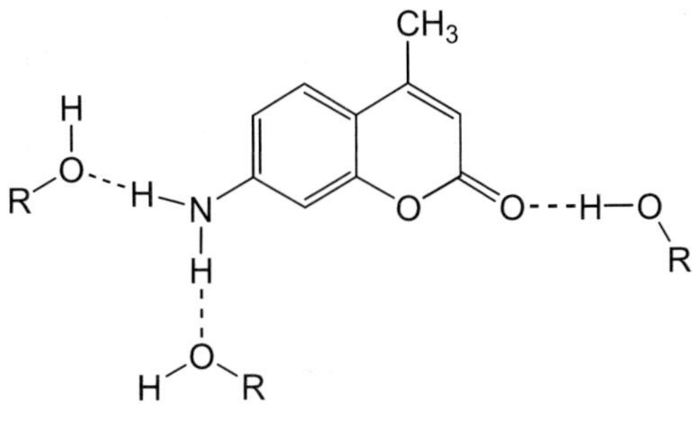

Scheme 7.1

the ground state. The resulting shifts of the fluorescence spectrum to lower energies explains the negative signs for both α and β coefficients in Eq. (7.2).

7.3
Photoinduced charge transfer (PCT) and solvent relaxation

The energy of the emitting state is often different from that of the Franck–Condon (FC) state; one of the reasons pertaining to the polarity effect is the process called *solvent relaxation*, the origin of which will be now explained. In most cases, the dipole moment of an aromatic molecule in the excited state μ_e differs from that in the ground state μ_g. In fact, absorption of a photon by a fluorophore occurs in a very short time ($\approx 10^{-15}$ s) with respect to the displacement of nuclei (Franck–Condon principle) but allows a redistribution of electrons, which results in an almost instantaneous change in the dipole moment. Most polarity probes undergo *intramolecular charge transfer* upon excitation so that $\mu_e > \mu_g$. Therefore, following excitation, the solvent cage undergoes a relaxation, i.e. a reorganization, leading to a relaxed state of minimum free energy (Figure 7.2). *The higher the polarity of the solvent, the lower the energy of the relaxed state and the larger the red-shift of the emission spectrum.*

It is important to note that the rate of solvent relaxation depends on the solvent viscosity. If the time required for the reorganization of solvent molecules around

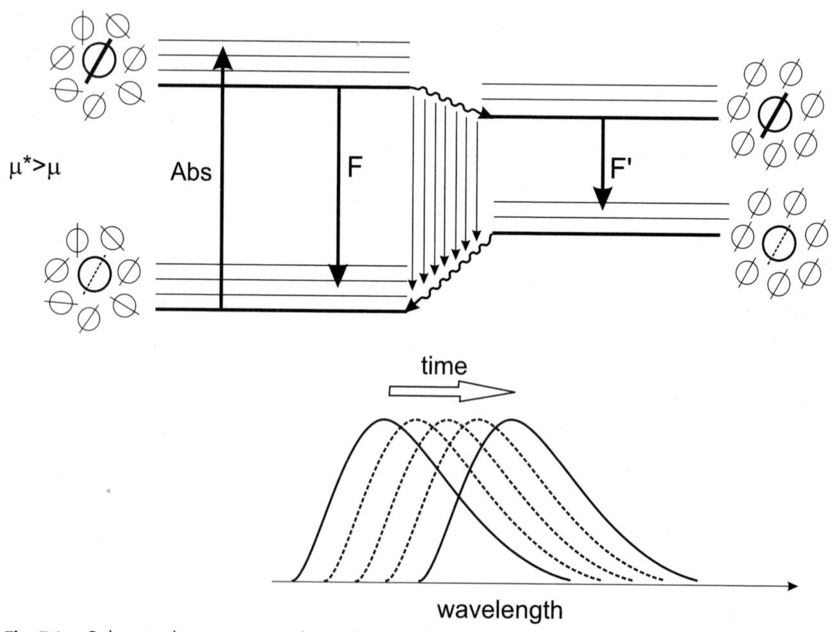

Fig. 7.2. Solvent relaxation around a probe that has a weak dipole moment in the ground state and a large dipole moment in the excited state.

the solute is short with respect to the excited-state lifetime, fluorescence will essentially be emitted from molecules in equilibrium with their solvation shell (F′ in Figure 7.2). Emission of a fluorescence photon being quasi-instantaneous, the solute recovers its ground state dipole moment and a new relaxation process leads to the most stable initial configuration of the system solute–solvent in the ground state.

In contrast, if the medium is too viscous to allow solvent molecules to reorganize, emission arises from a state close to the Franck–Condon state (FC) (as in the case of a nonpolar medium) and no shift of the fluorescence spectrum will be observed (F in Figure 7.2).

Finally, if the solvent reorganization time is of the order of the excited-state lifetime, the first emitted photons will correspond to wavelengths shorter than those emitting at longer times. In this case, the fluorescence spectrum observed under continuous illumination will be shifted but the position of the maximum cannot be directly related to the solvent polarity.

It should be recalled that, in polar rigid media, excitation on the red-edge of the absorption spectrum causes a red-shift of the fluorescence spectrum with respect to that observed on excitation in the bulk of the absorption spectrum (see the explanation of the red-edge effect in Section 3.5.1). Such a red-shift is still observable if the solvent relaxation competes with the fluorescence decay, but it disappears in fluid solutions because of dynamic equilibrium among the various solvation sites.

It is expected that, during the reorganization of solvent molecules, the time evolution of the fluorescence intensity depends on the observation wavelength, but once the equilibrium solute–solvent configuration is attained, the fluorescence decay only reflects the depopulation of the excited state. From the time-resolved fluorescence intensities recorded at various wavelengths, the fluorescence spectrum at a given time can be reconstructed, so that the time evolution of the fluorescence spectrum can be monitored during solvent relaxation. Fluorescence thus provides an outstanding tool for monitoring the response time of solvent molecules (or polar molecules of a microenvironment) following excitation of a probe molecule whose dipole moment is quasi-instantaneously changed by absorption of a photon.

The principle of the determination of time-resolved fluorescence spectra is described in Section 6.2.8. For solvent relaxation in the nanosecond time range, the single-photon timing technique can be used. The first investigation using this technique was reported by Ware and coworkers (1971). Figure 7.3 shows the reconstructed spectra of 4-aminophthalimide (4-AP) at various times after excitation. The solvent, propanol at −70 °C, is viscous enough to permit observation of solvent relaxation in a time range compatible with the instrument response (FWHM of 5 ns).

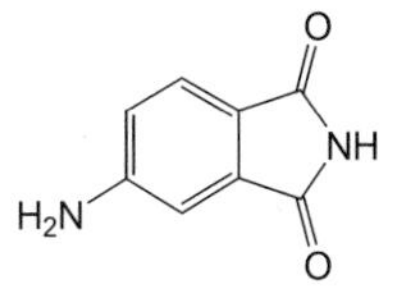

O
NH
H_2N
O

4-AP

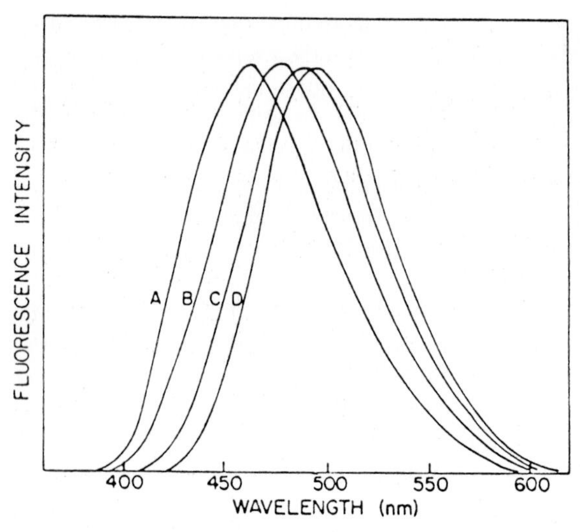

Fig. 7.3. Time-resolved fluorescence spectrum of 4-aminophthalimide at −70 °C in *n*-propanol. A: 4 ns; B: 8 ns; C: 15 ns; D: 23 ns (from Ware, 1971).

The shift of the fluorescence spectrum as a function of time reflects the reorganization of propanol molecules around the excited phthalimide molecules, whose dipole moment is 7.1 D instead of 3.5 D in the ground state (with a change in orientation of 20°). The time evolution of this shift is not strictly a single exponential.

In contrast, at room temperature, the reconstructed fluorescence spectra were found to be identical to the steady-state spectrum, which means that solvent relaxation occurs at times much shorter than 1 ns in fluid solution.

From a practical point of view, it should be emphasized that, if relaxation is not complete within the excited-state lifetime, this can lead to misinterpretation of the shift of the steady-state fluorescence spectrum in terms of polarity.

The technique of fluorescence up-conversion (see Chapter 11), allowing observations at the time-scale of picoseconds and femtoseconds, prompted a number of fundamental investigations on solvation dynamics that turned out to be quite complex (Barbara and Jarzeba, 1990) (see Box 7.1).

7.4
Theory of solvatochromic shifts

If solvent (or environment) relaxation is complete, equations for the dipole–dipole interaction solvatochromic shifts can be derived within the simple model of spherical-centered dipoles in isotropically polarizable spheres and within the assumption of equal dipole moments in Franck–Condon and relaxed states. The solvatochromic shifts (expressed in wavenumbers) are then given by Eqs (7.3) and (7.4) for absorption and emission, respectively:

Box 7.1 Solvation dynamics

To understand solvation dynamics, it is necessary to recall some aspects of dielectric relaxation in the framework of the simple continuum model, which treats the solvent as a uniform dielectric medium with exponential dielectric response.

When a constant electric field is suddenly applied to an ensemble of polar molecules, the orientation polarization increases exponentially with a time constant τ_D called the *dielectric relaxation time* or *Debye relaxation time*. The reciprocal of τ_D characterizes the rate at which the dipole moments of molecules orient themselves with respect to the electric field.

A single Debye relaxation time τ_D has been measured for a number of common liquids, called *Debye liquids*. However, for alcohols, three relaxation times ($\tau_{D1} > \tau_{D2} > \tau_{D3}$) are generally found:

- τ_{D1} is relevant to the dynamics of hydrogen bonds (formation and breaking) in aggregates of alcohol molecules;
- τ_{D2} corresponds to the rotation of single alcohol molecules;
- τ_{D3} is assigned to the rotation of the hydroxyl group around the C–O bond.

For example, for ethanol at room temperature $\tau_{D1} = 191$ ps, $\tau_{D2} = 16$ ps, $\tau_{D3} = 1.6$ ps.

Consequently, the time evolution of the center gravity (expressed in wavenumbers) of the fluorescence spectrum of a fluorophore in a polar environment should be written in the following general form:

$$\bar{\nu}(t) = \bar{\nu}(\infty) + (\bar{\nu}(0) - \bar{\nu}(\infty)) \sum_i \alpha_i \exp(-t/\tau_{S_i})$$

where $\bar{\nu}(0)$, $\bar{\nu}(t)$ and $\bar{\nu}(\infty)$ respectively are the wavenumbers of the center of gravity immediately after excitation, at a certain time t after excitation, and at a time sufficiently long to ensure the excited-state solvent configuration is at equilibrium. τ_{Si} represents the spectral relaxation times and $\sum_i \alpha_i = 1$.

From the above equation it appears convenient to characterize solvation dynamics by means of the *solvation time correlation function* $C(t)$, defined as[a]

$$C(t) = \frac{\bar{\nu}(t) - \bar{\nu}(\infty)}{\bar{\nu}(0) - \bar{\nu}(\infty)}$$

This function varies from 1 to 0 as time varies from 0 (instant of excitation) to ∞ (i.e. when the equilibrium solute–solvent interaction is attained). It is assumed that (i) the fluorescence spectrum is shifted without change in shape, (ii) there is no contribution of vibrational relaxation or changes in geometry to the shift, (iii) the possible change in excited-state dipole moment during solvent relaxation is neglected.

The correlation function $C(t)$ is purely phenomenological. Interpretation of its time evolution is often based on theory in which the *longitudinal relaxation time*, τ_L, is introduced. This time is a fraction of the Debye relaxation time:

$$\tau_L \simeq \frac{\varepsilon_\infty}{\varepsilon} \tau_D$$

where ε is the static dielectric constant (in the presence of a constant electric field) and ε_∞ is the dielectric constant in the presence of an electric field at high frequency, i.e. under conditions where the molecules do not have enough time to reorient, but deformation of the electronic cloud leads to electronic polarization. ε_∞ is often taken as the square of the index of refraction (measured in the visible spectral range).

Experiments in the picosecond time range show that $C(t)$ is non-exponential in most solvents with an average spectral relaxation time $\langle \tau_S \rangle$ greater than the longitudinal relaxation time τ_L and smaller than the Debye time τ_D.

However, picosecond resolution is insufficient to fully describe solvation dynamics. In fact, computer simulations have shown that in small-molecule solvents (e.g. acetonitrile, water, methyl chloride), the ultrafast part of solvation dynamics (< 300 fs) can be assigned to inertial motion of solvent molecules belonging to the first solvation layer, and can be described by a Gaussian function[b]. An exponential term (or a sum of exponentials) must be added to take into account the contribution of rotational and translational diffusion motions. Therefore, $C(t)$ can be written in the following form:

$$C(t) = \alpha_0 \exp(-\omega_S^2 t^2/2) + \sum_i \alpha_i \exp(-t/\tau_i)$$

where ω_S is called the solvation frequency. The $1/e$ time for the decay of a Gaussian of this form ($= 1.4\omega_S^{-1}$) represents the inertial time of the solvent.

With instruments based on fluorescence up-conversion (see Chapter 11) that offer the best time resolution (about 100 fs), such a fast inertial component was indeed detected in the fluorescence decay[c,d]. Using coumarin 153 as a solute (whose dipole moment increases from 6.5 to 15 D upon excitation), the inertial times of acetonitrile, dimethylformamide, dimethyl sulfoxide and benzonitrile were found to be 0.13, 0.20, 0.17 and 0.41 ps, respectively[d].

a) Barbara P. F. and Jarzeba W. (1990) *Adv. Photochem.* **15**, 1.

b) Maroncelli M., Kumar P. V., Papazyan A., Horng M. L., Rosenthal S. J. and Fleming G. R. (1994), in: Gauduel Y. and Rossky P. (Eds), *AIP Conference Proceedings* **298**, p. 310.

c) Rosenthal S. J., Xie X., Du M. and Fleming G. R. (1991) *J. Chem. Phys.* **95**, 4715.

d) Horng M. L., Gardecki J. A., Papazyan A. and Maroncelli M. (1995) *J. Phys. Chem.* **99**, 17311.

Tab. 7.2. Dielectric constant (at 20 °C), index of refraction (at 20 °C) and orientational polarizability Δf of some solvents $\left(\Delta f = \dfrac{\varepsilon - 1}{2\varepsilon + 1} - \dfrac{n^2 - 1}{2n^2 + 1}\right)$

Solvent	ε	n_D^{20}	Δf
Cyclohexane	2.023	1.4266	−0.001
Benzene	2.284	1.5011	0.003
Toluene	2.379	1.4961	0.013
Dioxan	2.218	1.4224	0.021
Chloroform	4.806	1.4459	0.149
Diethyl ether	4.335	1.3526	0.167
Butyl acetate	5.01	1.3941	0.171
Dichloromethane	9.08	1.4242	0.219
Dimethylsulfoxide	48.9	1.4770	0.265
N,N-Dimethylformamide	37.6	1.4305	0.275
Acetonitrile	38.8	1.3442	0.306
Ethanol	25.07	1.3611	0.290
Methanol	33.62	1.3288	0.309

$$\overline{\nu_a} = -\frac{2}{hc}\boldsymbol{\mu}_g \cdot (\boldsymbol{\mu}_e - \boldsymbol{\mu}_g)a^{-3}\Delta f + \text{const.} \tag{7.3}$$

$$\overline{\nu_f} = -\frac{2}{hc}\boldsymbol{\mu}_e \cdot (\boldsymbol{\mu}_g - \boldsymbol{\mu}_e)a^{-3}\Delta f + \text{const.} \tag{7.4}$$

where h is Planck's constant, c is the velocity of light, a is the radius of the cavity in which the solute resides and Δf is the *orientation polarizability* defined as

$$\Delta f = f(\varepsilon) - f(n^2) = \frac{\varepsilon - 1}{2\varepsilon + 1} - \frac{n^2 - 1}{2n^2 + 1} \tag{7.5}$$

Δf ranges from 0.001 in cyclohexane to 0.320 in water (Table 7.2).

Subtraction of Eqs (7.3) and (7.4) leads to the Lippert–Mataga equation:

$$\overline{\nu_a} - \overline{\nu_f} = \frac{2}{hc}(\mu_e - \mu_g)^2 a^{-3}\Delta f + \text{const.} \tag{7.6}$$

This expression of the Stokes shift depends only on the absolute magnitude of the charge transfer dipole moment $\Delta\mu_{ge} = \mu_e - \mu_g$ and not on the angle between the dipoles. The validity of Eq. (7.6) can be checked by using various solvents and by plotting $\overline{\nu_a} - \overline{\nu_f}$ as a function of Δf (Lippert's plot). A linear variation is not always observed because only the dipole–dipole interactions are taken into account and the solute polarizability is neglected[5]. By choosing solvents without hydrogen

5) In a similar approach, MacRae has included the solute polarizability, which is approximated to $a^3/2$. Equation (7.6) is still valid but Δf is replaced by

$$\Delta g = \frac{\varepsilon - 1}{\varepsilon + 2} - \frac{n^2 - 1}{n^2 + 2}.$$

bonding donor or acceptor capability, a linear behavior is often observed, which allows us to determine the increase in dipole moment $\Delta\mu_{ge}$ upon excitation, provided that a correct estimation of the cavity radius is possible. The uncertainty arises from estimation of the Onsager cavity and from the assumption of a spherical shape of this cavity. For elongated molecules, an ellipsoid form is more appropriate. In spite of these uncertainties, the Lippert–Mataga relation is widely used by taking the molecular radius as the cavity radius. Furthermore, Suppan (1983) derived another useful equation from Eqs (7.3) and (7.4) under the assumption that μ_g and μ_e are colinear but without any assumption about the cavity radius a or the form of the solvent polarity function. This equation involves the differences in absorption and emission solvatochromic shifts between two solvents 1 and 2:

$$\boxed{\frac{(\overline{\nu_f})_1 - (\overline{\nu_f})_2}{(\overline{\nu_a})_1 - (\overline{\nu_a})_2} = \frac{\mu_e}{\mu_g}} \tag{7.7}$$

Equations (7.6) and (7.7) provide a means of determining excited dipole moments together with dipole vector angles, but they are valid only if (i) the dipole moments in the FC and relaxed states are identical, (ii) the cavity radius remains unchanged upon excitation, (iii) the solvent shifts are measured in solvents of the same refractive index but of different dielectric constants.

When the emissive state is a charge transfer state that is not attainable by direct excitation (e.g. which results from electron transfer in a donor–bridge–acceptor molecule; see example at the end of the next section), the theories described above cannot be applied because the absorption spectrum of the charge transfer state is not known. Weller's theory for exciplexes is then more appropriate and only deals with the shift of the fluorescence spectrum, which is given by

$$\overline{\nu_f} = -\frac{2}{hc}\mu_e^2 a^{-3}\Delta f' + \text{const.} \tag{7.8}$$

where

$$\Delta f' = \frac{\varepsilon - 1}{2\varepsilon + 1} - \frac{n^2 - 1}{4n^2 + 2} \tag{7.9}$$

When a twisted intramolecular charge transfer (TICT) state is formed in the excited state (see Section 3.4.4 for the description of TICT formation), the molecular geometry and the charge distribution must be taken into account. Neglecting the ground state dipole moment with respect to that of the excited state and approximating the solute polarizability to $a^3/2$, the following formula is obtained (Rettig, 1982):

$$\overline{\nu_f} = -\frac{2}{hc}\mu_e^2 a^{-3}\Delta f'' + \text{const.} \tag{7.10}$$

Tab. 7.3. Ground state and excited state dipole moments of some fluorophores determined from solvatochromic shifts (except data from reference 6)

Molecule	μ_g *(D)*	μ_e *(D)*	$\Delta\mu$ *(D)*	*Ref.*
4-Aminophthalimide	3.5	7	3.5	1
4-Amino-9-fluorenone	4.9	12	7.1	2
DMANS[a)]	6.5	24	17.5	2
DMABN[b)]	5.5	10 (LE)	4.5	3
		20 (TICT)	14.5	
Coumarin 153	6.5	15	8.5	4
PRODAN[c)]	4.7	11.7	7	5
	5.2	9.6–10.2	4.4–5	6
DCM[d)]	6	26	20	7

a) 4-*N*,*N*-dimethylamino-4′-nitrostilbene
b) 4-*N*,*N*-dimethylaminobenzonitrile
c) 6-propionyl-2-(dimethylaminonaphthalene)
d) 4-dicyanomethylene-2-methyl-6-*p*-dimethylamino-styryl-4H-pyran
1) Suppan P. (1987) *J. Chem. Soc., Faraday Trans. 1* **83, 495.**
2) Hagan T., Pilloud D. and Suppan P. (1987) *Chem. Phys. Lett.* **139, 499.**
3) Suppan P. (1985) *J. Lumin.* **33, 29.**
4) Baumann W. and Nagy Z. (1993) *Pure Appl. Chem.* **65, 1729.**
5) Catalan J., Perez P., Laynez J. and Blanco F. G. (1991) *J. Fluorescence* **1, 215.**
6) Samanta A. and Fessenden R. W. (2000) *J. Phys. Chem. A* **104, 8972.**
7) Meyer M. and Mialocq J. C. (1985) *Opt. Commun.* **64, 264.**

where

$$\Delta f'' = \frac{\varepsilon - 1}{\varepsilon + 2} - \frac{n^2 - 1}{2n^2 + 4} \tag{7.11}$$

The assumptions made in theories of solvatochromic shifts, together with the uncertainty over the size and shape of the cavity radius, explain why the determination of excited-state dipole moments is not accurate. Examples of values of excited state dipole moments are given in Table 7.3.

7.5
Examples of PCT fluorescent probes for polarity

One of the most well known polarity probes is ANS (1-anilino-8-naphthalene sulfonate), discovered by Weber and Lawrence in 1954. It exhibits the interesting feature of being non-fluorescent in aqueous solutions and highly fluorescent in solvents of low polarity. This feature allows us to visualize only hydrophobic regions of biological systems without interference from non-fluorescent ANS molecules remaining in the surrounding aqueous environment.

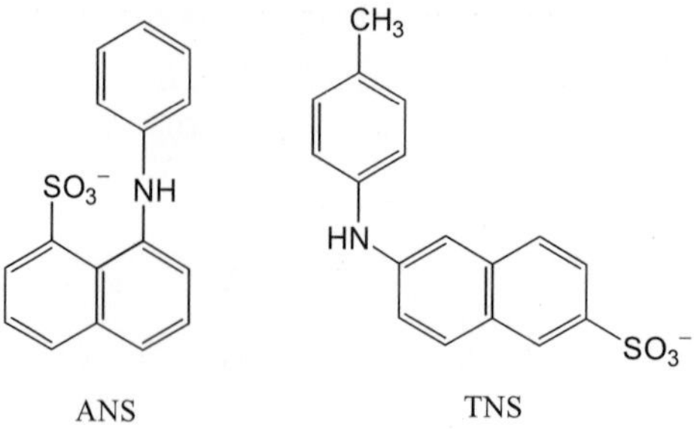

ANS TNS

ANS and its derivative TNS (*p*-toluidinyl-6-naphthalene sulfonate) have been extensively used for probing proteins, biological membranes and micellar systems. The reasons why these probes undergo large variations in fluorescence with solvent polarity have been the object of numerous investigations focusing on intramolecular charge transfer, specific solvent–solute interactions (see Section 7.6), change in molecular conformation, intersystem crossing to the triplet state, and monophotonic photoionization. Several of these processes may be competitive and it is hard to determine which one is predominant.

An ideal polarity probe based on photoinduced charge transfer and solvent relaxation should (i) undergo a large change in dipole moment upon excitation but without change in direction, (ii) bear no permanent charge in order to avoid contributions from ionic interactions, (iii) be soluble in solvents of various polarity, from the apolar solvents to the most polar ones.

In contrast to ANS, PRODAN [6-propionyl-2-(dimethylaminonaphthalene)] (see Figure 7.4), designed by Weber and Farris (1979), does not bear a charge and fulfills most of the requirements described above. Figure 7.5 shows the fluorescence spectra in various solvents. The increase in the Stokes shift is very large: from 4300 cm^{-1} in cyclohexane to 8640 cm^{-1} in water. Such an outstanding sensitivity to polarity explains the success of PRODAN in probing the polarity of various chemical and biological systems (protein binding sites, phospholipid membranes, etc.). The origin of this sensitivity was initially assigned to a large increase in the dipole moment upon excitation, but this explanation was found not to be true. The liter-

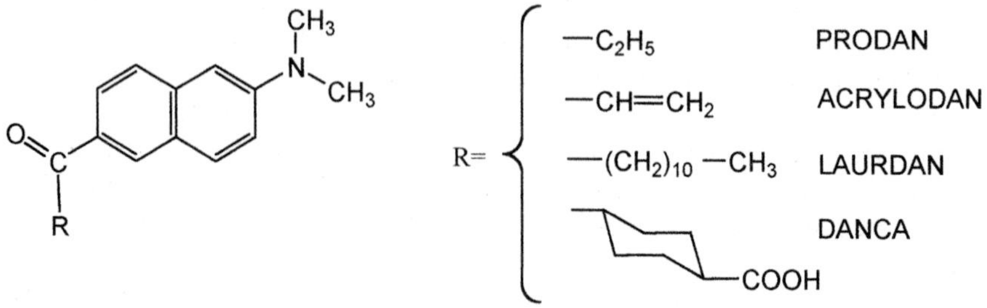

R= $-C_2H_5$ PRODAN; $-CH{=}CH_2$ ACRYLODAN; $-(CH_2)_{10}-CH_3$ LAURDAN; (cyclohexyl)–COOH DANCA

Fig. 7.4. Formulae of PRODAN and its derivatives.

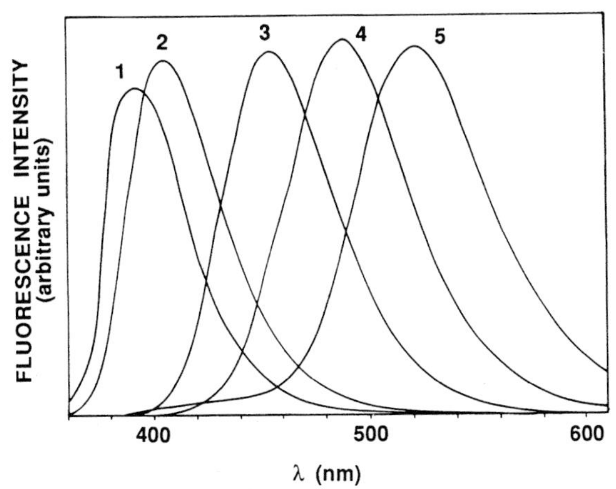

Fig. 7.5. Uncorrected fluorescence spectra of PRODAN in cyclohexane (1), chlorobenzene (2), dimethylformamide (3), ethanol (4) and water (5) (redrawn from Weber and Farris, 1979).

ature is quite confusing in regard to the photoinduced processes occurring in this molecule (see Box 7.2).

Various derivatives of PRODAN have been used. Their formulae are shown in Figure 7.4. DANCA [4-2′-(dimethylamino)-6′-naphthoylcyclohexanecarboxylic acid] was used to determine the polarity of the myoglobin haem pocket (Macgregor and Weber, 1986). DANCA indeed has a higher affinity for apomyoglobin than PRODAN. It was concluded that the pocket is actually a polar environment, and the polarity can be accounted for by peptide amide dipoles. However, as explained in Box 7.2, specific interactions with N–H groups of the protein binding site may also account for the observations.

Another chemical variant of PRODAN is ACRYLODAN [6-acryloyl-2-(dimethylamino)naphthalene], which covalently binds to protein–SH groups.

Replacement of the ethyl group of PRODAN by a C_{11} paraffinic chain yields LAURDAN, which is well suited to the study of phospholipid vesicles (Parasassi et al., 1998). Experiments carried out with LAURDAN and PRODAN in bilayers with different polar head composition and charge, i.e. different phospholipids at pH values between 4 and 10, showed that the spectral shifts do not depend on the polar head residue and on its charge, but only depend on the phase state of the bilayer. Therefore, the presence of a few water molecules in the bilayer at the level of the glycerol backbone is likely to be responsible for the microenvironmental relaxation around the probe. But this relaxation occurs only when the bilayer is in the liquid crystalline phase, and increasing the temperature causes gradual red-shift of the emission spectrum as a result of the increased concentration of water in the bilayer and to the increased molecular mobility. LAURDAN and PRODAN have also been used in studies of polarity in natural membranes, and fluorescence microscopy allows spatial resolution of the polarity microheterogeneity of these natural membranes.

TICT probes can be used as polarity probes (Rettig and Lapouyade, 1994). The classical TICT compound exhibiting dual fluorescence, DMABN (see Section 3.4.4),

Box 7.2 PRODAN – an ideal polarity probe?

The first point to be addressed is the increase in dipole moment, $\Delta\mu$, on excitation of PRODAN (Formula in Figure 7.4). The determinations of $\Delta\mu$ reported in the literature, apart from one, are based on solvatochromic shifts analyzed with the Lippert–Mataga equation. In the original paper by Weber and Farris[a] $\Delta\mu$ was estimated to be ~20 D, but this value was later recognized to be overestimated and recalculation led to a value of 8 D[b]. Another study yielded a consistent value of 7 D[c]. A completely different method based on transient dielectric loss measurement provided a somewhat lower value: 4.4–5.0 D[d]. From all these results, it can be concluded that the increase in dipole moment on excitation is not responsible for the high sensitivity of PRODAN to solvent polarity.

The second point concerns the nature of the emitting state of PRODAN, and the question arises as to whether the emitting state of PRODAN is or is not a TICT (twisted intramolecular charge transfer) state involving rotation of the dimethylamino group and/or the propanoyl group. From semi-empirical calculations based on various methods, controversial conclusions have been drawn. Nowak et al.[e] came to the conclusion that the TICT state could be the lowest state in polar media, whereas Ilich and Prendergast[f] found that specific electrostatic interactions were necessary for lowering the TICT state under the locally excited state. Parusel carried out calculations by several methods: (i) a gas phase *ab initio* calculation indicated a low-lying TICT state[g]; (ii) other calculations suggested that the emission originates from a planar intramolecular charge transfer state and a TICT state resulting from the rotation of the dimethylamino group[h]; (iii) the same author concluded that emission arises from a TICT state involving rotation of the propanoyl group for which the rotational barrier is significantly reduced[i]. These controversial conclusions exemplify the general difficulty (as already outlined) in assigning a TICT character to an emitting state.

The moderate increase in dipole moment on excitation (5–7 D) is not in favor of emission from a highly polar TICT state. At any rate, whatever the exact nature of the emitting state, such a modest value of $\Delta\mu$ cannot explain the outstanding solvatochromism of PRODAN. Specific interactions such as hydrogen bonding are then likely to play a role. In fact, PRODAN contains hydrogen bond acceptor groups (carbonyl and tertiary amine groups) and care should be taken when interpreting data from solvatochromic shifts of PRODAN in microenvironments containing hydrogen bond donor groups[d].

a) Weber G. and Farris F. J. (1979) *Biochemistry* **18**, 3075.
b) Balter A., Nowak W., Pavelkiewich W. and Kowalczyk A. (1988) *Chem. Phys. Lett.* **143**, 565.
c) Catalan J., Perez P., Laynez J. and Blanco F. G. (1991) *J. Fluorescence* **1**, 215.
d) Samanta A. and Fessenden R. W. (2000) *J. Phys. Chem. A* **104**, 8972.
e) Nowak W., Adamczak P., Balterm A. and Sygula A. (1986) *J. Mol. Struct. (THEOCHEM)* **139**, 13.
f) Ilich P. and Prendergast F. G. (1989) *J. Phys. Chem. A* **93**, 4441.
g) Parusel A., Köhler G. J. and Schneider F. W. (1997) *J. Mol. Struct. (THEOCHEM)* **398–399**, 341.
h) Parusel A., Nowak W., Grimme S. and Köhler G. J. (1998) *J. Phys. Chem. A* **102**, 7149.
i) Parusel A. (1998) *J. Chem. Soc. Faraday Trans.* **94**, 2923.

is of particular interest. The long-wavelength band (corresponding to emission from the TICT state) is red-shifted when the solvent polarity increases, with a concomitant increase in the ratio of the two bands, which offers the possibility of ratiometric measurements. For instance, DMABN has been used for probing the polarity inside cyclodextrin cavities and micelles. PIPBN and DMANCN exhibiting dual fluorescence can also be used as polarity probes.

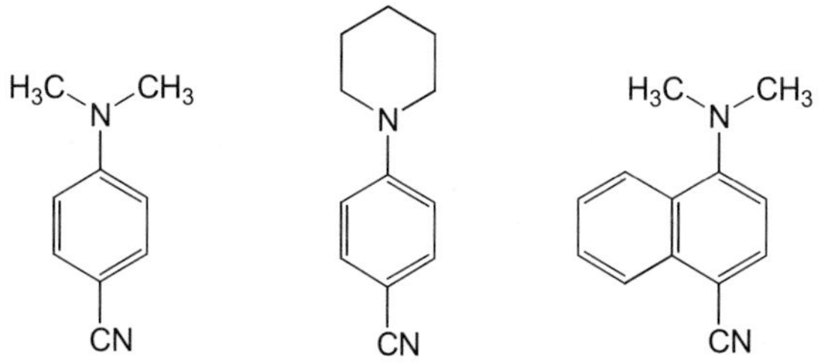

H_3C N CH_3 CN — DMABN

N CN — PIPBN

H_3C N CH_3 CN — DMANCN

Some bichromophoric systems, whose structure is based on the donor–bridge–acceptor principle, can undergo complete charge transfer, i.e. electron transfer. The resulting huge dipole moment in the excited state explains the very high sensitivity to solvent polarity of such molecules. An example is FP (1-phenyl-4-[(4-cyano-1-naphthyl)methylene]piperidine) (Hermant et al., 1990) in which photoinduced electron transfer occurs from the anilino group (donor) to the cyanonaphthalene moiety (acceptor).

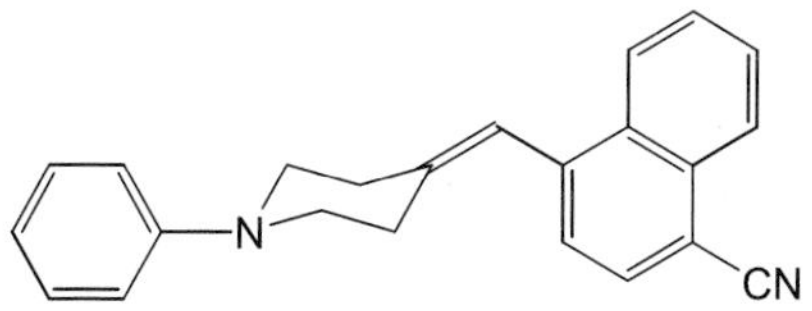

N CN

FP

The dipole moment in the excited state was estimated (by means of Eqs 7.8 and 7.9) to be 31.8 D. The fluorescence maximum is located at 407 nm in *n*-hexane and 697 nm in acetonitrile. Unfortunately, protic solvents cause complete quenching; therefore, this family of molecules cannot be used as polarity probes in protic microenvironments.

7.6 Effects of specific interactions

Several examples described above have shown that specific interactions such as hydrogen bonding interactions should be considered as one of the various aspects of polarity. This important point deserves further discussion because hydrogen bonding can lead in some cases to dramatic changes in absorption or fluorescence spectra.

7.6.1 Effects of hydrogen bonding on absorption and fluorescence spectra

In the case of $n \rightarrow \pi^*$ transitions, the electronic density on a heteroatom like nitrogen decreases upon excitation. This results in a decrease in the capability of this heteroatom to form hydrogen bonds. The effect on absorption should then be similar to that resulting from a decrease in dipole moment upon excitation, and a blue-shift of the absorption spectrum is expected; the higher the strength of hydrogen bonding, the larger the shift. This criterion is convenient for assigning a n–π^* band. The spectral shift can be used to determine the energy of the hydrogen bond.

It is easy to predict that the fluorescence emitted from a singlet state n–π^* will be always less sensitive to the ability of the solvent to form hydrogen bonds than absorption. In fact, if $n \rightarrow \pi^*$ excitation of a heterocycle containing nitrogen (e.g. in solution in methanol) causes hydrogen bond breaking (e.g. N...$HOCH_3$), the fluorescence spectrum will only be slightly affected by the ability of the solvent to form hydrogen bonds because emission arises from an n–π^* state without hydrogen bonds.

In the case of $\pi \rightarrow \pi^*$ transitions, it is often observed that the heteroatom of a heterocyle (e.g. N) is more basic in the excited state than in the ground state. The resulting excited molecule can thus be hydrogen bonded more strongly than the ground state. $\pi^* \rightarrow \pi$ fluorescence is thus more sensitive to hydrogen bonding than $\pi \rightarrow \pi^*$ absorption.

7.6.2 Examples of the effects of specific interactions

A remarkable observation illustrating the dramatic effect of specific interactions is the following. Addition of 0.2% of ethanol in cyclohexane causes a significant change in the fluorescence spectrum of 2-anilinonaphthalene, and for a 3% ethanol content, the spectrum is only slightly different from that observed in pure ethanol. Specific interaction via preferential solvation is undoubtedly responsible for these effects (Brand et al., 1971). This observation shows that a very small amount of a hydrogen bonding solvent may be sufficient to significantly change the fluorescence spectrum of a fluorophore, although the macroscopic properties of the solvent (dielectric constant, refractive index) are not significantly affected.

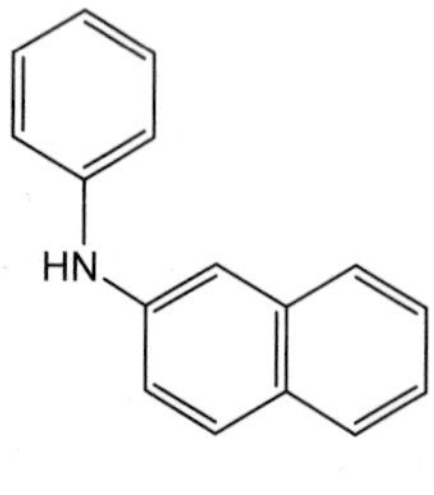

HN

2-anilinonaphthalene

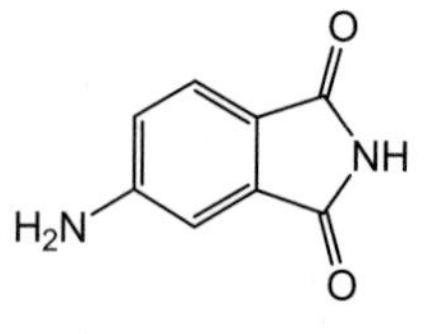

O NH H_2N O

4-AP

Another interesting example is 4-aminophthalimide (4-AP) (whose time-resolved spectra during solvent relaxation are described in Section 7.2.1). As a result of the increase in the dipole moment upon excitation of about 4 D, we cannot expect a large shift in the fluorescence spectrum with polarity. In fact, when going from nonpolar diethylether ($\varepsilon = 4.2$) to acetonitrile ($\varepsilon = 35.9$), a shift of only 33 nm of the fluorescence spectrum is observed, whereas the shift is much larger in hydrogen bond-donating solvents such as alcohols and water (e.g. 93 nm in methanol). Specific interaction is thus undoubtedly responsible for such a dramatic shift (Saroja et al., 1998). The fluorescence quantum yield of 4-AP is high in nonpolar or polar aprotic solvents but very low in protic solvents (0.1 in methanol and 0.01 in water) because hydrogen bonding presumably enhances intersystem crossing. In accordance with these changes in fluorescence quantum yields, the excited-state lifetime is much longer in aprotic solvents (14–15 ns) than in protic solvents ($\approx$ 1 ns in water).

4-AP has been used to probe micellar media (Saroja et al., 1998). The probe is located at the micellar interface and is well suited to monitoring micellar aggregation. In fact, the sharp change in the fluorescence intensity versus surfactant concentration allows the critical micellar concentration (CMC) to be determined. Excellent agreement with the literature values was found for anionic, cationic and nonionic surfactants. The electroneutrality of 4-AP and its small size are distinct advantages over ionic probes like ANS or TNS.

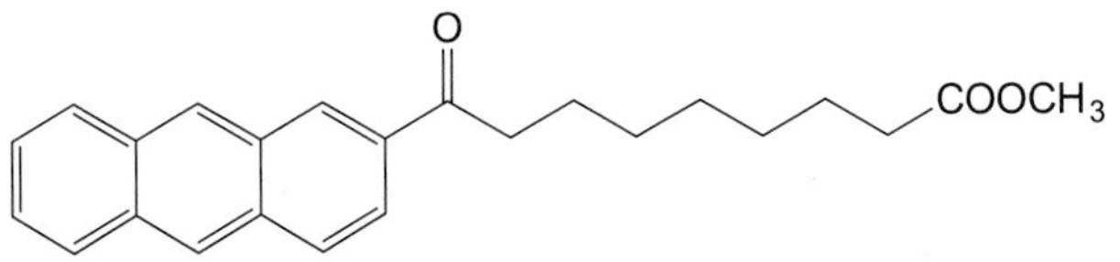

methyl 8-(2-anthroyl)-octanoate

A very marked effect of specific interactions can also be observed with anthroyl derivatives and in particular with methyl 8-(2-anthroyl)-octanoate. The fluorescence spectrum of this compound in hexane exhibits a clear vibrational structure, whereas in *N,N*-dimethylformamide and ethanol, the loss of vibrational structure is accompanied by a pronounced red-shift (Figure 7.6) (Pérochon et al., 1991).

When Stokes shifts are plotted as a function of the orientation polarizability Δf (Lippert's plot, see Section 7.2.2), solvents are distributed in a rather complex manner. A linear relationship is found only in the case of aprotic solvents of relatively low polarity. The very large Stokes shifts observed in protic solvents (methanol, ethanol, water) are related to their ability to form hydrogen bonds.

The 2-anthroyl fluorophore can be incorporated synthetically in phosphatidylcholine vesicles (anthroyl-PC), which provides an elegant tool for investigating the bilayers of egg phosphatidylcholine vesicles (Pérochon et al., 1992).

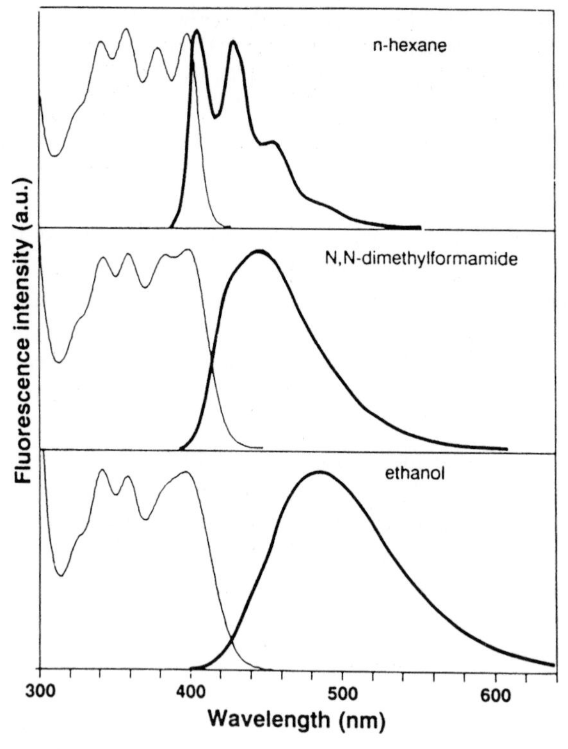

Fig. 7.6. Normalized fluorescence excitation spectra and fluorescence (bold lines) spectra of methyl 8-(2-anthroyl)-octanoate in *n*-hexane, *N,N*-dimethylformamide and ethanol at 20 °C. The maximum of emission is located at 430, 450 and 485 nm, respectively (excitation wavelength 360 nm) (reproduced with permission from Pérochon et al., 1991).

anthroyl-PC

The anthroyl fluorophore is located deep in the hydrophobic region of the lipid bilayer corresponding to the C_9–C_{16} segment of the acyl chains. The excited-state lifetime, associated with a non-structured fluorescence spectrum with a maximum at 460 nm (to be compared to those shown in Figure 7.6), can be accounted for by interaction of the fluorophore with water molecules that diffuse across the bilayer. Information is thus obtained on the permeability of lipid bilayers to water and its modulation by cholesterol.

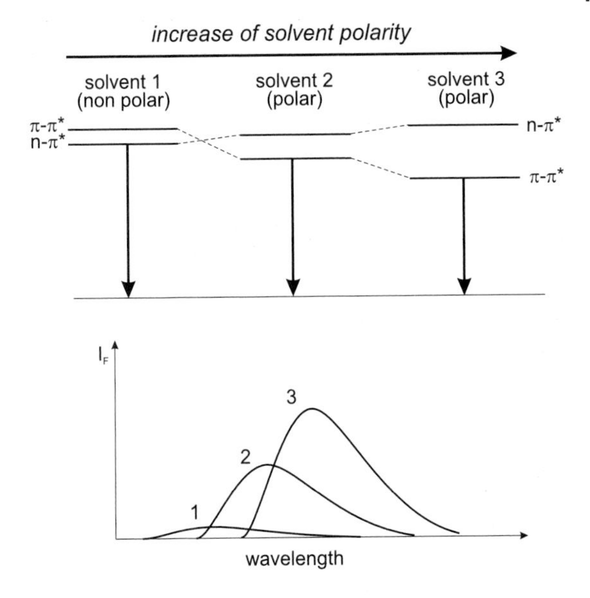

Fig. 7.7. The effects of polarity-induced inversion of n–π^* and π–π^* states.

7.6.3
Polarity-induced inversion of n–π^* and π–π^* states

A change in the ability of a solvent to form hydrogen bonds can affect the nature (n–π^* vs π–π^*) of the lowest singlet state. It was explained in Section 3.4.2.3 that some aromatic carbonyl compounds often have low-lying, closely spaced π–π^* and n–π^* states. Inversion of these two states can be observed when the polarity and the hydrogen-bonding power of the solvent increases, because the n–π^* state shifts to higher energy whereas the π–π^* state shifts to lower energy. This results in an increase in fluorescence quantum yield because radiative emission from n–π^* states is known to be less efficient than from π–π^* states (see Section 3.4.1). The other consequence is a red-shift of the fluorescence spectrum (Figure 7.7).

Pyrene-1-carboxaldehyde and 7-alkoxycoumarins belong to this class of polarity probes.

CHO

RO O O

pyrenecarboxaldehyde 7-alkoxycoumarins

The fluorescence quantum yield of pyrenecarboxaldehyde is very low in nonpolar solvents such as *n*-hexane (< 0.001) but it drastically increases in polar solvents

(0.15 in methanol). In nonpolar solvents, fluorescence is emitted from an n–π^* state. When the solvent polarity increases, the π–π^* state that lies slightly above the n–π^* state is brought below the n–π^* state by solvent relaxation during the lifetime of the excited state, and thus becomes the emitting state.

The fluorescence maximum of pyrenecarboxaldehyde varies linearly with the solvent dielectric constant over a broad range (10 to 80) (Kalyanasundaran and Thomas, 1977a). This molecule has been used for probing polarity at the micelle–water interface where it is located owing to the presence of polar carbonyl groups. For a given hydrocarbon chain length, the surface polarity depends on the nature of the head group, as expected: it increases on going from anionic to cationic to non-ionic micelles. Estimates of the polarity at the micelle–water interfaces were found to be in excellent agreement with ζ potential values for the micellar Stern layer, derived from the double layer theory.

7.7 Polarity-induced changes in vibronic bands. The Py scale of polarity

In some aromatic molecules that have a high degree of symmetry, i.e. with a minimum D_{2h} symmetry (e.g. benzene, triphenylene, naphthalene, pyrene, coronene), the first singlet absorption ($S_0 \rightarrow S_1$) may be symmetry forbidden[6] and the corresponding oscillator strength is weak. The intensities of the various forbidden vibronic bands are highly sensitive to solvent polarity (*Ham effect*). In polar solvents, the intensity of the 0–0 band increases at the expense of the others.

The relative changes in intensity of the vibronic bands in the pyrene fluorescence spectrum has its origin in the extent of vibronic coupling between the weakly allowed first excited state and the strongly allowed second excited state. Dipole-induced dipole interactions between the solvent and pyrene play a major role. The polarity of the solvent determines the extent to which an induced dipole moment is formed by vibrational distortions of the nuclear coordinates of pyrene (Karpovich and Blanchard, 1995).

The changes in the fluorescence spectrum of pyrene in solvents of different polarities (Figure 7.8) show that the polarity of an environment can be estimated by measuring the ratio of the fluorescence intensities of the third and first vibronic bands, I_I/I_{III} (Kalyanasundaram and Thomas, 1977b; Dong and Winnik, 1984). This ratio ranges from $\approx$0.6 in hydrocarbon media to $\approx$2 in dimethylsulfoxide (see Table 7.4). These values provide a polarity scale called the *Py scale*. When dividing solvents by class (aprotic aliphatics, protic aliphatics, aprotic aromatics), each class gives an excellent correlation between the Py scale and the π^* scale. The Py scale appears to be relatively insensitive to the hydrogen bonding ability of protic solvents.

6) This rule is not general and, in particular, it does not apply to anthracene and perylene whose symmetry is also D_{2h}.

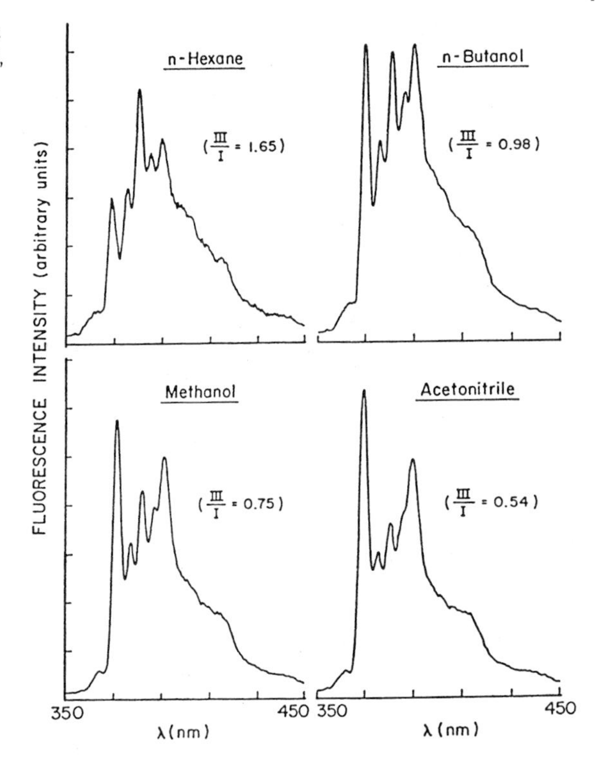

Fig. 7.8. Fluorescence spectra of pyrene in hexane, *n*-butanol, methanol and acetonitrile showing the polarity dependence of vibronic band intensities (excitation wavelength 310 nm) (reproduced with permission from Kalyanasundaran and Thomas, 1977b).

Tab. 7.4. Solvent dependence of the ratio I_I/I_{III} of the fluorescence intensities of the first and third vibronic bands in the fluorescence spectrum of pyrene.

Solvent	*Dielectric constant*	*Py scale* I_I/I_{III} [a)]
n-Hexane	1.89	0.58
Cyclohexane	2.02	0.58
Benzene	2.28	1.05
Toluene	2.38	1.04
Chloroform	4.80	1.25
Ethyl acetate	6.02	1.37
Acetic acid (glacial)	6.15	1.37
Tetrahydrofuran	7.6	1.35
Benzyl alcohol	13.1	1.24
Ethanol	25.1	1.18
Methanol	33.6	1.35
N,*N*-Dimethylformamide	36.7	1.81
Acetonitrile	37.5	1.79
Dimethyl sulfoxide	46.7	1.95

a) Data from Dong and Winnik, 1984.

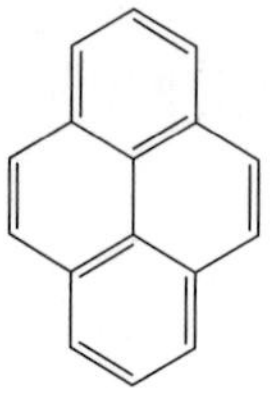

pyrene (Py) 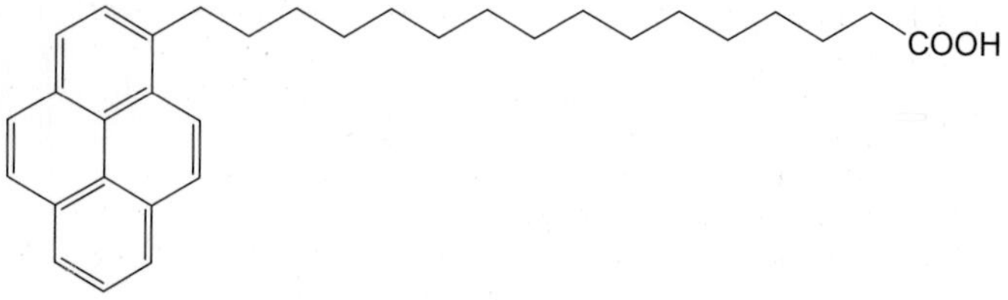 1-pyrenehexadecanoic acid

Pyrene has been used to investigate the extent of water penetration into micelles and to accurately determine critical micellar concentrations (Kalyanasundaram, 1987). Polarity studies of silica or alumina surfaces have also been reported. In lipid vesicles, measurement of the ratio I_I/I_{III} provides a simple tool for determination of phase transition temperatures and also the effect of cholesterol addition.

To investigate the extent of water penetration into the hydrocarbon region of lipid bilayers, pyrene linked to a carboxylic group via a paraffinic chain (e.g. 1-pyrenehexadecanoic acid) is best suited (L'Heureux and Fragata, 1987).

7.8 Conclusion

The concept of polarity covers all types of solute–solvent interactions (including hydrogen bonding). Therefore, polarity cannot be characterized by a single parameter. Erroneous interpretation may arise from misunderstandings of basic phenomena. For example, a polarity-dependent probe does not unequivocally indicate a hydrophobic environment whenever a blue-shift of the fluorescence spectrum is observed. It should be emphasized again that solvent (or microenvironment) relaxation should be completed during the lifetime of the excited state for a correct interpretation of the shift in the fluorescence spectrum in terms of polarity.

Attention should be paid to specific interactions, which should be taken into account in the interpretation of spectral shifts in relation to the polarity of a medium. Drastic changes in the fluorescence spectrum may indeed be induced by hydrogen bonding.

In contrast, the Py scale, based on the relative intensities of vibronic bands of pyrene, appears to be relatively insensitive to hydrogen bonding ability of solvents.

7.9 Bibliography

BARBARA P. F. and JARZEBA W. **(1990)** Ultrafast Photochemical Intramolecular Charge and Excited-Solvation, *Adv. Photochem.* 15, 1–68.

BRAND L., SELISKAR C. J. and TURNER D. C. **(1971)** The Effects of Chemical Environment on Fluorescence Probes, in: CHANCE B., LEE C. P. and BLAISIE J.-K. (Eds), *Probes of Structure and Function of Macromolecules and Membranes*, Academic Press, New York, pp. 17–39.

BUNCEL E. and RAJAGOPAL S. **(1990)** Solvatochromism and Solvent Polarity Scales, *Acc. Chem. Res.* 23, 226–231.

DONG D. C. and WINNIK M. A. **(1984)** The Py

Scale of Solvent Polarities, *Can. J. Chem.* 62, 2560–2565.

HERMANT R. M., BAKKER N. A. C., SCHERER T., KRIJNEN B. and VERHOEVEN J. W. **(1990)** Systematic Studies of a Series of Highly Fluorescent Rod-Shaped Donor–Acceptor Systems, *J. Am. Chem. Soc.* 112, 1214–1221.

KALYANASUNDARAN K. **(1987)** Photochemistry in Microheteregeneous Systems, Academic Press, New York, Chap. 2.

KALYANASUNDARAN K. and THOMAS J. K. **(1977a)** Solvent-Dependent Fluorescence of Pyrene-3-Carboxaldehyde and its Applications in the Estimation of Polarity at Micelle–Water Interfaces, *J. Phys. Chem.* 81, 2176–2180.

KALYANASUNDARAN K. and THOMAS J. K. **(1977b)** Environmental Effects on Vibronic Band Intensities in Pyrene Monomer Fluorescence and their Application in Studies of Micellar Systems, *J. Am. Chem. Soc.* 99, 2039–2044.

KAMLET M. J., ABBOUD J.-L. and TAFT R. W. **(1977)** The Solvatochromic Comparison Method. 6. The π^* Scale of Solvent Polarities, *J. Am. Chem. Soc.* 99, 6027–6038.

KAMLET M. J., DICKINSON C. and TAFT R. W. **(1981)** Linear Solvation Energy Relationships. Solvent Effects on Some Fluorescent Probes, *Chem. Phys. Lett.* 77, 69–72.

KAMLET M. J., ABBOUD J.-L., ABRAHAM M. H. and TAFT R. W. **(1983)** Linear Solvation Energy Relationships. 23. A Comprehensive Collection of the Solvatochromic Parameters, π^*, α, and β, and Some Methods for Simplifying the Generalized Solvatochromic Equation, *J. Org. Chem.* 48, 2877–2887.

KARPOVICH D. S. and BLANCHARD G. J. **(1995)** Relating the Polarity-Dependent Fluorescence Response of Pyrene to Vibronic Coupling. Achieving a Fundamental Understanding of the Py Polarity Scale, *J. Phys. Chem.* 99, 3951–3958.

L'HEUREUX G. P. and FRAGATA M. **(1987)** Micropolarities of Lipid Bilayers and Micelles, *J. Colloid Interface Sci.* 117, 513–522.

MACGREGOR R. B. and WEBER G. **(1986)** Estimation of the Polarity of the Protein Interior by Optical Spectroscopy, *Nature* 319, 70–73.

PARASASSI T., KRASNOWSKA E. K., BAGATOLLI L. and GRATTON E. **(1998)** *J. Fluorescence* 8, 365–373.

PÉROCHON E., LOPEZ A. and TOCANNE J. F. **(1991)** Fluorescence Properties of Methyl 8-(2-Anthroyl) Octanoate, a Solvatochromic Lipophilic Probe, *Chem. Phys. Lipids* 59, 17–28.

PÉROCHON E., LOPEZ A. and TOCANNE J. F. **(1992)** Polarity of Lipid Bilayers. A Fluorescence Investigation, *Biochemistry* 31, 7672–7682.

RAMAMURTHY V. (Ed.) **(1991)** Photochemistry in Organized and Constrained Media, VCH Publishers, New York.

REICHARDT C. **(1988)** *Solvent Effects in Organic Chemistry*, Verlag Chemie, Weinheim.

RETTIG W. **(1982)** Application of Simplified Microstructural Solvent Interaction Model to the Solvatochromism of Twisted Intramolecular Charge Transfer (TICT) States, *J. Mol. Struct.* 8, 303–327.

RETTIG W. and LAPOUYADE R. **(1994)** Fluorescent Probes Based on Twisted Intramolecular Charge Transfer (TICT) States and Other Adiabatic Reactions, in: LAKOWICZ J. R. (Ed.), *Topics in Fluorescence Spectroscopy, Vol. 4, Probe Design and Chemical Sensing*, Plenum Press, New York, pp. 109–149.

SAROJA G., SOUJANYA T., RAMACHANDRAM B. and SAMANTA A. **(1998)** 4-Aminophthalimide Derivatives as Environment-Sensitive Probes, *J. Fluorescence* 8, 405–410.

SUPPAN P. **(1983)** Excited-State Dipole Moments from Absorption/Fluorescence Solvatochromic Ratios, *Chem. Phys. Lett.* 94, 272–275.

SUPPAN P. **(1990)** Solvatochromic Shifts: The Influence of the Medium on the Energy of Electronic States, *J. Photochem. Photobiol.* A50, 293–330.

VALEUR B. **(1993)** Fluorescent Probes for Evaluation of Local Physical and Structural Parameters, in: SCHULMAN S. G. (Ed.), *Molecular Luminescence Spectroscopy. Methods and Applications: Part 3*, Wiley-Interscience, New York, pp. 25–84.

WARE W. R., LEE S. K., BRANT G. J. and CHOW P. P. **(1971)** Nanosecond Time-Resolved Emission Spectroscopy: Spectral Shifts due to Solvent-Excited Solute Relaxation, *J. Chem. Phys.* 54, 4729–4737.

WEBER G. and FARRIS F. J. **(1979)** Synthesis and Spectral Properties of a Hydrophobic Fluorescent Probe: 6-Propionyl-2-(Dimethylamino)naphthalene, *Biochemistry* 18, 3075–3078.

8
Microviscosity, fluidity, molecular mobility. Estimation by means of fluorescent probes

Un fluide a de la viscosité lorsque les molécules ont de l'adhésion entre elles.

[*A fluid has viscosity when the molecules adhere to each other.*]

A.-H. Paulian, 1773

We should first emphasize that viscosity is a macroscopic parameter which loses its physical meaning on a molecular scale. Therefore, the term 'microviscosity' should be used with caution, and the term 'fluidity' can be alternatively used to characterize, in a very general way, the effects of viscous drag and cohesion of the probed microenvironment (polymers, micelles, gels, lipid bilayers of vesicles or biological membranes, etc.).

8.1
What is viscosity? Significance at a microscopic level

The concept of viscosity was first introduced by Newton in the seventeenth century as the proportionality factor between the velocity gradient $\mathrm{d}v/\mathrm{d}x$ in the direction perpendicular to the flow direction (shear rate) and the force per unit area F/S required to maintain the flow (shear stress). We now call a fluid that obeys such a linear relation a *Newtonian fluid.*

In 1851, Stokes (well known for his pioneering work on luminescence; see Chapter 1) showed that the relation linking the force exerted by a fluid on a sphere to the viscosity η of the medium is $F = 6\pi\eta r v$, where r is the radius of the sphere and v its constant velocity. In this relation, the quantity $6\pi\eta r$ appears as a friction coefficient, i.e. the ratio of the viscous force to the velocity.

Einstein's work on the diffusion of particles (1906) led to the well known Stokes–Einstein relation giving the diffusion coefficient D of a sphere:

$$D = \frac{kT}{\xi} \tag{8.1}$$

Tab. 8.1. Main fluorescence techniques for the determination of fluidity (from Valeur, 1993)

Technique	*Measured fluorescence characteristics*	*Phenomenon*	*Comments*
Molecular rotors	fluorescence quantum yield and/or lifetime	internal torsional motion	very sensitive to free volume; fast experiment
Excimer formation	fluorescence spectra		fast experiment
1) intermolecular	ratio of excimer and monomer bands	translational diffusion	diffusion perturbed by microheterogeneities
2) intramolecular	ratio of excimer and monomer bands	internal rotational diffusion	more reliable than intermolecular formation
Fluorescence quenching	fluorescence quantum yield and/or lifetime	translational diffusion	addition of two probes; same drawback as intermolecular excimer formation
Fluorescence polarization	emission anisotropy	rotational diffusion of the whole probe	
1) steady state			simple technique but Perrin's Law often not valid
2) time-resolved			sophisticated technique but very powerful; also provides order parameters

where k is the Boltzmann constant, T is the absolute temperature and ξ is the friction coefficient. For translational and rotational diffusions, the diffusion coefficients are respectively

$$D_t = \frac{kT}{6\pi\eta r} \tag{8.2}$$

$$D_r = \frac{kT}{8\pi\eta r^3} = \frac{kT}{6\eta V} \tag{8.3}$$

where r is the hydrodynamic radius of the sphere and V its hydrodynamic volume.

In all fluorescence techniques permitting evaluation of the fluidity of a microenvironment by means of a fluorescent probe (see Table 8.1), the underlying physical quantity is a diffusion coefficient (either rotational or translational) expressing the viscous drag of the surrounding molecules. The major problem is then to relate the diffusion constant D to the viscosity η. This is indeed a difficult problem because Eqs (8.2) and (8.3) are valid only for a rigid sphere that is large compared to the molecular dimensions, moving in a homogeneous Newtonian fluid, and obeying the Stokes hydrodynamic law. Many other relations have been proposed but it should be pointed out that *there is no satisfactory relationship between diffusion and bulk viscosity for probes.* The main reason is that the size of probes is comparable to that of the surrounding molecules forming the microenvironment to be probed. Another difficulty arises in the case of organized assemblies such as micellar systems, biological membranes, etc. because the microenvironment is not isotropic.

In other words, viscosity is a macroscopic parameter and any attempt to get absolute values of the viscosity of a medium from measurements using a fluorescent probe is hopeless.

The term *microviscosity* is often used, but again no absolute values can be given, and the best we can do is to speak of an *equivalent viscosity*, i.e. the viscosity of a homogeneous medium in which the response of the probe is the same. But a difficulty arises as to the choice of the reference solvent because the rotational relaxation rate of a probe in various solvents of the same macroscopic viscosity depends on the nature of the solvent (chemical structure and possible internal order).

Various modifications of the Stokes–Einstein relation have been proposed to take into account the microscopic effects (shape, free volume, solvent–probe interactions, etc.). In particular, the diffusion of molecular probes being more rapid than predicted by the theory, the 'slip' boundary condition can be introduced, and sometimes a mixture of 'stick' and 'slip' boundary conditions is assumed. Equation (8.3) can then be rewritten as

$$D_r = \frac{kT}{6\eta Vsg} \tag{8.4}$$

where s is the coupling factor ($s = 1$ for 'stick' and $s < 1$ for 'slip' boundary conditions) and g is the shape factor.

In the case of charged molecules, an additional friction force should be introduced as a result of the induced polarization of the surrounding solvent molecules.

Another microscopic approach to the viscosity problem was developed by Gierer and Wirtz (1953) and it is worthwhile describing the main aspects of this theory, which is of interest because it takes account of the finite thickness of the solvent layers and the existence of holes in the solvent (*free volume*). The Stokes–Einstein law can be modified using a microscopic friction coefficient ξ_{micro}

$$\boxed{D = \frac{kT}{\xi_{micro}}} \tag{8.5}$$

and by introducing a microfriction factor f (< 1) defined as

$$\boxed{\xi_{micro} = \xi_{Stokes} f} \tag{8.6}$$

A solute molecule moves according to two diffusional processes: a viscous process with displacement of solvent molecules (*Stokes diffusion*) and a process associated

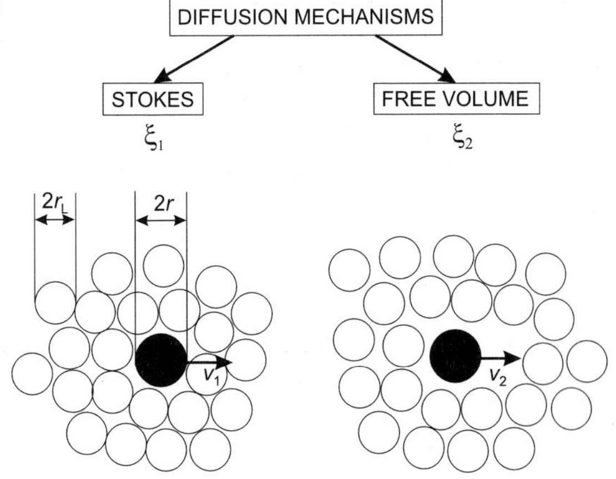

Fig. 8.1. Stokes and free volume translational diffusion processes. Black circles: solute molecules. White circles: solvent molecules.

with migration into holes of the solvent (*free volume diffusion*), as illustrated in Figure 8.1.

The total velocity of the solute molecule is given by $v = v_1 + v_2$. Because the friction coefficient is the ratio of the viscous force to the velocity ($\xi = F/v$), the microscopic friction coefficient ξ_{micro} consists of two parts:

$$\frac{1}{\xi_{\text{micro}}} = \frac{1}{\xi_1} + \frac{1}{\xi_2} \tag{8.7}$$

ξ_1, corresponding to the Stokes diffusional process, can be written as the product of the Stokes friction coefficient multiplied by a correcting factor f_t' taking into account the finite thickness of the solvent layers

$$\xi_1 = \xi_{\text{Stokes}} f_t' \tag{8.8}$$

with

$$f_t' = \left[2\frac{r_L}{r} + \frac{1}{1 + r_L/r}\right]^{-1} \tag{8.9}$$

where r and r_L are the radii of the solute and the solvent molecules, respectively, assuming that they are spherical.

Using Eqs (8.4) to (8.8), the microfriction factor f_t for translation is given by

$$f_t = \frac{\xi_{\text{micro}}}{\xi_{\text{Stokes}}} = \frac{1}{1/f_t' + \xi_{\text{Stokes}}/x_2} = \frac{f_t'}{1 + v_2/v_1} \tag{8.10}$$

The translational diffusion coefficient on the molecular scale is then

$$D_t = \frac{kT}{6\pi\eta r f_t} \tag{8.11}$$

For rotational diffusion, the correcting factor is

$$f_r' = \left[6\frac{r_L}{r} + \frac{1}{(1 + r_L/r)^3} \right]^{-1} \tag{8.12}$$

The importance of free volume effects in diffusional processes at a molecular level should be further emphasized. An empirical relationship between viscosity and free volume was proposed by Doolittle:

$$\boxed{\eta = \eta_0 \exp\left(\frac{V_0}{V_f}\right)} \tag{8.13}$$

where η_0 is a constant, and V_0 and V_f are the van der Waals volume and the free volume of the solvent, respectively.

These preliminary considerations should be borne in mind during the following discussion on the various methods of characterization of 'microviscosity'.

8.2
Use of molecular rotors

A *molecular rotor* – as a fluorescent probe – is a molecule that undergoes internal rotation(s) resulting in viscosity-dependent changes in its emissive properties. The possibilities of internal rotation associated with intramolecular charge transfer in various fluorescent molecules have already been examined in Section 3.4.4. In the present section, the viscosity dependence of the fluorescence quantum yield is further discussed, with special attention to the effect of free volume. Various examples of molecular rotors of practical interest are given in Figure 8.2: they belong to the family of diphenylmethane dyes (e.g. auramine O), triphenylmethane dyes (e.g. crystal violet), and TICT compounds that can form, upon excitation, a twisted intramolecular charge transfer state (e.g. DMABN) (the latter compounds possess an electron-donating group (e.g. dimethylamino group) conjugated to an electron-withdrawing group (e.g. cyano group)).

In solvents of medium and high viscosity, an empirical relation has been proposed (Loutfy and Arnold, 1982) to link the non-radiative rate constant for de-excitation to the ratio of the van der Waals volume to the free volume according to

$$\boxed{k_{nr} = k_{nr}^0 \exp\left(-x\frac{V_0}{V_f}\right)} \tag{8.14}$$

where k_{nr}^0 is the free-rotor reorientation rate and x is a constant for a particular probe. Because the fluorescence quantum yield is related to the radiative and non-

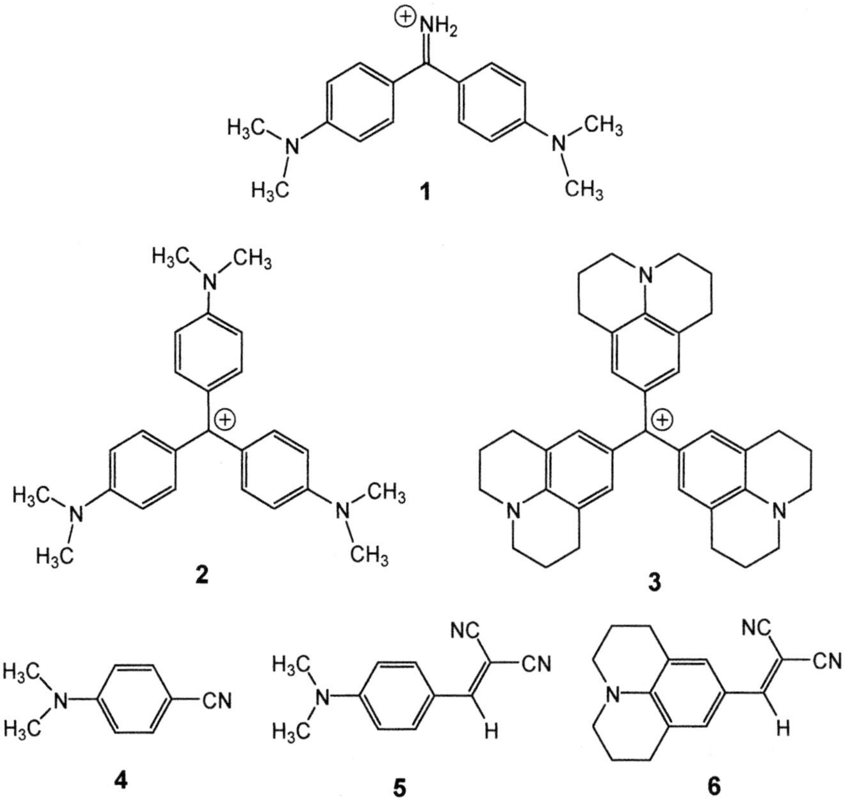

Fig. 8.2. Examples of molecular rotors. 1: auramine O. 2: crystal violet. 4: *p-N,N*-dimethylaminobenzonitrile (DMABN). 5: *p-N,N*-dimethylaminobenzylidenemalononitrile. 6: julolidinebenzylidenemalononitrile.

radiative rate constants by the expression $\Phi_F = k_r/(k_r + k_{nr})$, we obtain

$$\frac{\Phi_F}{1 - \Phi_F} = \frac{k_{nr}}{k_{nr}^0} \exp\left(x \frac{V_0}{V_f}\right) \tag{8.15}$$

Combination of this equation with the Doolittle equation (8.13) yields

$$\frac{\Phi_F}{1 - \Phi_F} = a\eta^x \tag{8.16}$$

For small fluorescence quantum yields, an approximate expression can be used:

$$\boxed{\Phi_F \cong a\eta^x} \tag{8.17}$$

When changes in viscosity are achieved by variations of temperature, this expression should be rewritten as

$$\boxed{\Phi_F \cong b(\eta/T)^x} \tag{8.18}$$

In most investigations in solvents of medium or high viscosity, or in polymers above the glass transition temperature, the fluorescence quantum yields were in fact found to be a power function of the bulk viscosity, with values of the exponent x less than 1 (e.g. for *p-N,N*-dimethylaminobenzylidenemalononitrile, $x = 0.69$ in glycerol and 0.43 in dimethylphthalate). This means that *the effective viscosity probed by a molecular rotor appears to be less than the bulk viscosity η because of free volume effects.*

Molecular rotors allow us to study changes in free volume of polymers as a function of polymerization reaction parameters, molecular weight, stereoregularity, crosslinking, polymer chain relaxation and flexibility. Application to monitoring of polymerization reactions is illustrated in Box 8.1.

8.3
Methods based on intermolecular quenching or intermolecular excimer formation

Dynamic quenching of fluorescence is described in Section 4.2.2. This translational diffusion process is viscosity-dependent and is thus expected to provide information on the fluidity of a microenvironment, but it must occur in a time-scale comparable to the excited-state lifetime of the fluorophore (experimental time window). When transient effects are negligible, the rate constant k_q for quenching can be easily determined by measuring the fluorescence intensity or lifetime as a function of the quencher concentration; the results can be analyzed using the Stern–Volmer relation:

$$\frac{\Phi_0}{\Phi} = \frac{I_0}{I} = 1 + k_q \tau_0[\mathrm{Q}] = 1 + K_{SV}[\mathrm{Q}] \tag{8.19}$$

where Φ_0 and Φ are the fluorescence quantum yields of the probe in the absence and in the presence of quencher, respectively.

Let us recall that, if the bimolecular process is diffusion-controlled, k_q is identical to the diffusional rate constant k_1. If k_1 is assumed to be independent of time (see Chapter 4), it can be expressed by the following simplified form (Smoluchowski relation):

$$k_1 = 4\pi N R_c D \tag{8.20}$$

where R_c is the distance of closest approach (in cm), D is the mutual diffusion coefficient (in cm^2 s^{-1}), N is equal to $N_a/1000$, N_a being Avogadro's number. The distance of closest approach is generally taken as the sum of the radii of the two

Box 8.1 Monitoring of polymerization reactions by means of molecular rotors[a)]

The variations in fluorescence intensity of compound 1 during the polymerization reaction of methyl methacrylate (MMA), ethyl methacrylate (EMA) and *n*-butyl methacrylate (*n*-BMA), initiated using AIBN at 70 °C, are shown in Figure B8.1.1.

1

At the beginning of the polymerization reaction, the viscosity of the medium is low, and efficient rotation (about the ethylenic double bond and the single bond linking ethylene to the phenyl ring) accounts for the low fluorescence quantum yield. After a lag period, when approaching the polymer glassy state, the fluorescence intensity increases rapidly as a result of the sharp increase in

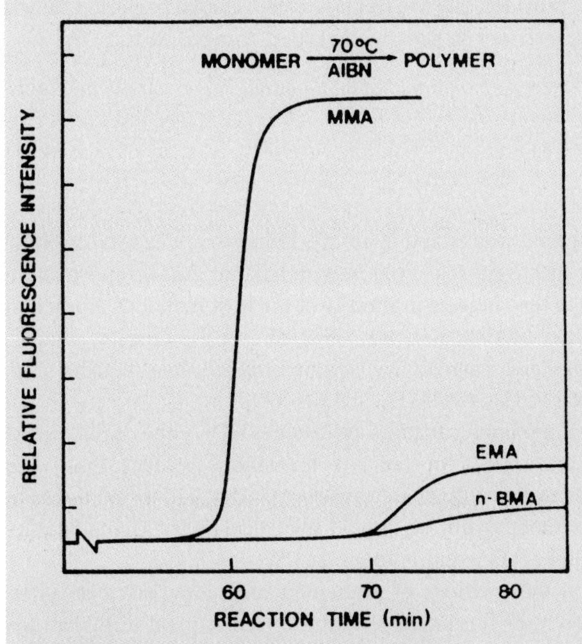

Fig. B8.1.1. Variation in the fluorescence intensity of compound 1 during the polymerization of methyl methacrylate (MMA), ethyl methacrylate (EMA), *n*-butyl methacrylate (*n*-BMA) (reproduced with permission from Loufty[a)]).

viscosity with drastic reduction of the free volume. Then, the fluorescence intensity levels off when the limiting conversion is reached.

It should be noted that the rate of change of fluorescence intensity and the fluorescence enhancement factor depend on the polymerization conditions (temperature, initiator concentration, monomer reactivity) and on the nature of the polymer formed.

Such an easy way to monitor on-line the progress of bulk polymerization, in order to prevent runaway reactions, is of great practical interest.

a) Loufty R. O. (1986) *Pure & Appl. Chem.* **58**, 1239.

molecules (R_M for the fluorophore and R_Q for the quencher). The mutual diffusion coefficient D is the sum of the translational diffusion coefficients of the two species, D_M and D_Q.

The method using intermolecular excimer formation is based on the same principle because this process is also diffusion-controlled. Excimers should, of course, be formed during the monomer excited-state lifetime. In Section 4.4.1, it was shown that the ratio I_E/I_M of the intensities of the excimer and monomer bands is proportional to k_1 provided that the transient term can be neglected. When the dissociation rate of the excimer is slow with respect to de-excitation, the relationship is

$$\frac{I_E}{I_M} = \frac{k_r'}{k_r}\tau_E k_1[M] \tag{8.21}$$

where τ_E is the excimer lifetime, and k_r and k_r' are the radiative rate constants of the monomer and excimer, respectively. It should be emphasized that erroneous conclusions may be drawn as to the changes in fluidity of the host matrix as a function of temperature if the excimer lifetime τ_E is not constant over the range of temperature investigated. Time-resolved fluorescence experiments are then required. The relevant equations are given in Chapter 4 (Eqs 4.43–4.47).

Changes in fluidity of a medium can thus be monitored via the variations of $I_0/I - 1$ for quenching, and I_E/I_M for excimer formation, because these two quantities are proportional to the diffusional rate constant k_1, i.e. proportional to the diffusion coefficient D. Once again, we should not calculate the viscosity value from D by means of the Stokes–Einstein relation (see Section 8.1).

The serious drawback of the methods of evaluation of fluidity based on intermolecular quenching or excimer formation is that the translational diffusion can be perturbed in constrained media. It should be emphasized that, in the case of biological membranes, problems in the estimation of fluidity arise from the presence of proteins and possible additives (e.g. cholesterol). Nevertheless, excimer formation with pyrene or pyrene-labeled phospholipids can provide interesting in-

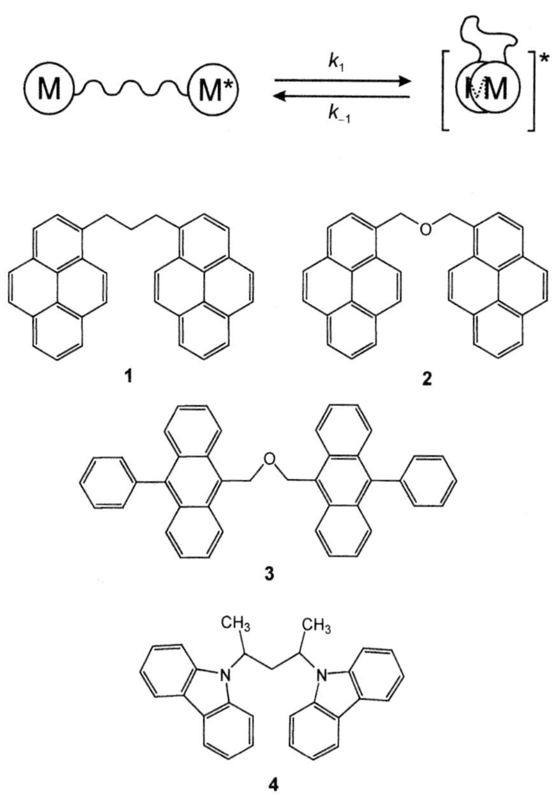

$$M \sim\sim\sim M^* \underset{k_{-1}}{\overset{k_1}{\rightleftharpoons}} [MM]^*$$

Fig. 8.3. Excimer-forming bifluorophores for the study of fluidity. 1: α,ω-di-(1-pyrenyl)-propane; 2: α,ω-di-(1-pyrenyl)-methylether; 3: 10,10′-diphenyl-bis-9-anthrylmethyloxide (DIPHANT); 4: meso-2,4-di-(*N*-carbazolyl)pentane.

formation on phospholipid bilayers and in particular on phase transition, lateral diffusion and lateral distribution (Vanderkoi and Callis, 1974; Hresko et al., 1986).

Molecular mobility in polymer films and bulk polymers can also be probed by excimer formation of pyrene (Chu and Thomas, 1990).

8.4 Methods based on intramolecular excimer formation

Bifluorophoric molecules consisting of two identical fluorophores linked by a short flexible chain may form an excimer. Examples of such bifluorophores currently used in investigations of fluidity are given in Figure 8.3.

In contrast to intermolecular excimer formation, this process is not translational but requires close approach of the two moieties through internal rotations during the lifetime of the excited state. Information on fluidity is thus obtained without the difficulty of possible perturbation of the diffusion process by microheterogeneity of the medium, as mentioned above for intermolecular excimer formation.

Moreover, the efficiency of excimer formation does not depend on the concentration of fluorophores, so that Eq. (8.21) should be rewritten as

$$\frac{I_E}{I_M} = \frac{k'_r}{k_r} \tau_E k_1 \tag{8.22}$$

As in the case of intermolecular excimer formation, it should be recalled that difficulties may arise from the possible temperature dependence of the excimer lifetime, when effects of temperature on fluidity are investigated. It is then recommended that time-resolved fluorescence experiments are performed. The relevant equations established in Chapter 4 (Eqs 4.43–4.47) must be used after replacing $k_1[M]$ by k_1.

The viscosity dependence of intramolecular excimer formation is complex. As in the case of molecular rotors (Section 8.2), most of the experimental observations can be interpreted in terms of free volume. However, compared to molecular rotors, the free volume fraction measured by intramolecular excimers is smaller. The volume swept out during the conformational change required for excimer formation is in fact larger, and consequently these probes do not respond in frozen media or polymers below the glass transition temperature.

Using again the Doolittle equation (8.13) and assuming that the rate constant for excimer formation is given by

$$k_1 = k_1^0 \exp\left(-x \frac{V_0}{V_f}\right) \tag{8.23}$$

(note the analogy with Eq. 8.14), we get

$$k_1 = \alpha \eta^{-x} \tag{8.24}$$

It is important to note that the response of a probe is not the same in two solvents of identical viscosity but of different chemical nature. For instance, the variations in I_E/I_M for α,ω-di-(1-pyrenyl)propane in ethanol/glycerol mixtures and in hexadecane/paraffin mixtures at 25 °C (either degassed or undegassed) show significant differences between I_E/I_M values for the same viscosity but with different solvent mixtures. A dependence on the chain length was also found. This effect of chain length and solvent can be explained in terms of the internal rotations involved in excimer formation. These rotations depend on the torsional potential of the various bonds and on the solvent nature; the solvent intervenes not only by the viscous drag, but also by its microstructure and possible interactions with the probe. The distribution of distances between the two fluorophores at the instant of excitation evolves during the excited-state lifetime of the monomer. Some of the bifluorophores with one excited fluorophore form an excimer because of favorable initial interchromophoric distance and favorable time evolution. Once the two chromophores are close to each other, the formation of the excited dimer (excimer) is very rapid, so that the overall process is diffusion-dependent. The initial distribution and its further evolution depends on chain length, solvent nature, viscosity and temperature (Viriot et al., 1983).

It should be noted that in organized assemblies, the local order may affect the internal rotations and the distribution of interchromophoric distances. Therefore, it is not surprising that the values of the *equivalent viscosity* may depend on the probe and, in particular, on the chain length. A comparison between degassed and undegassed solutions showed a strong effect of oxygen at low viscosities. The effect of oxygen leading to smaller values of the ratio I_E/I_M is simple to understand on the basis of Eq. (8.22): the ratio I_E/I_M is proportional to the excited-state lifetime of the excimer, which is reduced by oxygen quenching. The rate of oxygen quenching decreases with increasing viscosity, and its effect on excimer formation in dipyrenylalkanes becomes negligible at viscosities higher than 100 cP.

Intramolecular excimers have been used for probing bulk polymers, micelles, vesicles and biological membranes (Bokobza and Monnerie, 1986; Bokobza, 1990; Georgescauld et al., 1980; Vauhkonen et al., 1990, Viriot et al., 1983; Zachariasse et al., 1983). In particular, this method provides useful information on the local dynamics of polymer chains in the bulk (see Box 8.2).

In conclusion, the method of intramolecular excimer formation is rapid and convenient, but the above discussion has shown that great care is needed for a reliable interpretation of the experimental results. In some cases it has been demonstrated that the results in terms of equivalent microviscosity are consistent with those obtained by the fluorescence polarization method (described in Section 8.5), but this is not a general rule. Nevertheless, the relative changes in fluidity and local dynamics upon an external perturbation are less dependent on the probe, and useful applications to the study of temperature or pressure effects have been reported.

8.5 Fluorescence polarization method

In Chapter 5, devoted to fluorescence polarization, it was shown that information on the rotational motions of a fluorophore can be obtained from emission anisotropy measurements. Application to the evaluation of the fluidity of a medium, or molecular mobility, is presented below.

8.5.1 Choice of probes

The probes to be used in fluorescence polarization experiments should fulfil as far as possible the following requirements:

- minimum disturbance of the medium to be probed;
- symmetry and directions of the transition moments such that rotations can be considered as isotropic;

Box 8.2 Intramolecular excimer fluorescence for probing the mobility of bulk polymers

The method is based on the fact that the rate of conformational change required for excimer formation depends on the free volume induced by the segmental motions of the polymer occurring above the glass transition. DIPHANT (compound 3 in Figure 8.3) was used as an excimer-forming probe of three polymer samples consisting of polybutadiene, polyisoprene and poly(dimethylsiloxane).[a]

Figure B8.2.1 shows the fluorescence spectra of DIPHANT in a polybutadiene matrix. The I_E/I_M ratios turned out to be significantly lower than in solution, which means that the internal rotation of the probe is restricted in such a relatively rigid polymer matrix. The fluorescence intensity of the monomer is approximately constant at temperatures ranging from −100 to −20 °C, which indicates that the probe motions are hindered, and then decreases with a concomitant increase in the excimer fluorescence. The onset of probe mobility, detected by the start of the decrease in the monomer intensity and lifetime occurs at about −20 °C, i.e. well above the low-frequency static reference temperature T_g (glass transition temperature) of the polybutadiene sample, which is −91 °C (measured at 1 Hz). This temperature shift shows the strong dependence of the apparent polymer flexibility on the characteristic frequency of the experimental technique. This frequency is the reciprocal of the monomer excited-state

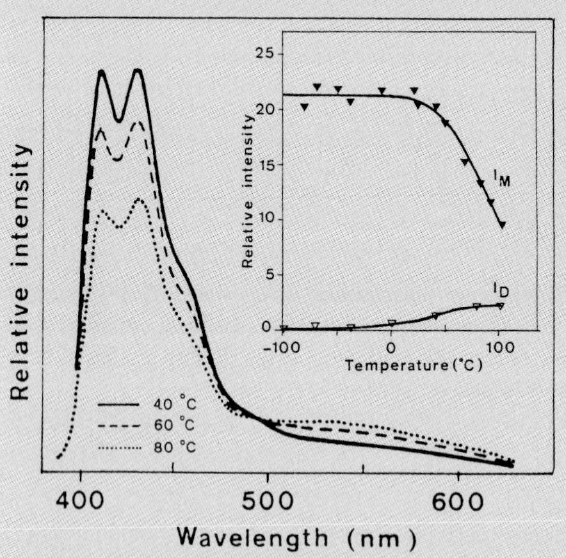

Fig. B8.2.1. Temperature dependence of the fluorescence spectrum of DIPHANT dispersed in polybutadiene. Insert: temperature evolution of the fluorescence intensity of monomer and excimer (reproduced with permission from Bokobza and Monnerie[a]).

lifetime (experimental window for excimer formation): $\sim 1.5 \times 10^8$ s^{-1} for DIPHANT, and $\sim 5 \times 10^6$ s^{-1} for pyrene-based bifluorophores (whose lifetimes are of the order of 200 ns).

The correlation time τ_c of the motions involved in intramolecular excimer formation is defined as the reciprocal of the rate constant k_1 for this process. Its temperature dependence can be interpreted in terms of the WLF equation[b)] for polymers at temperatures ranging from the glass transition temperature T_g to roughly $T_g + 100°$:

$$\ln \frac{\tau_c(T)}{\tau_c(T_g)} = -\frac{C_1(T - T_g)}{C_2 + (T - T_g)}$$

where C_1 and C_2 are two constants that depend on the chemical structure of the polymer. Their values were selected to be those given by Ferry[c)] for elastomers. The values of the rate constant k_1 were determined by time-resolved fluorescence measurements, and the values of log τ_c were plotted as a function of $T - T_g$ (Figure B8.2.2). The data reveal an important effect of the chemical structure of the matrix and a good agreement with the WLF equation in all cases.

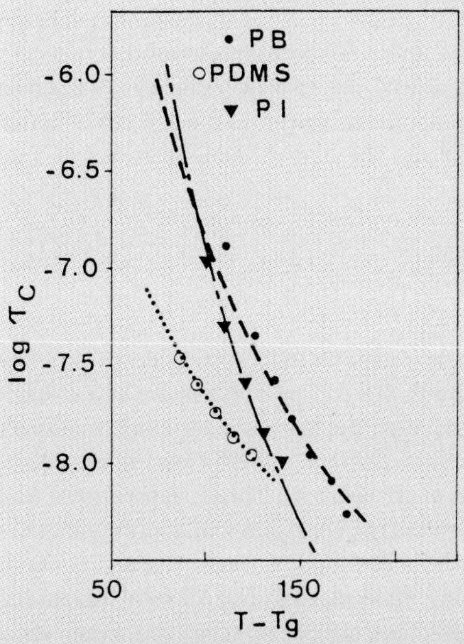

Fig. B8.2.2. Logarithmic plot of the correlation time versus $T - T_g$ for DIPHANT dispersed in polybutadiene (PB), polyisoprene (PI) and poly(dimethylsiloxane) (PDMS). The broken lines are the best fits with the WLF equation (reproduced with permission from Bokobza and Monnerie[a)]).

This investigation shows that it is indeed possible to study the flexibility of polymer chains in polymer matrices by means of excimer-forming probes and that the rotational mobility of these probes reflect the glass transition relaxation phenomena of the polymer host matrix, in agreement with the appropriate WLF equation.

a) Bokobza L. and Monnerie L. (1986), in: Winnik M. A. (Ed.), *Photophysical and Photochemical Tools in Polymer Science*, D. Reidel Publishing Company, New York, pp. 449–66.

b) Williams M., Landel R. F. and Ferry J. D. (1955) *J. Am. Chem. Soc.* **89**, 962. The WLF equation is commonly used for mechanical relaxation data analysis at low frequency ($< 10^6$ Hz).

c) Ferry J. D. (1970) *Viscoelastic Properties of Polymers*, Wiley, New York.

- minimum specific interactions with the surrounding molecules;
- minimum sensitivity of the excited-state lifetime to the microenvironment when only steady-state anisotropy is measured.

Figure 8.4 shows examples of some probes. Regarding disturbance of the medium, small molecules are to be preferred, but the measurement is then often sensitive to specific interactions with the surrounding molecules. Note that lipid-like probes such as parinaric acids offer minimum disturbance for the investigation of lipid bilayers. With large probes, these interactions have minor effects, as demonstrated in the case of the large probe BTBP [*N*,*N'*-bis(2,5-di-ter-butylpheny)-3,4,9,10-perylenetetracarboximide] by the linear relationship observed between rotational correlation time and bulk viscosity of the solvent (*n*-alkanes, *n*-alcohols, ethanol/glycerol mixtures, paraffin oil/dodecane mixtures) in the 0.5–150 cP range. However, such a large probe is not suitable for the study of confined media of small dimensions.

8.5.2 Homogeneous isotropic media

The case of probes undergoing isotropic rotations in a homogeneous isotropic medium will be examined first. Rotations are isotropic when the probe has a spherical shape, but it is difficult to find such probes[1] because most fluorescent molecules are aromatic and thus more or less planar. Nevertheless, when a probe interacts with the solvent molecules through hydrogen bonds, experiments have shown that in some cases the observed rotational behaviour can approach that of a sphere. In the case of rod-like probes whose direction of absorption and emission transition moments coincide with the long molecular axis (e.g. diphenylhexatriene; Figure 8.4), the rotations can be considered as isotropic because any rotation about this long axis has no effect on the emission anisotropy.

1) Fullerene C_{60} is perfectly spherical but for symmetry reasons, the emitted fluorescence is completely depolarized even in the absence of rotational motions. Moreover, its fluorescence quantum yield is very low ($\sim 2 \times 10^{-4}$).

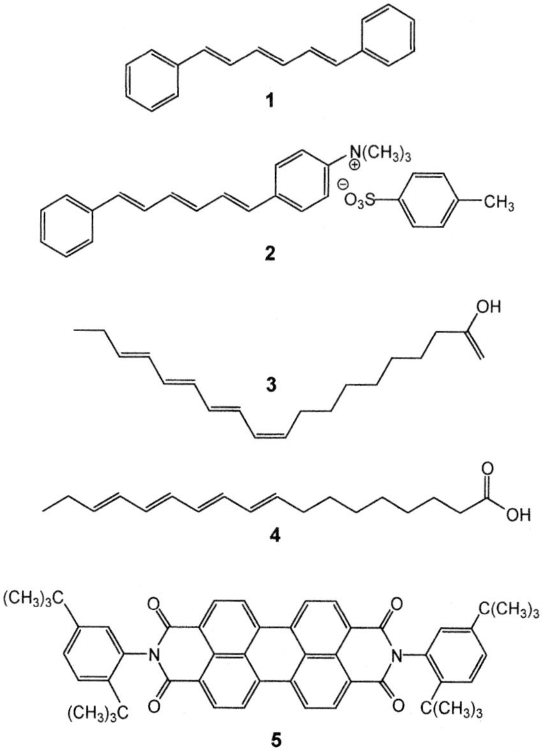

Fig. 8.4. Examples of probes used in fluorescence polarization experiments to evaluate fluidity and molecular mobility. 1: 1,6-diphenyl-1,2,5-hexatriene (DPH). 2: 1-(4-trimethyl-ammoniumphenyl)-6-phenyl 1,3,5-hexatriene, *p*-toluene sulfonate (TMA-DPH). 3: *cis*-parinaric acid. 4: *trans*-parinaric acid. 5. BTBP [*N*,*N*′-bis(2,5-di-ter-butylphenyl)-3,4,9,10-perylenetetracarboximide].

For isotropic motions in an isotropic medium, the values of the instantaneous and steady-state emission anisotropies are linked to the rotational diffusion coefficient D_r by the following relations (see Chapter 5):

$$r(t) = r_0 \exp(-6D_r t) \tag{8.25}$$

$$\frac{1}{\bar{r}} = \frac{1}{r_0}(1 + 6D_r\tau) \tag{8.26}$$

which allow determination of D_r. In principle, a value of the viscosity η $(= kT/6VD_r)$ could be calculated from the Stokes–Einstein relation (8.3) provided that the hydrodynamic volume V is known. In addition to the problem of validity of the Stokes–Einstein relation (see above), the hydrodynamic volume cannot be calculated on a simple geometrical basis but must also take into account the solvation shell.

Introducing the rotational correlation time $\tau_c = (6D_r)^{-1}$, Eqs (8.25) and (8.26) can be rewritten as

$$r(t) = r_0 \exp(-t/\tau_c) \tag{8.27}$$

$$\frac{1}{\bar{r}} = \frac{1}{r_0}\left(1 + \frac{\tau}{\tau_c}\right) \tag{8.28}$$

The changes in correlation time upon an external perturbation (e.g. temperature, pressure, additive, etc.) reflects well the changes in fluidity of a medium. It should again be emphasized that any 'microviscosity' value that could be calculated from the Stokes–Einstein relation would be questionable and thus useless.

8.5.3 Ordered systems

Equations (8.25) to (8.28) are no longer valid in the case of hindered rotations occurring in anisotropic media such as lipid bilayers and liquid crystals. In these media, the rotational motions of the probe are hindered and the emission anisotropy does not decay to zero but to a steady value r_∞ (see Chapter 5). For isotropic rotations (rod-like probe), assuming a single correlation time, the emission anisotropy can be written in the following form:

$$r(t) = (r_0 - r_\infty)\exp(-t/\tau_c) + r_\infty \tag{8.29}$$

Time-resolved emission anisotropy experiments provide information not only on the fluidity via the correlation time τ_c, but also on the order of the medium via the ratio r_∞/r_0. The theoretical aspects are presented in Section 5.5.2, with special attention to the *wobble-in-cone model* (Kinosita et al., 1977; Lipari and Szabo, 1980). Phospholipid vesicles and natural membranes have been extensively studied by time-resolved fluorescence anisotropy. An illustration is given in Box 8.3.

8.5.4 Practical aspects

From a practical point of view, the steady-state technique (continuous illumination) is far simpler than the time-resolved technique, but it can only be used in the case of isotropic rotations in isotropic media (Eqs 8.26 and 8.28) provided that the probe lifetime is known. Attention should be paid to the fact that the variations in steady-state anisotropy resulting from an external perturbation (e.g. temperature) may not be due only to changes in rotational rate, because this perturbation may also affect the lifetime.

The time-resolved technique is much more powerful but requires expensive instrumentation.

It is worth pointing out that many artifacts can alter the measurements of emission anisotropy. It is necessary to control the instrument with a scattering non-fluorescent solution (r close to 1) and with a solution of a fluorophore with a long lifetime in a solvent of low viscosity ($r \approx 0$). It is also recommended that the probe concentration is kept low enough to avoid interaction between probes.

Box 8.3 Investigation of the dynamics and molecular order of phosphatidylinositol incorporated into artificial and natural membranes[a)]

Lipid–protein interactions are of major importance in the structural and dynamic properties of biological membranes. Fluorescent probes can provide much information on these interactions. For example, van Paridon et al.[a)] used a synthetic derivative of phosphatidylinositol (PI) with a *cis*-parinaric acid (see formula in Figure 8.4) covalently linked on the *sn*-2 position for probing phospholipid vesicles and biological membranes. The emission anisotropy decays of this 2-parinaroyl-phosphatidylinositol (PPI) probe incorporated into vesicles consisting of phosphatidylcholine (PC) (with a fraction of 5 mol % of PI) and into acetylcholine receptor rich membranes from *Torpedo marmorata* are shown in Figure B8.3.1.

The curves were fitted using the following decay function, which must be considered as a purely mathematical model:

$$r(t) = a_1 \exp(-t/\tau_{c1}) + a_2 \exp(-t/\tau_{c2}) + r_\infty$$

where τ_{c1} and τ_{c2} are the correlation times for rotational diffusion of the probe and r_∞ is the residual anisotropy at long time with respect to the correlation times.

The parameters $a_1, a_2, \tau_{c1}, \tau_{c2}$, and r_∞ are obtained from the best fit of $I_\parallel(t)$ and $I_\perp(t)$ given by Eqs (5.7) and (5.8) of Chapter 5:

$$I_\parallel(t) = \frac{I(t)}{3}[1 + 2r(t)]$$

$$I_\perp(t) = \frac{I(t)}{3}[1 - r(t)]$$

where $I(t)$ is the total fluorescence intensity decay $[= I_\parallel(t) + 2I_\perp(t)]$ that can be satisfactorily fitted in this case by a sum of two exponentials.

Then, according to the *wobble-in-cone model* (see Section 5.6.2), the order parameter that is related to the half angle of the cone, and the wobbling diffusion constant (reflecting the chain mobility) can be determined.

For comparison, 2-parinaroyl-phosphatidylcholine (PPC) was also incorporated into the membrane preparations. For the *Torpedo* membranes, the acyl chain order measured by PPI was found to be lower than that by PPC, whereas the opposite was true for the vesicles. This inversion strongly suggests that PI has different interactions with certain membrane components compared to PC. In contrast, the correlation times of PPI and PPC were only slightly different in *Torpedo* membranes, and the values showed little difference with those measured in vesicles.

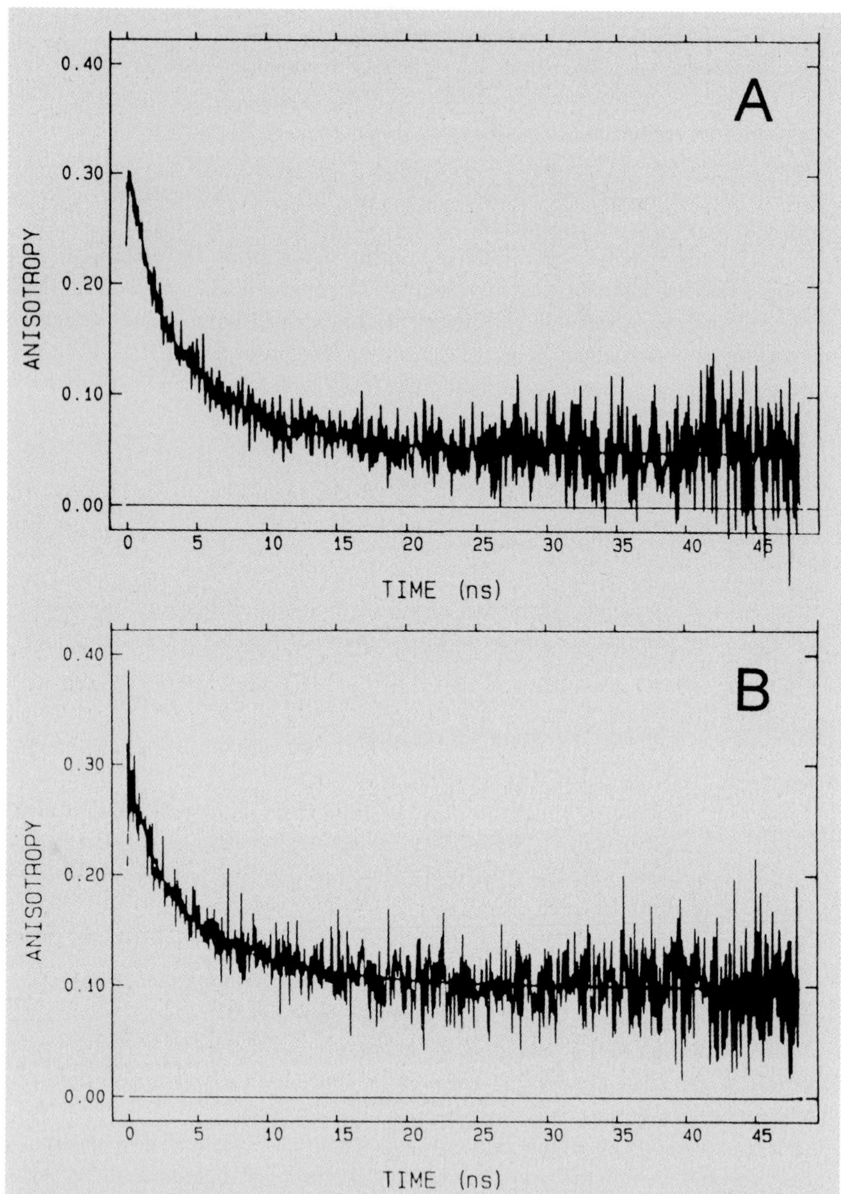

Fig. B8.3.1. Fluorescence anisotropy decays at 4 °C of PPI. A: in phospholipid vesicles (PC:PI, 95:5 mol %). B: in *Torpedo* membranes. From the best fit of the $I(t)$ and $I_{\perp}(t)$ components, and by using the wobble-in-cone model, the half angles of the cone are found to be 61° (A) and 45° (B), and the wobbling diffusion constants are 0.041 ns^{-1} (A) and 0.020 ns^{-1} (B) (reproduced with permission from van Paridon et al.[a]).

a) van Paridon P. A., Shute J. K., Wirtz K. W. A. and Visser A. J. W. G. (1988) *Eur. Biophys. J.* **16**, 53–63.

Otherwise, depolarization would also be a result of energy transfer between probe molecules. Because the transition moments of two interacting probes are unlikely to be parallel, this effect is indeed formally equivalent to a rotation. Moreover, artefacts may arise from scattering light that is not totally rejected in the detection system.

The fluorescence polarization technique is a very powerful tool for studying the fluidity and orientational order of organized assemblies: aqueous micelles, reverse micelles and microemulsions, lipid bilayers, synthetic nonionic vesicles, liquid crystals. This technique is also very useful for probing the segmental mobility of polymers and antibody molecules.

8.6 Concluding remarks

The various fluorescence-based methods for the determination of fluidity and molecular mobility are summarised in Table 8.1. Attention should be paid to the comments indicated in the last column.

The choice of method depends on the system to be investigated. The methods of intermolecular quenching and intermolecular excimer formation are not recommended for probing fluidity of microheterogeneous media because of possible perturbation of the translational diffusion process. The methods of intramolecular excimer formation and molecular rotors are convenient and rapid, but the time-resolved fluorescence polarization technique provides much more detailed information, including the order of an anisotropic medium.

Whatever the technique used, it is important to note that (i) only an equivalent viscosity can be determined, (ii) the response of a probe may be different in solvents of the same viscosity but of different chemical nature and structure, (iii) the measured equivalent viscosity often depends on the probe and on the fluorescence technique. Nevertheless, the relative variations of the diffusion coefficient resulting from an external perturbation are generally much less dependent on the technique and on the nature of the probe. Therefore, the fluorescence techniques are very valuable in monitoring changes in fluidity upon an external perturbation such as temperature, pressure and addition of compounds (e.g. cholesterol added to lipid vesicles; alcohols and oil added to micellar systems).

8.7 Bibliography

Al-Hassan K. A. and Rettig, W. **(1986)** Free Volume Sensing Fluorescent Probes, *Chem. Phys. Lett.* 126, 273–279.

Anwand D., Müller F. W., Strehmel B. and Schiller K. **(1991)** Determination of the Molecular Mobility and the Free Volume of Thin Polymeric Films with Fluorescence Probes. *Makromol. Chem.* 192, 1981–1991.

Balter, A. **(1997)** Probing Bioviscosity via Fluorescence, *J. Fluorescence* 7, S99–105.

Bokobza L. **(1990)** Investigation of Local Dynamics of Polymer Chains in the Bulk by the Excimer Fluorescence Method, *Prog. Polym. Sci.* 15, 337.

Bokobza L. and Monnerie L. **(1986)** Excimer Fluorescence as a Probe of Mobility in Polymer Melts, in: Winnik M. A. (Ed.), *Photophysical and Photochemical Tools in Polymer Science*, D. Reidel Publishing Company, Dordrecht, pp. 449–66.

Chu D. Y. and Thomas J. K. **(1990)** Photophysical Studies of Molecular Mobility in Polymer Films and Bulk Polymers. 3. Dynamic Excimer Formation of Pyrene in Bulk PDMS, *Macromolecules* 23, 2217–2222.

De Schryver F., Collart P., Vandendriessche J., Goedeweeck R., Swinnen A. and Van der Auweraer M. **(1987)** Intramolecular Excimer Formation in Bichromophoric Molecules Linked by a Short Flexible Chain, *Acc. Chem. Res.* 20, 159–166.

Georgescauld D., Desmasez J. P., Lapouyade R., Babeau A., Richard H. and Winnik M. **(1980)** Intramolecular Excimer Fluorescence: A New Probe of Phase Transitions in Synthetic Phospholipid Membranes, *Photochem. Photobiol.* 31, 539–545.

Gierer A. and Wirtz K. **(1953)** Molekulare Theorie der Mikroreibung, *Z. Naturforsch.* 8a, 532–538.

Hresko R. C., Sugar I. P., Barenholz Y. and Thompson T. E. **(1986)** Lateral Diffusion of a Pyrene-Labeled Phosphatidylcholine in Phosphatidylcholine Bilayers/Fluorescence Phase and Modulation Study, *Biochemistry* 25, 3813–3823.

Kinosita K., Kawato S. and Ikegami A. **(1977)** A Theory of Fluorescence Polarization Decay in Membranes, *Biophys. J.* 20, 289–305.

Lipari G. and Szabo A. **(1980)** Effect of Vibrational Motion on Fluorescence Depolarization and Nuclear Magnetic Resonance Relaxation in Macromolecules and Membranes, *Biophys. J.* 30, 489–506.

Loutfy R. O. **(1986)** Fluorescence Probes for Polymer Free Volume, *Pure & Appl. Chem.* 58, 1239–1248.

Loutfy R. O. and Arnold B. A. **(1982)** Effects of Viscosity and Temperature on Torsional Relaxation of Molecular Rotors, *J. Phys. Chem.* 86, 4205–4211.

Rettig W. and Lapouyade R. **(1994)** Fluorescence Probes Based on Twisted Intramolecular Charge Transfer (TICT) States and other Adiabatic Reactions, in: Lakowicz J. R. (Ed.), *Topics in Fluorescence Spectroscopy, Vol. 4, Probe Design and Chemical Sensing*, Plenum Press, New York, pp. 109–149.

Valeur B. **(1993)** Fluorescent Probes for Evaluation of Local Physical, Structural Parameters, in: Schulman S. G. (Ed.), *Molecular Luminescence Spectroscopy. Methods and Applications: Part 3*, Wiley-Interscience, New York, pp. 25–84.

Vanderkoi J. M. and Callis J. B. **(1974)** Pyrene. A Probe of Lateral Diffusion in the Hydrophobic Region of Membrane, *Biochemistry* 13, 4000–4006.

Vauhkonen M., Sassaroli M., Somerharju P. and Eisinger J. **(1990)** Dipyrenylphosphatidylcholines as Membrane Fluidity Probes. Relationship between Intramolecular and Intermolecular Excimer Formation Rates, *Biophys. J.* 57, 291–300.

Viriot M. L., Bouchy M., Donner M. and André J.-C. **(1983)** Kinetics of Partly Diffusion-Controlled Reactions. XII. Intramolecular Excimers as Fluorescent Probes of the 'Microviscosity' of Living Cells. *Photobiochem. Photobiophys.* 5, 293–306.

Zachariasse K. A., Kozankiewicz B. and Kühnle W. **(1983)** Micelles and Biological Membranes Studied by Probe Molecules, in: Zewail A. H. (Ed.), *Photochemistry and Photobiology*, Vol. II, Harwood, London, pp. 941–960.

9
Resonance energy transfer and its applications

En se propageant dans l'espace, cette cause active que nous nommons lumière, anéantit, pour ainsi dire, les distances, agrandit la sphère que nous habitons, nous montre des êtres dont nous n'aurions jamais soupçonné l'existence, et nous révèle des propriétés dont le sens de la vue pouvait seul nous donner la notion.

[*While propagating through space, the active cause that we call light annihilates, so to speak, distances, enlarges the sphere in which we live, shows us things, which we would have never suspected the existence of, and reveals to us properties which we could only grasp through our sense of sight.*]

Encyclopédie Méthodique. Physique
Monge, Cassini, Bertholon et al., 1819

9.1
Introduction

The mechanisms of resonance energy transfer (RET) between a donor and an acceptor have been described in Section 4.6.3. The aim of this chapter is to present further aspects of RET and its applications for probing matter or living systems. RET is particularly widely used to determine distances in biomolecules and supramolecular associations and assemblies.

Because this chapter will be mainly concerned with the Förster mechanism of transfer, the results of the Förster theory, given in Section 4.6.3, are recalled here for convenience. The rate constant for transfer between a donor and an acceptor at a distance r is:

$$k_T = \frac{1}{\tau_D^0}\left[\frac{R_0}{r}\right]^6 \tag{9.1}$$

where τ_D^0 is the excited-state lifetime of the donor in the absence of transfer and R_0 is the Förster critical radius (distance at which transfer and spontaneous decay of the excited donor are equally probable, i.e. $k_T = 1/\tau_D^0$). R_0 (in Å) is given by

$$R_0 = 0.2108\left[\kappa^2 \Phi_D^0 n^{-4} \int_0^\infty I_D(\lambda)\varepsilon_A(\lambda)\lambda^4 \, d\lambda\right]^{1/6} \tag{9.2}$$

where κ^2 is the orientational factor, which can take values from 0 (perpendicular transition moments) to 4 (collinear transition moments) (see Section 4.6.3 for more details on κ^2), Φ_D^0 is the fluorescence quantum yield of the donor in the absence of transfer, n is the average refractive index of the medium in the wavelength range where spectral overlap is significant, $I_D(\lambda)$ is the fluorescence spectrum of the donor normalized so that $\int_0^\infty I_D(\lambda)\, d\lambda = 1$, $\varepsilon_A(\lambda)$ is the molar absorption coefficient of the acceptor (in $dm^3\ mol^{-1}\ cm^{-1}$) and λ is the wavelength in nanometers. Values of R_0 in the range of 10–80 Å have been reported (see below).

The transfer efficiency is given by

$$\Phi_T = \frac{k_T}{1/\tau_D^0 + k_T} = \frac{1}{1 + (r/R_0)^6} \tag{9.3}$$

The sixth power dependence explains why resonance energy transfer is most sensitive to the donor–acceptor distance when this distance is comparable to the Förster critical radius.

It should be emphasized that the transfer rate depends not only on distance separation but also on the mutual orientation of the fluorophores. Therefore, it may be more appropriate in some cases to rewrite the transfer rate as follows:

$$k_T = \frac{3}{2}\frac{\kappa^2}{\tau_D^0}\left[\frac{\bar{R}_0}{r}\right]^6 \tag{9.4}$$

where $\bar{R}_0$ is the Förster radius for $\kappa^2 = 2/3$

$$\bar{R}_0 = 0.2108\left[\frac{2}{3}\Phi_D^0 n^{-4} \int_0^\infty I_D(\lambda)\varepsilon_A(\lambda)\lambda^4 \, d\lambda\right]^{1/6} \tag{9.5}$$

9.2
Determination of distances at a supramolecular level using RET

9.2.1
Single distance between donor and acceptor

The Förster resonance energy transfer can be used as a *spectroscopic ruler* in the range of 10–100 Å. The distance between the donor and acceptor molecules should be constant during the donor lifetime, and greater than about 10 Å in order to avoid the effect of short-range interactions. The validity of such a spectroscopic ruler has been confirmed by studies on model systems in which the donor and acceptor are separated by well-defined rigid spacers. Several precautions must be taken to ensure correct use of the spectroscopic ruler, which is based on the use of Eqs (9.1) to (9.3):

(i) The critical distance R_0 should be determined under the same experimental conditions as those of the investigated system because R_0 involves the quantum yield of the donor and the overlap integral, which both depend on the nature of the microenvironment. For this reason, the values of R_0 reported in Table 9.1 can only be used as a guide for selection of a pair. Tryptophan can be used as a donor for distance measurements in proteins; a few R_0 values are given in this table (for further values, see Berlman, 1973; Wu and Brand, 1994; and Van der Meer, 1994).

(ii) As regards the orientation factor κ^2, it is usually taken as 2/3, which is the isotropic dynamic average, i.e. under the assumptions that both donor and acceptor transient moments randomize rapidly during the donor lifetime and sample all orientations. However, these conditions may not be met owing to the constraints of the microenvironment of the donor and acceptor. A variation of κ^2 from 2/3 to 4 results in only a 35% error in r because of the sixth root of

Tab. 9.1. Examples of Förster critical radii

Donor	*Acceptor*	*R_0 (Å)*
Naphthalene	Dansyl	~22
Anthracene	Perylene	~31
Pyrene	Perylene	~36
Phenanthrene	Rhodamine B	~47
Fluorescein	Tetramethylrhodamine	~55
Fluorescein-5-isothiocyanate	Eosin maleimide	~60
Rhodamine 6G	Malachite green	~61
Tryptophan	Dansyl	~21
Tryptophan	ANS	~23
Tryptophan	Anthroyl	~25
Tryptophan	Pyrene	~28

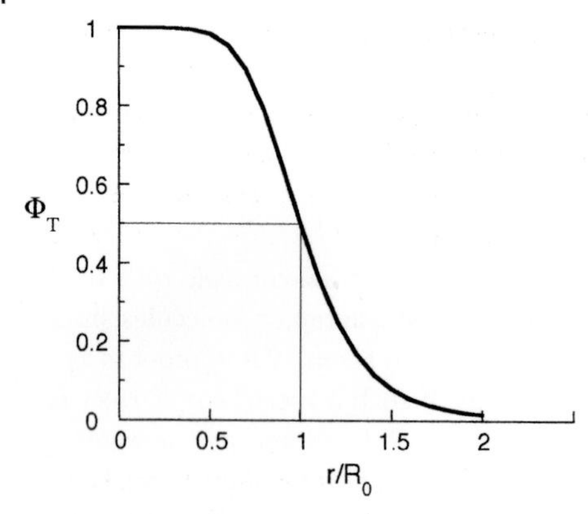

Fig. 9.1. Variations in the transfer efficiency as a function of the ratio donor–acceptor distance/Förster critical radius.

κ^2 appearing in R_0 (Eq. 9.2), but a variation from 2/3 to 0.01 results in a two-fold decrease in r. Information on the orientational freedom of donor and acceptor and their relative orientations can be obtained from fluorescence polarization experiments (Dale et al., 1975).

(iii) Because of the sixth power dependence, the variation in Φ_T (Eq. 9.3) is sharp around $r = R_0$, as shown in Figure 9.1: for $r < R_0/2$, the transfer efficiency is close to 1, and for $r > 2R_0$, it approaches 0; therefore the distance between donor and acceptor should be in the following range $0.5R_0 < r < 1.5R_0$.

(iv) The distance to be measured should be constant during the donor lifetime.

This section deals with a single donor–acceptor distance. Let us consider first the case where the donor and acceptor can freely rotate at a rate higher than the energy transfer rate, so that the orientation factor κ^2 can be taken as 2/3 (isotropic dynamic average). The donor–acceptor distance can then be determined by steady-state measurements via the value of the *transfer efficiency* (Eq. 9.3):

$$r = \left(\frac{1}{\Phi_T} - 1\right)^{1/6} R_0 \tag{9.6}$$

Three steady-state methods can be used to determine the energy transfer efficiency. In the following description of these methods, the fluorescence intensity is indicated with two wavelengths in parentheses: the first one is the excitation wavelength, and the second is the observation wavelength. Because the characteristics of the donor and/or acceptor are measured in the presence and in the absence of transfer, the concentrations of donor and acceptor and their microenvironments must be the same under both these conditions.

Steady-state method 1: decrease in donor fluorescence Transfer from donor to acceptor causes the quantum yield of the donor to decrease. The transfer efficiency is given by

$$\Phi_T = 1 - \frac{\Phi_D}{\Phi_D^0} \tag{9.7}$$

where Φ_D^0 and Φ_D are the donor quantum yields in the absence and presence of acceptor, respectively.

Because only the relative quantum yields are to be determined, a single observation wavelength is sufficient and the latter is selected so that there is no emission from the acceptor[1]. Then, Eq. (9.7) can be rewritten in terms of absorbances at the excitation wavelength λ_D and fluorescence intensities of the donor in the absence and presence of acceptor:

$$\Phi_T = 1 - \frac{A(\lambda_D)}{A_D(\lambda_D)} \frac{I_D(\lambda_D, \lambda_D^{em})}{I_D^0(\lambda_D, \lambda_D^{em})} \tag{9.8}$$

The factor A/A_D arises from the contribution of the acceptor moiety to the overall absorption at the excitation wavelength λ_D. Attention should be paid to the inner filter effect caused by absorption of the acceptor at the emission wavelength of the donor. Some correction may be necessary.

Steady-state method 2: comparison between the absorption spectrum and the excitation spectrum (through observation of the acceptor fluorescence) The corrected excitation spectrum is represented by

$$I_A(\lambda, \lambda_A^{em}) = C\Phi_A[A_A(\lambda) + A_D(\lambda)\Phi_T] \tag{9.9}$$

where C is a constant (instrumental factor). The absorption spectrum is defined as

$$A(\lambda) = A_A(\lambda) + A_D(\lambda) \tag{9.10}$$

In the case of total transfer ($\Phi_T = 1$), these two spectra are identical after normalization at the same height. But for any value of Φ_T less than 1, the excitation band corresponding to the donor is relatively lower than the absorption band. The comparison of the absorption and excitation spectra can be done at two wavelengths λ_D and λ_A corresponding to the absorption maxima of the donor and the acceptor, respectively. If there is no absorption of the donor at λ_A, we get

$$\frac{I_A(\lambda_A, \lambda_A^{em})}{A_A(\lambda_A)} = C\Phi_A \tag{9.11}$$

$$\frac{I_A(\lambda_D, \lambda_A^{em})}{A_A(\lambda_D)} = C\frac{\Phi_A[A_A(\lambda_D) + A_D(\lambda_D)\Phi_T]}{A_A(\lambda_D)} \tag{9.12}$$

The ratio of Eqs (9.12) and (9.11) yields:

$$\Phi_T = \frac{A_A(\lambda_A)}{A_D(\lambda_D)} \left[\frac{I_A(\lambda_D, \lambda_A^{em})}{I_A(\lambda_A, \lambda_A^{em})} - \frac{A_A(\lambda_D)}{A_A(\lambda_A)} \right] \tag{9.13}$$

1) Subtraction of the acceptor fluorescence is possible but would decrease the accuracy if it is large.

Steady-state method 3: Enhancement of acceptor fluorescence The fluorescence intensity of the acceptor is enhanced in the presence of transfer. Comparison with the intensity in the absence of transfer provides the transfer efficiency:

$$\Phi_T = \frac{A_A(\lambda_D)}{A_D(\lambda_D)}\left[\frac{I_A(\lambda_D, \lambda_A^{em})}{I_A^0(\lambda_D, \lambda_A^{em})} - 1\right] \tag{9.14}$$

It should be noted that $A_A(\lambda_D)$ is often small and thus difficult to measure accurately, which may lead to large errors in Φ_T.

Method 1 appears to be more straightforward than methods 2 and 3. However, it cannot be used in the case of very low donor quantum yields. It should also be noted that quenching of the donor by the acceptor may occur. This point can be checked by a complementary study based on observation of the acceptor fluorescence (method 2 or 3).

Time-resolved emission of the donor or acceptor fluorescence provides direct information on the transfer rate, without the difficulties that may result from inner filter effects.

Time-resolved method 1: decay of the donor fluorescence If the fluorescence decay of the donor following pulse excitation is a single exponential, the measurement of the decay time in the presence (τ_D) and absence (τ_D^0) of transfer is a straightforward method of determining the transfer rate constant, the transfer efficiency and the donor–acceptor distance, by using the following relations:

$$1/\tau_D = 1/\tau_D^0 + k_T \tag{9.15}$$

$$\Phi_T = 1 - \frac{\tau_D}{\tau_D^0} \tag{9.16}$$

$$r = \frac{R_0}{(\tau_D^0/\tau_D - 1)^{1/6}} \tag{9.17}$$

The advantage of this method is its ability to check whether the donor fluorescence decay in the absence and presence of acceptor is a single exponential or not. If this decay is not a single exponential in the absence of acceptor, this is likely to be due to some heterogeneity of the microenvironment of the donor. It can then be empirically modeled as a sum of exponentials:

$$i_D(t) = \sum_i \alpha_i \exp(-t/\tau_i) \tag{9.18}$$

Transfer efficiency can be calculated by using the average decay times of the donor in the absence and presence of acceptor, as follows:

$$\Phi_T = 1 - \frac{\langle\tau_D\rangle}{\langle\tau_D^0\rangle} \tag{9.19}$$

where the amplitude-averaged decay times are defined as[2)]

$$\langle\tau\rangle = \frac{\sum_i \alpha_i \tau_i}{\sum_i \alpha_i} \tag{9.20}$$

This approach is valid if the donor fluorescence decay is not too far from a single exponential, so that an average of the donor–acceptor distance can be estimated.

Time-resolved method 2: increase in the acceptor fluorescence The transfer rate constant can also be determined from the increase in the acceptor fluorescence following pulse excitation of the donor. The concentration of excited acceptors following δ-pulse excitation of the donor obeys the following differential equation:

$$\frac{\mathrm{d}[\mathrm{A}^*]}{\mathrm{d}t} = k_\mathrm{T}[\mathrm{D}^*] - (1/\tau_\mathrm{A}^0)[\mathrm{A}^*] \tag{9.21}$$

Assuming that there is no direct excitation of the acceptor and that the donor decay is single-exponential, the solution of this differential equation, with the initial condition $[\mathrm{D}^*] = [\mathrm{D}^*]_0$ at time $t = 0$, is

$$[\mathrm{A}^*] = \frac{[\mathrm{D}^*]_0 k_\mathrm{T}}{1/\tau_\mathrm{D} - 1/\tau_\mathrm{A}^0}[\mathrm{e}^{-t/\tau_\mathrm{A}^0} - \mathrm{e}^{-t/\tau_\mathrm{D}}] \tag{9.22}$$

where $1/\tau_\mathrm{D}$ is given by Eq. (9.15). The negative term corresponds to the increase in acceptor fluorescence with a time constant that is equal to the decay time of the donor τ_D because the excited acceptors are formed from excited donors.

In most cases, direct excitation of the acceptor at the excitation wavelength of the donor cannot be avoided, so that Eq. (9.22) should be rewritten as

$$[\mathrm{A}^*] = \left\{\frac{[\mathrm{D}^*]_0 k_\mathrm{T}}{1/\tau_\mathrm{D} - 1/\tau_\mathrm{A}^0} + [\mathrm{A}^*]_0\right\}\mathrm{e}^{-t/\tau_\mathrm{A}^0} - \frac{[\mathrm{D}^*]_0 k_\mathrm{T}}{1/\tau_\mathrm{D} - 1/\tau_\mathrm{A}^0}\mathrm{e}^{-t/\tau_\mathrm{D}} \tag{9.23}$$

where $[\mathrm{A}^*]_0$ is the concentration of directly excited acceptors at time $t = 0$. It should be noted that direct excitation of the acceptor affects only the pre-exponential factors but has no influence on the time constants.

In phase-modulation fluorometry, it is worth noting that the transfer rate constant can be determined from the phase shift between the fluorescence of the acceptor excited directly and via donor excitation.

As in method 1, an averaged value of the rise time may be considered if necessary.

2) For the calculation of transfer efficiency, it is incorrect to use the intensity-averaged decay time ($\sum \alpha_i \tau_i^2 / \sum \alpha_i \tau_i$) (see Section 6.2.1 for the definition of average decay times) because such an integrated intensity is not relevant to a dynamic process like energy transfer. In fact, the signal measured at a certain time after excitation is proportional to the number of donor molecules still excited at that time and able to transfer their energy to an acceptor molecule; therefore, the amplitude-averaged decay time should be used.

The above considerations apply to the case of the isotropic dynamic average (corresponding to an average orientation factor κ^2 of 2/3). However, the uncertainty in the average value of the orientation factor is one of the major sources of error in the estimation of average distances. Fluorescence polarization experiments can help to reduce the uncertainties in κ^2 (Dale et al., 1975) but may not provide complete information as to how the fluorophores orient. From analysis of time-resolved experiments, an 'apparent' distance distribution owing to the contribution from orientational heterogeneity can be obtained (Wu and Brand, 1992). For instance, in the case of static orientation of donor and acceptor with a single distance, the shape of the 'apparent' distribution fit depends on the ratio of Förster radius to average distance. A real distance distribution should have no such dependence.

The case of a real distribution of distances will now be examined.

9.2.2
Distributions of distances in donor–acceptor pairs

The donor–acceptor distance may not be unique, especially when the donor and acceptor are linked by a flexible chain (e.g. end-labeled oligomers of polymethylene or polyethylene oxide, oligopeptides, oligonucleotides, polymer chains).

The most commonly used method for the evaluation of a distribution of distances is based on the measurement of the donor fluorescence decay[3]. If the δ-response of the donor fluorescence is a single exponential in the absence of acceptor, it becomes multi-exponential in the presence of acceptor:

$$i_D(t) = i_D(0) \int_0^\infty f(r) \exp\left[-\frac{t}{\tau_D^0} - \frac{t}{\tau_D^0}\left(\frac{R_0}{r}\right)^6\right] \mathrm{d}r \tag{9.24}$$

where $f(r)$ is the distribution function of donor–acceptor distances.

If the δ-response of the donor fluorescence is not a single exponential in the absence of acceptor (and can be modeled by means of Eq. 9.18), Eq. (9.24) should be rewritten as

$$i_D(t) = \sum_{n=1}^{\infty} \alpha_i \int_0^\infty f(r) \exp\left[-\frac{t}{\tau_i^0} - \frac{t}{\tau_i^0}\left(\frac{R_0}{r}\right)^6\right] \mathrm{d}r \tag{9.25}$$

In the case of fluorophores linked by a flexible chain, the distance distribution is

3) In principle, steady-state fluorescence measurements can also provide information on distance distributions by using various levels of an external quencher of the donor: this causes R_0 to vary through changes in donor quantum yield. However, because of the sixth root dependence of R_0 on the donor quantum yield, a variation of this parameter by a factor of 10 leads to a variation of only 30% in R_0. Moreover, static quenching and transient diffusion phenomena cause problems. Therefore, such a steady-state method cannot be reliably used in practice.

Box 9.1 Recovery of distance distributions in a triantennary glycopeptide[a)]

The flexibility of the sugar linkages in glycoproteins results in multiple conformations that can be detected by time-resolved fluorescence experiments. Figure B9.1.1 shows an example of a triantennary glycopeptide labeled with a

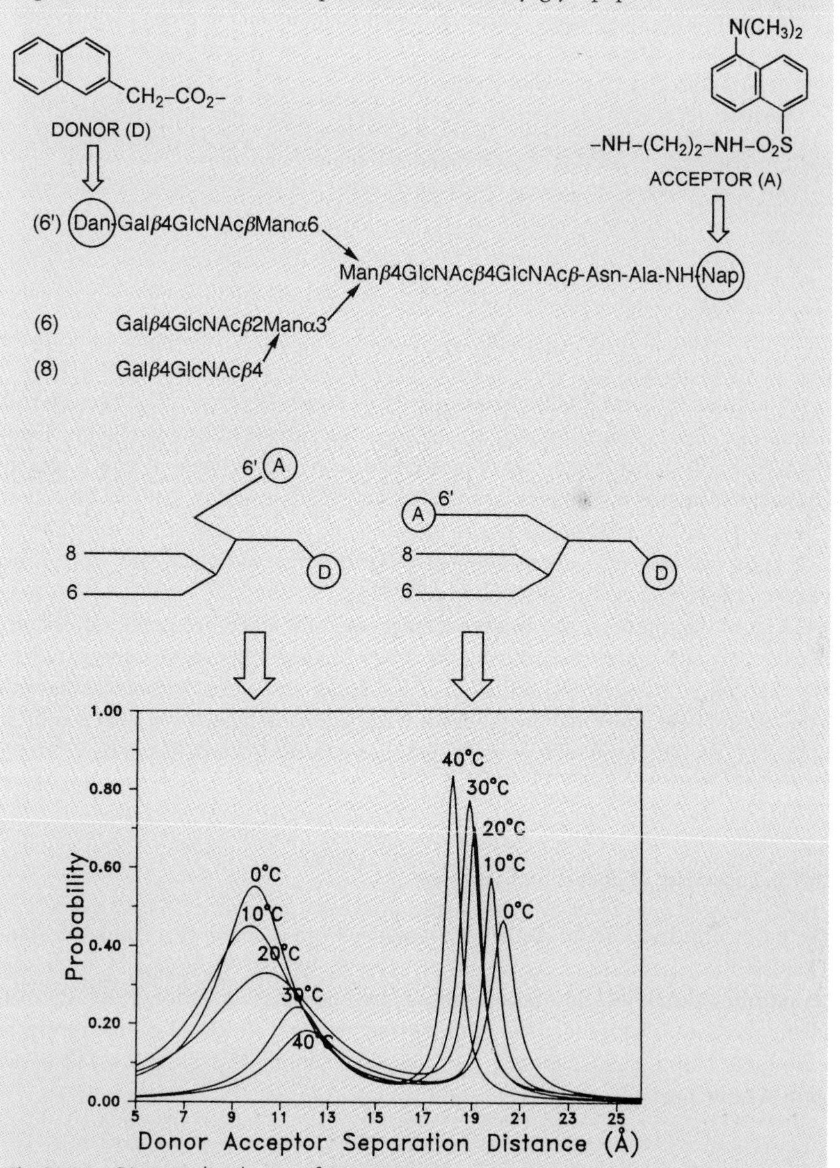

Fig. B9.1.1. Distance distribution of a triantennary glycopeptide reflecting the multiple conformations (adapted from Wu et al.[a)]).

naphthyl-2-acetyl donor and a dansylethylenediamine acceptor. The fluorescence decay curves were found to be satisfactorily fitted with a bimodal Lorentzian distribution, which suggests that the branch labeled with the acceptor can fold back towards the donor. The two conformations differ in donor–acceptor distance by about 10 Å.

Interestingly, one of the conformers obtained from data analysis corresponds to the bound conformation of the same chain in a biantennary oligosaccharide–lectin complex from X-ray diffraction.

Resonance energy transfer is thus a powerful tool for understanding the functions of glycoproteins.

a) Wu P. G., Rice K. G., Brand L. and Lee Y. C. (1991) *Proc. Nat. Acad. Sci. USA* **88**, 9355–9.

generally assumed to be Gaussian, but this assumption is no longer valid in the case of short chains.

As outlined in Section 9.2.1, orientational heterogeneity may affect the determination of a distance distribution, especially in the case of static orientation. Time-resolved fluorescence experiments provide an 'apparent' average distance and an 'apparent' distance distribution containing contributions from both distance and orientation (Wu and Brand, 1992).

A good example of a distribution of distances that was recovered from time-resolved measurements is presented in Box 9.1.

Distance distributions can be determined when the interchromophoric distance does not change significantly during the donor lifetime. Otherwise, energy transfer is enhanced by translational diffusion of the donor and acceptor moieties towards each other. Then, information on relative diffusion coefficients of chain ends in oligopeptides (Katchalski-Katzir et al., 1981) or polymers can be obtained.

9.3
RET in ensembles of donors and acceptors

We have considered so far the energy transfer from a donor to a single acceptor. Extension to ensembles of donor and acceptor molecules distributed at random in an infinite volume will now be considered, paying special attention to the viscosity of the medium. Then, the effect of dimensionality and restricted geometry will be examined. Homotransfer among the donors or among the acceptors will be assumed to be negligible.

9.3.1
RET in three dimensions. Effect of viscosity

The viscosity of the medium plays an important role in the dynamics of energy transfer because the mean distance diffused by the donor and acceptor relative to each other during the excited-state lifetime of the donor τ_D^0 is $(6D\tau_D^0)^{1/2}$[4], in which D is the mutual diffusion coefficient of the donor and acceptor (given by Eq. 4.12), depends on viscosity. This diffusion distance is thus to be compared with the mean distance r between donors and acceptors, so that the criterion will be $6D\tau_D^0/r^2$. Three regimes can be distinguished:

1) $6D\tau_D^0/r^2 \ll 1$: the donor and acceptor cannot significantly diffuse during the excited-state lifetime of the donor. This case is called the *static limit*. Distinction according to the mechanism of tranfer should be made.

(a) When energy transfer from a donor to nearby acceptors can occur via the exchange mechanism (see Section 4.6.3), the Perrin model provides an approximate description. As explained in Section 4.2.3, this model is based on the existence of a *sphere of effective quenching* of the donor in which transfer from the donor to the acceptor occurs with 100% efficiency. If the acceptor lies beyond this sphere, there is no transfer from the donor. Perrin's equation can then be written as

$$\frac{I_D^0}{I_D} = \exp(V_q N_a[A]) \qquad (9.26)$$

where V_q is the volume of the quenching sphere of the donor, and N_a is Avogadro's number. A plot of $\ln(I_D^0/I_D)$ versus [A] yields V_q.

(b) When energy transfer is possible via the Förster dipole–dipole mechanism, transfer occurs from a donor to acceptors that are at a distance less than about $2R_0$ (R_0: Förster critical radius). An analytical expression for the fluorescence decay of the donor in the presence of an ensemble of acceptors can be obtained under the following assumptions:

- the acceptors are randomly distributed in three dimensions;
- translational diffusion is slow compared to the rate of transfer;
- rotational motion of both donor and acceptor molecules is much faster than transfer and is unrestricted, so that the orientation factor is set equal to 2/3.

The survival probability $G^s(t)$ of the donor molecule (i.e. the probability that when excited at $t = 0$, it is still excited at time t) is obtained by summation over all possible rate constants k_T (given by Eq. 9.1), each corresponding to a given donor–acceptor distance r. For a donor molecule surrounded with n acceptor molecules distributed at random in a spherical volume whose radius is much larger than the Förster critical radius R_0, $G^s(t)$ is given by

$$\frac{dG^s}{dt} = -\left[\frac{1}{\tau_D^0} + \sum_{i=1}^{n} k_{Ti}\right] G^s = -\left[\frac{1}{\tau_D^0} + \frac{1}{\tau_D^0}\sum_{i=1}^{n}\left(\frac{R_0}{r_i}\right)^6\right] G^s \qquad (9.27)$$

4) According to classical equations of transla-tional Brownian motion in three dimensions, the mean diffusion distance of a particle during time t is $(6Dt)^{1/2}$.

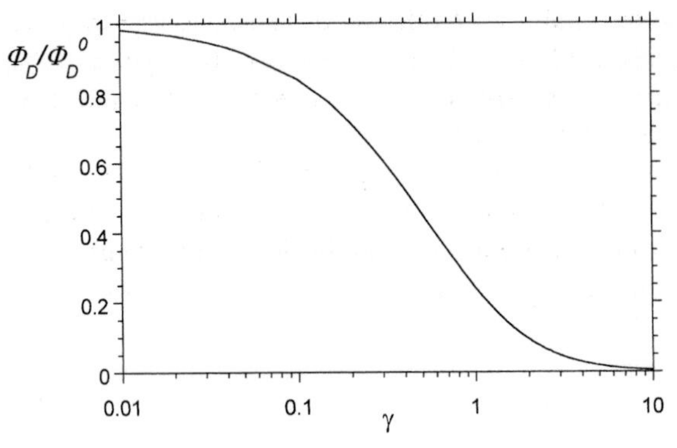

Fig. 9.2. Variations in the relative fluorescence intensity as a function of the parameter γ according to Eq. (9.31).

The solution of this differential equation, with the initial condition $G(0) = 1$, is

$$G^s(t) = \exp\left[-\frac{t}{\tau_D^0} - \frac{t}{\tau_D^0}\sum_{i=1}^{n}\left(\frac{R_0}{r_i}\right)^6\right]$$

$$= \exp\left(-\frac{t}{\tau_D^0}\right)\prod_{n=1}^{n}\exp\left[-\frac{t}{\tau_D^0}\left(\frac{R_0}{r_i}\right)^6\right] \tag{9.28}$$

The fluorescence of the donor is proportional to the average $\langle G^s(t)\rangle$ of this quantity. Förster obtained in this way the following relation:

$$i_D(t) = i_D(0)\exp\left[-\frac{t}{\tau_D^0} - 2\gamma\left(\frac{t}{\tau_D^0}\right)^{1/2}\right] \tag{9.29}$$

where γ is given by

$$\gamma = \frac{\sqrt{\pi}}{2}C_A\frac{4}{3}\pi R_0^3 \tag{9.30}$$

C_A is the concentration of acceptors (expressed in number of molecules per Å^3). $C_A\frac{4}{3}\pi R_0^3$ represents the average number of acceptor molecules in a sphere of radius R_0. Calculation of the integral of $\langle G^s(t)\rangle$ leads to the following expression for the ratio of the fluorescence quantum yields in the presence and absence of energy transfer:

$$\frac{\Phi_D}{\Phi_D^0} = 1 - \sqrt{\pi}\gamma\exp(\gamma^2)[1 - \mathrm{erf}(\gamma)] \tag{9.31}$$

where $\mathrm{erf}(\gamma)$ is the error function $\mathrm{erf}(\gamma) = \frac{2}{\sqrt{\pi}}\int_0^{\gamma} e^{-u^2}\,du$.

The variation of Φ_D/Φ_D^0 versus γ are shown in Figure 9.2.

2) $6D\tau_D^0/r^2 \gg 1$: the mean distance diffused by the donor and acceptor relative to each other during the excited-state lifetime of the donor is much larger than the mean distance *R* between donors and acceptors. Two cases are to be considered according to the mechanism of energy transfer:

(a) In the case of the exchange mechanism, energy transfer occurs via a collisional process and thus obeys Stern–Volmer kinetics, the transfer rate constant being that of the mutual diffusional rate constant k_1 of the donor and acceptor (see Section 4.2.1):

$$\frac{I_D^0}{I_D} = \frac{\tau_D^0}{\tau_D} = 1 + k_1 \tau_D^0 [A] \tag{9.32}$$

where [A] is the acceptor concentration.

(b) In the case of the Förster mechanism, energy transfer from donors occurs not only to acceptors that are within a distance about $2R_0$ at the instant of excitation but also to acceptors that come within this distance during the donor excited-state lifetime τ_D^0. The limit corresponding to $6D\tau_D^0/r^2 \gg 1$ is called the *rapid diffusion limit.*

For an ensemble of donor and acceptor molecules distributed at random in an infinite volume, it is easy to calculate the sum of the rate constants for transfer from donor to all acceptors because all donors of this ensemble are identical in the rapid diffusion limit:

$$k_T = C_A \int_{R_c}^{\infty} \frac{1}{\tau_D^0} \left(\frac{R_0}{r}\right)^6 4\pi r^2 \, dr = C_A \frac{4}{3}\pi R_0^3 \frac{1}{\tau_D^0} \left(\frac{R_0}{R_c}\right)^3 \tag{9.33}$$

where C_A is the concentration of acceptors (molecules/Å^3) and R_c is the distance of closest approach.

It should be emphasized that, because small molecules in usual solvents have diffusion coefficients $<10^{-5}$ cm^2 s^{-1}, the rapid diffusion limit can be attained only for donors with lifetimes of ~1 ms. This is the case for lanthanide ions; for instance, the lifetime of Tb^{3+} chelated to dipicolinate is 2.2 ms. Stryer and coworkers (1978) showed that using Tb^{3+} as a donor and rhodamine B as an acceptor, the concentration of rhodamine B resulting in 50% transfer was 6.7×10^{-6} M, which is three orders of magnitude less than the concentration corresponding to 50% transfer in the static limit.

It is interesting to note that energy transfer in the rapid diffusion limit is sensitive to the distance of closest approach of the donor and acceptor. Based on this observation, some interesting applications in biology have been described, such as the measurement of the distance at which an acceptor is buried in biological macromolecules and membrane systems.

3) $6D\tau_D^0/r^2 \approx 1$: the mean distance diffused by the donor and acceptor relative to each other during the excited-state lifetime of the donor is comparable to the mean

distance r between donors and acceptors. This case is very complex. Among the various approaches, one of the most successful is that due to Gösele and coworkers (1975) (see Box 4.1 of Chapter 4) who obtained the following approximate solution:

$$i_D(t) = i_D(0) \exp\left[-\frac{t}{\tau_D^0} - 2B\gamma\left(\frac{t}{\tau_D^0}\right)^{1/2}\right] \tag{9.34}$$

where the parameter B is given by

$$B = \left[\frac{1 + 5.4x + 4.00x^2}{1 + 3.34x}\right]^{3/4} \tag{9.35}$$

with $x = D(R_0^6/\tau_D^0)^{-1/3}t^{2/3}$ where D is the mutual diffusion coefficient and γ is given by Eq. (9.30).

Finally, as described in Box 4.1 of Chapter 4, an exact numerical solution of the diffusion equation (based on Fick's second law with an added sink term that falls off as r^{-6}) was calculated by Butler and Pilling (1979). These authors showed that, even for high values of R_0 (≈ 60 Å), large errors are made when using the Förster equation for diffusion coefficients $> 10^{-5}$ cm^2 s^{-1}. Equation (9.34) proposed by Gösele et al. provides an excellent approximation.

9.3.2
Effects of dimensionality on RET

Equation (9.29) for Förster kinetics is valid for randomly distributed acceptors in an infinite volume, i.e. in three dimensions. If the dimension is not 3, but 1, or 2, Eq. (9.29) must be rewritten in a more general form:

$$i_D(t) = i_D(0) \exp\left[-\frac{t}{\tau_D^0} - 2\gamma\left(\frac{t}{\tau_D^0}\right)^{d/6}\right] \tag{9.36}$$

where d is the Euclidian dimension (1, 2 or 3) and γ is given by

$$\gamma = \frac{\Gamma(1 - d/6)}{2} C_A V_d R_0^d \tag{9.37}$$

In this expression, V_d is the volume of the unit sphere of dimension d ($V_1 = 2$; $V_2 = \pi$; $V_3 = 4\pi/3$). $C_A V_d R_0^d$ represents the number of acceptors in the sphere of dimension d and of radius R_0. Γ is the gamma function: for $d = 3$, $\Gamma(1/2) = \sqrt{\pi}$; for $d = 2$, $\Gamma(2/3) = 1.35417\ldots$; for $d = 1$, $\Gamma(5/6) = 1.12878$.

Figure 9.3 shows the effect of dimensionality on the decay of the donor using Eqs (9.36) and (9.37).

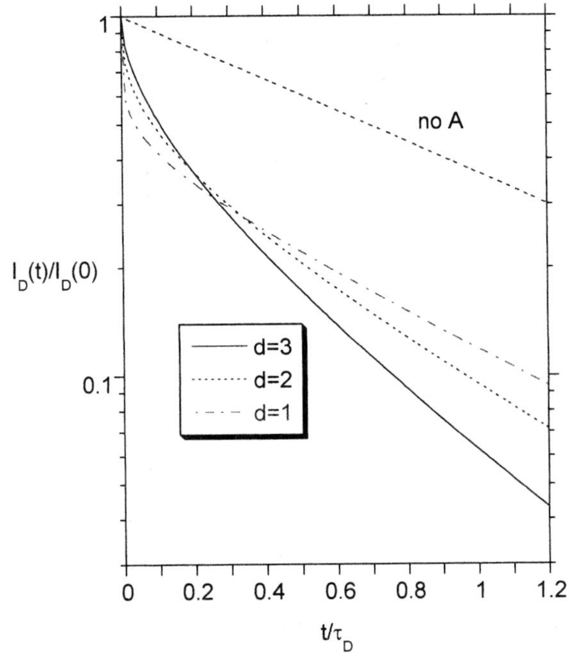

Fig. 9.3. Effect of dimensionality on the fluorescence decay of the donor. The value taken for γ (Eq. 9.37) is $\Gamma(1 - d/6)$, i.e. for $C_A V_d R_0^d = 1$ (average of 1 acceptor in the sphere of dimension d and of radius R_0).

Only a few studies of RET in one dimension – between dyes intercalated into DNA – have been reported.

RET in two dimensions has been studied in monolayers, Langmuir–Blodgett films and phospholipid bilayers of vesicles. In particular, RET provides a very useful tool for the investigation of biological membranes, as exemplified in Box 9.2.

In media of fractal structure, non-integer d values have been found (Dewey, 1992). However, it should be emphasized that a good fit of donor fluorescence decay curves with a stretched exponential leading to non-integer d values have been in some cases improperly interpreted in terms of fractal structure. An 'apparent fractal' dimension may not be due to an actual self-similar structure, but to the effect of restricted geometries (see Section 9.3.3). Another cause of non-integer values is a non-random distribution of acceptors.

9.3.3
Effects of restricted geometries on RET

Analytical expressions for the fluorescence decay in the case of RET between donor and acceptor molecules randomly distributed in various models of restricted geometries (spheres, cylinders, etc.) that mimic simple pores have been established by Klafter and Blumen (1985). The donor decay can be written in a form similar to that of Eq. (9.36):

Box 9.2 RET in phospholipid vesicles as models of biological membranes[a,b]

RET can be used to investigate the lateral organization of phospholipids (range of ~100 Å) in gel and fluid phases. Indeed, information can be obtained on the probe heterogeneity distribution: the donors sense various concentrations of acceptor according to their localization. A continuous probability function of having donors with a mean local concentration C_A of acceptors in their surroundings should thus be introduced in Eq. (9.36) written in two dimensions:

$$I_D(t) = I_D(0)\exp\left(-\frac{t}{\tau_D^0}\right)\int_{C_{min}}^{C_{max}} f(C)\exp\left[-C\left(\frac{t}{\tau_D^0}\right)^{1/3}\right]\mathrm{d}C$$

where C is proportional to the concentration C_A of acceptors:

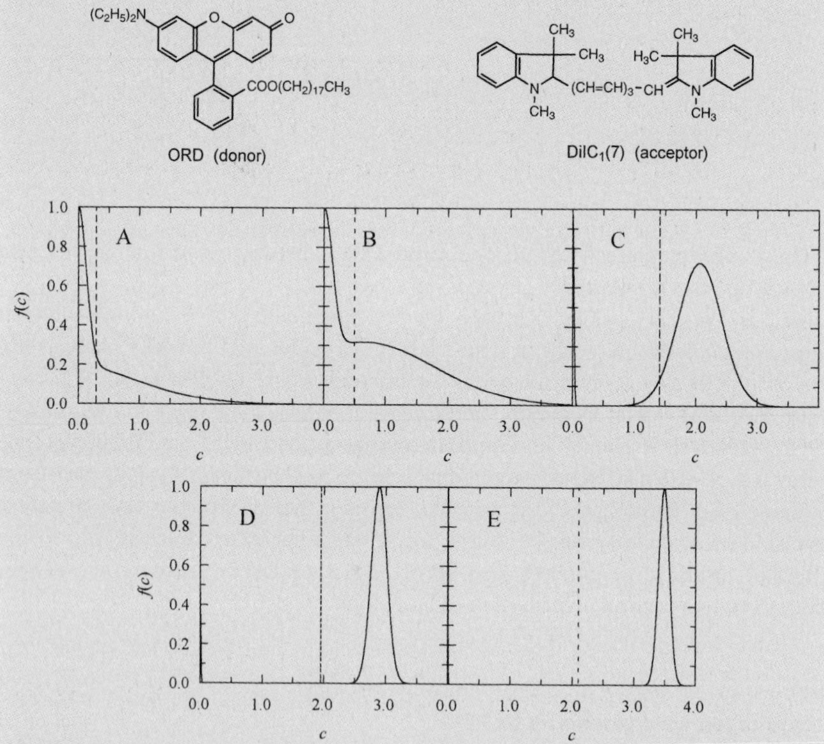

Fig. B9.2.1. Concentration distributions of acceptor [DiIC$_1$(7)] obtained from the analysis of the donor (ORB) decay curves in the gel phase DPPC LUV (25 °C) in the presence of different amounts of acceptor (ratio DiIC$_1$(7) to outer leaflet DPPC: A, 0.0022; B, 0.0044; C, 0.0082; D, 0.0126; E, 0.0171). The vertical broken lines represent the concentration values recovered with a model of discrete exponentials. C_2 is set at 0 (isolated donors). (Reproduced with permission from Lowa et al.[a])

$$C = \Gamma(2/3) C_A \pi R_0^2$$

The recovery of $f(C)$ is an ill-conditioned problem and it is reasonable to take a Gaussian distribution. Moreover, when probes experience two distinct environments (as expected for instance for gel/fluid heterogeneity), a sum of two Gaussian curves should be adequate for data analysis[a]:

$$f(C) = A\left[\exp\frac{(-C - C_1)^2}{2\sigma_1^2} + h\exp\frac{(-C - C_2)^2}{2\sigma_2^2}\right]$$

An investigation of large unilamellar vesicles of dipalmitoylphosphatidylcholine using octadecyl rhodamine B (ORB) (donor) and 1,1′,3,3,3′,3′-hexamethylindotricarbocyanine [$DiIC_1(7)$] (acceptor) was carried out by Prieto and coworkers[b]. In the fluid phase (50 °C), for moderate (1%) acceptor concentrations, donors and acceptors are essentially randomly distributed, and data analysis shows that the acceptor concentration distribution curves are well described by a single narrow Gaussian. In contrast, for the gel phase (25 °C), a sum of two Gaussians is required to fit the decay curves, as shown in Figure B9.2.1. Consequently, in the gel phase, acceptors are partially segregated into a pseudo (defect) phase and donors are more randomly dispersed in the gel phase. These conclusions are supported by Monte Carlo simulations and additional photophysical measurements (steady-state energy transfer, fluorescence self-quenching in steady and transient states, and excitation energy transport).

a) Loura L. M. S., Fedorov A. and Prieto M. (2000) *J. Phys. Chem. B* **104**, 6920.
b) Loura L. M. S. and Prieto M. (2000) *J. Phys. Chem. B* **104**, 6911.

$$i_D(t) = i_D(0)\exp\left[-\frac{t}{\tau_D^0} - A_0\Gamma\left(1 - \frac{\bar{d}}{6}\right)\left(\frac{t}{\tau_D^0}\right)^{\bar{d}/6}\right] \tag{9.38}$$

but $\bar{d}$ is now an 'effective' or 'apparent' dimension and A_0 is related to the surface concentration of acceptor.

Deviations from the Förster decay (Eq. 9.29) arise from the geometrical restrictions. In the case of spheres, the restricted space results in a crossover from a three-dimensional Förster-type behavior to a time-independent limit. In an infinite cylinder, the cylindrical geometry leads to a crossover from a three-dimensional to a one-dimensional behavior. In both cases, the geometrical restriction induces a slower relaxation of the donor.

RET provides a spectroscopic method of probing local pore geometries. Levitz et al. (1988) critically evaluated the application of RET in probing the morphology of porous solids (e.g. silica gels).

9.4
RET between like molecules. Excitation energy migration in assemblies of chromophores

We have considered so far non-radiative energy transfer from a donor to an acceptor that is a different molecule (heterotransfer). Energy transfer between like molecules is also possible (homotransfer) if there is some overlap between the absorption and fluorescence spectra.

$$D^* + D \rightarrow D + D^*$$

In an assembly of chromophores, the process can repeat itself so that the excitation migrates over several molecules; it is called *excitation transport*, or *energy migration*.

Homotransfer does not cause additional de-excitation of the donor molecules, i.e. does not result in fluorescence quenching. In fact, the probability of de-excitation of a donor molecule does not depend on the fact that this molecule was initially excited by absorption of a photon or by transfer of excitation from another donor molecule. Therefore, the fluorescence decay of a population of donor molecules is not perturbed by possible excitation transport among donors. Because the transition dipole moments of the molecules are not parallel (except in very rare cases), the polarization of the emitted fluorescence is affected by homotransfer and information on the kinetics of excitation transport is provided by the decay of emission anisotropy.

9.4.1
RET within a pair of like chromophores

For pairs of like chromophores at a fixed distance and with random and uncorrelated static orientations, the decay of emission anisotropy of the indirectly excited chromophore varies with time, tending to zero (Berberan-Santos and Valeur, 1991) in contradiction to earlier works where it was reported to be 4% of that of the directly excited chromophore. Therefore, because the probability that emission arises from the directly excited chromophore is $1/2$, the decay of emission anisotropy of the latter levels off at $r_0/2$. This can be generalized to an ensemble of n chromophores (with random and uncorrelated static orientations): the decay of emission anisotropy of the directly excited chromophore levels off at r_0/n.

Consequently, homotransfer is fully characterized by the survival probability $G^s(t)$ representing the average probability that an initially excited molecule is still excited at time t. From these relations, the emission anisotropy can be directly expressed as $r(t) = r_0 G^s(t)$.

9.4.2
RET in assemblies of like chromophores

For assemblies of like chromophores in three dimensions in an infinite volume, many theories have provided various expressions of the survival probability $G^s(t)$. Only one of them will be given here, owing to its simplicity and good accuracy. Huber (1981) obtained the following relationship:

$$G^s(t) = \frac{r(t)}{r_0} = \exp\left[-2\gamma'\left(\frac{t}{\tau_0}\right)^{1/2}\right] \tag{9.39}$$

where τ_0 is the intrinsic excited-state lifetime and γ' is given by

$$\gamma' = \frac{\gamma}{\sqrt{2}} = \frac{\sqrt{\pi}}{2\sqrt{2}} C \frac{4}{3} \pi R_0^3 \tag{9.40}$$

where C is the concentration of molecules (number of molecules/Å^3), and R_0 is the critical Förster radius (with the assumption of orientation-averaged interaction).

Investigations of excitation energy transport among chromophores are of major interest in understanding how nature harvests sunlight. In fact, excitation energy migrates among chlorophyll molecules within the antennae pigments and is eventually trapped in the reaction center, where it is converted into chemical energy. Considerable effort has been devoted to the understanding of energy migration in photosynthetic units with the help of fluorescence techniques (Van Grondelle et al., 1996; Pullerits and Sundström, 1996).

Excitation energy transport has also been studied in artificial systems such as polymers having chromophores substituted at intervals along the chains (Guillet, 1985; Webber, 1990), polynuclear complexes (Balzani, 1992) and multi-chromophoric cyclodextrins (Berberan-Santos et al., 1999). These studies are of interest not only in providing a better understanding of the light-harvesting step of photosynthesis, but also in designing photomolecular devices for conversion of light (Balzani and Scandola, 1990).

9.4.3
Lack of energy transfer upon excitation at the red-edge of the absorption spectrum (Weber's red-edge effect)

Section 3.5.1 described the various effects observed upon excitation at the red-edge of the absorption spectrum. In particular, a lack of energy transfer was first observed by G. Weber (1960) (and is called Weber's effect for this reason). This effect can be explained in terms of inhomogeneous broadening of spectra. In a rigid polar solution of fluorophores that are close enough to undergo non-radiative en-

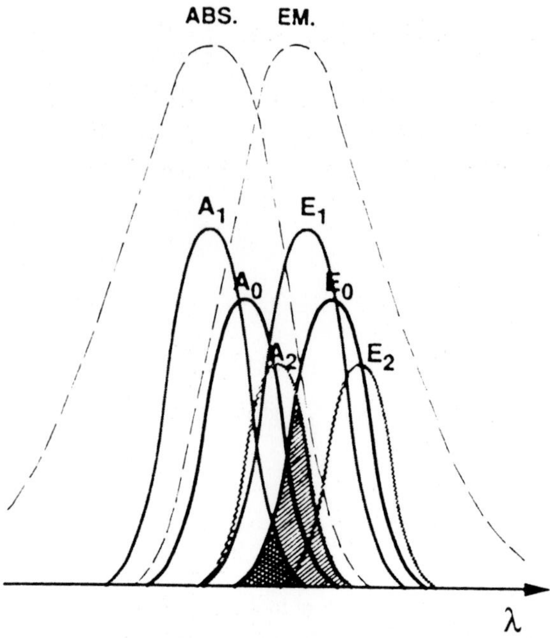

Fig. 9.4. Inhomogeneous spectral broadening responsible for directed energy transfer. The spectral overlap between the emission spectrum E_0 of an excited species (whose absorption spectrum is A_0) and the absorption spectrum A_2 of a solvate absorbing at higher wavelengths (cross-hatched area + double cross-hatched area) is larger than the spectral overlap with the absorption spectrum A_1 of a solvate absorbing at lower wavelengths (double cross-hatched area).

ergy transfer, the probability of transfer from a chromophore in a given configurational state to a chromophore whose configurational state corresponds to a lower electronic transition frequency is greater than the probability of transfer to a chromophore with a higher electronic transition frequency. This is because of the smaller overlap in the latter case between the donor fluorescence and acceptor absorption spectra, as illustrated in Figure 9.4.

In both mechanisms of transfer by dipole–dipole interaction or exchange interaction, the efficiency of transfer is indeed related to the spectral overlap. Therefore, the transfer from 'blue' solvates to 'red' ones is more probable than the reverse transfer. As back transfer has a lower probability, we can speak of *directed non-radiative energy transfer*. This explains the lack of energy transfer upon red-edge excitation: as the excitation wavelength increases beyond the absorption maximum, energy hopping from the directly excited chromophore becomes less and less probable because the proportion of 'blue' partners to which transfer is weak or impossible drastically increases. In other words, energy hopping is spectrally selective. Such a directed energy hopping may be of major importance in energy transport from the antenna to the reaction center in photosynthetic membranes. An example of a very pronounced red-edge effect is given in Box 9.3.

Box 9.3 Red-edge effect in multi-chromophoric cyclodextrins[a)]

A β-cyclodextrin bearing seven 2-naphthoyloxy chromophores, CD7(6), is a good model for studying the effect of the excitation wavelength on energy hopping among chromophores in well-defined positions, as in photosynthetic antennae.

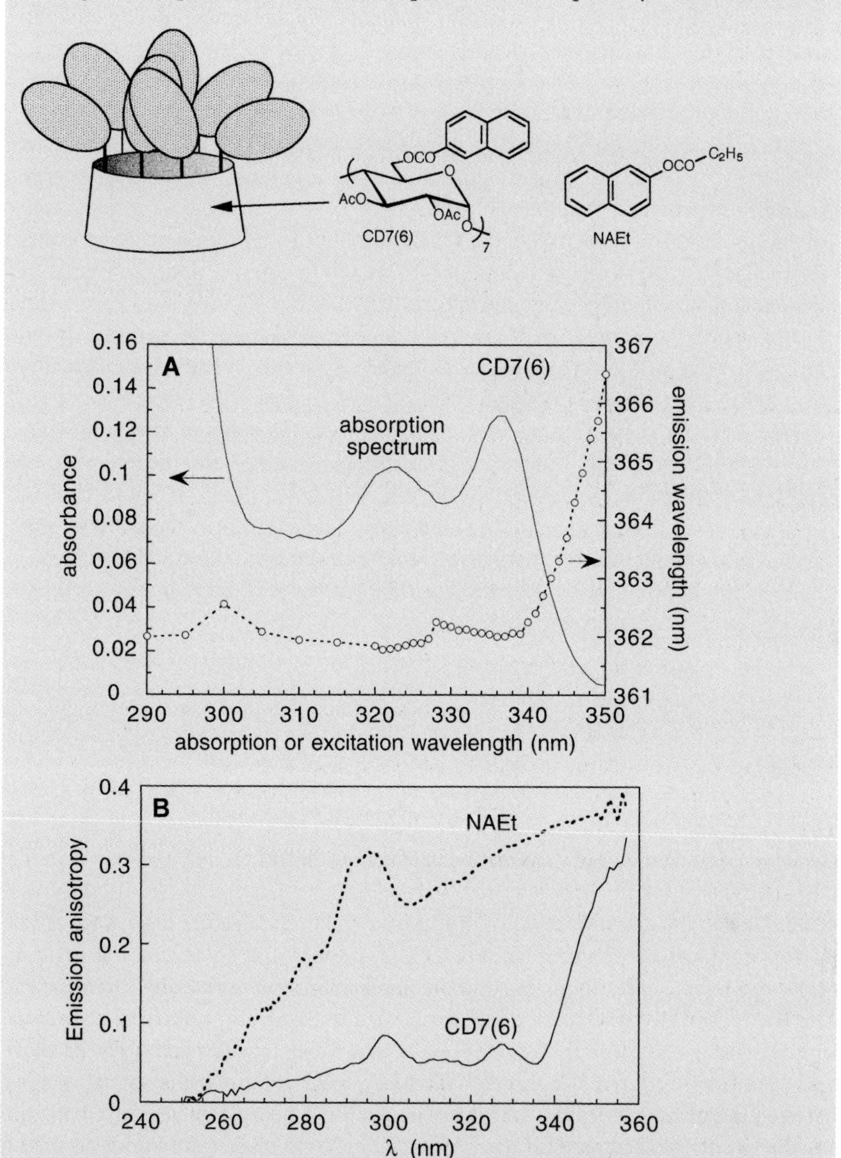

Fig. B9.3.1. A: absorption spectrum of the multi-chromophoric cyclodextrin CD7(6) and variations in the emission maximum as a function of the excitation wavelength (broken line). B: excitation polarization spectra of the model compound NAEt and CD7(6). Solvent: mixture (9:1 v/v) of propylene glycol and 1,4-dioxane at 200 K (adapted from Berberan-Santos et al.[a)]).

Absorption spectra, emission spectra and excitation polarization spectra were recorded in a propylene glycol–dioxane glass at 200 K. Comparison was made with the reference chromophore 2-ethylnaphthoate (NAEt).

Because the rotational motions of the chromophores are frozen in the rigid glass, the depolarization effect observed with CD7(6) as compared with NAEt (see Figure B9.3.1) is solely caused by non-radiative electronic energy transfer, i.e. energy hopping between chromophores. In fact, an emitting chromophore that is indirectly excited by energy transfer is likely to have its emission transition moment oriented differently from that of the directly excited one.

When increasing the excitation wavelength beyond 335 nm, the emission anisotropy drastically increases, thus indicating a gradual decrease in energy transfer efficiency. At the extreme red-edge ($\approx$ 350 nm) there is a complete lack of energy transfer (Weber's effect). It is also noticed that the increase in emission anisotropy at wavelengths ranging from 320 to 328 nm (longwave edge of the second vibronic band) is significantly steeper for CD7(6) than for NAEt. Furthermore, the maximum of emission anisotropy around 300 nm (i.e. at the longwave edge of the second electronic band) is located at higher wavelengths for CD7(6) than for NAEt. We can thus speak of a *vibronic red-edge effect.*

Figure B9.3.1 shows the parallelism between the increase in emission spectrum displacement and fluorescence anisotropy observed for the red-edge of most vibronic bands and especially for the 0–0 one. It can be interpreted in terms of inhomogeous spectral broadening due to solvation heterogeneity. The decrease in energy transfer that is observed upon red-edge excitation is evidence that energy hopping is not chaotic but directed toward lower energy chromophores, as in photosynthetic antennae.

a) Berberan-Santos M. N., Pouget J., Valeur B., Canceill J., Jullien L. and Lehn J.-M. (1993) *J. Phys. Chem.* **97**, 11376–9.

9.5
Overview of qualitative and quantitative applications of RET

In addition to the determination of distances at a supramolecular level, RET can be used to demonstrate the mutual approach of a donor and an acceptor at a supramolecular level as a result of aggregation, association, conformational changes, etc. The donor and acceptor molecules are generally covalently linked to molecular, macromolecular or supramolecular species that move toward each other or move away. From the variations in transfer efficiency, information on the spatial relation between donor and acceptors can thus be obtained. Because of its simplicity, the steady-state RET-based method has been used in many diverse situations as shown below[5].

5) Leading references can be found in the reviews by Cheung (1991), Clegg (1996), Valeur (1993), Webber (1990), Wu and Brand (1994) and the books by Guillet (1985), Van der Meer et al. (1994).

Box 9.4 Formation of hairpin structures as studied by RET[a)]

Single-stranded DNA or RNA may adopt hairpin structures in which the distance between two sequences is much shorter than in the absence of hairpin. Figure B9.4.1 shows two synthetic targets, both containing 45 nucleotides, but only the first one is able to form a hairpin via a loop of four thymines. The second one is used as a control. Both contain the complementary sequences for ethidium–13-mer and 11-mer–coumarin separated by the same number of bases. The efficiency of energy transfer from coumarin to ethidium is close to zero for the control, whereas it is about 25% in the hairpin structure. This value is low but the spatial conformation of this particular three-way junction is only partially known, and the transfer efficiency depends on the relative orientation and/or distance between coumarin and ethidium.

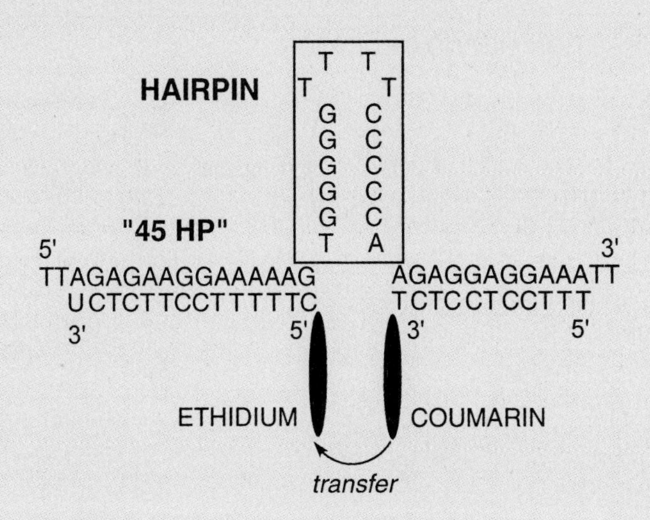

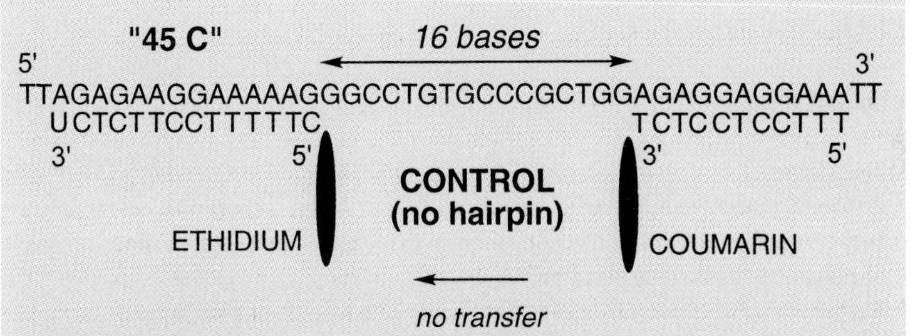

Fig. B9.4.1. Sequences of the 45-mer containing a hairpin structure and the control 45-mer (redrawn from Mergny et al.[a)]).

a) Mergny J. L. et al. (1994) *Nucl. Acids Research* 22, 920–8.

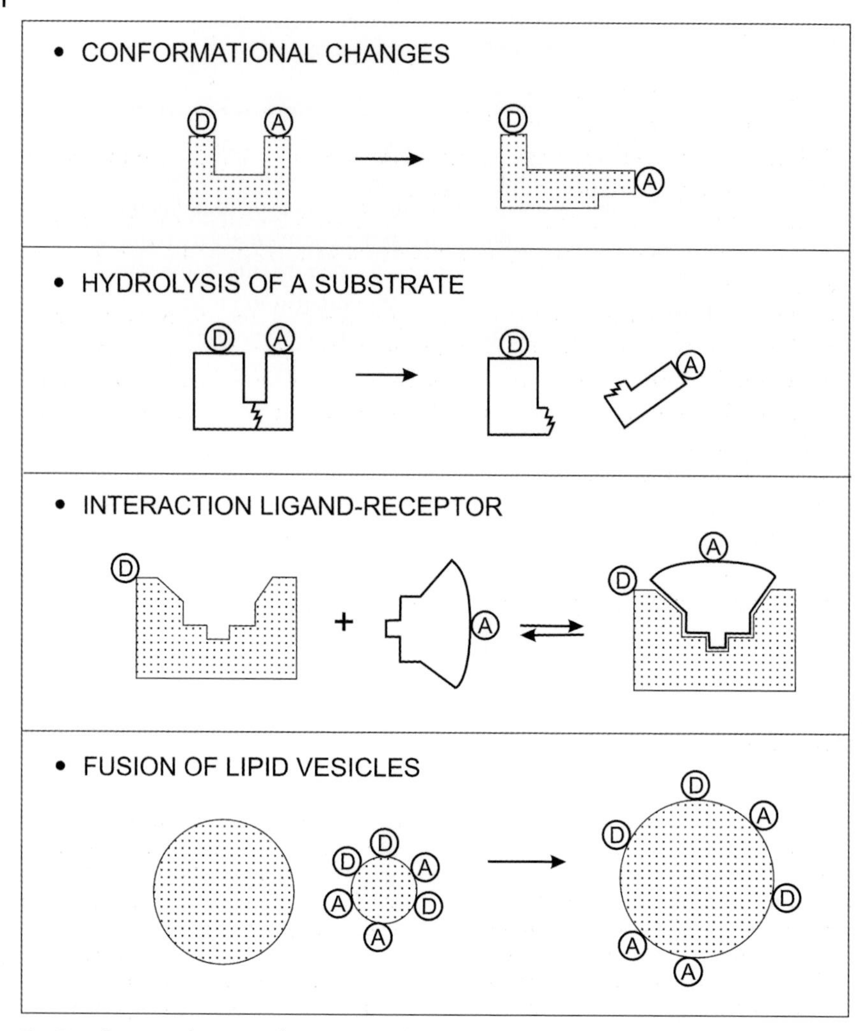

Fig. 9.5. Some applications of RET.

Chemical sciences

- Scintillators
- Polymers: interpenetration of polymer chains, phase separation, compatibility between polymers, interdiffusion of latex particles, interface thickness in blends of polymers, light-harvesting polymers, etc.
- Supramolecular systems: molecular devices, artificial photosynthesis, antenna effect, etc.
- Restricted media and fractals. Morphology of porous solids. Determination of dimensionality.
- Chemical sensors

Life sciences

- Ligand–receptor interactions
- Conformational changes of biomolecules
- Proteins: *in vivo* protein–protein interactions, protein folding kinetics, protein subunit exchange, enzyme activity assay, etc.
- Membranes and models: membrane organization (e.g. membrane domains, lipid distribution, peptide association, lipid order in vesicles, membrane fusion assays, etc.)
- Nucleic acid structures and sequences: primary and secondary structure of DNA fragments, translocation of genes between two chromosomes, detection of nucleic acid hybridization, formation of hairpin structures (see Box 9.4), interaction with drugs, DNA triple helix, DNA–protein interaction, automated DNA sequencing, etc.
- Nucleic acid–protein interaction
- Immunoassays
- Biosensors

Some examples of these applications are shown in Figure 9.5.

9.6 Bibliography

Andrews D. L. and Demidov A. A. (Eds) **(1999)** *Resonance Energy Transfer*, Wiley & Sons, New York.

Baumann J. and Fayer M. D. **(1986)** Excitation Energy Transfer in Disordered Two-Dimensional and Anisotropic Three-Dimensional Systems: Effects of Spatial Geometry on Time-Resolved Observables, *J. Chem. Phys.* 85, 4087–4107.

Balzani V. and Scandola F. **(1990)** *Supramolecular Photochemistry*, Horwood, New York.

Balzani V., Campagna S., Denti G. and Serroni S. **(1992)** Supramolecular Photochemistry: Antenna Effect in Polynuclear Metal Complexes, in: Kochanski E. (Ed.), *Photoprocesses in Transition Metal Complexes, Biosystems and Other Molecules. Experiment and Theory*, Kluwer Academic Publishers, Dordrecht, p. 233.

Berberan-Santos M. N. and Valeur B. **(1991)** Fluorescence Polarization by Electronic Energy Transfer in Donor–Acceptor Pairs of Like and Unlike Molecules, *J. Chem. Phys.* 95, 8048–8055.

Berberan-Santos M. N., Choppinet P., Fedorov A., Jullien L., Valeur B. **(1999)** Multichromophoric Cyclodextrins. 6. Investigation of Excitation Energy Hopping by Monte-Carlo Simulations and Time-Resolved Fluorescence Anisotropy, *J. Am. Chem. Soc.* 121, 2526–2533.

Berlman I. D. **(1973)** Energy Transfer Parameters of Aromatic Compounds, Academic Press, New York.

Butler P. R. and Pilling M. J. **(1979)** The Breakdown of Förster Kinetics in Low Viscosity Liquids. An Approximate Analytical Form for the Time-Dependent Rate Constant *Chem. Phys.* 41, 239–243.

Cheung H. C. **(1991)** Resonance Energy Transfer, in: Lakowicz, J. R. (Ed.), *Topics in Fluorescence Spectroscopy, Vol. 2, Principles*, Plenum Press, New York, pp. 127–176.

Clegg R. M. **(1996)** Fluorescence Resonance Energy Transfer, in: Wang X. F. and Herman B. (Eds), *Fluorescence Imaging*

Spectroscopy and Microscopy, Wiley & Sons, New York, pp. 179–252.

Dale R. E., Eisinger J. and Blumberg W. E. **(1975)** The Orientational Freedom of Molecular Probes. The Orientation Factor in Intramolecular Energy Transfer, *Biophys. J.* 26, 161–194.

Dale R. E. and Eisinger J. **(1975)** Polarized Excitation Transfer, in: Chen R. F. and Edelhoch H. (Eds), *Biochemical Fluorescence: Concepts*, Vol. 1, Marcel Dekker, New York, p. 115.

Dewey T. G. **(1992)** Fluorescence Resonance Energy Transfer in Fractals, *Acc. Chem. Res.* 25, 195–200.

Gösele U., Hauser M., Klein U. K. A. and Frey R. **(1975)** Diffusion and Long-Range Energy Transfer, *Chem. Phys. Lett.*, 34, 519–522.

Guillet J. E. **(1985)** *Polymer Photophysics and Photochemistry*, Cambridge University Press, Cambridge.

Huber D. L. **(1981)** Dynamics of Incoherent Transfer, *Top. Appl. Phys.* 49, 83–111.

Klafter J. and Blumen A. **(1985)** Direct Energy Transfer in Restricted Geometries, *J. Luminescence* 34, 77–82.

Klafter, J. and Drake, J. M. (Eds) **(1989)** *Molecular Dynamics in Restricted Geometries*, Wiley & Sons, New York.

Katchalski-Katzir E., Haas E. and Steinberg I. A. **(1981)** Study of Conformation and Mobility of Polypeptides in Solution by a Novel Fluorescence Method, *Ann. N.Y. Acad. Sci.* 36, 44–61.

Levitz P., Drake J. M. and Klafter J. **(1988)** Critical Evaluation of the Applications of Direct Energy Transfer in Probing the Morphology of Porous Solids, *J. Chem. Phys.* 89, 5224–5236.

Morawetz H. **(1999)** On the Versatility of Fluorescence Techniques in Polymer Research, *J. Polym. Sci.* A 37, 1725–1735.

Pullerits T. and Sundström V. **(1996)** Photosynthetic Light-Harvesting Pigment–Protein Complexes: Toward Understanding How and Why, *Acc. Chem. Res.* 29, 381–389.

Scholes G. D. and Ghiggino K. P. **(1994)** Electronic Interactions and Interchromophore Excitation Transfer, *J. Phys. Chem.* 98, 4580–4590.

Speiser S. **(1996)** Photophysics and Mechanisms of Intramolecular Electronic Energy Transfer in Bichromophoric Molecular Systems: Solution and Supersonic Jet Studies, *Chem. Rev.* 96, 1953–1976.

Stryer L. **(1978)** Fluorescence Energy Transfer as a Spectroscopic Ruler, *Ann. Rev. Biochem.* 47, 819–846.

Thomas D. D., Carlsen W. F. and Stryer L. **(1978)** Fluorescence Energy Transfer in the Rapid-Diffusion Limit, *Proc. Nat. Acad. Sci.* 75, 5746–5750.

Valeur B. **(1989)** Intramolecular Excitation Energy Transfer in Bichromophoric Molecules, in: Jameson D. and Reinhart G. D. (Eds), *Fluorescent Biomolecules*, Plenum Press, New York, pp. 269–303.

Valeur B. **(1993)** Fluorescent Probes for Evaluation of Local Physical, Structural Parameters, in: Schulman S. G. (Ed.), *Molecular Luminescence Spectroscopy. Methods and Applications*, Part 3, Wiley-Interscience, New York, pp. 25–84.

Van der Meer B. W., Coker G. III and Chen S.-Y. S. Chen **(1994)** *Resonance Energy Transfer. Theory and Data*, VCH, New York.

Van Grondelle R., Monshouwer R. and Valkunas L. **(1996)** Photosynthetic Antennae. Photosynthetic Light-Harvesting, *Ber. Bunsenges. Phys. Chem.* 100, 1950–1957.

Webber S. E. **(1990)** Photon Harvesting Polymers, *Chem. Rev.* 90, 1469–1482.

Weber G. **(1960)** Fluorescence Polarization Spectrum and Electronic Energy Transfer in Tyrosine, Tryptophan and Related Compounds, *Biochem. J.* 75, 335–345.

Weber G. and Shinitsky M. **(1970)** Failure of Energy Transfer between Identical Aromatic Molecules on Excitation at the Long Wave Edge of the Absorption Spectrum, *Proc. Nat. Acad. Sci. USA* 65, 823–830.

Wu P. and Brand L. **(1992)** Orientation Factor in Steady-State and Time-Resolved Resonance Energy Transfer Measurements, *Biochemistry* 31, 7939–7947.

Wu P. and Brand L. **(1994)** Resonance Energy Transfer: Methods and Applications, *Anal. Biochem.* 218, 1–13.

10
Fluorescent molecular sensors of ions and molecules

La part de l'imagination dans le travail scientifique est la même que dans le travail du peintre ou de l'évrivain. Elle consiste à découper le réel, et à recombiner les morceaux pour créer quelque chose de neuf.

[*The part played by the imagination in scientific work is the same as in the work of the painter or the writer. It consists of cutting up reality, and recombining the pieces to create something new.*]

François Jacob, 1981

10.1
Fundamental aspects

The design of fluorescent sensors is of major importance because of the high demand in analytical chemistry, clinical biochemistry, medicine, the environment, etc. Numerous chemical and biochemical analytes can be detected by fluorescence methods: cations (H^+, Li^+, Na^+, K^+, Ca^{2+}, Mg^{2+}, Zn^{2+}, Pb^{2+}, Al^{3+}, Cd^{2+}, etc.), anions (halide ions, citrates, carboxylates, phosphates, ATP, etc.), neutral molecules (sugars, e.g. glucose, etc.) and gases (O_2, CO_2, NO, etc.). There is already a wide choice of fluorescent molecular sensors for particular applications and many of them are commercially available. However, there is still a need for sensors with improved selectivity and minimum perturbation of the microenvironment to be probed. Moreover, there is the potential for progress in the development of fluorescent sensors for biochemical analytes (amino acids, coenzymes, carbohydrates, nucleosides, nucleotides, etc.).

The success of fluorescent sensors can be explained by the distinct advantages offered by fluorescence detection in terms of sensitivity, selectivity, response time, local observation (e.g. by fluorescence imaging spectroscopy). Moreover, remote sensing is possible by using optical fibers. The great improvement in the sensitivity and the spatial or temporal resolution of instruments is also a factor in their success.

Many terms are used in the field of fluorescence sensing: fluorescent sensors, fluorosensors, fluorescent chemosensors, fluorescent molecular sensors, luminescent sensor molecules, luminescent sensors, fluorescent biosensors, fluorescent optical sensors, etc. It is important to make a clear distinction between the *analyte-responsive (supra)molecular moiety*, involving a fluorophore that signals the presence of an analyte by changes in its fluorescence characteristics, and the complete *optical sensing device*, i.e. the light source, the analyte-responsive (supra)molecular moiety properly immobilized (e.g. in plastified polymers, sol–gel matrices, etc.), the optical system (involving an optical fiber or not) and the light detector (photomultiplier or photodiode) connected to appropriate electronics for displaying the signal. In principle, a *fluorescent sensor* is the complete device, but in many papers, authors assign the term *fluorescent sensor* to the fluorescent analyte-responsive (supra)-molecular moiety as well. In order to avoid confusion, it is recommended in the latter case to use the term *fluorescent molecular sensor*.

Another distinction should be made (independently of the fluorescence aspects) between *chemical sensors* (also called *chemosensors*) and *biosensors*. In the former, the analyte-responsive moiety is of abiotic origin, whereas it is a biological macromolecule (e.g. protein) in the latter.

In fluorescent molecular sensors, the fluorophore is the *signaling* species, i.e. it acts as a signal transducer that converts the information (presence of an analyte) into an optical signal expressed as the changes in the photophysical characteristics of the fluorophore. In contrast, in an electrochemical sensor, the information is converted into an electrical signal.

The present chapter is restricted to fluorescent molecular sensors, for which three classes can be distinguished (Figure 10.1):

- Class 1: fluorophores that undergo quenching upon collision with an analyte (e.g. O_2, Cl^-).
- Class 2: fluorophores that can reversibly bind an analyte. If the analyte is a proton, the term *fluorescent pH indicator* is often used. If the analyte is an ion, the term *fluorescent chelating agent* is appropriate. Fluorescence can be either quenched upon binding (CEQ type: Chelation Enhancement of Quenching), or enhanced (CEF type: Chelation Enhancement of Fluorescence). In the latter case, the compound is said to be *fluorogenic* [e.g. 8-hydroxyquinoline (oxine)].
- Class 3: fluorophores linked, via a spacer or not, to a receptor. The design of such sensors, which are based on molecule or ion recognition by a receptor, requires special care in order to fulfil the criteria of affinity and selectivity. These aspects are relevant to the field of *supramolecular chemistry*. The changes in photophysical properties of the fluorophore upon interaction with the bound analyte are due to the perturbation by the latter of photoinduced processes such as electron transfer, charge transfer, energy transfer, excimer or exciplex formation or disappearance, etc. These aspects are relevant to the field of *photophysics*. In the case of ion recognition, the receptor is called an *ionophore*, and the whole molecular sensor is

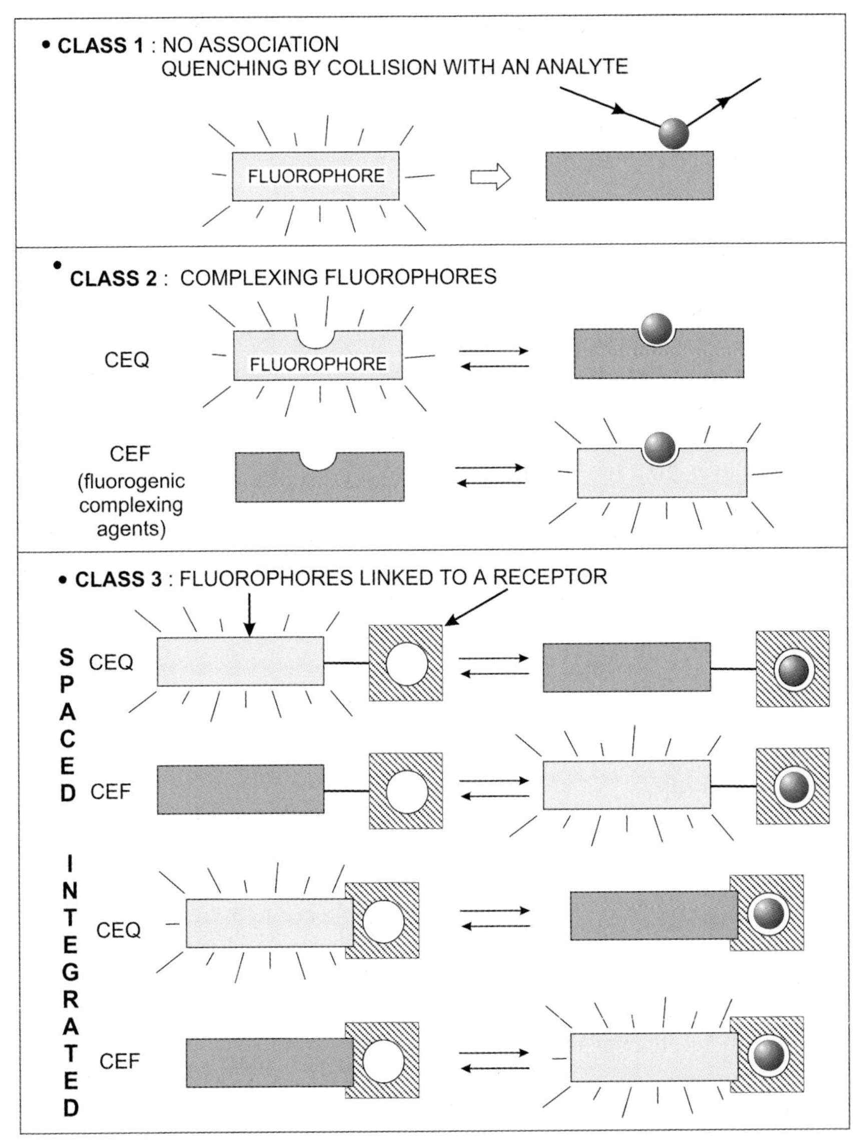

CEF : Chelation or Complexation Enhancement of Fluorescence
CEQ : Chelation or Complexation Enhancement of Quenching

Fig. 10.1. Main classes of fluorescent molecular sensors of ions or molecules.

called a *fluoroionophore*. Again, fluorescence can be quenched (CEQ) or enhanced (CEF).

In cases where fluorescence sensing is accompanied by binding of the analyte (classes 2 and 3), the dissociation constant of the complex should match the expected range of analyte concentration, which can vary greatly according to the field of application.

Several books and many reviews have been devoted to fluorescent molecular sensors (see Bibliography at the end of this chapter). This chapter will present only selected examples to help the reader understand the fundamental aspects and the principles.

10.2 pH sensing by means of fluorescent indicators

Fluorescent pH indicators offer much better sensitivity than the classical dyes such as phenolphthalein, thymol blue, etc., based on color change. They are thus widely used in analytical chemistry, bioanalytical chemistry, cellular biology (for measuring intracellular pH), medicine (for monitoring pH and pCO_2 in blood; pCO_2 is determined via the bicarbonate couple). Fluorescence microscopy can provide spatial information on pH. Moreover, remote sensing of pH is possible by means of fiber optic chemical sensors.

10.2.1 Principles

First, it should be recalled that the value of pH is defined by the activity of protons H^+ (simplified notation for $H_3O^+_{aq}$) – and not by their concentration – via the following relationship:

$$pH = -\log a_{H^+} \tag{10.1}$$

A glass electrode yields a response that is directly linked to the activity of protons. This is not true for optical methods (spectrophotometry and spectrofluorometry) using pH indicators. In fact, these methods are based on the optical determination of the concentrations of the acidic form, [A], and the basic form, [B], of the indicator[1)] and use of the well known Henderson–Hasselbach equation

$$\boxed{pH = pK_a + \log \frac{[B]}{[A]}} \tag{10.2}$$

For fluorometric titrations, this equation can be rewritten as

$$\boxed{pH = pK_a + \log \frac{I - I_A}{I_B - I}} \tag{10.3}$$

1) The acidic and basic forms can also be written as HA and A^-, respectively, or BH^+ and B.

where I is the fluorescence intensity at a given wavelength, I_A and I_B are the fluorescence intensities measured at the same wavelength when the indicator is only in the acidic form or only in the basic form, respectively (see Appendix A).

However, the Henderson–Hasselbach equation results from an oversimplification that deserves special attention. In fact, the equilibrium constant K characterizing the acid–base equilibrium must be written with activities:

$$K = \frac{a_{H^+} a_B}{a_A} \tag{10.4}$$

K is the true value of the acidity constant, which depends only on temperature. Using the definition of pH (Eq. 10.1), Eq. (10.4) becomes

$$\mathrm{pH} = \mathrm{p}K + \log \frac{a_B}{a_A} \tag{10.5}$$

or

$$\boxed{\mathrm{pH} = \mathrm{p}K + \log \frac{[\mathrm{B}]}{[\mathrm{A}]} + \log \frac{f_B}{f_A}} \tag{10.6}$$

where f_B and f_A are the activity coefficients relative to molarities. The activity coefficients are close to 1 only in very dilute aqueous solutions (reference state: solute at infinite dilution).

By comparison with Eq. (10.2), we obtain

$$\mathrm{p}K_a = \mathrm{p}K + \log \frac{f_B}{f_A} \tag{10.7}$$

Therefore, K_a should be considered as an 'apparent' constant that depends not only on temperature, but also on factors able to modify the activity coefficients:

- ionic strength (if A and/or B are ionic). Great care should be taken in the case of media containing high concentrations of salts. Highly charged indicators like pyranine are very sensitive to ionic strength (see Box 10.1). For less highly charged indicators, the effect is less significant. For instance, the pK_a of fluorescein is almost independent of ionic strength.
- specific interactions depending on the chemical nature of the indicator and the surrounding medium (e.g. buffer constituents, composition of the physiological medium, vicinity of interfaces in microheterogeneous media, etc.).
- structural changes of the medium. In particular, when probing microheterogeneous media (e.g. the vicinity of interfaces of micelles or lipid bilayers), the pK_a value differs significantly from the true pK value measured in dilute aqueous solutions.

Box 10.1 Effect of ionic strength on the pK_a of pyranine

Pyranine (1-hydroxypyrene-3,6,8-trisulfonic acid trisodium salt, Figure 10.2) has a pK of 7.2 in a medium at infinite dilution. Such a dye is chosen here because it is particularly sensitive to ionic strength owing to the number of negative charges it bears: three in the acidic form and four in the basic form. The evolution of the absorption spectrum in a buffer at pH 7.2 as a function of the ionic strength (addition of KCl) indicates that the ratio $[PyO^-]/[PyOH]$ does not remain constant[a]. Because pK is a constant and the pH is kept constant by the buffer, Eq. (10.8) shows that the variations of $[PyO^-]/[PyOH]$ compensate for the variations in the ratio of the activity coefficients, f_{PyO^-}/f_{PyOH}, as a function of ionic strength. Once the ratio $[PyO^-]/[PyOH]$ is determined, the apparent pK value, pK_a, can be calculated by means of the following equation:

$$pK_a = pH - \log \frac{[PyO^-]}{[PyOH]}$$

- At ionic strengths $I < 1$ M: the ratio $[PyO^-]/[PyOH]$ increases with increasing I. The resulting decrease in f_{PyO^-}/f_{PyOH} is in agreement with the Debye–Huckel theory: the activity coefficient of PyO^- (four charges) is indeed expected to decrease more than that of PyOH (three charges). At an ionic strength of 1 M, pK_a is equal to 6.7, i.e. 0.5 unit lower than the value in pure water.
- At ionic strengths $I > 1$ M: the spectra do not exhibit isosbestic points any more and an increase, instead of a decrease, in pK_a with increasing I is now observed. In this range, the specific interactions and the structural changes are predominant.

The effect of various charged species on the pK_a of pyranine is shown below[b].

Solution	*pK_a*
Water with 0.0066 M phosphate buffer	7.51 ± 0.03
Water with 6% albumin	7.83 ± 0.06
Water with cetyl trimethylammonium bromide (6.6 mM)	7.69 ± 0.03
Water with sodium dodecylsulfate (50 mM)	7.43 ± 0.06

All these data show that the ionic strength can induce a change in pK_a up to 0.5–0.6 unit. Because the ionic strength effect causes spectral changes indistinguishable from those caused by pH, a correction method is desirable. A method based on a double pH indicator system, i.e. pyranine and 4-methylumbelliferone, was proposed by Opitz and Lübbers[c].

a) Valeur B. and Bardez E. (1989), Proton Transfer in Reverse Micelles and Characterization of the Acidity in the Water Pool, in: Pileni M. P. (Ed.), *Structure and Reactivity in Reverse Micelles*, Elsevier, New York, p. 103.

b) Wolfbeis O. S., Fürlinger E., Kroneis H. and Marsoner H. (1983) *Fresenius Z. Anal. Chem.* **314**, 119.

c) Opitz N. and Lübbers D. W. (1984) *Adv. Exp. Med. Biol.* **169**, 907.

In almost all applications, *fluorescent pH indicators are employed in a pH range around the ground state pK_a (even if the excited state pK is different)*. Therefore, the absorption (and excitation) spectrum depends on pH in the investigated range. These indicators can be divided into three classes (see formulae in Figure 10.2) on the basis of the elementary processes (photoinduced proton transfer or electron transfer) that are involved.

- Class A: *Fluorophores that undergo photoinduced proton transfer but not electron transfer* (e.g. pyranine, hydroxycoumarins; see Sections 10.2.2.1 and 10.2.2.2). Most of these fluorophores are much more acidic in the excited state than in the ground state (case discussed in Chapter 4, Section 4.5), i.e. the pK in the excited state is much lower than that in the ground state. Therefore, in a pH range around the ground state pK, the emitting form is always the basic form because excitation of the acidic form is followed by excited-state deprotonation. The fluorescence spectrum is thus unchanged, in contrast to the excitation spectrum.
- Class B: *Fluorophores that undergo neither photoinduced proton transfer nor photoinduced electron transfer* (e.g. fluorescein, eosin Y, benzo[*c*]xanthene dyes). The evolution of the fluorescence spectrum versus pH should be similar to that of the absorption spectrum. In other words, when increasing the pH, the absorption and emission bands of the acidic form should decrease with a concomitant increase in the absorption and emission bands of the basic form. This is indeed observed for benzo[*c*]xanthene dyes that have been developed for physiological applications; these dyes can be described as semi-naphthofluoresceins (SNAFL) and semi-naphthorhodafluors (SNARF) (see Section 10.2.2.4). In contrast, the currently used xanthene dyes like fluorescein derivatives and eosin Y show little change in fluorescence bandshape, but much more significant changes in absorption or excitation spectra (see Section 10.2.2.3).
- Class C: *Fluorophores that undergo no photoinduced proton transfer but only photoinduced electron transfer.* The fluorescence quantum yield of these fluorophores is very low when they are in the non-protonated form because of internal quenching by electron transfer. Protonation (which suppresses electron transfer) induces a very large enhancement of fluorescence (see Section 10.2.2.5). The bandshapes of the excitation and fluorescence spectra are independent of pH.

From a practical point of view, the pH dependence of the emission and/or excitation spectra is first examined in a preliminary experiment. The evolution of the fluorescence intensity for an appropriate couple of excitation and emission wavelengths is then recorded under experimental conditions (in particular composition of the medium, ionic strength, etc.) as close as possible to the medium in which pH must be determined. The resulting curve can be used as a calibration curve.

It is recommended, whenever possible, to carry out ratiometric measurements, i.e. to determine the ratio of the fluorescence intensities measured at two wavelengths according to the following methods:

- *one emission and two excitation wavelengths* (see Figure 10.3A): this excitation ratio method is possible for most indicators (class A and class B) and is used in conventional fluorescence microscopy.

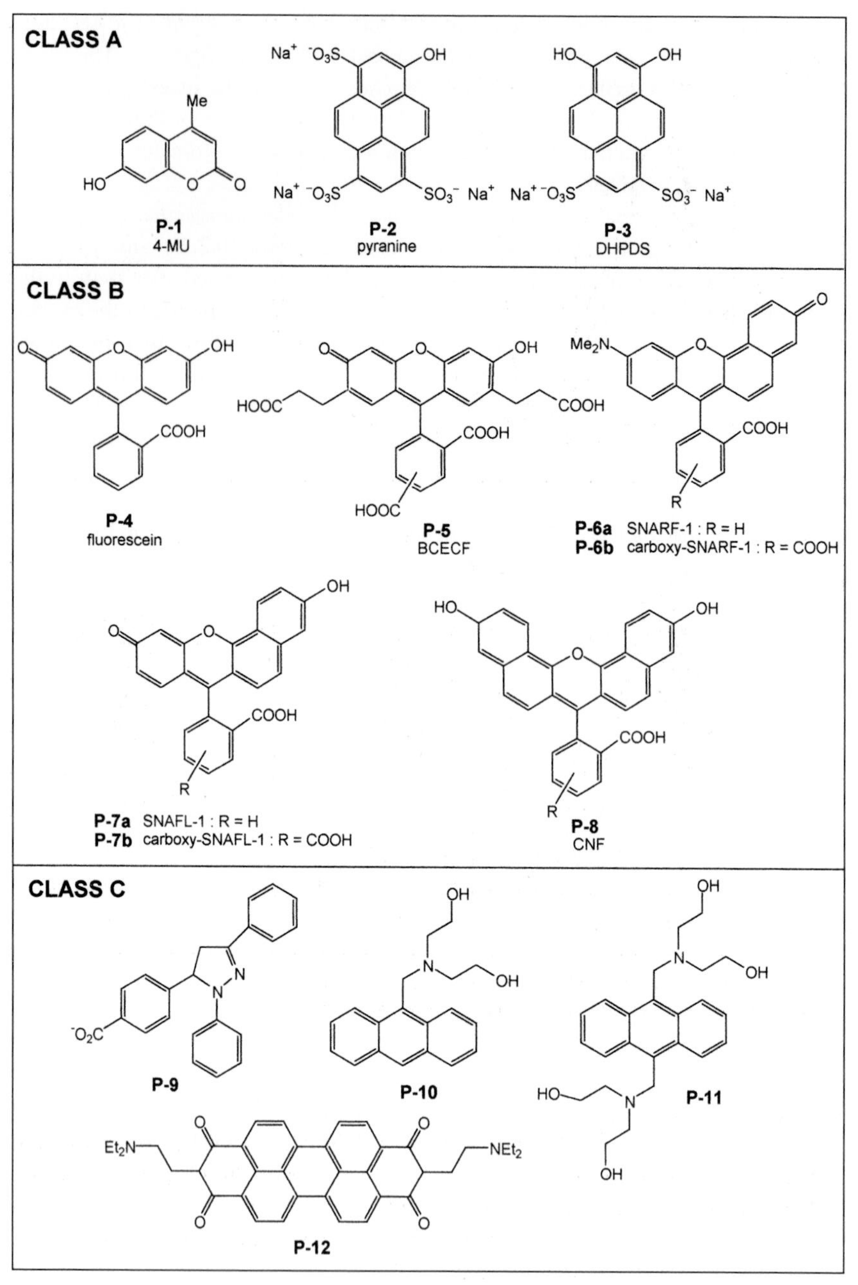

Fig. 10.2. Various classes of fluorescent pH indicators. (P-1 to P-8: Haugland R. P., *Handbook of Fluorescent Probes and Research Chemicals*, 6th edn, Molecular Probes, Inc., Eugene, OR. P-9: de Silva A. P. et al. (1989) *J. Chem. Soc., Chem. Commun.* 1054. P-10 and P-11: Bissell R. A. et al. (1992) *J. Chem. Soc., Perkin Trans. 2.* 1559. P-12: Daffy L. M. et al. (1998) *Chem. Eur. J.* **4**, 1810).

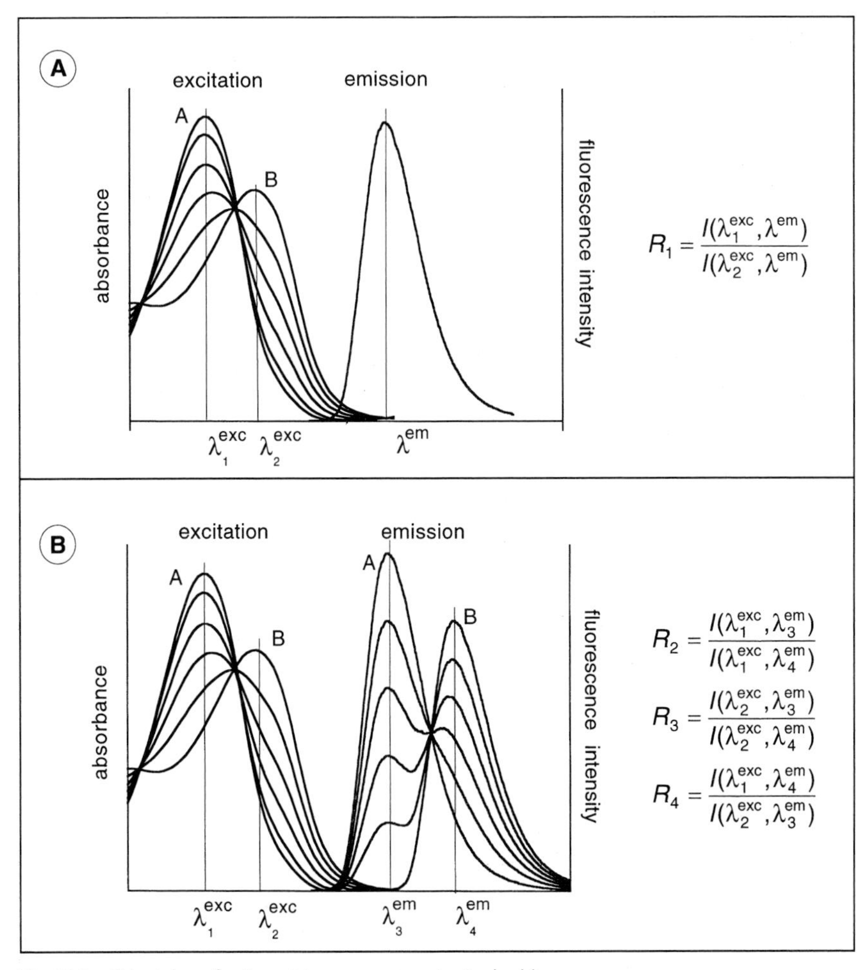

Fig. 10.3. Principles of ratiometric measurements. A: double-excitation measurements. B: double-emission measurements.

- *one excitation and two emission wavelengths* (Figure 10.3B, ratios R_2 and R_3): this emission ratio method is applicable only to indicators exhibiting dual emission (emission of the acidic and basic forms at different wavelengths) (part of class B). The changes in fluorescence spectrum vs pH follow those of the absorption spectrum. This method is preferred for flow cytometry and confocal microscopy, and it allows emission ratio imaging.
- *two excitation and two emission wavelengths* (Figure 10.3B, ratio R_4): this method is also possible for indicators exhibiting dual emission. The main advantage of this method is that the emission intensities are much greater because both the acidic and basic forms are excited near their absorption maxima, resulting in stronger emission from both species.

Tab. 10.1. Examples of fluorescent pH indicators allowing ratiometric measurements. It should be recalled that the values of pK_a (given here at room temperature for dilute solutions) can be more or less affected by changes in ionic strength and temperature

Fluorophore	pK_a	*Type of measurement*
4-MU	7.8	excitation ratio 365/335 nm
pyranine	7.2	excitation ratio 450/400 nm
fluorescein	2.2, 4.4, 6.7	excitation ratio 490/435 nm
BCECF	7.0	excitation ratio 505/439 nm
SNAFL dyes	7.0–7.8	excitation ratio 490/540 nm or emission ratio 540/630 nm
SNARF dyes	7.0–7.8	emission ratio 580/630 nm
CNF	7.5	excitation ratio 600/510 nm or emission ratio 550/670 nm

Abbreviations: 4-MU: 4-methylumbelliferone; BCECF: 2′,7′-bis(carboxyethyl)-5 (or 6)-carboxyfluorescein; SNAFL: semi-naphthofluoresceins; SNARF: semi-naphthorhodafluors; CNF: carboxynaphthofluorescein.

The ratiometric measurements are preferable because the ratio of the fluorescence intensities at two wavelengths is in fact independent of the total concentration of the dye, photobleaching, fluctuations of the source intensity, sensitivity of the instrument, etc. The characteristics of some fluorescent pH indicators allowing ratiometric measurements are given in Table 10.1.

Measurements at one excitation and one emission wavelength are always possible of course (this is the sole possibility for class C), but they do not offer the advantages of the ratiometric methods described above. In this case, the calibration curve may be successfully fitted to Eq. (10.3).

For ratiometric measurements, the following equation may be used to fit the calibration curve (see Appendix A):

$$\mathrm{pH} = \mathrm{p}K_a + \log \frac{R - R_A}{R_B - R} + \log \frac{I_A(\lambda_2)}{I_B(\lambda_2)} \tag{10.8}$$

where R is the ratio $I(\lambda_1)/I(\lambda_2)$ of the fluorescence intensities at two excitation wavelengths (or two emission wavelengths) λ_1 and λ_2. R_A and R_B are the values of R when only the acidic form or the basic form is present, respectively. $I_A(\lambda_2)/I_B(\lambda_2)$ is the ratio of the fluorescence intensity of the acidic form alone to the intensity of the basic form alone at the wavelength λ_2 chosen for the denominator of R. It should be emphasized that R, R_A and R_B are very sensitive to the wavelengths chosen and also to instrumental settings.

The preceding relationships show that the usual working range is about 2 pH units around $\mathrm{p}K_a$. However, several 2 pH unit ranges are juxtaposed when the indicator possesses more than one $\mathrm{p}K_a$ (e.g. fluorescein).

10.2.2 The main fluorescent pH indicators

10.2.2.1 Coumarins

A derivative of coumarin that has been extensively used for intracellular pH measurement is 4-methylumbelliferone (4-methyl-7-hydroxycoumarin) because of its pK_a value of 7.8, the relatively large variation in its fluorescence intensity versus pH, and its low toxicity. Excitation ratio measurements at 365 and 334 nm with observation at 450 nm permit a six- to ten-fold increase over the pH range from 6 to 8.

4-Chloromethyl-7-hydroxycoumarin and 6,7-dihydroxy-4-methylcoumarin have been also used as pH indicators.

10.2.2.2 Pyranine

Pyranine (1-hydroxypyrene-3,6,8-trisulfonic acid trisodium salt) has already been described in Chapter 4 and its sensitivity to ionic strength is outlined in Box 10.1. Its ground state pK_a is 7.2 (in a medium of low ionic strength). In the pH range of practical interest for intracellular measurements, i.e. around 7, the acidic form of pyranine undergoes deprotonation upon excitation, which explains why the fluorescence spectrum is unchanged in this pH range and corresponds to the emission of the basic form (nevertheless, a weak emission of the acidic form appears as a result of geminate recombination; see Chapter 4, Section 4.5.3). In contrast, the large changes in the absorption and excitation spectra reflect the varying proportions of the acidic and basic forms in the ground state (Figure 10.4). Excitation ratio measurements, 450/400 nm, are possible.

The pyranine derivative DHPDS (1,3-dihydroxypyrene-6,8-disulfonic acid) retains the high pH sensitivity of pyranine but its acido–basic properties are much less sensitive to ionic strength. This indicator has two pK_a values: 7.33 ± 0.04 and 8.55 ± 0.2. Ratiometric measurements in both excitation and emission are possible.

10.2.2.3 Fluorescein and its derivatives

Fluorescein is one of the most well-known fluorescent dyes, discovered by A. von Baeyer in 1871. It has been used in numerous applications. There are nine different forms of fluorescein; six of them exist in aqueous solutions with pK_a values of 2.2, 4.4 and 6.7 (Figure 10.5).

The exceptionally broad range of pH response (from 1 to 8) can be explained by the existence of two consecutive transitions: neutral form–monoanion and monoanion–dianion. The pH dependence of the emission and excitation spectra is shown in Figure 10.5.

The cationic form has a higher fluorescence quantum yield than the neutral form, which allows the excitation ratio (490/435 nm) measurement to be extended down to pH 2.

With the aim of increasing the apparent pK_a to a value more suitable for intracellular pH measurements, fluorescein derivatives have been designed: CF (5(or 6)-

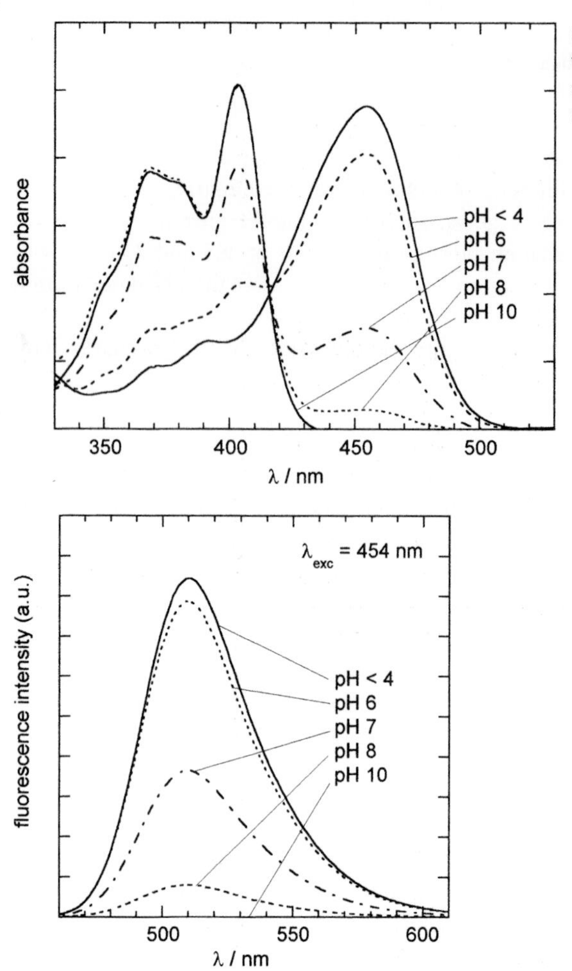

Fig. 10.4. pH dependence of the excitation and emission spectra of pyranine around neutral pH.

carboxyfluorescein) and BCECF (2′,7′-bis(carboxyethyl)-5(or 6)-carboxyfluorescein). BCECF, whose pK_a is close to 7, has become the most widely used intracellular fluorescent pH indicator (see Table 10.1).

10.2.2.4 SNARF and SNAFL

SNARF and SNAFL indicators are benzo[*c*]xanthene dyes that can be described as semi-naphthofluoresceins and semi-naphthorhodafluors, respectively, depending on whether the benzo[*c*]xanthene ring is substituted at the 10-position with oxygen or with nitrogen, respectively (Whitaker et al., 1991). These indicators, whose pK_a values are in the physiological range, exhibit distinct emission bands for the protonated and deprotonated forms so that emission ratio measurements are possible. In SNAFL, the acidic form is more fluorescent, whereas in SNARF, the basic form is more fluorescent.

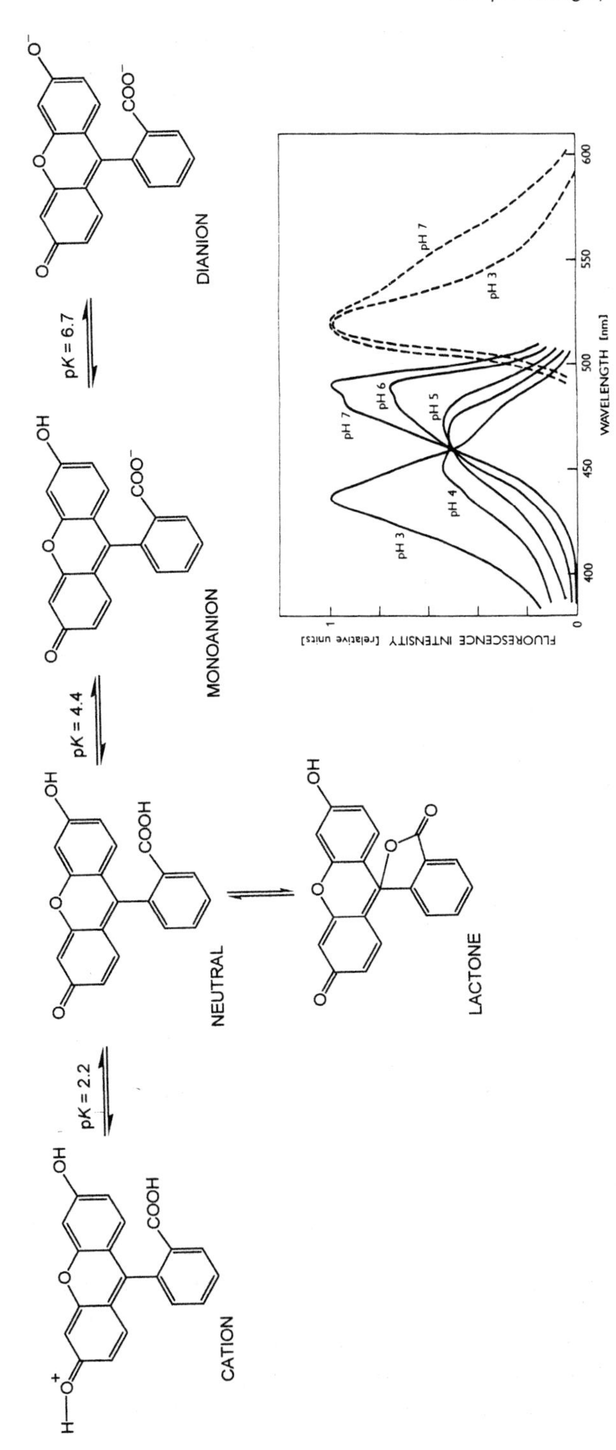

Fig. 10.5. Various forms of fluorescein in aqueous solution and pH dependence of its emission and excitation spectra. The emission spectra are normalized to the same height at the maximum (spectra from Slavik, 1994).

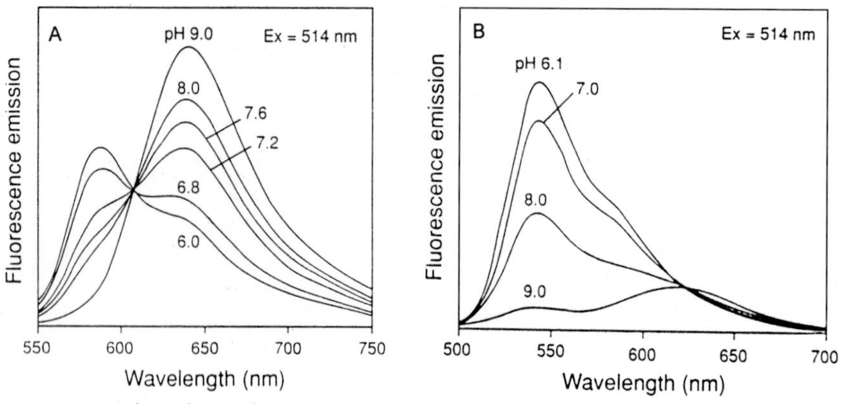

Fig. 10.6. pH dependence of the emission spectra of carboxy-SNARF-1 (A) and carboxy-SNAFL-1 (B) (from Haugland R. P., *Handbook of Fluorescent Probes and Research Chemicals*, 6th edn, Molecular Probes, Inc., Eugene, OR).

Carboxylated derivatives have been designed for improving retention by cells. The pH dependence of the fluorescence spectrum of carboxy SNARF-1 and carboxy-SNAFL-1 are shown in Figure 10.6.

Introduction of an additional phenyl ring in the SNAFL structure leads to carboxynaphthofluorescein (CNF), whose absorption and fluoresence spectra are further red-shifted. CNF has a pK_a of 7.5 and can be used in either excitation or emission ratiometric measurements.

10.2.2.5 **PET (photoinduced electron transfer) pH indicators**

Most fluorescent PET molecular sensors, including pH indicators of this type, consist of a fluorophore linked to an amine moiety via a methylene spacer. Photoinduced electron transfer (see Chapter 4, Section 4.3), which takes place from amino groups to aromatic hydrocarbons, causes fluorescence quenching of the latter. When the amino group is protonated (or strongly interacts with a cation), electron transfer is hindered and a very large enhancement of fluorescence is observed.

Figure 10.7 illustrates the mechanism in terms of molecular orbitals. Upon excitation of the fluorophore, an electron of the highest occupied molecular orbital (HOMO) is promoted to the lowest unoccupied molecular orbital (LUMO), which enables PET from the HOMO of the donor (proton-free amine or cation-free receptor) to that of the fluorophore, causing fluorescence quenching of the latter. Upon protonation (or cation binding), the redox potential of the donor is raised so that the relevant HOMO becomes lower in energy than that of the fluorophore; consequently, PET is no longer possible and fluorescence quenching is suppressed.

Examples of PET pH indicators are given in Figure 10.2 (class C). The pK_a values of compounds P-9, P-10 and P-11 are 4.4, 7.2 and 5.9, respectively, in water–methanol (4:1, v/v), and the value is 9.6 for compound P-12 in water–ethanol (1:1, v/v). The latter offers the advantage of long wavelength excitation ($\sim$ 500 nm).

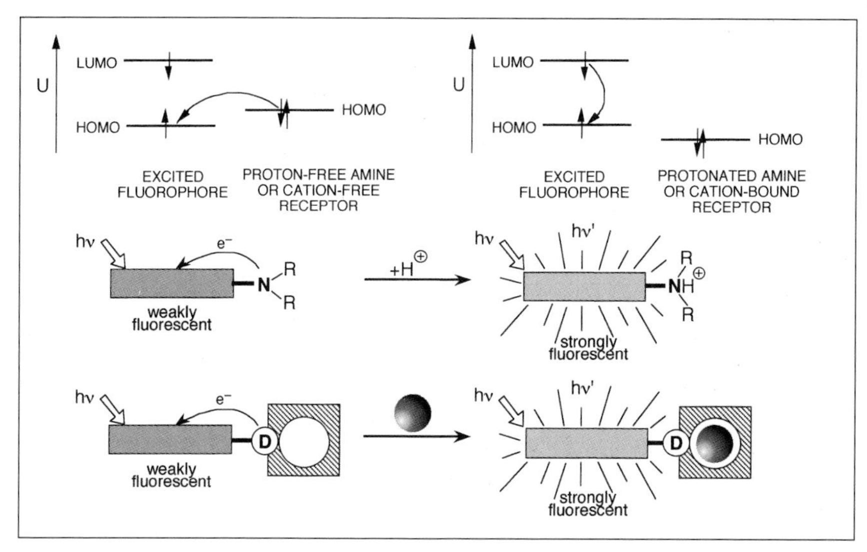

Fig. 10.7. Principles of H^+ or metal ion sensing by fluorescent PET sensors.

10.3
Fluorescent molecular sensors of cations

10.3.1
General aspects

Detecting cations is of great interest to many scientists, including chemists, biologists, clinical biochemists and environmentalists. Sodium, potassium, magnesium and calcium are involved in biological processes such as transmission of nerve impulses, muscle contraction, regulation of cell activity, etc. Zinc is an essential component of many enzymes (e.g. in carbonic anhydrase and zinc finger proteins); it plays a major role in enzyme regulation, gene expression, neurotransmission, etc.

In medicine, monitoring of metal ions (e.g. Na^+, K^+, Mg^{2+}, Ca^{2+}, Li^+) in blood and urine is of major importance in diagnosis. The normal concentrations are reported in Table 10.2. The values of the concentrations of Na^+ and K^+ show how difficult the selective detection of Na^+ in blood is where K^+ is at a concentration about 30 times higher. Note also that the concentration of Ca^{2+} in blood plasma and urine is in the millimolar range whereas inside a living cell it is in the micromolar range. Therefore, the well-known calcium sensors Indo-1, Fura-2, etc. (see Section 10.3.3.2) are suitable for cellular biology but not for clinical diagnosis. In medicine, it is also important to control lithium levels in the serum for patients under treatment for manic depression, and potassium in the case of high blood pressure.

Regarding the toxicity of some metal ions, it is well known that mercury, lead and cadmium are toxic to organisms, and early detection in the environment is

Tab. 10.2. Normal concentrations of some cations in blood and urines

Cation	*Blood* $pH = 7.35–7.42$	*Urine* $pH = 6–7$
Na^+	143 mM	125 mM
K^+	5 mM	65 mM
Mg^{2+}	1 mM	4 mM
Ca^{2+}	1.5 mM	4 mM

desirable. Aluminum is also potentially toxic: it is probably at the origin of some diseases such as osteomalacia, anemia, neurodegenerative or bone diseases. Control of aluminum content is thus necessary in farm produce and in the pharmaceutical industry.

In chemical oceanography, it has been demonstrated that some nutrients required for the survival of microorganisms in seawater contain zinc, iron and manganese as enzyme cofactors.

Colorimetric determination of cations based on changes in color on complexation by dye reagents started to be popular a long time ago, especially in the case of alkaline earth metal ions, which are efficiently chelated by agents of the EDTA type (Figure 10.8). Fluorimetric techniques being more sensitive than photometric ones, numerous fluorogenic chelating reagents were studied and applied to practical cases (Fernandez-Gutierrez and Muños de la Peña, 1985). Among them, oxine (8-hydroxyquinoline) (Figure 10.8) and many of its derivatives occupy an important place in analytical chemistry but they are not very specific. In contrast, fluorescent molecular sensors of the EDTA type exhibit high selectivity for calcium with respect to the other ions present in living cells. Examples are given below.

The discovery of crown ethers and cryptands (see examples in Figure 10.8) in the late 1960s opened up new possibilities for cation recognition with improvement of selectivity, especially for alkali metal ions for which there is a lack of selective chelators. Then, the idea of coupling these ionophores to chromophores or fluorophores, leading to so-called *chromoionophores* and *fluoroionophores*, respectively, emerged some years later (Löhr and Vögtle, 1985). As only fluorescent sensors are considered in this chapter, chromoionophores will not be described.

In the design of a fluoroionophore, much attention is paid to the characteristics of the ionophore moiety and to the expected changes in fluorescence characteristics of the fluorophore moiety upon binding (Figure 10.9). It should first be recalled that the stability of a complex between a given ligand and a cation depends on many factors: nature of the cation, nature of the solvent, temperature, ionic strength, and pH in some cases. In ion recognition, complex selectivity (i.e. the preferred complexation of a certain cation when other cations are present) is of major importance. In this regard, the characteristics of the ionophore, i.e. the ligand topology and the number and nature of the complexing heteroatoms or groups, should match the characteristics of the cation, i.e. ionic diameter, charge density, coordination number (Table 10.3), intrinsic nature (e.g. hardness of metal

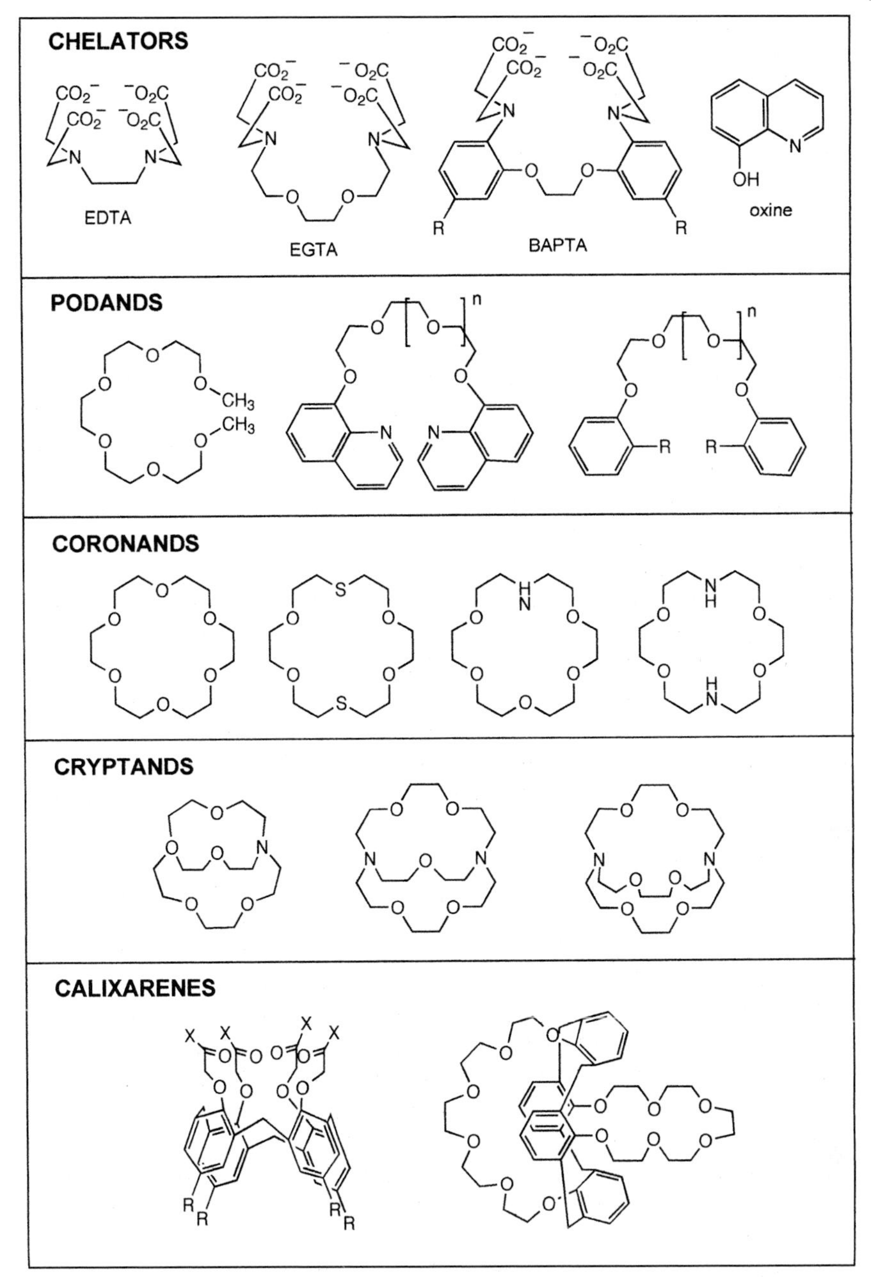

Fig. 10.8. Examples of artificial receptors of cations.

cations, nature and structure of organic cations, etc.) according to the general principles of supramolecular chemistry. The ionophore can be a chelator, an open-chain structure (podand), a macrocycle (coronand, e.g. crown ether), a macrobicycle (cryptand), a calixarene derivative, etc. (Figure 10.8). The very high stability of

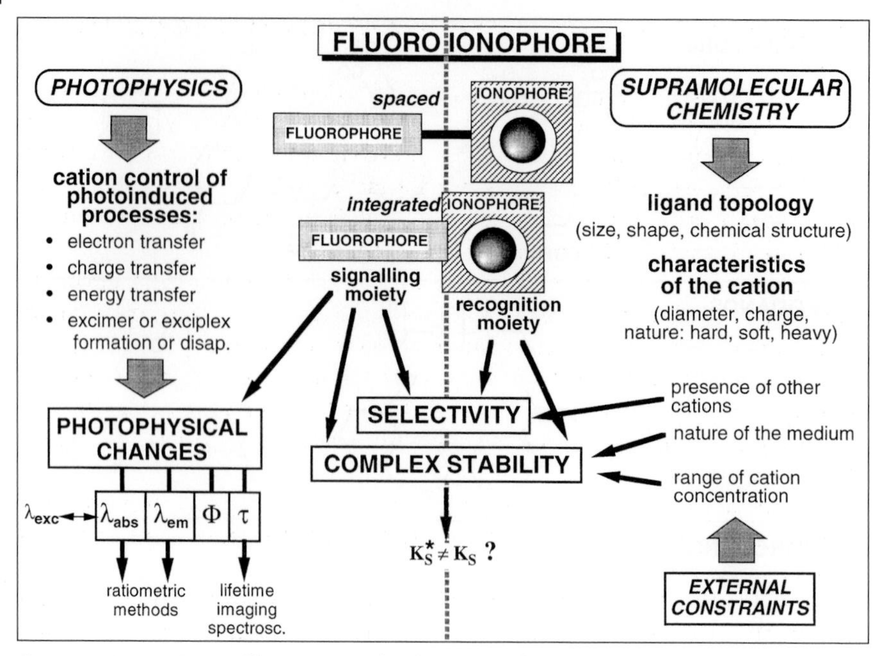

Fig. 10.9. Main aspects of fluorescent molecular sensors for cation recognition (fluoroionophores).

Tab. 10.3. Ionic diameter, charge density and coordination number of some cations

Cation	*Ionic diameter (Å)*	*Charge density (q Å^{-1})*	*Coordination number*
Li^+	1.36	1.47	6
Na^+	1.94	1.03	6
K^+	2.66	0.75	6
Mg^{2+}	1.32	3.03	4 or 5
Ca^{2+}	1.98	2.02	7 or 9
Ba^{2+}	2.68	1.49	9
Ni^{2+}	1.38	2.9	4 or 6
Cu^{2+}	1.44	2.78	4 or 6
Zn^{2+}	1.48	2.70	4 or 6
Cd^{2+}	1.94	2.06	4 or 6
Pb^{2+}	2.40	1.67	4 or 6

cryptates results from the three-dimensional encapsulation, and the complexation selectivity is also usually higher because of their inability to be deformed. In the case of complexes with coronands and cryptands, the most stable complexes are formed with ions having an ionic diameter close to that of the ligand cavity. Another principle generally applicable in chemistry predicts that hard oxygen centers combine with hard alkali metal ions, and soft sulfur or nitrogen centers with soft transition metal ions.

Medium effects are significant in both stabilities and selectivities of complexation with cations. The main factors are (i) the difference between ligand coordination energy and solvation energy, i.e. the solvating power of the ligand compared to that of the solvent, (ii) the difference in interaction with the ligand shell and the dielectric medium outside the first solvation shell. In the case of aqueous solutions, pH and ionic strength of course play an important role.

The connection between the ionophore and the fluorophore is a very important aspect of sensor design, bearing in mind the search for the strongest perturbation of the photophysical properties of the fluorophore by the cation. The ionophore may be linked to the fluorophore via a spacer, but in many cases some atoms or groups participating in the complexation belong to the fluorophore. Therefore, the selectivity of binding often results from the whole structure involving both signalling and recognition moieties.

More than one ionophore and/or more than one fluorophore may be involved in the structure of fluoroionophores. Figure 10.10 illustrates some of the structures that have been designed.

Attention should be paid to the possible existence of several complexes having different stoichiometries. A necessary preliminary experiment consists of recording the fluorescence and/or excitation spectra under experimental conditions (nature of the solvent, composition of the medium, ionic strength, pH (if it has an effect on the stability constant), etc.) as close as possible to the medium in which a cation must be detected. The variations in the fluorescence intensity for an appropriate couple of excitation and emission wavelengths (or for several emission or excitation wavelengths) as a function of cation concentration must be analyzed in order to determine the stoichiometry and the stability constant of the complexes (Appendix B). As in the case of pH determination (see Section 10.2.1), ratiometric measurements are recommended.

This section will focus on the design of fluorescent molecular sensors for cations. Many of them are poorly soluble in water and their complexing ability will thus be presented in organic solvents. Applications for probing aqueous solutions, and in particular aqueous biological media, are not possible with these molecular sensors. However, they are useful in monitoring extraction processes, e.g. for evaluation of the cation concentration in the organic phase. Moreover, the hydrophobic character of these sensors is not a drawback but an advantage (in terms of leaching) when used for doping the sensitive part (polymer or sol–gel film) of optical sensor devices for biomedical applications or for monitoring species in the environment.

The fluorescent molecular sensors will be presented with a classification according to the nature of the photoinduced process (mainly photoinduced electron or charge transfer, and excimer formation) that is responsible for photophysical changes upon cation binding. Such a classification should help the reader to understand the various effects of cation binding on the fluorescence characteristics reported in many papers. In most of these papers, little attention is often paid to the origin of cation-induced photophysical changes.

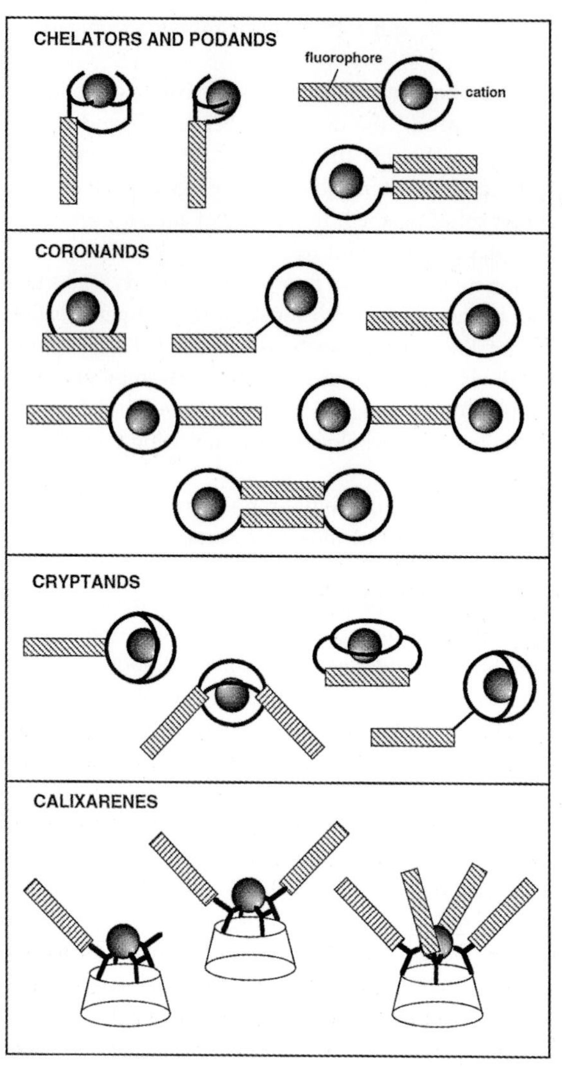

Fig. 10.10. Various topologies of fluoroionophores.

10.3.2
PET (photoinduced electron transfer) cation sensors

10.3.2.1 Principles

PET fluorescent sensors for cations have been highly developed (Bissell et al., 1993). The principle is the same as for PET pH indicators, described in Section 10.2.2.5.

Most PET fluorescent sensors for cations are based on the principle displayed in Figure 10.7, but other photoinduced electron transfer mechanisms can take place with transition metal ions (Fabbrizzi et al., 1996; Bergonzi et al., 1998). In fact, 3d metals exhibit redox activity and electron transfer can occur from the fluorophore

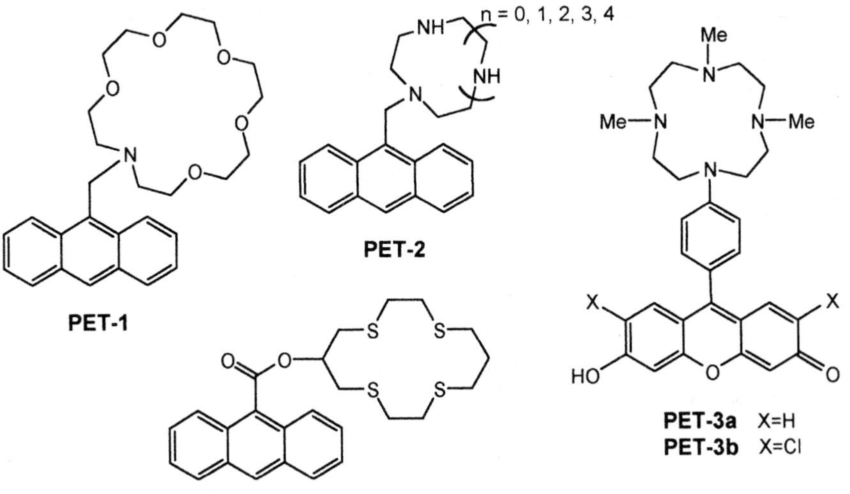

Fig. 10.11. Crown-containing PET sensors (PET-1: de Silva A. P. and de Silva S. A. (1986) *J. Chem. Soc., Chem. Commun.* 1709. PET-2: Akkaya E. U. et al. (1990) *J. Am. Chem. Soc.* **112**, 3590. PET-3: Hirano T. et al. (2000) *Angew. Chem. Int. Ed.* **39**, 1052. PET-4: De Santis G. et al. (2000) *Inorg. Chim. Acta* **257**, 69).

to the bound metal ion, or vice versa. In some cases, electron exchange is possible, which results in quenching of the fluorophore by non-radiative energy transfer according to the Dexter mechanism.

Various examples of PET sensors will be now presented; these are classified according to the chemical structure of the recognition moiety.

10.3.2.2 **Crown-containing PET sensors**

Examples of PET sensors containing various kinds of crowns are given in Figure 10.11. PET-1 is the first and simplest coronand PET sensor. Its fluorescence quantum yield increases from 0.003 to 0.14 upon binding of K^+ in methanol.

PET-2, containing a water-soluble polyazamacrocycle, was designed for the recognition of soft metal ions like Zn^{II}. This compound is sensitive to pH because of the protonability of the aliphatic nitrogen atoms. At pH 10, the fluorescence intensity of this compound increases 14-fold upon addition of Zn^{II}, whereas at neutral pH, protonation of the nitrogen atoms suppresses the PET process, which precludes applications in physiological media. If the macrocyclic polyamine is directly bound to the phenyl group of a fluorophore, as in PET-3a and PET-3b, the pK_a is lowered and the intensity changes are comparable at pH 10 and pH 7. The Zn^{II} complexes are thus fluorescent at physiological pH. Moreover, the PET-3a and PET-3b (containing fluorescein and dichlorofluorescein moieties) are excitable in the visible and are thus more appropriate for biological applications.

In PET-4, the crown does not contain nitrogen atoms but four sulfur atoms and is known for its strong affinity towards Cu^{II}. This sensor is also based on the PET

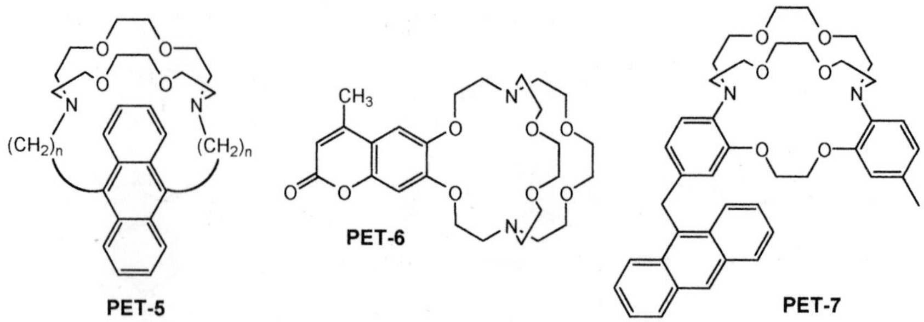

Fig. 10.12. Cryptand-based PET sensors (PET-5: Fages F. et al. (1989) *J. Am. Chem. Soc.* **111**, 8672. PET-6: Golchini K. et al. (1990) *Am. J. Physiol.* **258**, F438. PET-7: de Silva A. P. et al. (1990) *Tetrahedron Lett.* **31**, 5193).

principle, but in a different way to compounds PET-1 and PET-2. Quenching of fluorescence upon Cu^{II} binding arises from a photoinduced electron transfer from the fluorophore to the metal center and involves the Cu^{II}/Cu^{I} couple. It is remarkable that the other transition metal ions like Mn^{II}, Fe^{II}, Co^{II} and Ni^{II} have negligible effect in the ethanolic solutions where the studies were carried out. The crown can also efficiently bind Ag^{I}, but complexation is not signaled by a change in fluorescence intensity because the poor redox activity of this non-transition cation precludes electron transfer.

10.3.2.3 **Cryptand-based PET sensors**

PET-5, PET-6 and PET-7 are examples of macrobicyclic structures (cryptands) (Figure 10.12). The cavity of PET-6 and PET-7 fits well the size of K^+. PET-6 has been successfully used for monitoring levels of potassium in blood and across biological membranes, but pH must be controlled because of pH sensitivity of this compound via protonation of the nitrogen atoms. This difficulty has been elegantly overcome in benzannelated cryptand PET-7, in which the aromatic nitrogens have lower pK_a than those of aliphatic amines.

An interesting feature of PET-5 is its ability to form exciplexes characterized by an additional band at higher wavelengths, thus allowing ratiometric measurements at two different observation wavelengths.

10.3.2.4 **Podand-based and chelating PET sensors**

Podands PET-8 and PET-9 contain polyamine chains (Figure 10.13) and were thus aimed at Zn^{II} because this soft cation has a strong affinity for the soft nitrogen atoms. However, they operate in a very limited pH range and the affinity for Cu^{II} is also strong. PET-10 can bind Cu^{II} and Ni^{II} and favor oxidation of these cations to the trivalent state. Fluorescence quenching of anthracene upon binding should be ascribed in this case to an electron transfer from the reducing divalent metal center.

PET-11 (Zinpyr-1), containing fluorescein as a fluorophore, and bis(2-pyridylmethyl)amine as a chelating moiety, has been designed for probing Zn^{II} in living cells. This compound is indeed cell-permeable and it has essentially no

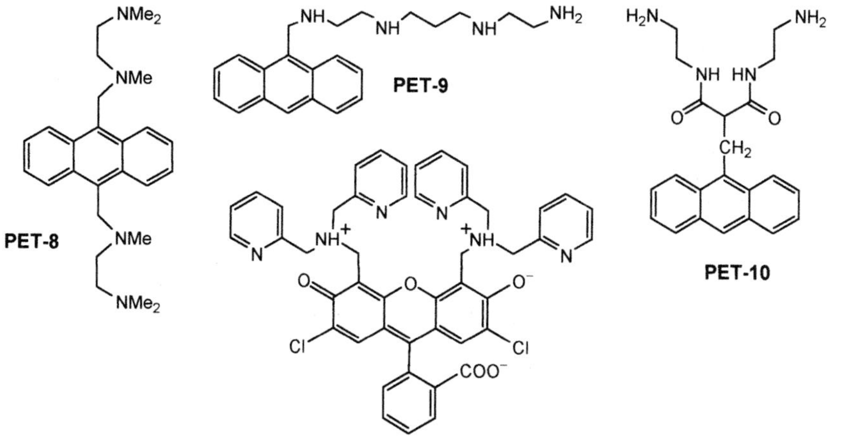

PET-8 PET-9 PET-10 PET-11 Zinpyr-1

Fig. 10.13. Polyamine-based PET sensors (PET-8: Huston M. E. et al. (1988) *J. Am. Chem. Soc.* **110**, 4460. PET-9: Fabbrizzi L. et al. (1996) *Inorg. Chem.* **35**, 1733. PET-10: Fabbrizzi L. et al. (1994) *Angew. Chem., Int. Ed. Engl.* **33**, 1975. PET-11: Walkup G. K. et al. (2000) *J. Am. Chem. Soc.* **122**, 5644).

measurable affinity for Ca^{2+} or Mg^{2+}. Upon addition of Zn^{II} under physiological conditions, the fluorescence quantum yield increases from 0.39 to 0.87. Concentrations in the nanomolar range can be determined (stability constant: 1.4×10^9).

Chelators with carboxylic groups are known to efficiently bind divalent hard cations like Ca^{2+} and Mg^{2+}. Many selective calcium probes have been designed by Tsien and coworkers for applications in cellular biology, i.e. for probing calcium concentrations in the micromolar range. The so-called BAPTA recognition moiety resembles EDTA, but the nitrogen atoms are linked to a phenyl group in order to avoid pH sensitivity in physiological media. Most of them are not based on the PET principle but on photoinduced charge transfer (see Section 10.3.3.2).

Figure 10.14 shows examples of chelating PET sensors. PET-12 to PET-14 are selective for calcium. In PET sensors, the changes in fluorescence quantum yield are accompanied by proportional changes in excited-state lifetime. Therefore, compounds PET-12 to PET-14 were found to be suitable for fluorescence lifetime imaging of calcium.

The same design principles apply to the magnesium sensors PET-15 and PET-16, in which the recognition moiety has a smaller 'cavity'.

10.3.2.5 Calixarene-based PET sensors

PET-17 has been designed for selective recognition of sodium (Figure 10.15). It contains four carbonyl functions, two of them being linked to pyrene and nitrobenzene at opposite sites on the calixarene lower rim. Complexation with Na^+ prevents close approach of pyrene and nitrobenzene and thus reduces the probability of PET. The fluorescence quantum yield increases from 0.0025 to 0.016.

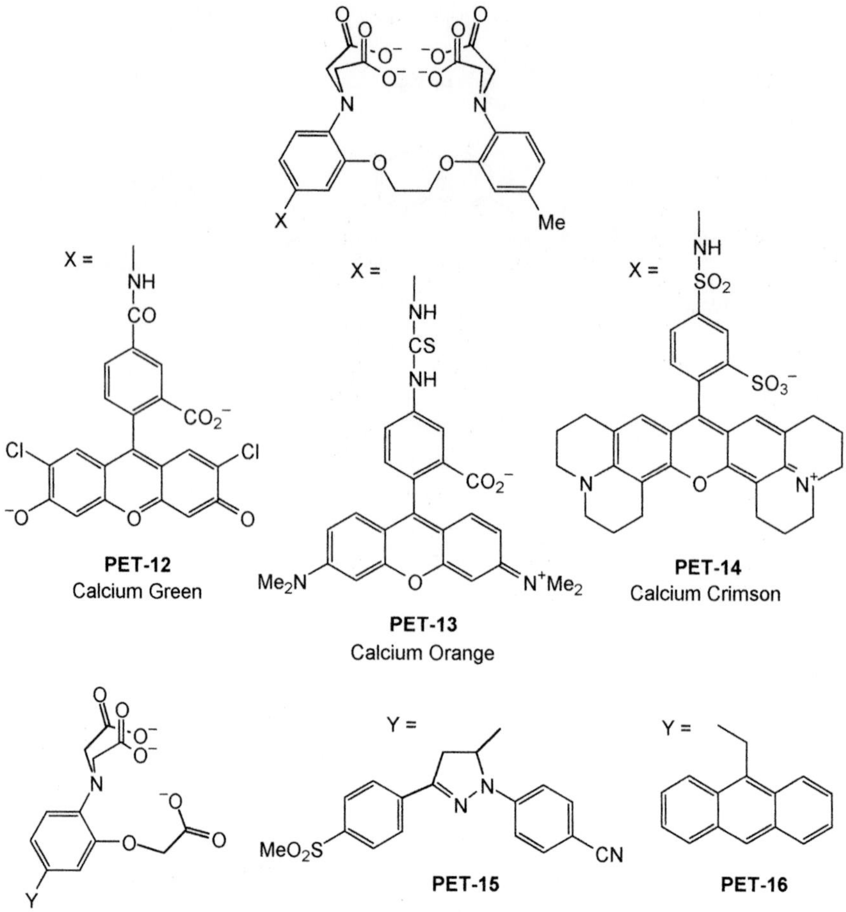

X = NH–CO (PET-12), X = NH–CS–NH (PET-13), X = NH–SO2 (PET-14)

PET-12
Calcium Green

PET-13
Calcium Orange

PET-14
Calcium Crimson

PET-15

PET-16

Fig. 10.14. Chelating PET sensors (PET-12, PET-13, PET-14: Kuhn M. A. (1993), in: Czarnik A. W. (Ed.), *Fluorescent Chemosensors for Ion and Molecule Recognition*, ACS Symposium Series 358. p. 147. PET-15, PET-16: de Silva A. P. et al. (1994) *J. Chem. Soc., Chem. Commun.* 1213).

Calixarene containing a dioxotetraaza unit, PET-18, is responsive to transition metal ions like Zn^{2+} and Ni^{2+}. Interaction of Zn^{2+} with the amino groups induces a fluorescence enhancement according to the PET principle. In contrast, some fluorescence quenching is observed in the case of Ni^{2+}. PET from the fluorophore to the metal ion is a reasonable explanation but energy transfer by electron exchange (Dexter mechanism) cannot be excluded.

10.3.2.6 **PET sensors involving excimer formation**

PET-19 (Figure 10.16) consists of a diazacrown ether with two pendant pyrene groups. As expected, cation binding results in a large change in the monomer/

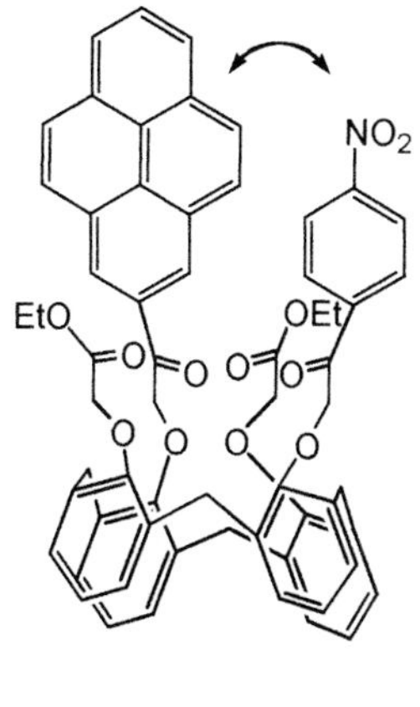

NO2 EtO OEt O O O O O O O O CH2 O O NH HN O NH HN O O OH HO O

PET-17 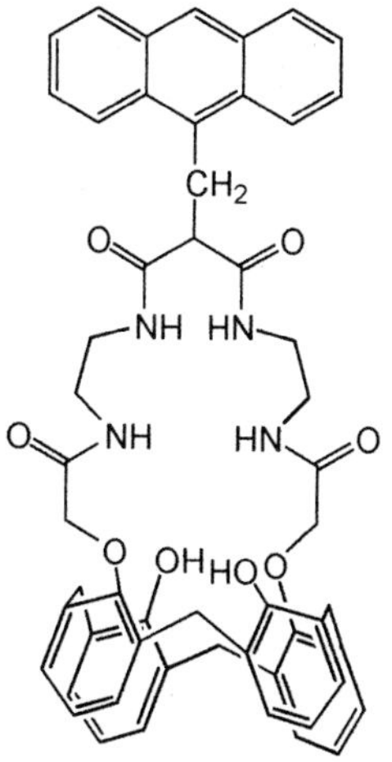 **PET-18**

Fig. 10.15. Calixarene-based PET sensors (PET-17: Aoki I. et al. (1992) *J. Chem. Soc., Chem. Commun.* 730. PET-18: Unob F. et al. (1998) *Tetrahedron Lett.* **39**, 2951).

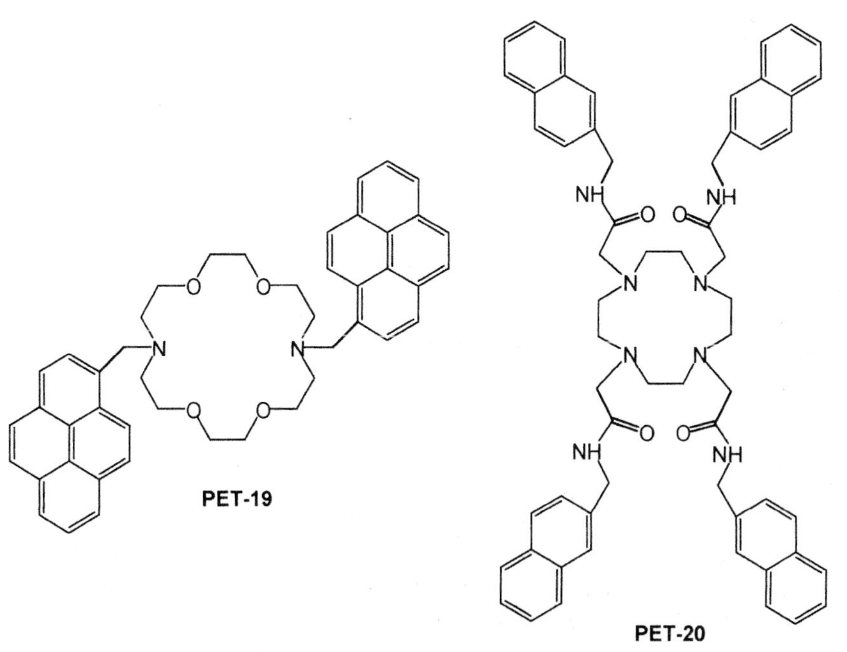

Fig. 10.16. PET sensors involving excimer formation (PET-19: Kubo K. et al. (1997) *Bull. Chem. Soc. Jpn.* **70**, 3041. PET-20: Parker D. and Williams J. A. G. (1995) *J. Chem. Soc., Perkin 2*, 1305.

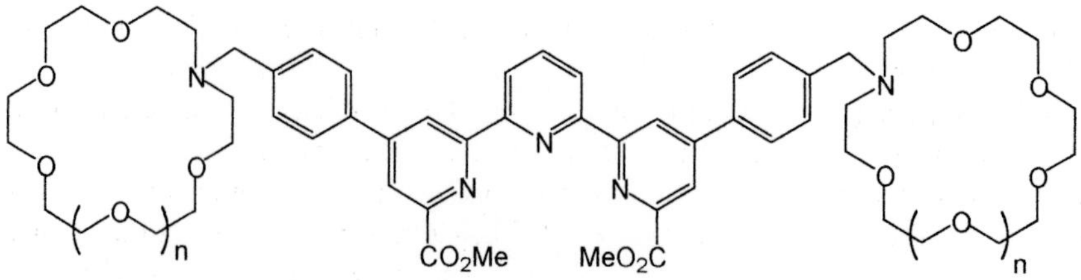

PET-21

Fig. 10.17. PET sensor involving energy transfer (PET-21: de Silva A. P. et al. (1997) *J. Chem. Soc., Chem. Commun.* 1891).

excimer ratio. There is a concomitant increase in the overall fluorescence emission as a result of the reduction in PET from the nitrogen atom to the pyrenyl groups. The monomer/excimer ratio was found to be strongly dependent on the nature of the metal ion. Among the investigated metal ions, the larger stability constants of the complexes were obtained for K^+ and Ba^{2+}, in accordance with the size of these cations with respect to the crown diameter.

In the same way, the emission spectrum of PET-20, existing as the protonated form in acetonitrile, exhibits an excimer band whose intensity decreases upon binding of Cd^{2+} and Pb^{2+}.

10.3.2.7 **Examples of PET sensors involving energy transfer**

PET-21 (Figure 10.17) contains a terpyridyl diester that can strongly bind Eu^{III}. Excitation energy transfer from this type of ligand to Eu^{III} is known to occur via the triplet state and should result in luminescence from Eu^{III}. However, when the crown is empty, only weak luminescence is detected because of quenching due to PET from the nitrogen atom of the crown. Binding of K^+ causes a very large enhancement of the luminescence quantum yield, as expected from cation-induced reduction of the PET efficiency. PET-21 provides the first example of metal-triggered metal-centered emission. Thanks to the long lifetime of Eu^{III} (hundreds of microseconds), time-delayed detection of the luminescence is possible which enables eradication of the fast parasitic fluorescence of biological samples.

10.3.3
Fluorescent PCT (photoinduced charge transfer) cation sensors

10.3.3.1 **Principles**

When a fluorophore contains an electron-donating group (often an amino group) conjugated to an electron-withdrawing group, it undergoes intramolecular charge transfer from the donor to the acceptor upon excitation by light. The consequent change in dipole moment results in a Stokes shift that depends on the micro-environment of the fluorophore; polarity probes have been designed on this basis (see Chapter 7). It can thus be anticipated that cations in close interaction with the donor or the acceptor moiety will change the photophysical properties of the fluorophore because the complexed cation affects the efficiency of intramolecular charge transfer.

When a group (like an amino group) playing the role of an electron donor within the fluorophore interacts with a cation, the latter reduces the electron-donating character of this group; owing to the resulting reduction in conjugation, a blue-shift of the absorption spectrum is expected, together with a decrease in the molar absorption coefficient. Conversely, a cation interacting with the acceptor group enhances the electron-withdrawing character of this group; the absorption spectrum is thus red-shifted and the molar absorption coefficient is increased. The fluorescence spectra are in principle shifted in the same direction as the absorption spectra. In addition to these shifts, changes in quantum yields and lifetimes are often observed. All these photophysical effects are obviously dependent on the charge and the size of the cation, and selectivity of these effects is expected.

The photophysical changes on cation binding can also be described in terms of charge dipole interaction. Let us consider only the case where the dipole moment in the excited state is larger than that in the ground state. Then, when the cation interacts with the donor group, the excited state is more strongly destabilized by the cation than the ground state, and a blue-shift of the absorption and emission spectra is expected (however the fluorescence spectrum undergoes only a slight blue-shift in most cases; this important observation will be discussed below). Conversely, when the cation interacts with the acceptor group, the excited state is more stabilized by the cation than the ground state, and this leads to a red-shift of the absorption and emission spectra (Figure 10.18).

10.3.3.2 PCT sensors in which the bound cation interacts with an electron-donating group

Crown-containing PCT sensors Many fluoroionophores have been designed according to the following principle: the cation receptor is an azacrown containing a nitrogen atom that is conjugated to an electron-withdrawing group (Figure 10.19). Compounds PCT-1 to PCT-4 exhibit a common feature: the blue-shift of the absorption spectrum is much larger than that of the emission spectrum on cation binding. An example is given in Figure 10.20. Such a small shift of the fluorescence spectrum – which at first sight is surprising – can be interpreted as follows. The photoinduced charge transfer reduces the electron density on the nitrogen atom of the crown, and this nitrogen atom becomes a non-coordinating atom because it is positively polarized. Therefore, excitation induces a photodisruption of the interaction between the cation and the nitrogen atom of the crown. The fluorescence spectrum is thus only slightly affected because most of the fluorescence is emitted from species in which the interaction between the cation and the fluorophore does not exist any more or is much weaker.

This interpretation is supported by a thorough study of the photophysics of PCT-1 and its complexes with Li^+ and Ca^{2+} (Martin et al., 1994). In particular, subpicosecond pump-probe spectroscopy provided compelling evidence for the disruption of the link between the crown nitrogen atom and the cation. A photo-

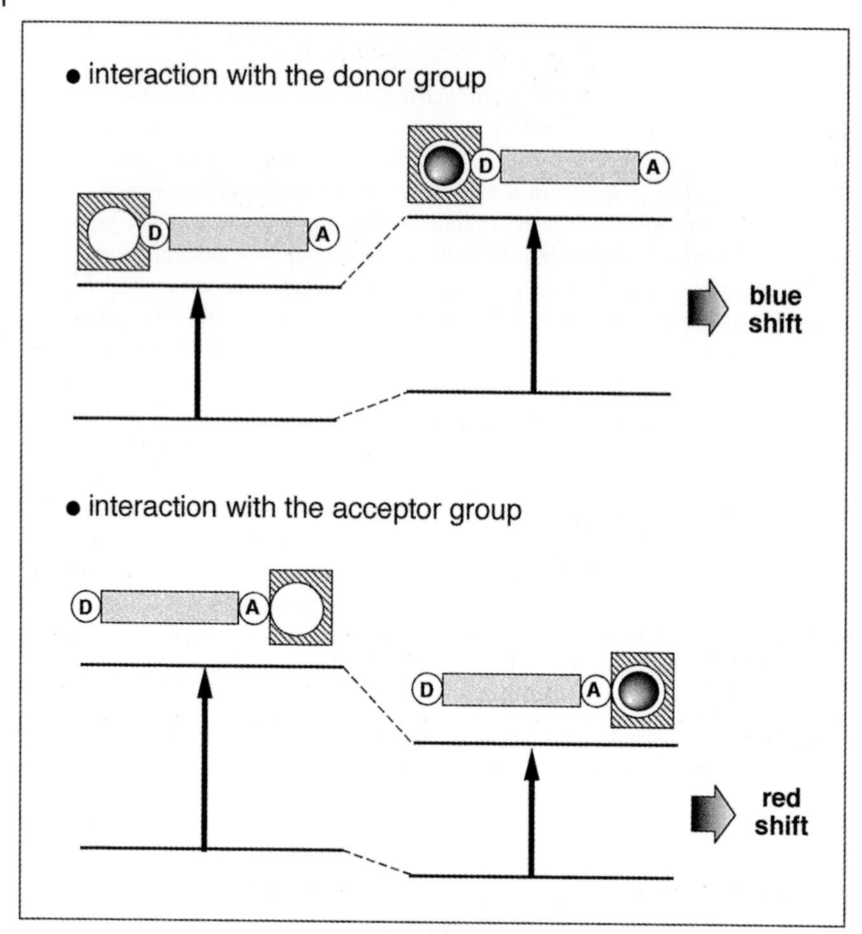

Fig. 10.18. Spectral displacements of PCT sensors resulting from interaction of a bound cation with an electron-donating or electron-withdrawing group.

disruption was also demonstrated in complexes of PCT-2 and PCT-3. The cation-induced spectral changes in PCT-4 were interpreted in the same way.

Such a photodisruption results in a lower stability of the complexes in the excited state. Therefore, excitation of these complexes by an intense pulse of light is expected to cause some cations to leave the crown and diffuse away, provided that the time constant for total release of the cation from the crown is shorter than the lifetime τ of the excited state (for a discussion on this point, see Valeur et al., 1997).

Intramolecular charge transfer in conjugated donor–acceptor molecules may be accompanied by internal rotation leading to TICT (twisted intramolecular charge transfer) states. A dual fluorescence may be observed as in PCT-5 (Létard et al., 1994) (which resembles the well-known DMABN (see section 3.4.4) containing a dimethylamino group instead of the monoaza-15-crown-5): the short-wavelength

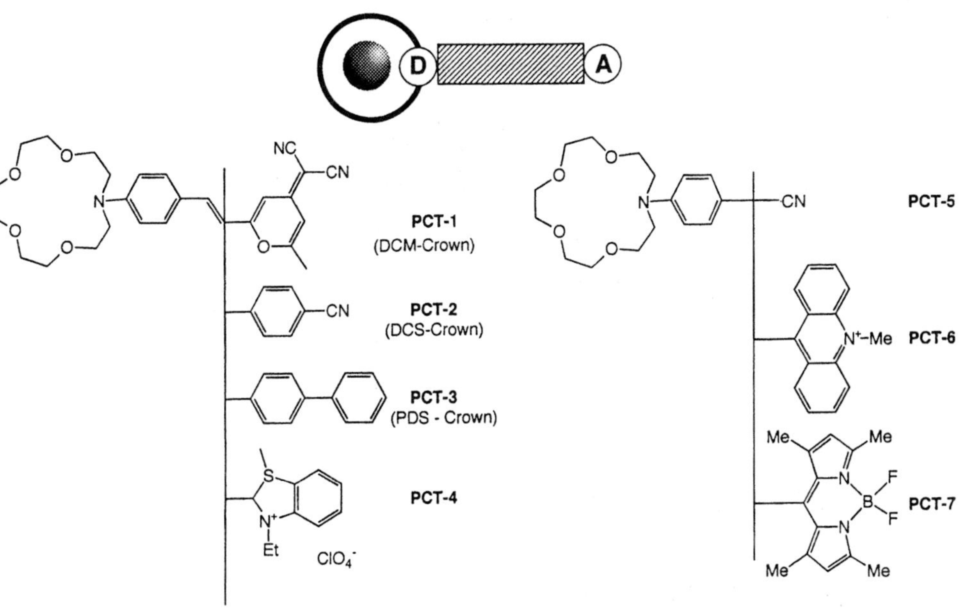

Fig. 10.19. Crown-containing PCT sensors in which the bound cation interacts with the donor group (PCT-1: Bourson J. and Valeur B. (1989) *J. Phys. Chem.* **93**, 3871. PCT-2 and PCT-3: Létard J. F. et al. (1993) *Pure Appl. Chem.* **65**, 1705. PCT-4: Ushakov E. N. et al. (1997) *Izv. Akad. Nauk, Ser. Khim.* 484 [*Russ. Chem. Bull.* **46**, 463]. PCT-5: Létard J. F. et al. (1995) *Rec. Trav. Chim. Pays-Bas* **114**, 517. PCT-6: Jonker S. A. et al. (1990) *Mol. Cryst. Liq. Cryst.* **183**, 273. PCT-7: Kollmannsberger M. et al. (1998) *J. Phys. Chem.* **102**, 10211).

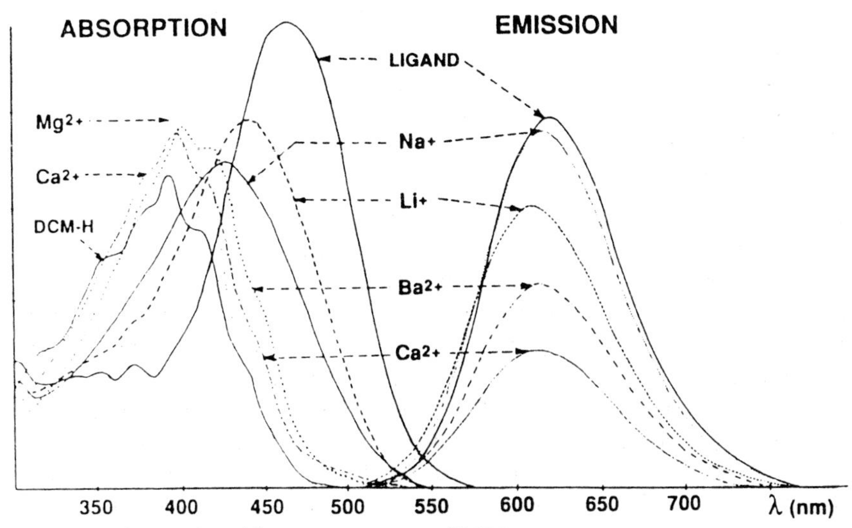

Fig. 10.20. Absorption and fluorescence spectra of PCT-1 (DCM-crown) and its complexes in acetonitrile.

band corresponds to the fluorescence from the locally excited state and the long-wavelength band arises from a TICT state. The fluorescence intensity of the latter decreases upon cation binding because interaction between a bound cation and the crown nitrogen disfavors the formation of a TICT state, which leads to a concomitant increase in the short-wavelength band.

The formation of a TICT state is often invoked even if no dual fluorescence is observed. For donor–acceptor stilbenes (PCT-2 and PCT-3), the proposed kinetic scheme contains three states: the planar state E^* reached upon excitation can lead to state P^*(non-fluorescent) by double-bond twist, and to TICT state A^* by single-bond twist, the latter being responsible for most of the emission.

The absence of fluorescence of PCT-6 may be due to the formation of a non-fluorescent TICT state, and an acridinium-type fluorescence is recovered upon binding of H^+ and Ag^+. The same explanation may hold for the low fluorescence of PCT-7, whose electron-withdrawing group, boron-dipyrromethen, must be twisted due to steric interactions with the phenyl ring: the fluorescence enhancement factor varies from 90 for Li^+ to 2250 for Mg^{2+}. Both PCT-6 and PCT-7 compounds undergo much larger fluorescence enhancements than most PCT molecular sensors (factors of ~2–5 are generally observed). They can be considered as limiting cases closely resembling PET sensors but with a virtual spacer.

The crown-containing fluoroionophores described above are of great interest for the understanding of cation–fluorophore interactions. They offer a wide variety of photophysical changes on cation binding that can be used for cation recognition. However, the selectivity of azacrowns towards metal ions is not good enough when stringent discrimination between cations of the same chemical family is required.

Improvement of selectivity can be achieved by the participation of external groups, as shown in Figure 10.21. In PCT-8 (PBFI) and PCT-9 (SBFI), the oxygen atom of the methoxy substituent of the fluorophore can interact with a cation; binding efficiency and selectivity are thus better than those of the crown alone. SBFI has been designed for probing intracellular sodium ions and PBFI for potassium ions. In both compounds, the photophysical changes are likely to be due to the reduction of the electron-donating character of the nitrogen atoms of the diazacrown by the complexed cation. Further improvement of the selectivity towards K^+ with respect to Na^+ is desirable. PCT-10 resembles PBFI but it shows greater selectivity for potassium over sodium than PBFI.

Chelating PCT sensors In light of the preceding considerations on crowned charge-transfer compounds, it is worth examining the photophysical properties of well-known chelators in the BAPTA series used for the recognition of cytosolic cations and in particular calcium ions (Grynkiewicz et al., 1985). For instance PCT-11 (Indo-1) and PCT-12 (Fura-2) (Figure 10.22) are widely used as calcium indicators. In these compounds, the fluorophore is a donor–acceptor molecule with an amino group as the electron-donating group that participates in the complexation, the ionophore being a chelating group of the BAPTA type.

Upon complexation by Ca^{2+} in water, the absorption spectrum of Fura-2 is blue-shifted, whereas there is almost no shift of the fluorescence spectrum. The same

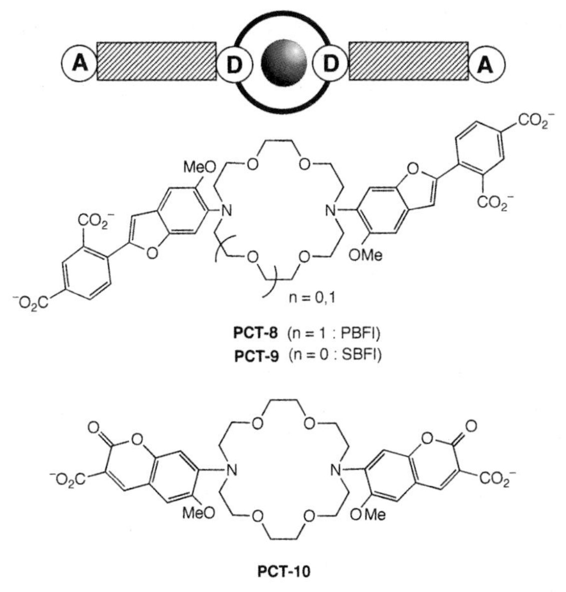

Fig. 10.21. Crown-containing bifluorophoric PCT sensors (PCT-8: Kasner S. E. and Ganz M. B. (1992) *Am. J. Physiol.* **262**, F462. PCT-9: Minta A. and Tsien R. Y. (1989) *J. Biol. Chem.* **264**, 19449. PCT-10: Crossley R. et al. (1994) *J. Chem. Soc., Perkin Trans. 2*, 513).

interpretation as for the above-described crown-ether-linked compounds can be proposed: the electron density of the nitrogen atom conjugated with the electron-withdrawing group of the fluorophore is reduced on excitation and might even become positively polarized – this causes disruption of the interaction between this nitrogen atom and a bound cation. Consequently, fluorescence emission closely resembles that of the free ligand.

In contrast to Fura-2, the photoinduced charge transfer in Indo-1 may not be sufficient to cause nitrogen–Ca^{2+} bond breaking. This interpretation is consistent with the fact that the fluorescence maximum of free Indo-1 is located at a shorter wavelength than Fura-2 by ~30 nm, thus indicating a less polar charge-transfer state.

By keeping the same fluorophores, but reducing the 'cavity' size of the ionophore, we obtain PCT-13 (Mag-Indo1) and PCT-14 (Mag-Fura2) (Figure 10.22), which are selective for magnesium. PCT-8, PCT-9 and PCT-11 to PCT-14 are commercially available in the non-fluorescent acetoxymethylester form so that they are cell permeant and they recover their fluorescence upon hydrolysis by enzymes (Molecular Probes, Inc., Eugene, OR).

Cryptand-based PCT sensors The above-described chelators are well suited to the detection of alkaline earth cations but not alkali cations. In contrast, cryptands are very selective towards the latter.

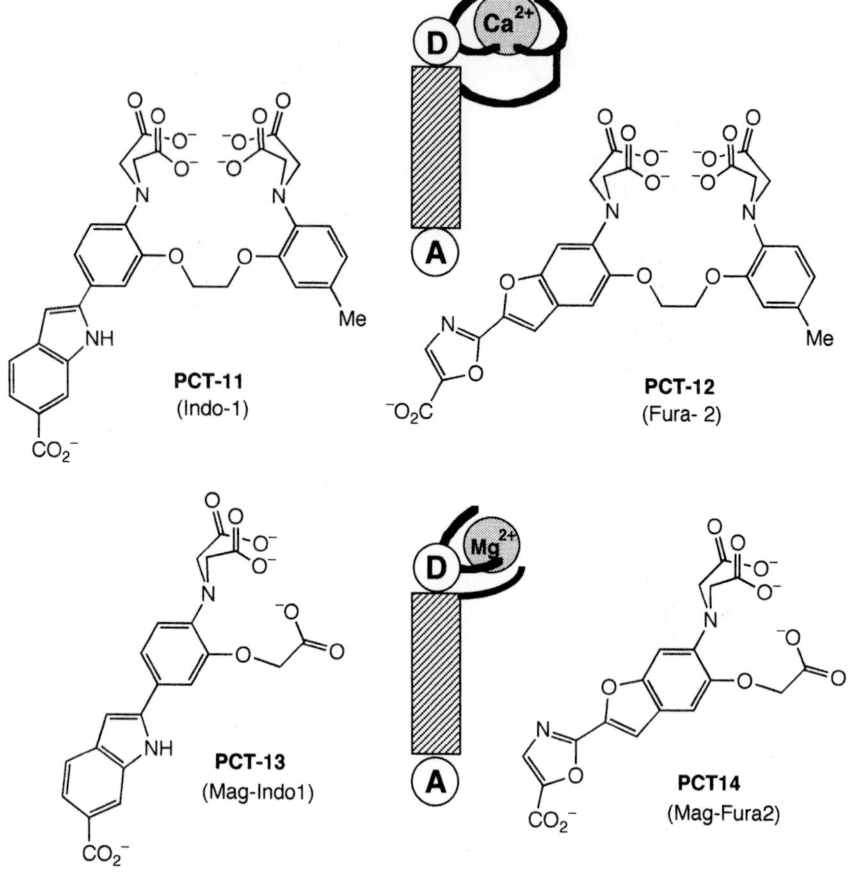

Fig. 10.22. Chelating PCT sensors for calcium and magnesium ions (PCT-11 and PCT-12: Grynkiewicz G. et al. (1985) *J. Biol. Chem.* **260**, 3440. PCT-13 and PCT-14: Haugland R. P., *Handbook of Fluorescent Probes and Research Chemicals*, 6th edn, Molecular Probes, Inc., Eugene, OR).

PCT-15 (FCryp-2) (Figure 10.23) is a nice example of a fluorescent signaling receptor in which the ionophore moiety has been specially designed to determine intracellular free sodium concentration. An indole derivative acts as the fluorophore: on sodium binding, the emission maximum shifts from 460 nm to 395 nm and the fluorescence intensity increases 25-fold. The origin of these photophysical changes has not yet been studied. The large Stokes shift of the free ligand may be accounted for by photoinduced charge transfer with concomitant internal rotation in the excited state leading to a TICT state. The blue-shift of the emission spectrum on sodium binding is likely to be due to the reduction in the electron-donating character of the dye-bound nitrogen atom of the cryptand.

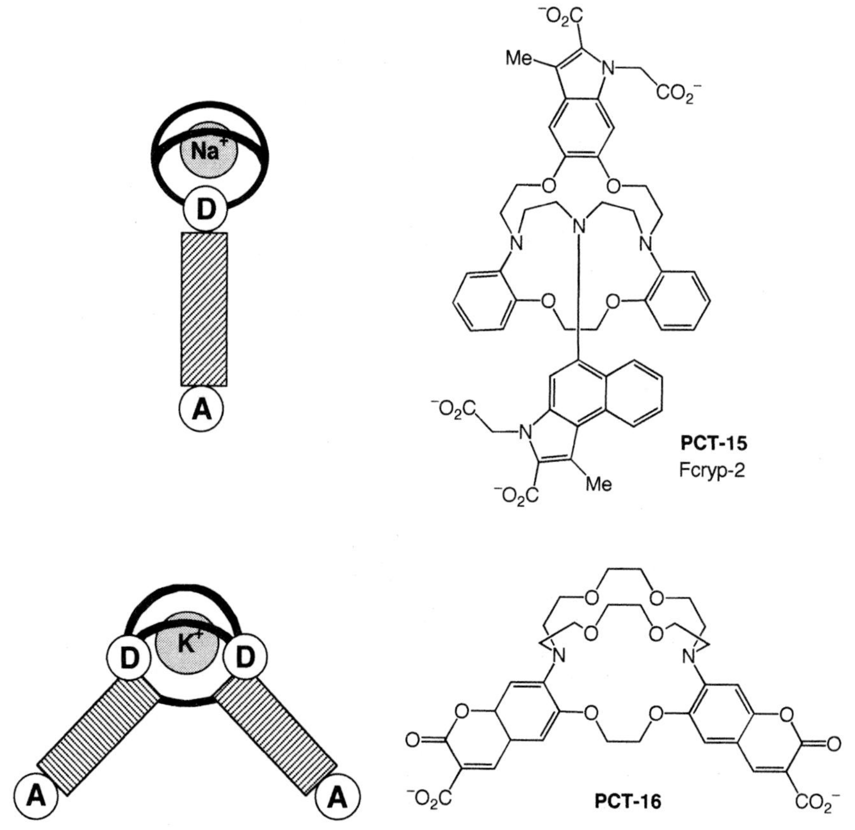

Fig. 10.23. Cryptand-based PCT sensors (PCT-15: Smith G. A. et al. (1988) *Biochem. J.* **250**, 227. PCT-16: Grossley R. et al. *J. Chem. Soc. Perkin Trans.* **2**, 1615).

Another example of a cryptand is PCT-16 (Figure 10.23), which has potential applications as an extracellular probe of potassium. The dissociation constant in water is 1 mmol dm^{-3}, i.e. slightly lower than that of the coumarocryptand PET-5 (see Section 10.3.2).

10.3.3.3 PCT sensors in which the bound cation interacts with an electron-withdrawing group

In contrast to the above-described systems, there are only few systems in which the bound cation can interact with the acceptor part of charge-transfer probes. The case of coumarins linked to crowns (Figure 10.24) is of special interest because the cation interacts directly with the electron-withdrawing group, i.e. the carbonyl group, in spite of the spacer between the fluorophore and the crown. An important consequence is the increase in stability constant of the complexes with respect to the same crown without external complexing atoms.

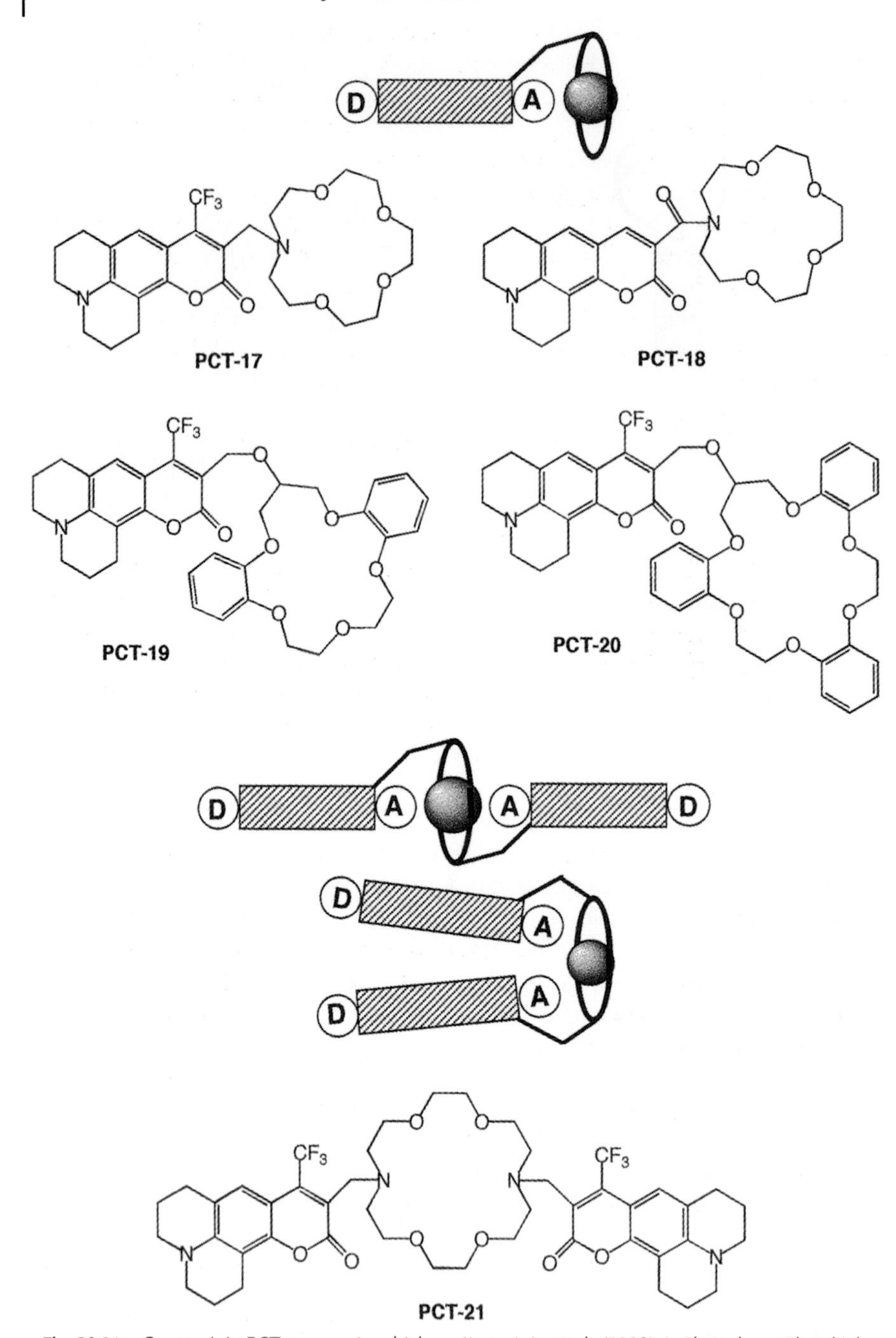

Fig. 10.24. Coumarinic PCT sensors in which the bound cation interacts with the acceptor group (PCT-17 and PCT-21: Bourson J. et al. (1993) *J. Phys. Chem.* **97**, 4552. PCT-18: Habib Jiwan J.-L. et al. (1998) *J. Photochem. Photobiol. A. Chemistry* **116**, 127. PCT-19 and PCT-20: Leray I. et al. (2000) *J. Photochem. Photobiol. A. Chemistry* **135**, 163).

Regarding the cation-induced photophysical changes, it should be kept in mind that the dipole moment of aminocoumarins in the excited state is larger than in the ground state because of the photoinduced charge transfer occurring from the nitrogen atom of the julolidyl ring to the carbonyl group. Therefore, when a cation is coordinated with the carbonyl group, the excited state is more stabilized than the ground state, so that both the absorption and emission spectra are red-shifted.

The selectivity of PCT-17 was in general found to be poor, because of the flexibility of the crown as well as around the spacer between the dye and the crown. Moreover, the nitrogen atom of the crown is easily protonable. The replacement of the methylene bridge of PCT-17 by an amide bridge in PCT-18 precludes pH sensitivity and leads to an improvement of the selectivity towards alkaline earth metal ions with respect to alkaline cations because the amide bridge is more rigid. Another way of improving the selectivity is to use a more rigid crown containing phenyl groups like dibenzocrown or tribenzocrown. This can be achieved by replacing monoaza-15-crown-5 by the more rigid dibenzocrown to give PCT-19. In acetonitrile, this system was found to be selective for some cations, i.e. the selectivity expressed as the ratio of the stability constants is 12 500 for Ca^{2+}/Mg^{2+} and 16 for Na^+/K^+. The spectral shifts with cations are larger with PCT-19 compared to those observed with PCT-17 and PCT-18 and could be explained by the length of the bridge. We might expect that PCT-19 adopts a conformation where the bound cation is closer to the carbonyl group compared to the other systems.

A tribenzocrown has been covalently linked to coumarine to give PCT-20. This complexing moiety is expected to be selective for K^+. The K^+/Na^+ selectivity is only 2 in acetonitrile, but 20 in ethanol.

These examples show that the most important parameters responsible for selectivity in these crowned coumarins are (i) the rigidity of the link beween the fluorophore and the crown, (ii) the rigidity of the crown itself, and (iii) the size of the crown.

Additional photophysical effects can be observed when the fluoroionophore contains two fluorophores. In PCT-21 (Figure 10.24), the carbonyl groups of the two coumarin moieties participate in the complexes: direct interaction between these groups and the cation explains the high stability constants and the photophysical changes. In addition to the shifts of the absorption and emission spectra, an interesting specific increase in the fluorescence quantum yield upon binding of K^+ and Ba^{2+} ions has been observed: in the complexes with these ions that fit best into the crown cavity, the carbonyl groups of the two coumarins are preferentially on the opposite sides with respect to the cation and self-quenching is thus partially or totally suppressed, whereas with cations smaller than the cavity size of the crown, the preferred conformation of the relevant complexes may be such that the two carbonyls are on the same side and the close approach of the coumarin moieties accounts for static quenching.

Calixarene-based compounds PCT-22 and PCT-23 (Figure 10.25) containing one or four appended naphthalenic fluorophores, respectively, exhibit outstanding fluorescence enhancements upon cation binding and are very selective for Na^+ (see Box 10.2).

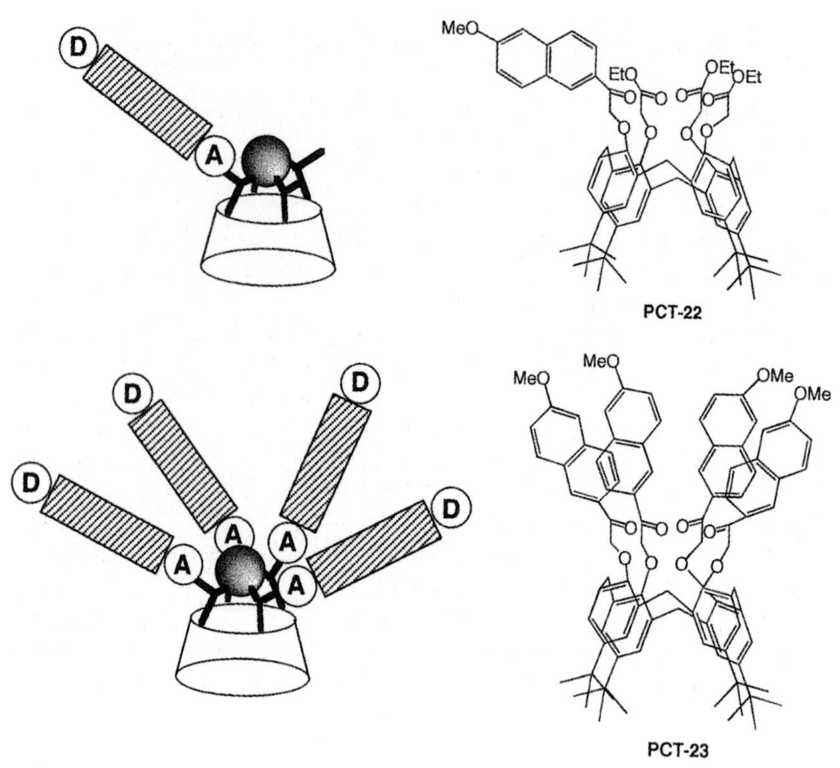

Fig. 10.25. Calixarene-based PCT sensor in which the bound cation interacts with the acceptor group (PCT-22: Leray I. et al. (1999) *Chem. Commun.* 795. PCT-23: Leray I. et al., *Chem. Eur. J.* **7**, 4590-8).

10.3.4
Excimer-based cation sensors

When a fluoroionophore contains two fluorophores whose mutual distance is affected by cation complexation, recognition of this cation can be monitored by the monomer/excimer fluorescence-intensity ratio (see Chapter 4, Section 4.4.1 for excimer formation). Cation binding may favor or hinder excimer formation. In any case, such a ratiometric method allowing self-calibration measurement is of great interest for practical applications.

The bisanthraceno-crown ether E-1 (Figure 10.26) exhibits a fluorescence spectrum composed of the characteristic monomer and excimer bands. Gradual addition of sodium perchlorate to a solution in methanol induces a decrease in the monomer band and an increase in the excimer band. Complexation is indeed expected to bring closer together the two anthracene units, which favors excimer formation. A 2:1 (metal:ligand) complex is formed with Na^+ in methanol and acetonitrile with a positive cooperative effect (see Appendix B). Interestingly, the overall stability constant obtained from absorption data was found to be lower than that

Box 10.2 Calixarene-based fluorescent molecular sensors for sodium ions

In general, PCT sensors undergo moderate changes in fluorescence intensities upon cation binding. Compound PCT-22 (Figure 10.25) is a noticeable exception. The fluorescence quantum yield of the free ligand is very low ($\sim 10^{-3}$ in acetonitrile). In addition to the red-shift of the emission spectrum expected from the interaction of a cation with the carbonyl group of the naphthalenic fluorophore, a considerable enhancement of the fluorescence quantum yield was observed, as shown in Figure B10.2.1[a].

Such an enhancement of the fluorescence quantum yield can be explained in terms of the relative locations of the singlet π–π^* and n–π^* states. In the absence of cation the lowest excited states has n–π^* character, which results in an efficient intersystem crossing to the triplet state and consequently a low fluorescence quantum yield. In the presence of cation, which strongly interacts with the lone pair of the carbonyl group, the n–π^* state is likely to be shifted to higher energy so that the lowest excited state becomes π–π^*. An outstanding selectivity of Na^+ versus K^+ was found: the ratio of the stability constants is 1300 in a mixture of ethanol and water (60:40 v/v).

In the case of PCT-23 (Figure 10.25) containing four appended naphthalenic fluorophores, similar photophysical effects induced by complexation of alkali and alkaline earth metal ions (i.e. red-shift of the absorption and the emission

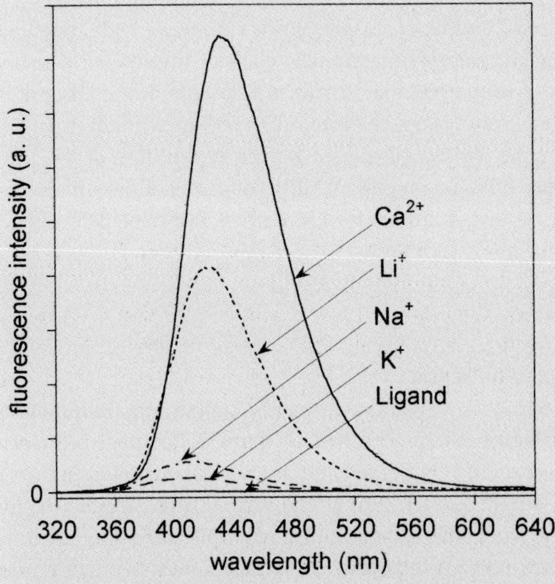

Fig. B10.2.1. Corrected fluorescence spectra of PCT-22 and its complexes with various cations in acetonitrile (excitation: 320 nm). (Redrawn from Leray L. et al.[a])

spectra, and enhancement of the fluorescence quantum yield) were observed. However, a broadening of the emission band or the appearance of a new band at higher wavelength results from complexation with alkali metal ions[b]. This is likely to be due to excimer formation between adjacent naphthalenic groups. The relative intensities of the monomer and excimer bands are related to the size of the complexed cation ($Li^+ > Na^+ > K^+$). The smaller the size of the cation, the smaller the average naphthalene–naphthalene distance, and the higher the probability of intramolecular excimer formation.

a) Leray I., O'Reilly F., Habib Jiwan J.-L., Soumillion J.-Ph. and Valeur B. (1999) *Chem. Commun.* 795.

b) Leray I., Lefèvre J. P., Delouis J. F., Delaire J. and Valeur B. (2001) *Chem. Eur. J.* 7, 4590-8.

from fluorescence data, which means that the complexing ability is greater in the excited state. E-1 forms only a 1:1 complex with K^+.

The non-cyclic ethers E-2 (Figure 10.26), with two pyrenes linked at both ends of the chain, show strong intramolecular excimer formation. Addition of alkaline earth metal ions leads to an increase in monomer emission at the expense of the excimer band. The helical structure of the 1:1 complexes is supported by ^{1}NMR spectra. Thanks to the pseudocyclic structure, the stability constants of the complexes with Ca^{2+}, Sr^{2+} and Ba^{2+} in acetonitrile are quite high (10^6–10^7 for $n = 5$), but the selectivity is poor as a consequence of the flexibility of the oxyethylene chain.

E-3 (Figure 10.26) is the first example of an ionophoric calixarene with appended fluorophores, demonstrating the interest in this new class of fluorescent sensors. The lower rim contains two pyrene units that can form excimers in the absence of cation. Addition of alkali metal ions affects the monomer versus excimer emission. According to the same principle, E-4 was designed for the recognition of Na^+; the Na^+/K^+ selectivity, as measured by the ratio of stability constants of the complexes, was indeed found to be 154, while the affinity for Li^+ was too low to be determined.

10.3.5 Miscellaneous

10.3.5.1 Oxyquinoline-based cation sensors

Various fluorescent chelators, podands and coronands containing oxyquinoline fluorophores are shown in Figure 10.27. The first of them is 8-hydroxyquinoline M-1 (8-HQ), often called oxine, and its derivatives, mainly 8-hydroxyquinoline-5-sulfonic acid M-2 (8-HQS). 8-HQ is considered as the second most important chelating agent after EDTA. The most interesting feature of 8-HQ and 8-HQS is their fluorogenic character, i.e. their very low quantum yield in aqueous or organic solutions and the fluorescence enhancement arising from cation binding. The non-fluorescent character of 8-HQ can be satisfactorily explained by the occurrence of excited-state proton transfer reactions coupled to an intramolecular electron

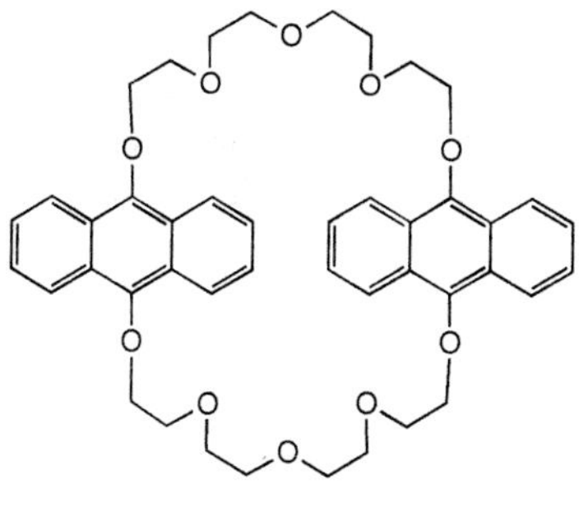

E-1

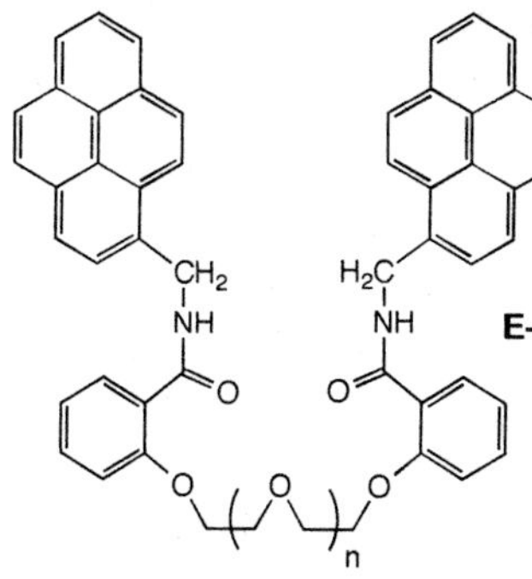

E-2 (n=2 to 5)

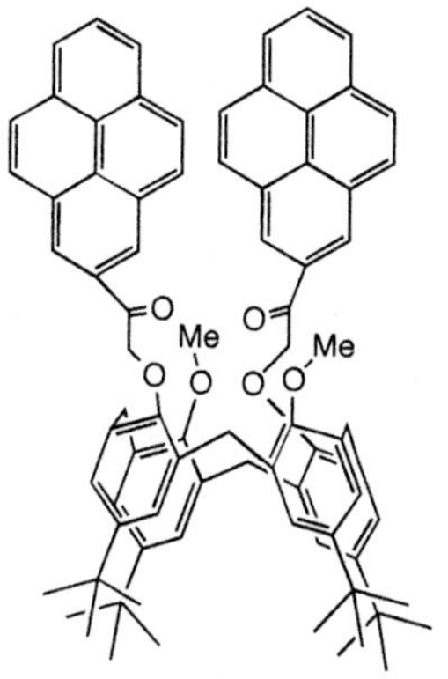

E-3

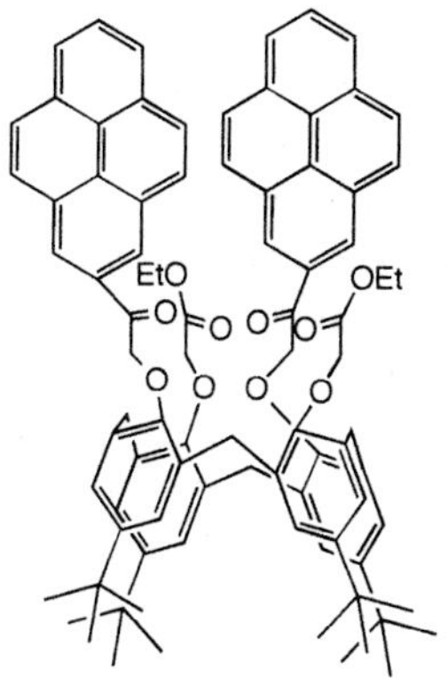

E-4

Fig. 10.26. Excimer-forming cation sensors (E-1: Bouas-Laurent H. et al. (1986) *J. Am. Chem. Soc.* **108**, 315. Marquis D. and Desvergne J.-P. (1994) *Chem. Phys. Lett.* **230**, 131. E-2: Suzuki Y. et al. (1998) *J. Phys. Chem.* **102**, 7910. E-3: Aoki I. et al. (1991) *Chem. Commun.* 1771. E-4: Jin T. et al. (1992) *Chem. Commun.* 499).

transfer. In fact, photoinduced deprotonation of the –OH group and protonation of the heterocyclic nitrogen atom can occur either with surrounding water molecules or intramolecularly, depending on the possible existence of H-bonding between the two functions. De-excitation of the resulting tautomer (cetonic form) occurs mainly via a non-radiative pathway.

In contrast, many metal chelates of 8-HQ and 8-HQS (e.g. Cd, Zn, Mg, Al, Ga, In) exhibit intense yellow–green fluorescence because the above-described photoprocesses are impaired by a bound cation. However, the changes in electronic distribution upon excitation are likely to weaken the bond between the oxygen atom and the metal ion, thus allowing some charge transfer from the phenolate $-O^-$ to the nitrogen atom of the adjacent ring. Weakening of this bond should be favored by the presence of water. It has indeed been observed that the metal chelates with 8-HQ or 8-HQS are more fluorescent in micellar media than in hydro-organic solvents and even more fluorescent in reverse micellar systems at low water content.

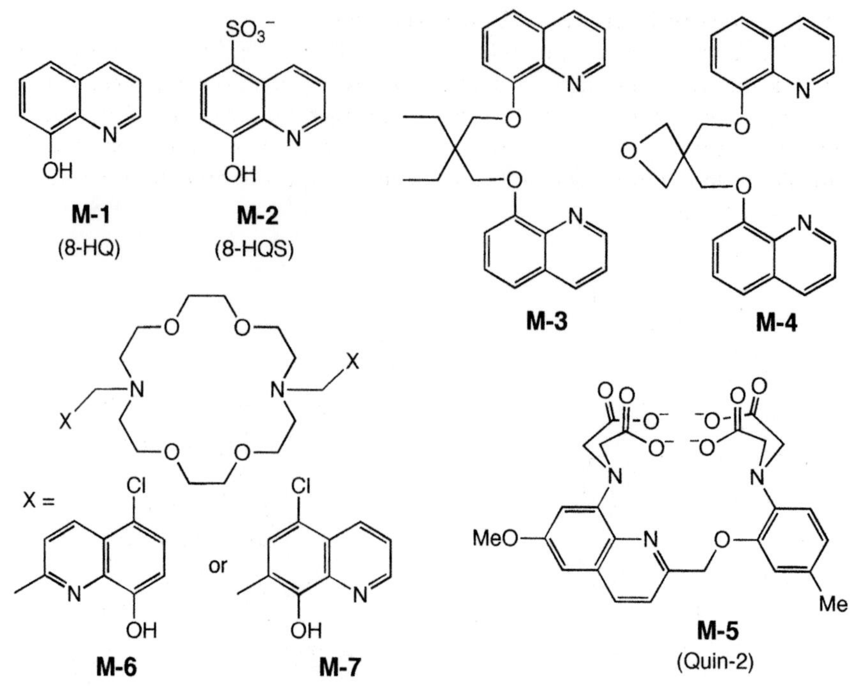

Fig. 10.27. Oxyquinoline-based chelators, podands and coronands (M-1 and M-2: K. Soroka et al. (1987) *Anal. Chem.* **59**, 629. Bardez E. et al. (1997) *J. Phys. Chem. B* **101**, 7786. M-3 and M-4: Hiratani K. (1987) *Chem. Commun.* 960. M-5: Tsien R. Y. and Pozzan T. (1989) *Methods in Enzymology* **172**, 230. M-6: Zhang X. X. et al. (1995) *J. Am. Chem. Soc.* **117**, 11507. M-7: Prodi L. et al. (1998) *Tetrahedron Lett.* **39**, 5451).

The selectivity of 8-HQ and 8-HQS is poor, but can be improved by appropriate substitution on the oxygen atom to form acyclic polyethers containing two oxyquinoline fluorophores, as in compounds M-3 and M-4 shown in Figure 10.27. The geometrical constraints in these compounds explain the excellent selectivity for the small lithium ion.

M-5 (Quin-2) was the first practical fluorescent indicator for cytosolic calcium with a simple 6-methoxyquinoline as its fluorophore. Ca^{2+}-binding increases the fluorescence intensity about six-fold (without spectral displacement, in contrast to Fura 2: see Section 10.3.3). The fluorescence lifetime of Quin-2 is highly sensitive to calcium concentration; Quin-2 can thus be used as a probe in the technique of fluorescence lifetime imaging.

Diaza-18-crown-6 substituted with 5-chloro-8-hydroxyquinoline exhibits very interesting complexing properties. For instance, M-6 is very selective for Ba^{2+} over other alkaline earth cations and for K^+ over Na^+ in methanol. Unfortunately, investigation of fluorogenic effects with other cations has not been reported. On the other hand, the fluorescence intensity of M-7 was shown to increase by a factor of 1000 in the presence of Mg^{2+} (in a mixture of methanol/water 1:1 v/v) whereas other alkaline earth ions have no effect.

10.3.5.2 **Further calixarene-based fluorescent sensors**

In calixarene-based compound M-8 (Figure 10.28), bearing four anthracene moieties on the lower rim, some changes in fluorescence intensity were observed on binding of alkali metal ions but no excimer emission was detected. Quenching of the fluorescence by Na^+ may arise from interaction of four anthracene residues brought in closer proximity to one another; enhancement of fluorescence by K^+ is difficult to explain.

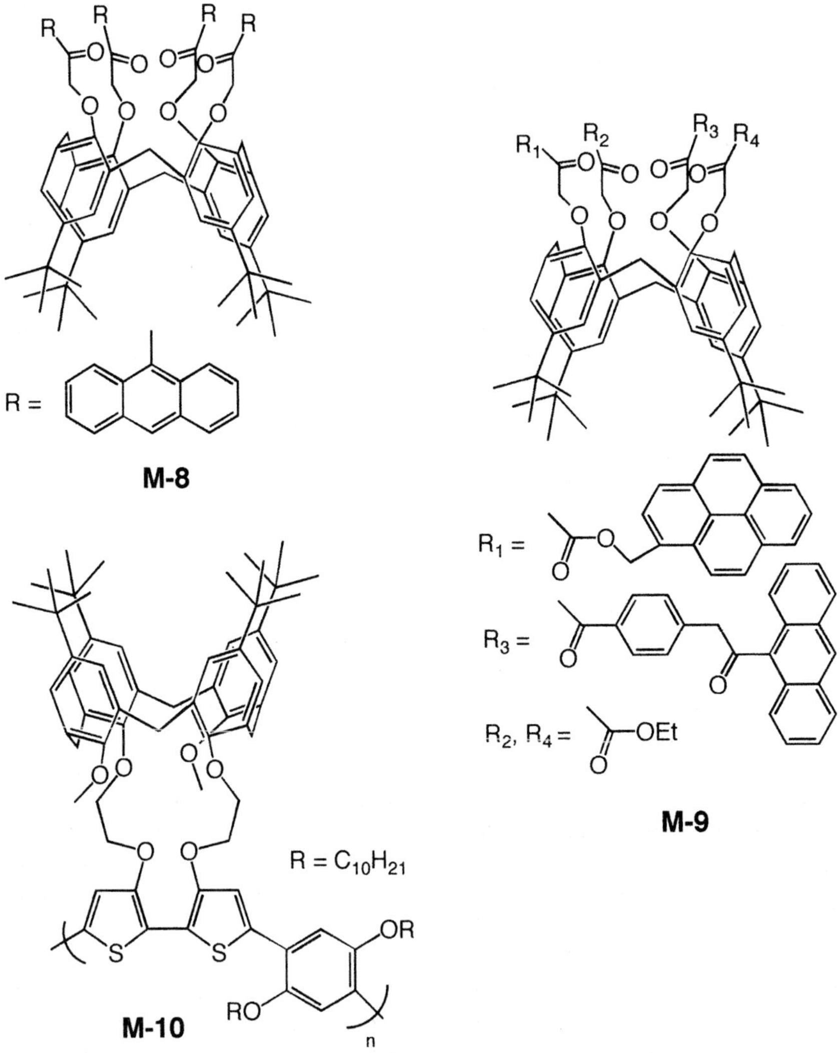

Fig. 10.28. Calixarene-based sensors (M-8: Perez-Jimenez C. et al. (1993) *Chem. Commun.* 480. M-9: Jin T. (1999) *Chem. Commun.* 249. M-10: Crawford K. et al. (1998) *J. Am. Chem. Soc.* **120**, 5187.).

The Na^+ sensor M-9 has a structure analogous to that of compound E-4, but instead of two identical pyrene fluorophores, it contains two different fluorophores with a pyrene group and an anthroyloxy group. Resonance energy transfer (see Chapter 9) from the former to the latter is then possible because of the spectral overlap between the fluorescence spectrum of the pyrene moiety and the absorption spectrum of the anthroyloxy moiety. Upon addition of Na^+ to a solution of M-9 in a mixture of MeOH and THF (15:1 v/v), the fluorescence of the anthroyloxy group increases significantly compared with that of the pyrene group, which permits a ratiometric measurement.

An original fluoroionophore M-10, consisting of a conjugated poly(phenylene bithiophene) linked to calixarene, was designed and tested with various cations. An outstanding selectivity towards Na^+ was observed, whereas Li^+, K^+ or Ca^{2+} induce negligible effects. The blue-shift of the emission spectrum upon sodium binding appeared to be much more pronounced than with a monomeric model compound and no emissive point was observed. This interesting behavior may be due to migration of excitation energy to regions of the polymer that do not have bound Na^+ and can relax to lower energy conformations.

10.3.6
Concluding remarks

The examples described above illustrate the immense variety of fluoroionophores that have been designed for cation recognition. Emphasis was put on the understanding of cation-induced photophysical changes, which should help the user and the designer of this kind of sensor.

A distinct advantage of PET sensors is the very large change in fluorescence intensity usually observed upon cation binding, so that the expressions 'off–on' and 'on–off' fluorescent sensors are often used. Another characteristic is the absence of shift of the fluorescence or excitation spectra, which precludes the possibility of intensity-ratio measurements at two wavelengths. Furthermore, PET often arises from a tertiary amine whose pH sensitivity may affect the response to cations.

In PCT sensors, the changes in fluorescence quantum yield on cation complexation are generally not very large compared to those observed with PET sensors. Nevertheless, exceptions can be found (see PCT-7 and PCT-22). However, the absorption and fluorescence spectra are shifted upon cation binding so that an appropriate choice of the excitation and observation wavelengths often allows us to observe quite large changes in fluorescence intensity. Moreover, ratiometric measurements are possible: the ratio of the fluorescence intensities at two appropriate emission or excitation wavelengths provides a measure of the cation concentration, which is independent of the probe concentration (provided that the ion is in excess) and is insensitive to the intensity of incident light, scattering, inner filter effects and photobleaching. Ratiometric measurements are also possible with excimer-based sensors.

It should be emphasized that the selectivity and efficiency of binding towards a given cation in a specific range of concentration can be very different according to the application. In this respect, the choice of the recognition moiety is of major importance but it is also important to note that the fluorophore itself often participates in the complexation and thus plays a role in the selectivity. Cryptands are known to be very selective to alkali ions but they often contain a tertiary amine that is pH sensitive. Calixarenes with appropriate appended groups including fluorophores have great potential in terms of molecular design.

10.4 Fluorescence molecular sensors of anions

Anions play key roles in chemical and biological processes. Many anions act as nucleophiles, bases, redox agents or phase transfer catalysts. Most enzymes bind anions as either substrates or cofactors. The chloride ion is of special interest because it is crucial in several phases of human biology and in disease regulation. Moreover, it is of great interest to detect anionic pollutants such as nitrates and phosphates in ground water. Design of selective anion molecular sensors with optical or electrochemical detection is thus of major interest, however it has received much less attention than molecular sensors for cations.

The methods of anion detection based on fluorescence involve quenching, complex formation, redox reactions and substitution reactions (Fernandez-Gutierrez and Muñoz de la Peña, 1985). This chapter will be restricted to anion molecular sensors based on collisional quenching (in general, they exhibit a poor selectivity) and on recognition by an anion receptor linked to a fluorophore (fluoroionophore).

10.4.1 Anion sensors based on collisional quenching

Many fluorescent molecular sensors for halide ions (except F^-) are based on collisional quenching of a dye. In particular, the determination of chloride anions in living cells is done according to this principle. Examples of halide ion sensors are given in Figure 10.29.

The drawback of these molecular sensors is their lack of selectivity, as shown by the Stern–Volmer constants (Table 10.4). For instance A-1, 6-methoxy-*N*-(3-sulfopropyl)quinolinium (SPQ) is mainly used as a Cl^--sensitive fluorescent indicator, but its fluorescence is also quenched by several other anions (I^-, Br^- and SCN^-, but not by NO_3^-).

Another feature is that the absence of spectral change precludes ratiometric measurements. However, dual-wavelength Cl^- sensors have been constructed. For instance, in compound A-6 (Figure 10.30), 6-methoxyquinolinium (MQ) as the Cl^--sensitive fluorophore (blue fluorescence) is linked to 6-aminoquinolinium (AQ) as the Cl^--insensitive fluorophore (green fluorescence), the spacer being either rigid or flexible.

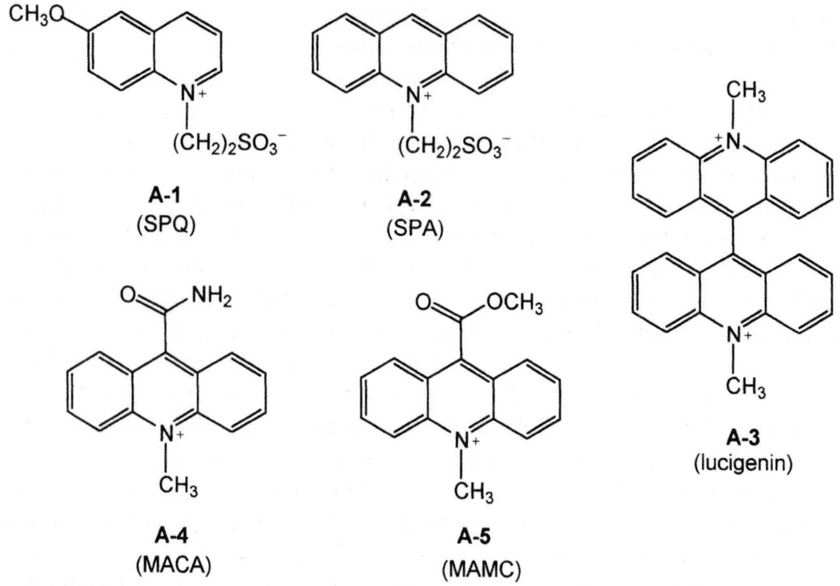

Fig. 10.29. Halide ion sensors (A-1 to A-5: Biwersi J. et al. (1994) *Anal. Biochem.* **219**, 139.

Tab. 10.4. Stern–Volmer constants (M^{-1}) of halide molecular sensors in aqueous solutions (see chemical formulae in Figure 10.29) (data from Biwersi et al., 1994)

Compound	Cl^-	Br^-	I^-	SCN^-
SPQ	118	175	276	211
SPA	5	224	307	255
Lucigenin	390	585	750	590
MACA	225	480	550	480
MAMC	160	250	267	283

Abbreviations: SPQ: 6-methoxy-*N*-(sulfopropyl)quinolinium; SPA: *N*-(sulfopropyl)acridinium; lucigenin: bis-*N*-methylacridinium nitrate; MACA: 10-methylacridinium-9-carboxamide; MAMC: *N*-methylacridinium-9-methyl carboxylate.

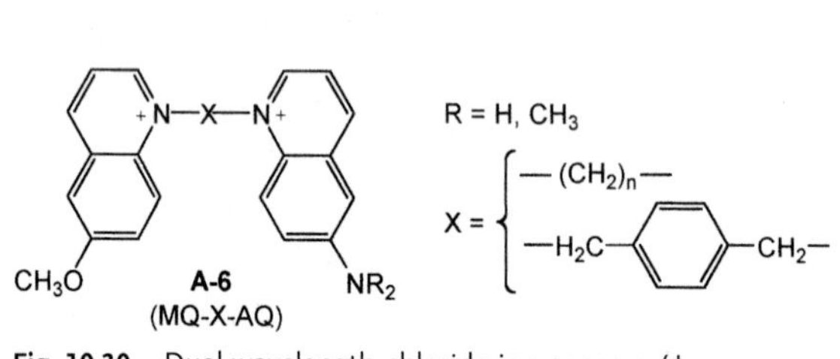

Fig. 10.30. Dual-wavelength chloride ion sensors (Jayaraman S. et al. (1999) *Am. J. Physiol.* **276**, C747).

10.4.2
Anion sensors containing an anion receptor

There are a limited number of fluorescent sensors for anion recognition. An outstanding example is the diprotonated form of hexadecyltetramethylsapphyrin (A-7) that contains a pentaaza macrocyclic core (Figure 10.31): the selectivity for fluoride ion was indeed found to be very high in methanol (stability constant of the complex $\sim 10^5$) with respect to chloride and bromide (stability constants $\leq 10^2$). Such selectivity can be explained by the fact that F^- (ionic radius $\sim$ 1.19 Å) can be accommodated within the sapphyrin cavity to form a 1:1 complex with the anion in the plane of the sapphyrin, whereas Cl^- and Br^- are too big (ionic radii 1.67 and 1.82 Å, respectively) and form out-of-plane ion-paired complexes. A two-fold enhancement of the fluorescent intensity is observed upon addition of fluoride. Such enhancement can be explained by the fact that the presence of F^- reduces the quenching due to coupling of the inner protons with the solvent.

Phosphate groups have attracted much attention because of their biological relevance. They can be recognized by anthrylpolyamine conjugate probes A-8 (Figure 10.32). The choice of pH is crucial: at pH 6, a fraction of 70% of A-8 exists as a triprotonated form, the nitrogen atom close to the anthracene moiety being unprotonated. The very low fluorescence of this compound is due to photoinduced electron transfer from the unprotonated amino group to anthracene. This trication can bind a complementary structure like monohydrogenophosphate whose three oxygen atoms interact with the three positive charges; the remaining phosphate OH group is in a favorable position to undergo intracomplex proton transfer to the unprotonated amino group, which eliminates intramolecular quenching. Then, binding is accompanied by a drastic enhancement of fluorescence. A-8 can also bind ATP, citrate and sulfate. This mode of recognition is conceptually very interesting but the stability of the complexes is low.

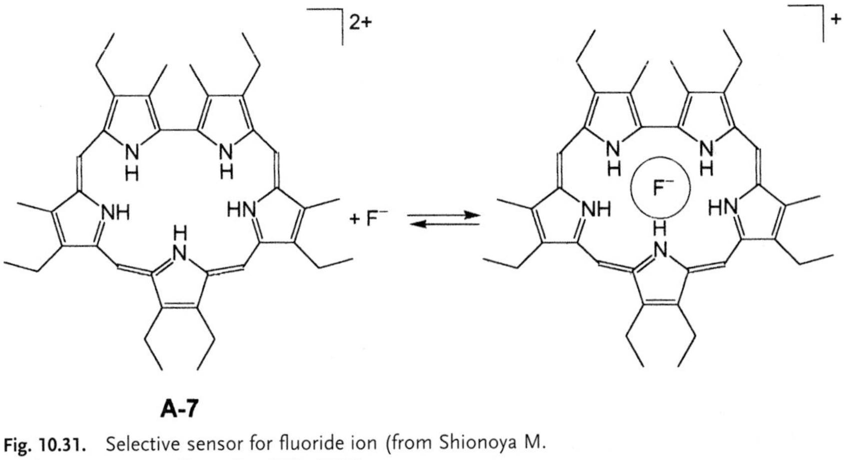

Fig. 10.31. Selective sensor for fluoride ion (from Shionoya M. et al. (1992) *J. Am. Chem. Soc.* **111**, 8735).

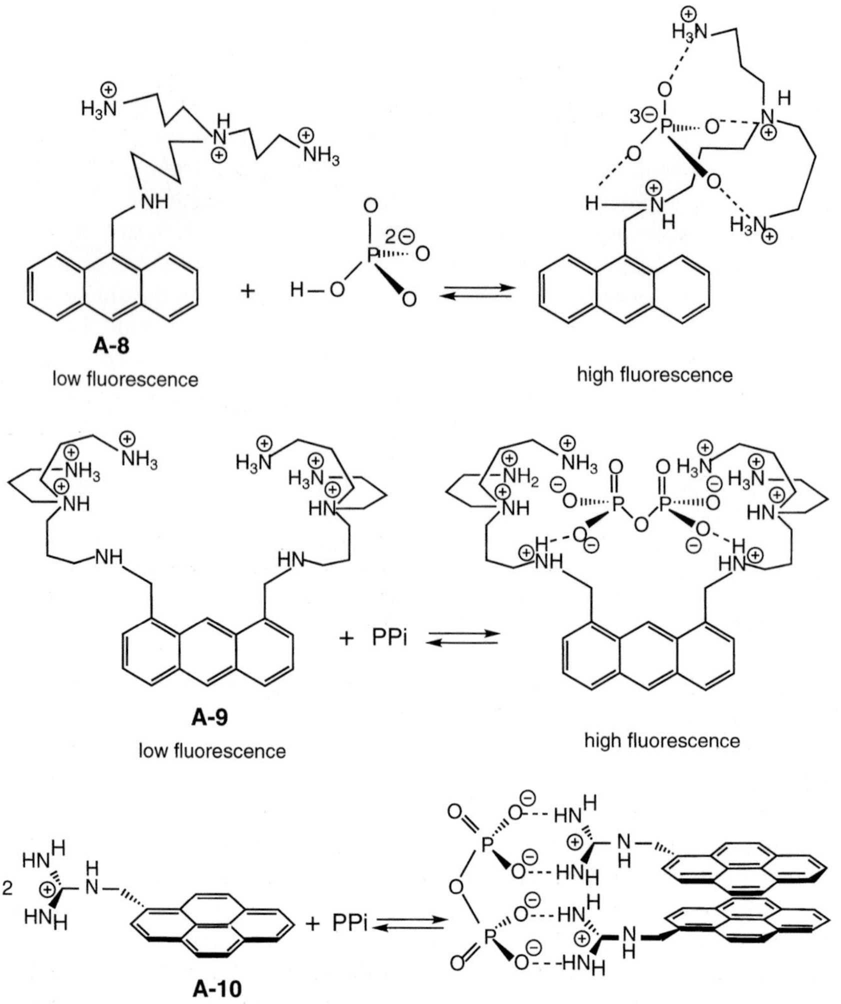

A-8
low fluorescence

high fluorescence

A-9
low fluorescence

high fluorescence

A-10

Fig. 10.32. Sensors for phosphate groups. (A-8: Huston M. E. et al. (1989) *J. Am. Chem. Soc.* **111**, 8735. A-9: Vance D. H. and Czarnik A. W. (1994) *J. Am. Chem. Soc.* **116**, 9397. A-10: Nishizawa S. et al. (1999) *J. Am. Chem. Soc.* **121**, 9463).

The same strategy has been applied to the recognition of pyrophosphate ions $P_2O_7^{4-}$ (PPi). A-9 (Figure 10.32) binds these ions over 2000 times more tightly than phosphate ions, permitting the real-time monitoring of pyrophosphate hydrolysis.

Detection of pyrophosphate has also been demonstrated by a simple self-assembling system A-10 (Figure 10.32) with a pyrene-functionalized monoguanidinium receptor. This receptor was found to self-assemble to form a 2:1 (host:-guest) complex with high selectivity for biologically important pyrophosphate ions in methanol. A sandwich-like ground-state pyrene dimer is formed. The character-

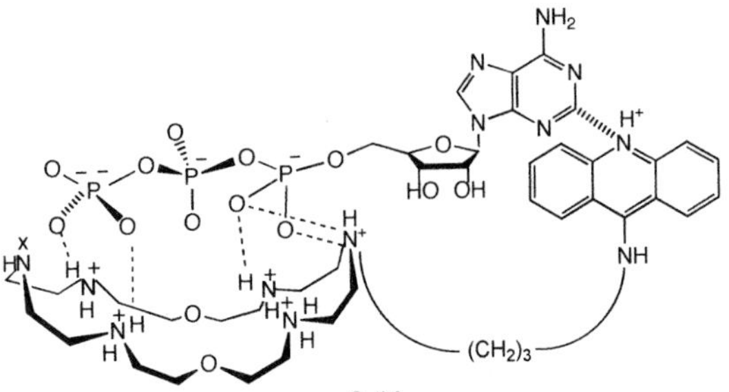

Fig. 10.33. Anion sensors based on protonated polyazamacrocycles and polyazamacrobicycles (A-11: Hosseini M. W. et al. (1988) *J. Chem. Soc., Chem. Commun.* 596. A-12: Dhaenens M. et al. (1993) *J. Chem. Soc., Perkin Trans. 2,* 1379. A-12: Teulade-Fichou M.-P. et al. (1996) *J. Chem. Soc., Perkin Trans. 2,* 2169).

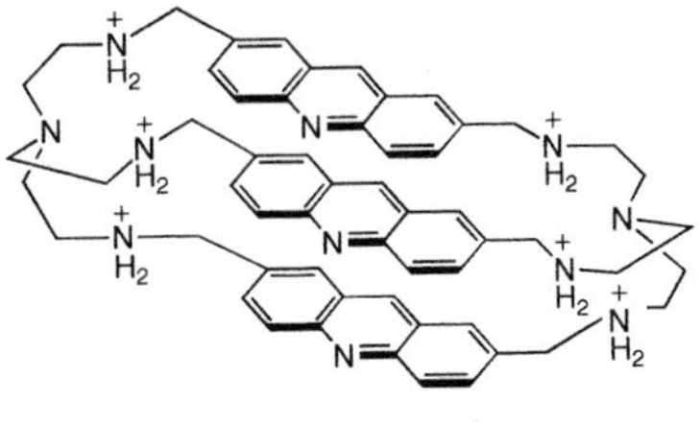

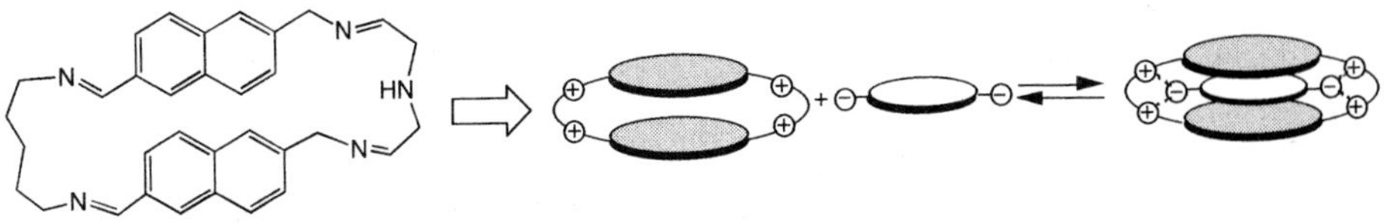

istic excimer fluorescence band appears upon pyrophosphate binding so that calibration via ratiometry is possible.

A better efficiency and selectivity is expected with probes based on macrocyclic and macropolycyclic polycations that are capable of forming strong and selective complexes with inorganic anions and with negatively charged functional groups, especially phosphate and carboxylate groups (Figure 10.33). In fact, protonated polyazamacrocycles and polyazamacrobicycles can complex anions. Compound A-11 is a good example of a fluorogenic anion receptor specially designed for nucleotide recognition and ATP hydrolysis: it contains a macrocyclic polyamine as a receptor of the triphosphate moiety and an acridine group for stacking interaction with the nucleic base. The fluorescence of the acridine group is significantly enhanced upon binding. In addition, amino groups of the protonated macrocyclic hexamine catalyse the hydrolysis of ATP.

The water-soluble bis-intercaland-type receptor molecule A-12 contains two naphthalene rings situated at a distance suitable for the intercalation of planar anionic substrates such as aromatic carboxylates and nucleotides. The four protonated nitrogen atoms (at pH = 6) prevent the collapse of the cavity (because of repulsive interactions between the positive charges) and precludes fluorescence quenching by photoinduced electron transfer from the nitrogen atoms to the naphthalene moieties. Complexation with dianionic substrates causes more or less efficient fluorescence quenching. Both stacking and electrostatic factors contribute to the stability of the complexes, which increases with the number of negative charges in the substrate. Moreover, interesting selectivities have been observed; in particular, among the nucleobases, guanine-containing species are preferentially bound.

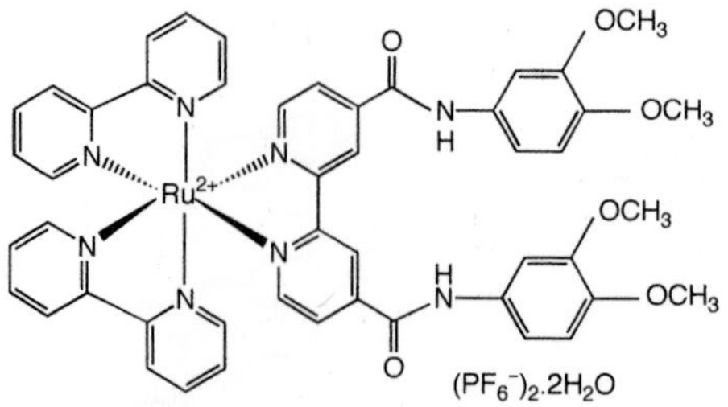

Fig. 10.34. Sensors based on acyclic, macrocyclic and calixarene ruthenium–bipyridyl (from Beer, P. D. (1996) *Chem. Commun.* 689).

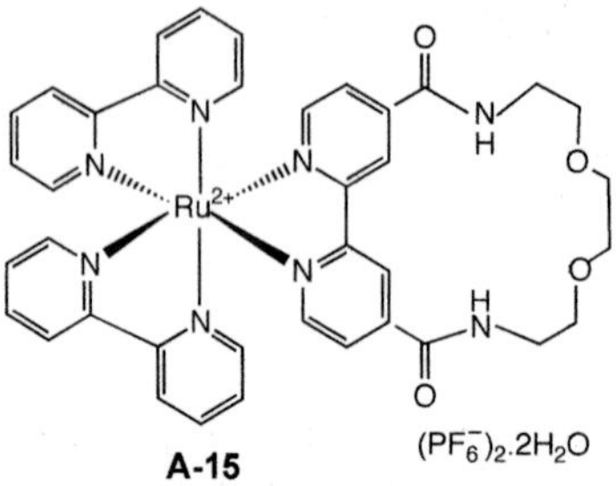

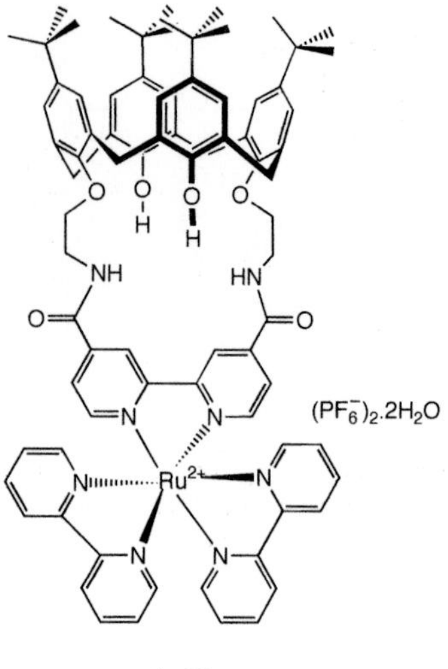

The fluorescence spectrum of the tris-acridine cryptand A-13 shows the characteristic monomer and excimer bands. Upon complexation with various organic anions (carboxylates, sulfonates, phosphates), the monomer band increases at the expense of the excimer band. The stability of the complexes depends on the contribution of the electrostatic and hydrophobic forces and on the structural complementarity. Stability constants of the complexes ranging from 10^3 to 10^7 have been measured. In particular, A-13 binds tightly to mono- and oligonucleotides, and it can discriminate by its optical response between a pyridimic and a purinic sequence.

Another interesting class of anion receptors (Beer, 1996) consists of acyclic, macrocyclic and lower-rim calix[4]arene structures in which the Lewis-acidic redox and photoactive ruthenium(II) bipyridyl moiety is introduced. These sensors offer the dual capability of detection of anionic species via either electrochemical or optical methods. Examples are given in Figure 10.34. Determination of the stability constants of the complexes in dimethyl sulfoxide reveals a high selectivity of the calixarene-based receptor for $H_2PO_4^-$. It is worth noting that hydrogen bonding plays an important role in the stabilization of the complexes. Six hydrogen bonds stabilize the Cl^- anion (two amide and four C–H groups) in the complex with A-14. Three hydrogen bonds (two amide and one calix[4]arene hydroxy) permit stabilization of $H_2PO_4^-$ in the complex with A-16. The fluorescence spectrum of the ligands undergoes significant blue-shift upon anion binding (16 nm for addition of

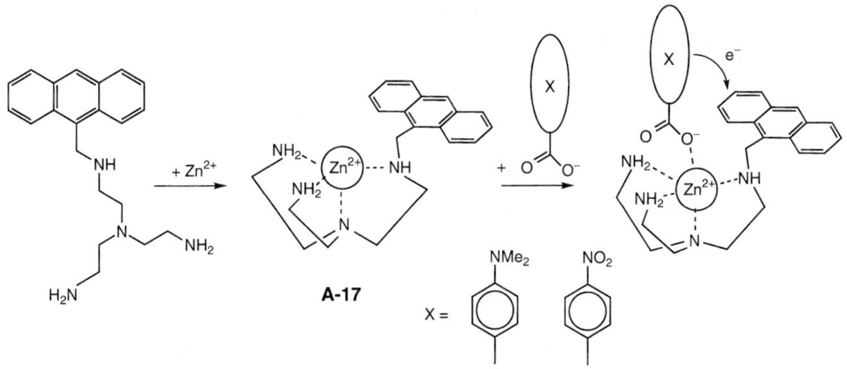

Fig. 10.35. Recognition of carboxylate ions by a complex of an anthrylamine with Zn^{II} (De Santis G. et al. (1996) *Angew. Chem. Int. Ed.* **35**, 202).

$H_2PO_4^-$ to A-16) with a concomitant large increase in fluorescence quantum yield. Such an increase may be due to the rigidification of the receptor by the bound anion, which decreases the efficiency of non-radiative de-excitation.

Carboxylate ions can be recognized and sensed by a complex of an anthrylamine (analogous to A-8) with Zn^{II} (Figure 10.35). The resulting four-coordinate metal center (A-17) has a vacant site for coordination of an anion to give a trigonal-bipyramidal arrangement. Affinity towards anions bearing a carboxylate group is strong. Recognition is signaled via fluorescence quenching of the appended fluorophore as a result of intramolecular electron transfer, e.g. from a bound 4-*N*,*N*-dimethylaminobenzoate to the excited anthracene moiety. Such a transfer is favored by the stacking of benzoate and anthracene. The selectivity is essentially determined by the energy of the metal–anion coordinative interaction; moreover, only the interactions with anions displaying distinctive electron donor or electron acceptor tendencies cause fluorescence quenching of anthracene. For instance, NO_3^- and SCN^- do not affect the anthracene emission and do not compete with dimethylaminobenzoate for the binding to the metal center, whereas Cl^- causes an intensity decrease of less than 5% and competes for binding. The acetate ion behaves in a similar way to Cl^-.

According to the same strategy, the cooperative action of boronic acid and zinc chelate was used in the design of the fluorescent sensor A-18 for recognition of uronic and sialic acid salts (Figure 10.36). The zinc-phenanthroline moiety is the fluorophore whose fluorescence is quenched by the tertiary amine in the free ligand state. Upon addition of uronic and sialic acids at pH 8, the carboxylate functions of these acids can ligate the zinc atom while the hydroxyl functions of the saccharide ring form a boronate ester at the saccharide recognition site. Because the interaction between boronic acid and amine is intensified, the PET process is suppressed and the fluorescence intensity increases.

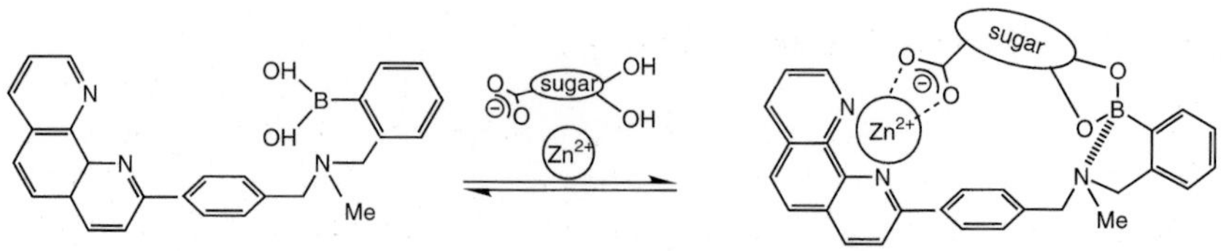

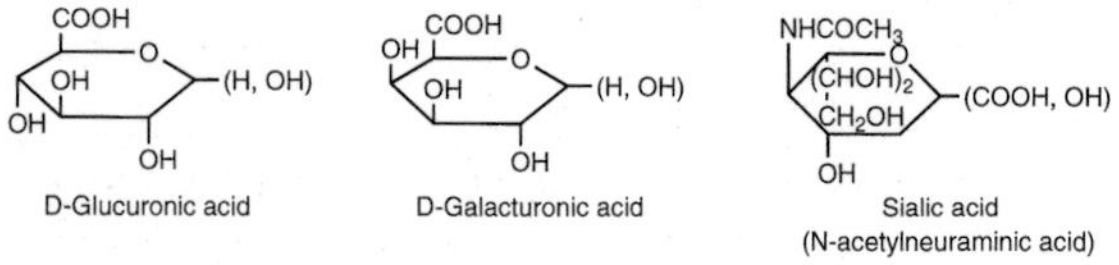

Fig. 10.36. Recognition of uronic and sialic acid salts by the cooperative action of boronic acid and zinc chelate (Yamamoto M. et al. (1996) *Tetrahedron* **54**, 3125).

An interesting practical application is the detection of the citrate anion in soft drinks, as shown in Box 10.3. The strategy is quite different from that of the preceding examples because the anion receptor is not linked to a fluorophore. The latter simply acts in competition with the citrate anion in a fashion that resembles fluorescence-based immunoassays.

There a great need for selective anion sensors, but the number of available sensors is rather limited because of difficulties in their design. However, new selective sensors are expected because of the considerable progress made in the synthesis of anion receptors.

10.5 Fluorescent molecular sensors of neutral molecules and surfactants

Recognition of neutral organic molecules in solution is a much greater challenge than recognition of ionic species because the involved interactions (Van der Waals interactions, hydrogen bonds) are much weaker than those existing with charged species and the electronic changes induced by complexation are smaller.

A few examples will be presented in this section. Among them, detection of steroid molecules (e.g. cortisone, hydrocortisone, progesterone, etc.) is of particular interest because of their biological relevance. Because saccharides play a significant role in the metabolic pathways of living organisms, it is necessary to detect the presence and to measure the concentration of biologically important sugars (glucose, fructose, galactose, etc.) in aqueous solutions. Determination of enantiomeric purity of synthetic drugs and monitoring of fermentation processes are examples of applications.

Another example of practical interest is the detection of surfactants that are extensively used in domestic and industrial applications; their slow degradation poses a severe problem of environmental pollution.

Box 10.3 Detection of citrate in beverages[a)]

At neutral pH, citrate bears three negative charges and is thus quite distinctive from interfering species like mono- and dicarboxylates, phosphates, sugars and simple salts. This observation led Metzger and Anslyn to design the A-19 receptor, consisting of three guanidinium groups that form hydrogen bonds and ion pairs with carboxylate groups (Figure B10.3.1). The positively charged groups are arranged on one face of a benzene ring. This conformation leads to good binding of citrate in water ($\log K_S = 3.83$).

None of the involved species are fluorescent. Therefore, for fluorescence signaling of citrate recognition, carboxyfluorescein is first added to the medium because binding to the receptor in the absence of citrate is possible and causes deprotonation of carboxyfluorescein, which results in high fluorescence. Citrate is then added, and because it has a better affinity for the receptor than carboxyfluorescein, it replaces the latter, which emits less fluorescence in the bulk solvent as a result of protonation. Note that this molecular sensor operates in a similar fashion to antibody-based biosensors in immunoassays. It was successfully tested on a variety of soft drinks.

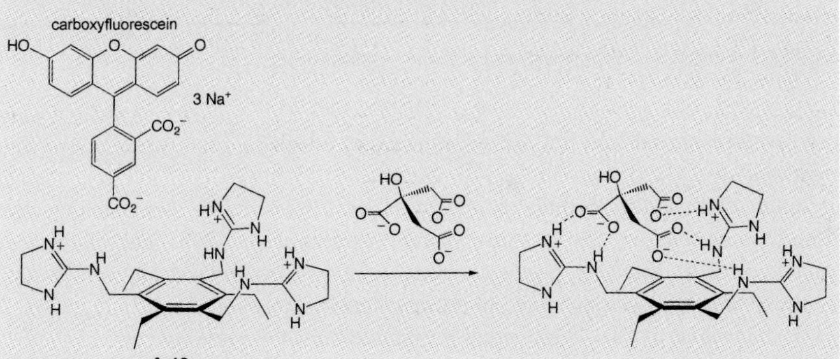

carboxyfluorescein

3 Na+

A-19

Fig. B10.3.1. Formation of a complex between the A-19 receptor and citrate ion[a)].

a) Metzger A. and Anslyn E. V. (1998) *Angew. Chem. Int. Ed.* 37, 649.

10.5.1
Cyclodextrin-based fluorescent sensors

α-, β- and γ-cyclodextrins (CDs) are toroidal molecules containing six, seven and eight glucopyranose units, respectively. The internal diameters of the cavities are approximately 5, 6.5 and 8.5 Å, respectively, and the depth is about 8 Å (Figure 10.37). Their ability to form inclusion complexes with various organic compounds in aqueous solutions is of major interest for molecular recognition. Numerous modified CDs have been designed for improving the selectivity of binding. CDs

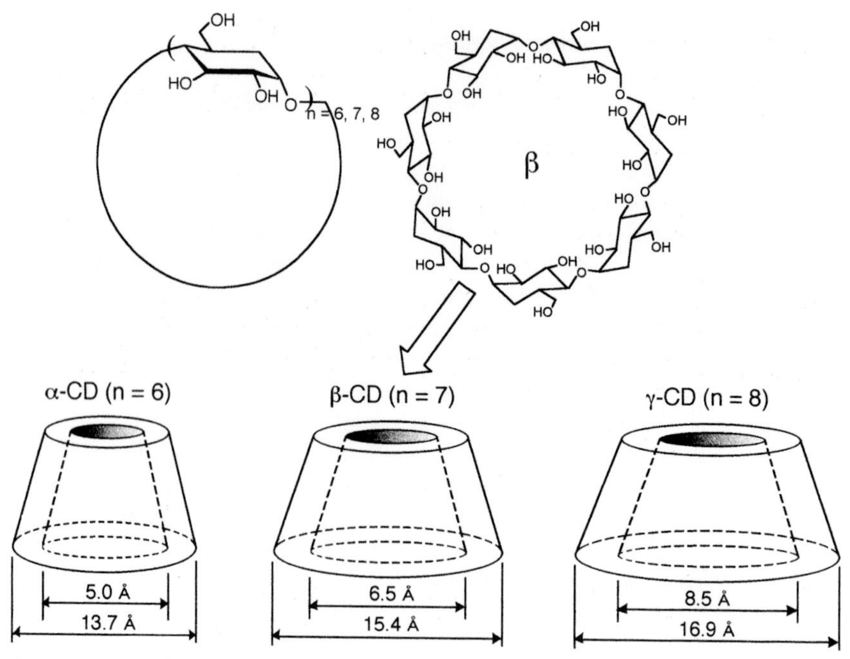

Fig. 10.37. Formulae and dimensions of α, β and γ-cyclodextrins.

can be transformed into fluorescent sensors by attaching one, two or more fluorophores.

Figure 10.38 shows modified β- and γ-cyclodextrins with two identical appended fluorophores that are able to form excimers (Ueno et al., 1997). They have been studied in 10% ethylene glycol aqueous solutions. β-cyclodextrins with two 2-naphthylsulfonyl moieties linked to the smaller rim (compounds β-CD1, β-CD2, β-CD3), have a cavity that is too small to include both fluorophores; one of them is outside the cavity and the other is inside. The latter can be excluded from the cavity upon inclusion of a guest molecule. Therefore, the excimer band in the fluorescence spectrum increases upon guest inclusion.

In contrast, the fluorescence spectra of the parent γ-cyclodextrins (compounds γ-CD1, γ-CD2, γ-CD3, γ-CD4) exhibit both monomer and excimer bands in the absence of guests because the cavity is large enough to accommodate both fluorophores (Figure 10.38). The ratio of excimer and monomer bands changes upon guest inclusion. The ratio of the intensities of the monomer and excimer bands was used for detecting various cyclic alcohols and steroids (cyclohexanol, cyclododecanol, *l*-borneol, 1-adamantanecarboxylic acid, cholic acid, deoxycholic acid and parent molecules, etc.).

Various CDs with a single appended fluorophore were also designed (Ueno et al., 1997). For β-CDs, the principle is the following: in the absence of guest, the fluorophore is encased in the cavity and exhibits photophysical properties that are characteristic of such a nonpolar restricted microenvironment. Upon addition of a

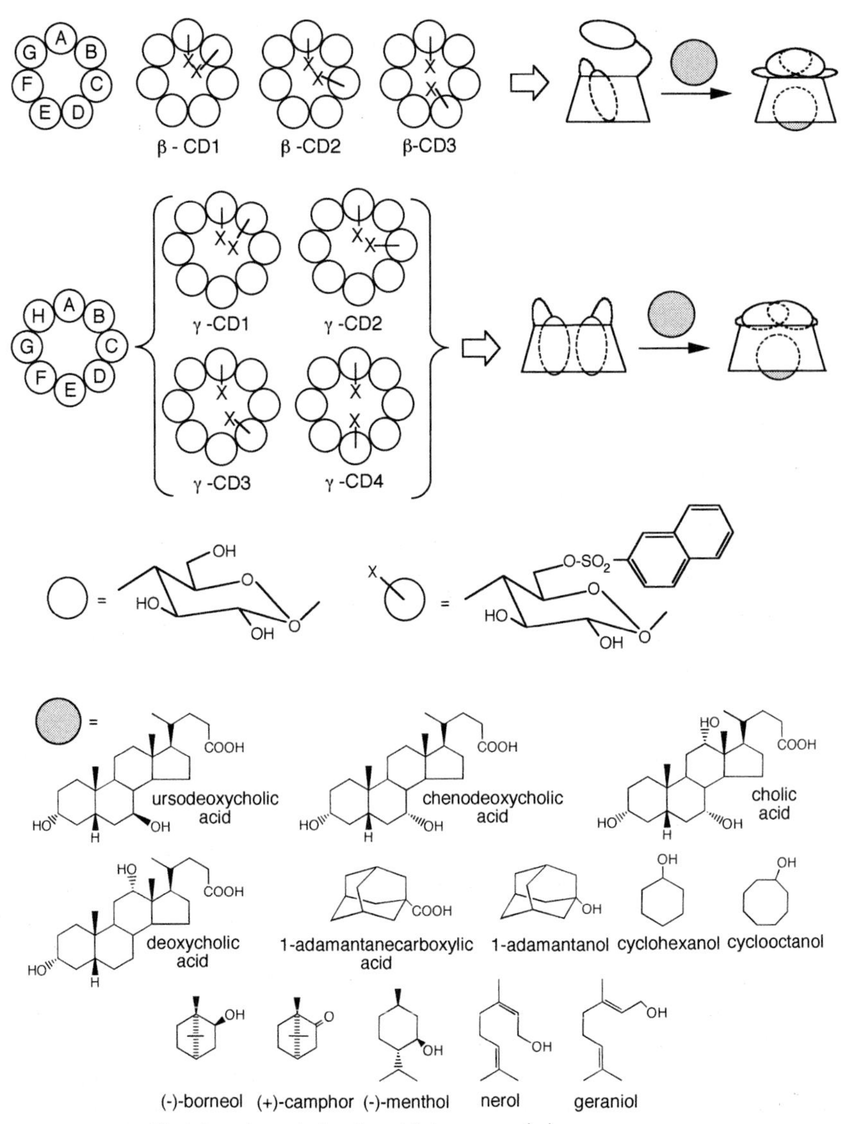

Fig. 10.38. Modified β- and γ-cyclodextrins with two appended 2-naphthylsulfonyl fluorophores (adapted from Minato S. et al. (1991) *Photochem. Photobiol.* **54**, 593).

guest molecule (like those presented in Figure 10.38), exclusion of the fluorophore is accompanied by changes in photophysical properties because the fluorophore moves to a free aqueous medium (Figure 10.39).

Dansyl (dimethylaminonaphthalene sulfonyl) fluorophore is a good candidate as an appended fluorophore because it is very sensitive to solvent polarity (see Chapter 7). Dansyl emits strong fluorescence in a hydrophobic environment, but weak

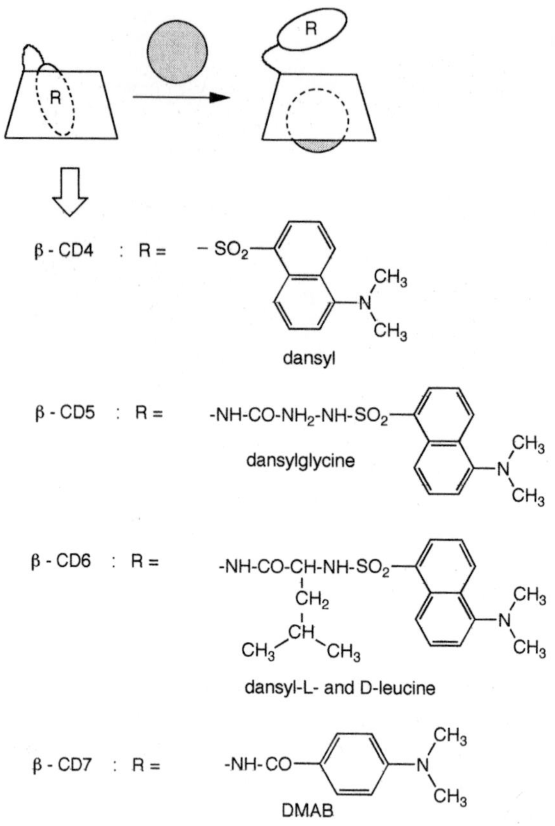

Fig. 10.39. Modified β-CDs with a single appended fluorophore (adapted from Ueno et al., 1997).

fluorescence in aqueous solutions. A guest-induced decrease in fluorescence intensity of compound β-CD4 with an appended dansyl moiety was indeed observed and the extent of this decrease parallels the affinity of the guest for the cavity. A remarkable result is the very large stability constant of the inclusion complex with ursodeoxycholic acid ($> 10^6$). Dansylglycine-, dansyl-L-leucine- and dansyl-L-leucine-modified β-CDs (β-CD5 and β-CD6) have also been prepared and show significantly different responses in the presence of the organic molecules mentioned above.

β-CDs bearing an appended *p*-dimethylaminobenzoyl (DMAB) moiety (called β-CD7) are of special interest. This fluorophore can form a twisted intramolecular charge transfer (TICT) in the excited state, like *p*-dimethylaminobenzonitrile (see Chapter 3, Section 3.4.4), and exhibits dual fluorescence: it can emit fluorescence from both the locally excited state and the TICT state, and the ratio of the intensities of the two bands depends on the microenvironment and in particular on solvent polarity. Circular dichroism experiments show that the DMAB moiety of β-CD7 is excluded from the cavity by forming an intermolecular inclusion complex with a guest molecule. Inclusion of the guest is accompanied by a decrease in the intensity of the TICT emission, changing the environment around the DMAB

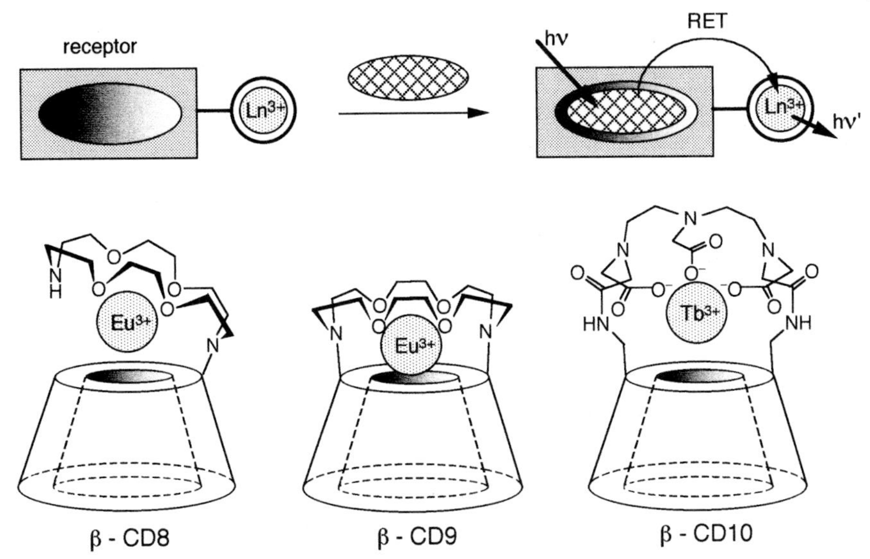

Fig. 10.40. Modified β-CDs with an appended crown containing a lanthanide ion (adapted from Hartmann et al., 1997).

moiety from the hydrophobic interior to the polar water environment. It is interesting to note that the TICT emission depends on the size, shape and polarity of the guest molecules. Among the guests presented in Figure 10.38, the stability constants range from 2000 for cyclohexanol to 2.2×10^5 for 1-adamantanecarboxylic acid.

Compounds β-CD8, β-CD9 and β-CD10 (Figure 10.40) consisting of a β-CD with an appended azacrown able to complex a lanthanide ion have been designed for sensing small organic molecules that absorb in the UV and form inclusion complexes with β-CDs. The principle is that in the absence of sensitizers, lanthanide ions exhibit very weak luminescence because their molar absorption coefficients are very low ($< 1\ M^{-1}\ cm^{-1}$). But when an organic π-electron system is located at proximity, absorption of a UV photon by this species is followed by energy transfer to the lanthanide ion, which emits long-lived luminescence. Such a sensitized luminescence is well known in various lanthanide complexes. In the present case, the analyte bound to the receptor is excited by UV light and the excitation energy is transferred to the lanthanide ion, which acts as a signal transducer.

Addition of benzene to a solution of β-CD8 causes a two-fold enhancement of Eu^{3+} luminescence intensity. The stability constant of the complex between β-CD8 and benzene (~ 200) is comparable to that of the complex with the native β-CD. In contrast, benzoic and naphthoic acids show much stronger associations with β-CD8 than with native β-CD because association is assisted by the interaction between the carboxylic groups and the metal ion. Moreover, the enhancement factor of the luminescence is larger.

Surprisingly, no significant increase in luminescence intensity was observed

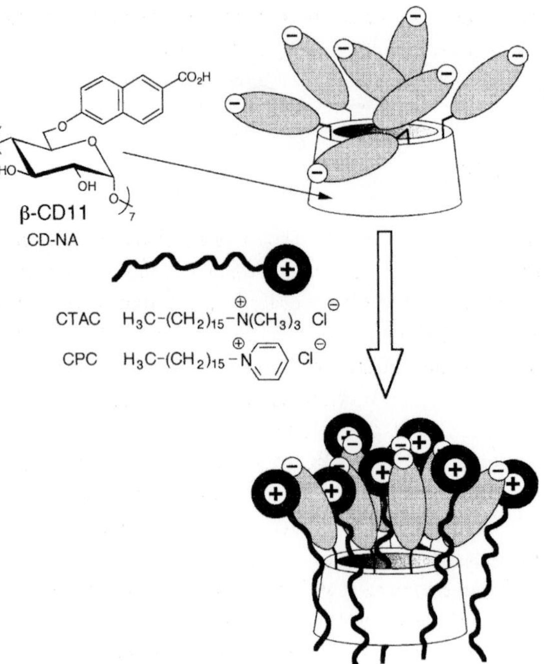

Fig. 10.41. Detection of cationic surfactants by a β-cyclodextrin with seven appended naphthoate fluorophores (Choppinet P. et al. (1996) *J. Chem. Soc. Perkin Trans. 2*, 249).

upon addition of benzene to β-CD9. This is because the association of benzene is defavored by the 3+ charge on the europium ion at the entrance of the cavity, which makes the interior of the latter less hydrophobic. This difficulty is overcome by the three carboxylate groups of β-CD10 that neutralize the 3+ charge of the lanthanide ion. But Tb^{3+} must be used instead of Eu^{3+} in order to avoid low-lying carboxylate ligand-to-metal charge transfer excited states than can interfere with the energy transfer process when the metal ion is Eu^{3+}. Large enhancement factors have been observed, in particular with biphenyl and durene.

Modified cyclodextrins can also be used for the detection of cationic surfactants in the environment, which is of major interest; the high toxicity of these surfactants is linked to a slow biodegradation owing to their bactericidal nature. Very sensitive detection is possible by means of a β-cyclodextrin derivative CD-NA, bearing seven negatively charged naphthoate fluorophores (Figure 10.41). Interaction with a cationic surfactant leads to a drop in excimer emission. The ratio of the fluorescence intensities of the monomer and excimer bands is directly related to the concentration of the surfactant. In the case of electroactive surfactants such as cetylpyridinium chloride (CPC), the fluorescence quenching arising from photo-induced electron transfer can be additionally used for sensing. CD-NA permits detection of cetyltrimethylammonium chloride (CTAC) and cetylpyridinium chloride in an aqueous solution at concentrations as low as a few micromoles per liter and up to about 50 micromoles per liter. Interaction between CD-NA and cationic surfactants can be interpreted by a micellization process (Figure 10.41) induced by

CD-NA rather than by the formation of 1:1 inclusion complexes. It should be noted that addition of the anionic surfactant sodium dodecylsulfate does not induce any photophysical effect.

10.5.2 Boronic acid-based fluorescent sensors

Receptors containing diboronic acids can precisely recognize saccharide molecules. In fact, one boronic acid can reversibly form a boronate ester with two OH groups (one diol group), and one diboronic acid can immobilize two suitably positioned diol units to form a saccharide-containing macrocycle. Selectivity can be achieved by controlling the relative spatial position of the two boronic acids in relation to the cis-diol moieties on the saccharide. A given monosaccharide possesses at least two binding sites that differ from other monosaccharides.

Appropriate combinations of boronic acid and fluorophores lead to a remarkable class of fluorescent sensors of saccharides (Shinkai et al., 1997, 2000, 2001). The concept of PET (photoinduced electron transfer) sensors (see Section 10.2.2.5 and Figure 10.7) has been introduced successfully as follows: a boronic acid moiety is combined intramolecularly with an aminomethylfluorophore; consequently, PET from the amine to the fluorophore causes fluorescence quenching of the latter. In the presence of a bound saccharide, the interaction between boronic acid and amine is intensified, which inhibits the PET process (Figure 10.42). S-1 is an outstanding example of a selective sensor for glucose based on this concept (see Box 10.4).

S-2, in which the spacer between the two boronic acids is flexible, has the additional capability of forming excimers. The 1:1 binding of a saccharide leads to an increase in the monomer fluorescence intensity. This increase has two origins: the decrease in excimer formation, and the increase in fluorescence quantum yield resulting from suppression of the PET process. The 1:1 complex is formed at low saccharide concentrations, but increasing the concentration leads to the formation of the 1:2 complex, as revealed by the increase in the ratio of the intensities of the excimer band to the monomer band. The selectivity of S-2 was found to be similar to that of S-1.

Both steric and electronic factors are used for chiral recognition of saccharides by the R and S forms of S-3. A difference in PET efficiency is created by the asymmetric immobilization of the amine groups relative to the binaphthyl moiety upon 1:1 complexation of saccharides by D- or L-isomers. For instance, D-fructose is recognized by the R form of S-2 with a large fluorescence enhancement.

10.5.3 Porphyrin-based fluorescent sensors

Porphyrins are very attractive as a 'platform' for the design of sensors for molecule recognition. The flat and relatively rigid structure of porphyrins offers the possibility of constructing various types of recognition sites. Numerous functionalized

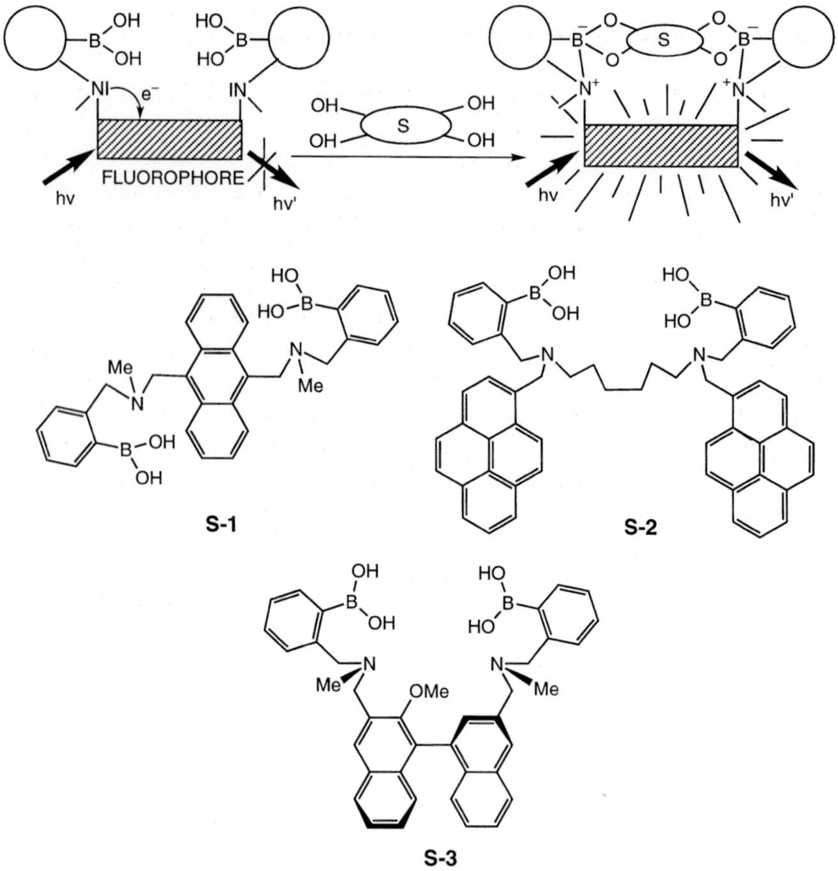

Fig. 10.42. Fluorescent sensors of saccharides based on boronic acids (adapted from James T. D. et al. (1996) *Chem. Commun.* 281).

porphyrins have been designed for the recognition of hydrophobic molecules, aminoacids, nucleobases, etc. Most of the investigations employ UV–vis spectrophotometry and/or NMR to demonstrate complex formation. This section will be limited to selected examples in which fluorescence is used for signaling the recognition event.

In line with the previous section, P-1 (Figure 10.43) associates a boronic acid moiety for recognition of diols (and in particular saccharides) and a tin-metallated porphyrin as a signaling moiety. This PET sensor changes its fluorescence intensity in an 'on–off' manner instead of the usual 'off–on' manner. In the absence of analyte, the boronic acid moiety is positioned within a sterically crowded cleft of the porphyrin. The weak fluorescence emitted by P-1 can be explained by the residual boron–nitrogen interaction that prevents complete quenching by electron transfer from the intramolecular amino group to the porphyrin moiety. Upon binding of diols, the boronic moiety is pulled away from the cleft and the diol is suspended

Box 10.4 A fluorescent molecular sensor selective for glucose

Compound S-1, designed by Shinkai and coworkers[a,b], can form different complexes with saccharides in methanol/water[c] buffer at pH 7.77, as shown in Figure B10.4.1: a non-cyclic 1:1 complex, a 1:1 cyclic complex and a 2:1 (saccharide:ligand) complex. The non-cyclic 1:1 complex could not be detected by fluorescence because both binding sites must be occupied in order to prevent fluorescence quenching by PET.

In the cyclic 1:1 complex, glucose is held close to the anthracene aromatic face, as represented in Scheme B10.4.1. In fact, the ^{1}H-NMR spectrum exhibits a very large paramagnetic shift for the H3 proton (−0.3 ppm), which thus points towards the π-electrons of the anthracene moiety[a,b].

Norrild and coworkers[d] showed that this structure is only valid as an initial complex formed under completely non-aqueous conditions. In the presence of water, a rapid rearrangement from the α-D-glucopyranose form to the α-D-glucofuranose occurs. In the latter form, all five free hydroxy groups of glucose are covalently bound to the sensor molecule (Figure B.10.4.2).

It should be emphasized that binding of glucose by S-1 occurs at very low concentrations. Detection at physiological levels is possible, as demonstrated by

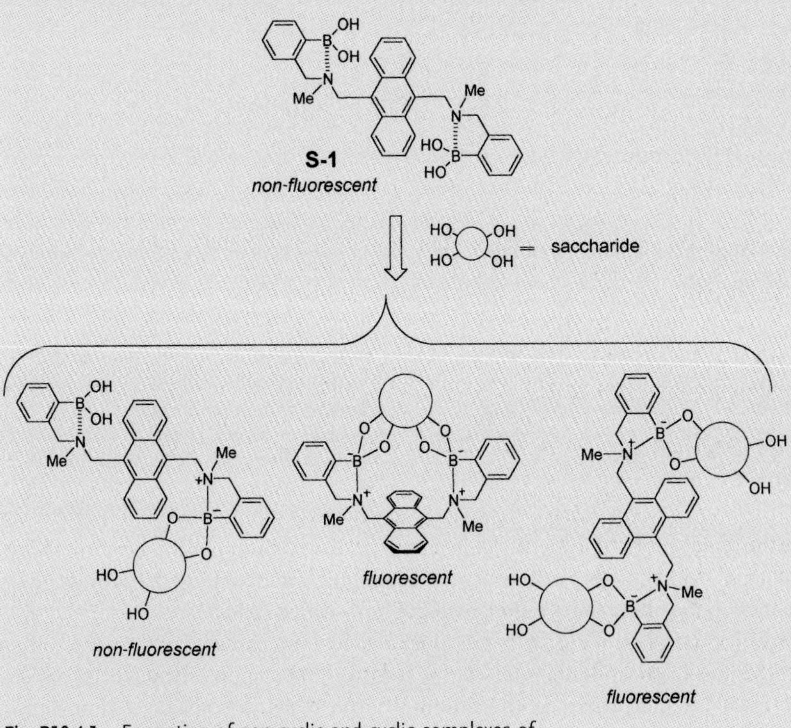

Fig. B10.4.1. Formation of non-cyclic and cyclic complexes of saccharides with the diboronic receptor S-1[a,b].

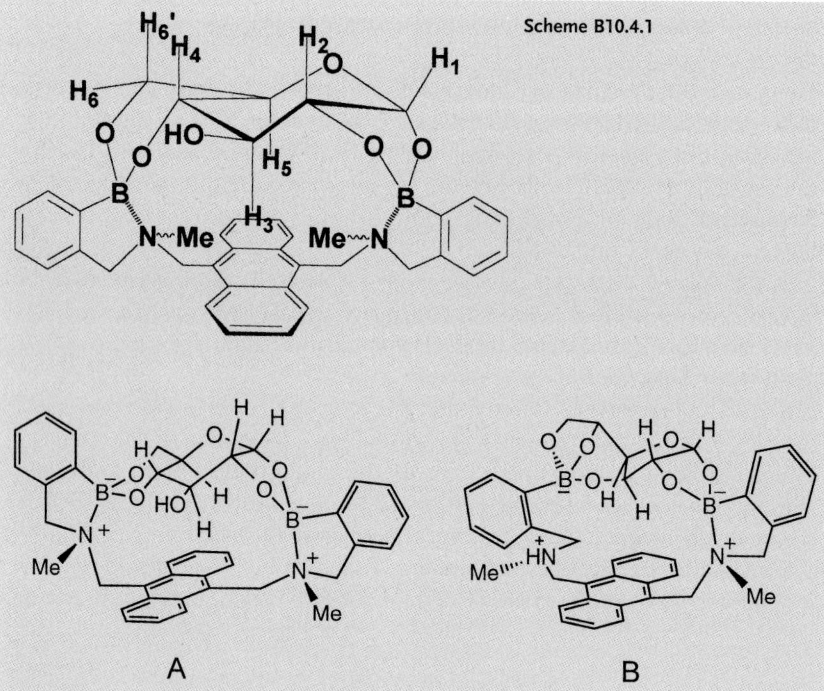

Fig. B10.4.2. Complexes of S-1 with glucose in the α-D-glucopyranose form (A) and the α-D-glucofuranose form (B)[d].

competitive binding studies with the three main monosaccharides present in human blood: D-glucose (0.3–1.0 mM), D-fructose ($\leq$ 0.1 mM) and D-galactose ($\leq$ 0.1 mM). When D-glucose is titrated in the presence of 0.1 mM D-fructose or D-galactose, the observed stability constant is the same as that obtained with D-glucose alone[a].

a) James T. D., Sandanayake K. R. A. S. and Shinkai S. (1994) *Angew. Chem. Int. Ed. Engl.* **33**, 2207.

b) James T. D., Sandanayake K. R. A. S., Iguchi R. and Shinkai S. (1994) *J. Am. Chem. Soc.* **117**, 8982.

c) The presence of methanol avoids any complications arising from precipitation. Detection at low saccharide concentation is possible in a water-only buffer.

d) Bielecki M., Eggert H. and Norrild J. C. (1994) *J. Chem. Soc. Perkin Trans 2* 449.

over the plane of the porphyrin. This results in a weakening of the boron–nitrogen interaction and thus in an increase in quenching efficiency by the amino group. The most bulky diols induce the greatest fluorescence reduction.

The zinc porphyrin P-2 with an appended hydroquinone can form complexes with quinones via H-bonds and charge-transfer interactions. In such complexes, PET from the porphyrin to the quinone is possible, and efficient fluorescence quenching was indeed observed upon binding with substituted benzoquinones, naphthaquinones and anthraquinones.

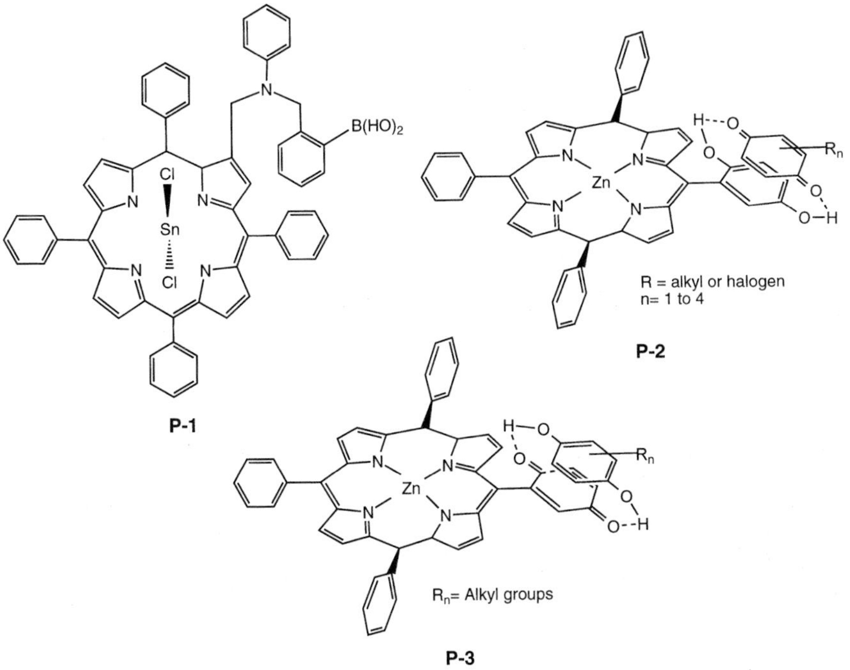

Fig. 10.43. Porphyrin-based fluorescent sensors of neutral molecules (P-1: Kijima H. et al. (1999) *Chem. Commun.* 2011. P-2: D'Souza F. (1996) *J. Am. Chem. Soc.* **118**, 923. P-3: D'Souza F. (1997) *Chem. Commun.* 533).

P-3, consisting of a free-base porphyrin with an appended quinone, is weakly fluorescent due to PET. Complexation with substituted hydroquinones induces fluorescence enhancement because of the unfavorable conditions for PET from the porphyrin to the quinone–hydroquinone entity.

10.6 Towards fluorescence-based chemical sensing devices

From a general point of view, a chemical sensor is a device capable of continuously monitoring the concentration of an analyte. The two main classes are *electrochemical sensors* and *optical chemical sensors*. The latter are based on the measurement of changes in an optical quantity: refractive index, light scattering, reflectance, absorbance, fluorescence, chemiluminescence, etc. For remote sensing, an optical fiber is used, and the optical sensor is then called an *optode*[2] because of

2) This term comes from the Greek (οπτος = visible, οδος = way) and is preferred to the term *optrode* that results from the combination of the words 'optical' and 'electrode'.

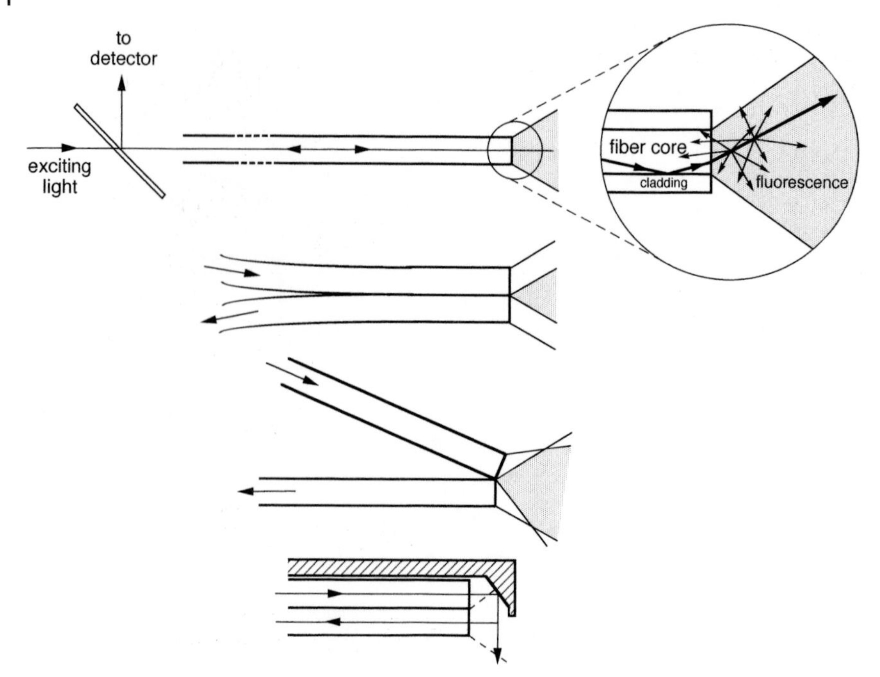

Fig. 10.44. Examples of optical configurations for passive optical sensors.

the analogy with the use of electrodes. However, the operating principles of optodes and electrodes are quite different.

Electrochemical sensors have several disadvantages with respect to optical sensors: (i) they are based on electrodes and require a reference electrode; (ii) the liquid–liquid junction is easily perturbed by external factors; (iii) they are sensitive to electrical interferences; (iv) miniaturization is not easy and their cost is relatively high. However, optical sensors also have some disadvantages: (i) ambient light can interfere; (ii) the range over which the concentration of an analyte can be accurately measured is often limited; (iii) they have generally limited long-term stability.

Among optical sensors, those based on fluorescence are of major interest because of their ability to use spectral and temporal information: multi-wavelength measurements allow simultaneous detection of two or more analytes, and discrimination between analytes is possible by time-resolved measurements. Multiplex capabilities represent the main advantage of such sensors compared to electrochemical devices.

Very different techniques are used in the design of fluorescence-based optical sensors. Passive and active modes of operation should be distinguished.

• In the *passive mode*, the optical device measures the variation in fluorescence characteristics (intensity, lifetime, polarization) of an intrinsically fluorescent analyte. The optical device can have different optical configurations involving in most cases an optical fiber (*passive optode*) (Figure 10.44).

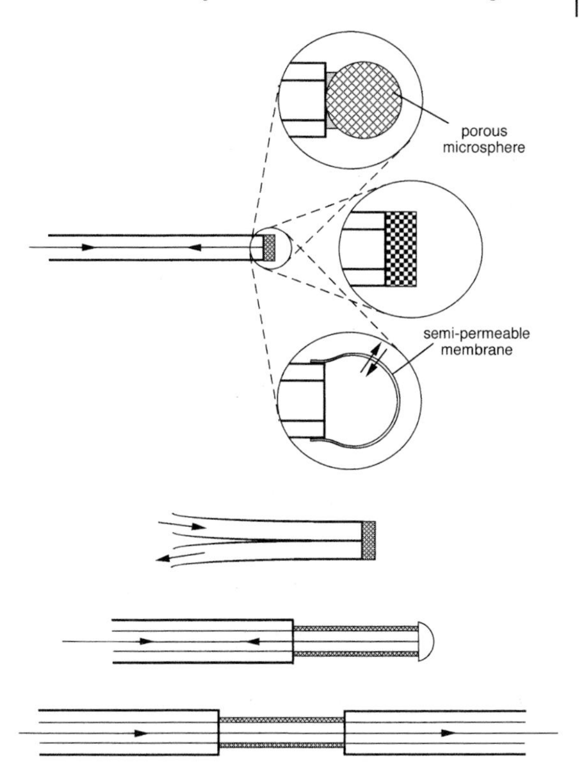

Fig. 10.45. Examples of optical configurations for active optical sensors.

• In the *active mode*, the optical device uses for optical transduction the changes in the fluorescence characteristics of a fluorescent molecular sensor (as described in the preceding sections) resulting from the interaction with an analyte[3]. The two main optical configurations are:

1. In most cases, the fluorescent molecular sensor is immobilized at the tip of an optical fiber (*active optode*) in a matrix that permits diffusion of the analyte (Figure 10.45). Plasticized polymers (e.g. polyvinyl chloride, PVC) and sol–gel glasses (e.g. SiO_2 or TiO_2) are often used as matrices. The advantage of the latter is the rapidity of the response. Various other methods have been used: grafting on silica, dextrins, cellulose, polymers (e.g. polystyrene, polyacrylamide), adsorption on membranes (e.g. polytetrafluoroethylene, PTFE), electrostatic coupling (with ion exchange membranes).
2. The second possibility is to make use of the continuous *evanescent field* that exists at the surface of an optical fiber. The cladding layer is replaced by a layer containing the fluorescent molecular sensor in a short portion of the optical fiber (Figure 10.45). At each reflection at the surface of the fiber core, the eva-

3) Alternatively, in some cases, the fluorophore chemically reacts with the analyte – these are outside the scope of this chapter.

nescent wave penetrates into the layer within a distance of the order of the wavelength of the propagating light. Fluorescence emitted within this evanescent region can be reciprocally coupled back into the core and conveyed towards the detector. Sensors using evanescent waves in planar waveguides have been also designed.

Various pH sensors have been built with a fluorescent pH indicator (fluorescein, eosin Y, pyranine, 4-methylumbelliferone, SNARF, carboxy-SNAFL) immobilized at the tip of an optical fiber. The response of a pH sensor corresponds to the titration curve of the indicator, which has a sigmoidal shape with an inflection point for $pH = pK_a$; but it should be emphasized that the effective pK_a value can be strongly influenced by the physical and chemical properties of the matrix in which the indicator is entrapped (or of the surface on which it is immobilized) without forgetting the dependence on temperature and ionic strength. In solution, the dynamic range is restricted to approximately two pH units, whereas it can be significantly extended (up to four units) when the indicator is immobilized in a microheterogeneous microenvironment (e.g. a sol–gel matrix).

The same principle can be applied to cation sensors by using an immobilized fluoroionophore. An alternative is the use of neutral ion carriers coupled to a fluorescent co-reagent in a hydrophobic layer (e.g. plasticized PVC). The principle is based on the electroneutrality in the membrane, as illustrated in Figure 10.46. The first method employs a hydrophobic pH indicator dye as a co-reagent: when a cation is pulled into the hydrophobic layer by the neutral carrier (e.g. valinomycin for K^+), the same number of H^+ must be released from the hydrophobic layer, and deprotonation of the pH indicator (whose role is to achieve proton exchange to maintain electroneutrality) is accompanied by a change in absorbance and fluorescence. In the second method, the co-reagent is an amphiphilic indicator possessing a hydrocarbon tail and a polar head that is a positively charged fluorescent group sensitive to polarity. Upon cation complexation in the layer by the neutral carrier, electroneutrality is maintained by partitioning out of the hydrophobic layer the polar head into the aqueous phase at the interface, which results in a change in fluorescence. The hydrophobic tail of the indicator prevents leaching.

Fluorescence-based gas sensors have been extensively developed. In particular, oxygen sensors are based on the dynamic quenching of a fluorophore (e.g. pyrene, pyrene butyric acid, complexes with Ru, Re, Pt) that is either entrapped in a matrix (e.g. porous glass, polymers like PVC, silicone layer) or adsorbed on a resin, or grafted on silica. For sensing acidic and basic gases such as ammonia, carbon dioxide, hydrogen cyanide or nitrogen oxide, a solution of a pH indicator (e.g. pyranine, fluorescein) is separated from the sample solution by a gas-permeable membrane (made of silicone or PTFE). The acidic or basic gas can cross the membrane and enter the solution containing the indicator, which undergoes proton tranfer. For further details on fluorescence-based chemical sensing devices, the reader is referred to specialized books and reviews (e.g. Arnold, 1992; Fuh et al., 1991; Janata, 1992; Seitz, 1984; Wolfbeis et al., 1988, 1991, 1996). It is beyond the scope of this chapter to describe biosensors.

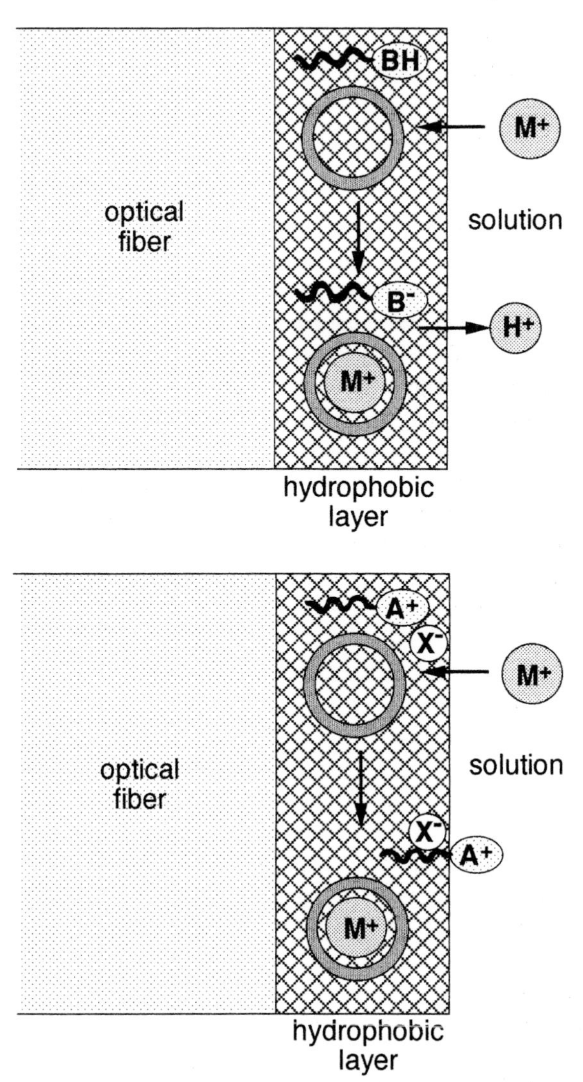

Fig. 10.46. Principles of optodes using a neutral carrier (e.g. valinomycin for K^+) coupled to an amphiphilic fluorophore: (A) an amphiphilic pH indicator (Seiler K. et al. (1989) *Anal. Sci.* **5**, 557). (B) a fluorescent cationic surfactant (e.g. dodecyl acridine orange) (Kawabata Y. et al. (1990) *Anal. Chem.* **62**, 2054).

Appendix A. Spectrophotometric and spectrofluorometric pH titrations

Single-wavelength measurements

The concentration of the indicator is kept low enough that the linear relationship between aborbance or fluorescence intensity and concentration is fulfilled.

At a given wavelength, where both acidic form A and basic form B absorb or emit, the absorbance or the fluorescence intensity, denoted Y, are given by

$$Y = a[\mathrm{A}] + b[\mathrm{B}] \tag{A.1}$$

In spectrophotometry, Y is the absorbance $A(\lambda)$ of the solution at the chosen wavelength λ. According to the Beer–Lambert law, a and b are the products of the absorption pathlength by the molar absorption coefficients of A and B, respectively.

In spectrofluorometry, Y is the fluorescence intensity $I_\mathrm{F}(\lambda_\mathrm{E}, \lambda_\mathrm{F})$ of the solution for a couple of excitation and observation wavelengths. a and b are proportional to the molar absorption coefficients (at the excitation wavelength) and the fluorescence quantum yields of A and B, respectively.

When the indicator is only in the acidic form or only in the basic form, the values of Y are, respectively,

$$Y_\mathrm{A} = ac_0 \tag{A.2}$$

$$Y_\mathrm{B} = bc_0 \tag{A.3}$$

where c_0 is the total concentration of indicator such that

$$c_0 = [\mathrm{A}] + [\mathrm{B}] \tag{A.4}$$

Combination of the preceding relations yields

$$\frac{[\mathrm{B}]}{[\mathrm{A}]} = \frac{Y - Y_\mathrm{A}}{Y_\mathrm{B} - Y} \tag{A.5}$$

The Henderson–Hasselbach equation (10.2) can thus be written as

$$\boxed{\mathrm{pH} = \mathrm{p}K_\mathrm{a} + \log \frac{Y - Y_\mathrm{A}}{Y_\mathrm{B} - Y}} \tag{A.6}$$

The value of $\mathrm{p}K_\mathrm{a}$ can thus be determined from the plot of $\log[(Y - Y_\mathrm{A})/(Y_\mathrm{B} - Y_\mathrm{A})]$ vs pH.

Dual-wavelength measurements

The absorbances or fluorescence intensities measured at two wavelengths λ_1 and λ_2 can be written in a form analogous to Eq. (A.1):

$$Y(\lambda_1) = a_1[\mathrm{A}] + b_1[\mathrm{B}] \tag{A.7}$$

$$Y(\lambda_2) = a_2[\mathrm{A}] + b_2[\mathrm{B}] \tag{A.8}$$

A ratiometric measurement consists of monitoring the ratio $Y(\lambda_1)/Y(\lambda_2)$ given by

$$R = \frac{a_1[\mathrm{A}] + b_1[\mathrm{B}]}{a_2[\mathrm{A}] + b_2[\mathrm{B}]} \tag{A.9}$$

When the indicator is only in the acidic form or only in the basic form, the values of R are, respectively,

$$R_\mathrm{A} = \frac{a_1}{a_2} \tag{A.10}$$

$$R_\mathrm{B} = \frac{b_1}{b_2} \tag{A.11}$$

Taking account of Eq. (A.4), Eq. (A.5) is replaced by

$$\frac{[\mathrm{B}]}{[\mathrm{A}]} = \frac{a_2}{b_2}\frac{R - R_\mathrm{A}}{R_\mathrm{B} - R} \tag{A.11}$$

Hence,

$$\mathrm{pH} = \mathrm{p}K_\mathrm{a} + \log\frac{R - R_\mathrm{A}}{R_\mathrm{B} - R} + \log\frac{a_2}{b_2} \tag{A.12}$$

The ratio a_2/b_2 represents the ratio of the absorbances or fluorescent intensities of the acidic form alone and the basic form alone at the wavelength λ_2:

$$\boxed{\mathrm{pH} = \mathrm{p}K_\mathrm{a} + \log\frac{R - R_\mathrm{A}}{R_\mathrm{B} - R} + \log\frac{Y_\mathrm{A}(\lambda_2)}{Y_\mathrm{B}(\lambda_2)}} \tag{A.13}$$

Appendix B. Determination of the stoichiometry and stability constant of metal complexes from spectrophotometric or spectrofluorometric titrations

Definition of the equilibrium constants

The ligand (whose absorbance or fluorescence intensity is measured) will be denoted L and the metal ion M[4]. Let us consider first the formation of a 1:1 complex, assuming that the equilibrium does not involve protons, which is the case for investigations in organic solvents or in buffered aqueous solutions:

$$\mathrm{M} + \mathrm{L} \rightleftharpoons \mathrm{ML}$$

The stability of the complex ML is characterized by the equilibrium constant, for which various terms are used: *stability constant, binding constant, association constant, affinity constant* (K_s or β) or *dissociation constant* (K_d, reciprocal of K_s). From

4) M may also represent an anion or a neutral molecule.

the thermodynamic point of view, the true equilibrium constant (which depends only on temperature) must be written with activities:

$$K_s = \frac{a_{ML}}{a_M a_L} \tag{B.1}$$

When the solution is dilute enough to approximate the activity coefficients to 1 (reference state: solute at infinite dilution), activities can then be replaced by molar fractions (dimensionless quantities), but in solution they are generally replaced by molar concentrations:

$$K_s = \frac{[ML]}{[M][L]} \tag{B.2}$$

However, *the equilibrium constant must still be considered as pure and dimensionless numbers* (according to the classical relation $-\Delta G^0 = RT \ln K_s$). All molar concentrations in the expression of K_s should thus be interpreted as molar concentrations relative to a standard state of 1 mol dm^{-3}: i.e. they are the numerical values of the molar concentrations[5]. If the solution is not dilute enough, the equilibrium constants can still be written with concentrations but they must be considered as *apparent stability constants.*

When a second complex of stoichiometry 2:1 (metal:ligand) is formed, the following equilibria should be considered, assuming stepwise binding

$$M + L \rightleftharpoons ML$$

$$ML + M \rightleftharpoons M_2L$$

The stepwise binding constants are defined as

$$K_{11} = \frac{[ML]}{[M][L]} \tag{B.3}$$

$$K_{21} = \frac{[M_2L]}{[ML][M]} \tag{B.4}$$

It is possible to write the formation of the 2:1 complex directly from the cation and ligand as follows:

$$2M + L \rightleftharpoons M_2L$$

and to define an overall binding constant

5) In spite of these thermodynamic requirements, most papers report equilibrium constants with dimensions for convenience (e.g. $dm^3\ mol^{-1}$ for the stability constant of a 1:1 complex or mol dm^{-3} for its dissociation constant).

$$\beta_{21} = \frac{[M_2L]}{[M]^2[L]} \quad (B.5)$$

Generalization to a complex M_mL_n can easily be made[6]:

$$mM + lL \rightleftharpoons M_mL_l$$

with

$$\beta_{ml} = \frac{[M_mL_l]}{[M]^m[L]^l} \quad (B.6)$$

Preliminary remarks on titrations by spectrophotometry and spectrofluorometry

In the following considerations, it will be assumed that the linear relationship between absorbance or fluorescence intensity and concentration is always fulfilled. Moreover, we will consider only the case where *the ligand absorbs light or emits fluorescence but not the cation.* In a titration experiment, the concentration of the ligand is kept constant and the metal salt is gradually added. The absorption spectrum or fluorescence spectrum is recorded as a function of cation concentration. Changes in these spectra upon complexation allow us to determine the stability constant of the complexes. Data can be processed using several wavelengths simultaneously. This may turn out to be necessary when several complexes are formed with overlap of their existence domains. This requires special software (a few are commercially available). However, when only one or two complexes exist, it may be sufficient to monitor the variations in absorbance or fluorescence intensity at an appropriate wavelength (chosen so that the changes are as large as possible), or the variations in the ratio of absorbances or fluorescence intensities at two wavelengths. This is the subject of the following sections.

Formation of a 1:1 complex (single-wavelength measurements)

$$M + L \rightleftharpoons ML \quad K_s = \frac{[ML]}{[M][L]} \quad (B.7)$$

Let Y_0 be the absorbance or the fluorescence intensity of the free ligand. In fluorometric experiments, the absorbance at the excitation wavelength should be less than ~ 0.1. Then, at a given wavelength, Y_0 is proportional to the concentration c_L:

$$Y_0 = ac_L \quad (B.8)$$

6) A more general equilibrium should be considered if protons are involved:

$$mM + lL + hH \rightleftharpoons M_mL_lH_h$$

with the global equilibrium constant

$$\beta_{mlh} = \frac{[M_mL_lH_h]}{[M]^m[L]^l[H]^h}$$

and in the presence of an excess of cation such that the ligand is fully complexed, Y reaches the limiting value Y_{lim}:

$$Y_{lim} = bc_L \tag{B.9}$$

In spectrophotometry, a and b are the products of the absorption pathlength by the molar absorption coefficients of the ligand and the complex, respectively. In spectrofluorometry, a and b are proportional to the molar absorption coefficients (at the excitation wavelength) and the fluorescent quantum yields of the ligand and the complex, respectively.

After addition of a given amount of cation at a concentration c_M, the absorbance or the fluorescence intensity becomes

$$Y = a[L] + b[ML] \tag{B.10}$$

Mass balance equations for the ligand and cation are

$$c_L = [L] + [ML] \tag{B.11}$$

$$c_M = [M] + [ML] \tag{B.12}$$

From Eqs (B.7) to (B.12), it is easy to derive the usual relation

$$\boxed{\frac{Y - Y_0}{Y_{lim} - Y} = K_s[M]} \tag{B.13}$$

This relationship can be used to determine K_s *under the condition that the concentration in free cation* $[M]$ *can be approximated to the total concentration* c_M: $(Y - Y_0)/(Y_{lim} - Y)$ is plotted as a function of c_M, and the plot should be linear and the slope yields K_s. Once K_s is known, the concentration of free cation $[M]$ can be determined by means of Eq. (B.13).

When Y_{lim} is not measurable because full complexation cannot be attained at a reasonable concentration of cation, it is better to use the following relation:

$$\boxed{\frac{Y_0}{Y - Y_0} = \frac{\alpha}{K_s[M]} + \alpha} \tag{B.14}$$

where $\alpha = a/(b - a)$. It is convenient to plot $Y_0/(Y - Y_0)$ versus $1/c_M$ provided that the approximation $[M] \approx c_M$ is valid. The ratio of the ordinate at the origin to the slope yields K_s. Such a plot is called a *double-reciprocal plot* or a *Benesi-Hildebrand plot.*

The approximation $[M] \approx c_M$ is made in most cases (Connors, 1987), and surprisingly little attention has been paid to cases where it is not valid (i.e. when $K_s c_L \gg 1$, see below). Yet an explicit expression of Y in the case of the formation of a 1:1 complex can be easily derived from the preceding equations without approximation. In fact, appropriate combinations of these equations lead to the following

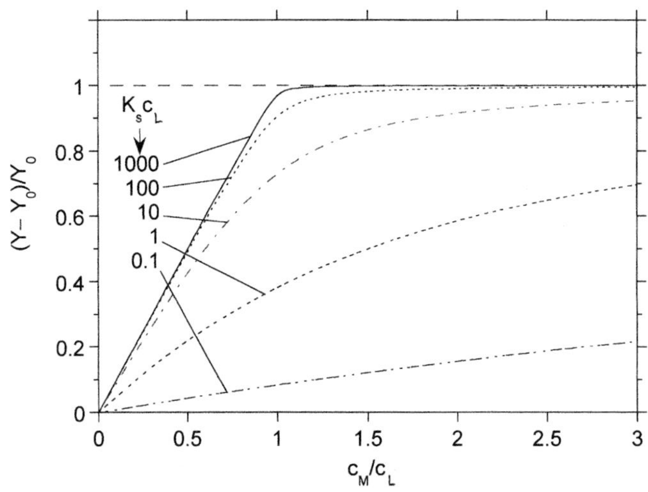

Fig. 10.B.1. Spectrophotometric or spectrofluorimetric titration curves for a complex 1:1 according to Eq. (B.16). Y_{lim} is chosen to be equal to $2Y_0$.

second-order equation:

$$c_L x^2 - (c_L + c_M + 1/K_s)x + c_M = 0 \tag{B.15}$$

where

$$x = \frac{Y - Y_0}{Y_{\mathrm{lim}} - Y_0}$$

Hence,

$$Y = Y_0 + \frac{Y_{\mathrm{lim}} - Y_0}{2}\left\{1 + \frac{c_M}{c_L} + \frac{1}{K_s c_L} - \left[\left(1 + \frac{c_M}{c_L} + \frac{1}{K_s c_L}\right)^2 - 4\frac{c_M}{c_L}\right]^{1/2}\right\} \tag{B.16}$$

K_s can thus be obtained by a nonlinear least squares analysis of Y versus c_M or c_M/c_L. If Y_{lim} cannot be accurately determined, it can be left as a floating parameter in the analysis. Figure 10.B.1 shows the variations of $(Y - Y_0)/Y_0$ versus c_M/c_L for $Y_{\mathrm{lim}} = 2Y_0$. Attention should be paid to the concentration of ligand c_L with respect to $1/K_s$. When $K_s c_L \gg 1$, the determination of K_s will be inaccurate because the titration curve, Y versus c_M/c_L, will essentially consist of two portions of straight lines. In fact, when $K_s c_L \gg 1$, Eq. (B.16) reduces to $Y = Y_0 + (Y_{\mathrm{lim}} - Y_0)c_M/c_L$ for $c_M/c_L < 1$ and $Y = Y_{\mathrm{lim}}$ for $c_M/c_L \geq 1$. Another restriction for the ligand concentration should be recalled when fluorescence intensities are measured: the absorbance at the excitation wavelength should be less than ~ 0.1. In contrast, in spectrophotometric experiments, absorbance can be measured up to 2–3. Figure 10.B.1

shows also that the smaller the value of $K_s c_L$, the larger the excess of cation to be added to reach Y_{lim}. It is often preferable to leave this parameter floating in the analysis.

Formation of a 1:1 complex (dual-wavelength measurements)

The absorbances or fluorescence intensities measured at two wavelengths λ_1 and λ_2 can be written in a form analogous to Eq. (B.10):

$$Y(\lambda_1) = a_1[\mathrm{L}] + b_1[\mathrm{ML}] \tag{B.17}$$

$$Y(\lambda_2) = a_2[\mathrm{L}] + b_2[\mathrm{ML}] \tag{B.18}$$

The ratiometric measurement consists of monitoring the ratio $Y(\lambda_1)/Y(\lambda_2)$ given by

$$R = \frac{a_1[\mathrm{L}] + b_1[\mathrm{ML}]}{a_2[\mathrm{L}] + b_2[\mathrm{ML}]} \tag{B.19}$$

For the free ligand and at full complexation, the values of R are, respectively,

$$R_0 = \frac{a_1}{a_2} \tag{B.20}$$

$$R_{lim} = \frac{b_1}{b_2} \tag{B.21}$$

Taking into account Eqs (B.11) and (B.12), we obtain the following equation:

$$\boxed{\frac{R - R_0}{R_{lim} - R}\frac{a_2}{b_2} = K_s[\mathrm{M}]} \tag{B.22}$$

Note that a_2/b_2 represents the ratio of the absorbances or fluorescent intensities of the free ligand and the complex at the wavelength λ_2: $a_2/b_2 = Y_0(\lambda_2)/Y_{lim}(\lambda_2)$.

Equation (B.22) can be used for the determination of K_s only if the concentration in free cation [M] can be approximated to the total concentration c_M.

When K_s is known, the concentration of free cation [M] can be determined by means of Eq. (B.22).

Formation of successive complexes ML and M_2L

Let us consider a ligand that can bind successively two cations according to the equilibria

$$\mathrm{M} + \mathrm{L} \rightleftharpoons \mathrm{ML} \quad K_{11} = \frac{[\mathrm{ML}]}{[\mathrm{M}][\mathrm{L}]} \tag{B.23}$$

$$ML + M \rightleftharpoons M_2L \quad K_{21} = \frac{[M_2L]}{[ML][M]} \tag{B.24}$$

The absorbance or the fluorescence intensity Y_0 of the free ligand is

$$Y_0 = ac_L \tag{B.25}$$

and after addition of a given amount of cation at a concentration c_M, the absorbance or the fluorescence intensity becomes

$$Y = a[L] + b[ML] + c[M_2L] \tag{B.26}$$

where a, b and c include the molar absorption coefficients (and fluorescence quantum yields) of M, ML and M_2L, respectively.

In the presence of an excess of cation so that the complex M_2L only is present, Y reaches the limiting value Y_{lim}:

$$Y_{lim} = cc_L \tag{B.27}$$

Mass balance equations for the ligand and cation are

$$c_L = [L] + [ML] + [M_2L] \tag{B.28}$$

$$c_M = [M] + [ML] + 2[M_2L] \tag{B.29}$$

From Eqs (B.23)–(B.29), the following expression for Y can be obtained:

$$Y = \frac{Y_0 + c_M b K_{11}[M] + Y_{lim}\beta_{21}[M]^2}{1 + K_{11}[M] + \beta_{21}[M]^2} \tag{B.30}$$

where $\beta_{21} = K_{11}K_{21}$. If the approximation $[M] \approx c_M$ is valid, K_{11} and β_{21} can be determined by a nonlinear least-squares analysis of Y versus c_M. Y_{lim} can also be left as a floating parameter in the analysis, if necessary. It should be noted that there is no explicit expression for Y versus c_M if the approximation $[M] \approx c_M$ is not valid.

Cooperativity

When a ligand can complex more than one cation, the question arises of possible cooperative binding. There are many definitions of cooperativity but they are all consistent with the following criterion (Connors, 1987). A system is

- *non-cooperative* if the ratio $K_{(i+1)1}/K_{i1}$ is equal to the statistical value calculated when all binding sites are identical and independent. These statistical values are given in Table 10.B.1.
- *positively cooperative* if the ratio $K_{(i+1)1}/K_{i1}$ is larger than the statistical value.

Tab. 10.B.1. Relative values of the stability constants in the case of n identical and independent binding sites (Connors, 1987)

n	K_{11}	K_{21}	K_{31}	K_{41}	K_{51}	K_{61}
2	2	1/2				
3	3	1	1/3			
4	4	3/2	2/3	1/4		
5	5	2	1	1/2	1/5	
6	6	5/2	4/3	3/4	2/5	1/6

- *negatively cooperative* (*or anti-cooperative*) if the ratio $K_{(i+1)1}/K_{i1}$ is smaller than the statistical value.

In particular, for a ditopic receptor that can bind successively two cations (see previous section), the criterion for cooperativity is $K_{21}/K_{11} > 1/4$, i.e. complexation of a second cation is made easier by the presence of a bound cation. For instance, a cooperative effect was observed with fluoroionophore E-1 (see Section 10.3.4).

Determination of the stoichiometry of a complex by the method of continuous variations (Job's method)

An assumed 1:1 stoichiometry for a complex can be confirmed or invalidated by the fit of the titration curves described above for this case. If the fit is not satisfactory, a model of formation of two successive complexes can be tried.

Information on the stoichiometry of a complex can also be obtained from the continuous variation method (see Connors, 1987). Let us consider a complex M_mL_l formed according to the equilibrium

$$m\mathrm{M} + l\mathrm{L} \rightleftharpoons \mathrm{M}_m\mathrm{L}_l$$

with

$$\beta_{ml} = \frac{[\mathrm{M}_m\mathrm{L}_l]}{[\mathrm{M}]^m[\mathrm{L}]^l} \tag{B.31}$$

The principle of the method as follows: the absorbance or the fluorescence intensity Y is measured for a series of solutions containing the ligand and the cation such that *the sum of the total concentrations of ligand and cation is constant.*

$$c_\mathrm{L} + c_\mathrm{M} = C = \text{constant} \tag{B.32}$$

The position of the maximum of Y is then related to the ratio m/l, as shown below.

It is convenient to use the following dimensionless quantity (which is analogous to a molar fraction but not strictly):

$$x = \frac{c_M}{c_M + c_L} = \frac{c_M}{C} \tag{B.33}$$

Mass balance equations are

$$c_L = [L] + l[M_mL_l] \tag{B.34}$$

$$c_M = [M] + m[M_mL_l] \tag{B.35}$$

These equations can be rewritten as

$$C(1-x) = [L] + l[M_mL_l] \tag{B.36}$$

$$Cx = [M] + m[M_mL_l] \tag{B.37}$$

Combination of Eqs (B.31)–(B.37) gives

$$\beta_{ml}\{Cx - m[M_mL_l]\}^m\{C(1-x) - l[M_mL_l]\}^n = [M_mL_l] \tag{B.38}$$

Taking the logarithm of this expression, then differentiating with respect to x, and finally setting $d[M_mL_l]/dx = 0$, we obtain

$$\boxed{\frac{m}{l} = \frac{x_{max}}{1 - x_{max}}} \tag{B.39}$$

This treatment assumes that a single complex is present, but this assumption may not be valid. When only one complex is present, the value of x_{max} is independent of the wavelength at which the absorbance or fluorescence intensity is measured. A dependence on wavelength is an indication of the presence of more than one complex.

For a 1:1 complex, $x_{max} = \frac{1}{2}$, according to Eq. (B.39). To illustrate the shape of Job's plot, the following equation can be derived with the same notations as above:

$$Y = aC(1-x) + \frac{(b-a)C}{2}\left\{1 + \frac{1}{K_sC} - \left[\left(1 + \frac{1}{K_sC}\right)^2 - 4x(1-x)\right]^{1/2}\right\} \tag{B.40}$$

where a and b have the same meaning as in Eqs (B.8) and (B.9). The product aC is equal to Y_0, i.e. the value of Y when no cation is added ($x = 0$).

When plotting the variations in absorbance or fluorescence intensity versus x, it is convenient to subtract the absorbance or fluorescence intensity that would be measured in the absence of cation at every concentration, i.e. $Y_0(1-x)$. In this way, the plot of $Y - Y_0(1-x)$ versus x starts from 0 for $x = 0$, goes through a maximum, and returns to 0 for $x = 1$. Equation (B.40) can thus be rewritten as

$$Y - Y_0(1-x) = \frac{(b/a - 1)Y_0}{2}\left\{1 + \frac{1}{K_sC} - \left[\left(1 + \frac{1}{K_sC}\right)^2 - 4x(1-x)\right]^{1/2}\right\} \tag{B.41}$$

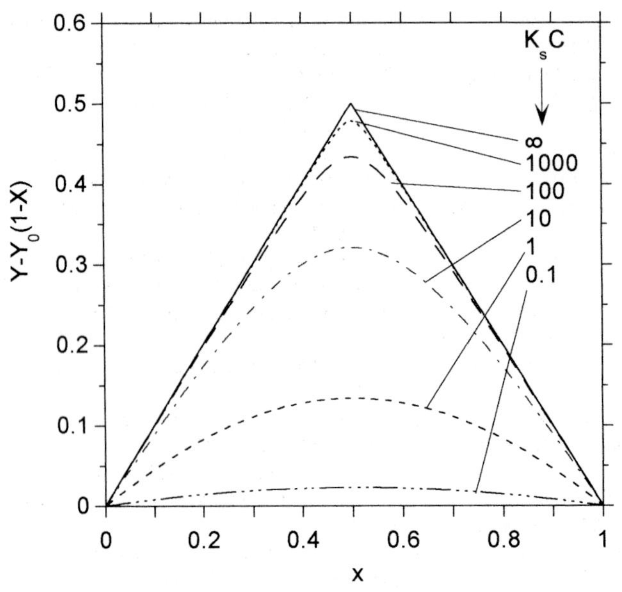

Fig. 10.B.2. Job's plots for a 1:1 complex according to Eq. (B.41). The ratio b/a is chosen to be equal to 2.

Plots of $y/y_0 - (1 - x)$ versus x are drawn in Figure 10.B.2 for $b/a = 2$ and various values of $K_S C$. The sharpness of the maximum, and therefore the accuracy with which x_{max} is located, depends on the magnitude of this parameter. Curve fitting with Eq. (B.41) yields K_S.

10.7 Bibliography

Alfimov M. V. and Gromov S. P. **(1999)** Fluorescence Properties of Crown-Containing Molecules, in: Rettig W. et al. (Eds), *Applied Fluorescence in Chemistry, Biology and Medicine*, Springer-Verlag, Berlin, pp. 161–78.

Arnold M. A. **(1992)** Fiber-Optic Chemical Sensors, *Anal. Chem.* 64, A1015–25.

Beer P. D. **(1996)** Anion Selective Recognition and Optical/Electrochemical Sensing by Novel Transition-Metal Receptor Systems, *Chem. Commun.* 689–96.

Beer P. D. and Cadman J. **(2000)** Electrochemical and Optical Sensing of Anion Systems by Transition Metal Based Receptors, *Coord. Chem. Rev.* 205, 131–55.

Bergonzi R., Fabbrizzi L., Lichelli M. and Mangano C. **(1998)** Molecular Switches of Fluorescence Operating Through Metal Centred Redox Couples, *Coord. Chem. Rev.* 170, 31–46.

Bissell R. A., de Silva A. P., Gunaratne H. Q. N., Lynch P. L. M., Maguire G. E. M., McCoy C. P. and Sandanayake K. R. A. S. **(1993)** Fluorescent PET (Photoinduced Electron Transfer) Sensors, *Top. Curr. Chem.* 168, 223–64.

Biwersi J., Tulk B. and Verlman A. S. **(1994)** Long-Wavelength Chloride-Sensitive Fluorescent Indicators, *Anal. Biochem.* 219, 139–43.

Connors K. A. **(1987)** Binding Constants. The Measurement of Molecular Complex Stability, John Wiley & Sons, New York.

Czarnik A. W. (Ed.) **(1993)** *Fluorescent Chemosensors for Ion and Molecule Recognition*, ACS Symposium Series 358, American Chemical Society, Washington, DC.

Czarnik A. W. **(1994)** Chemical Communication in Water Using Fluorescent Chemosensors, *Acc. Chem. Res.* 27, 302–8.

de Silva A. P., Gunaratne H. Q. N., Gunnlaugsson T., Huxley A. J. M., McCoy C. P., Rademacher J. T. and Rice T. E. **(1997)** Signaling Recognition Events with Fluorescent Sensors and Switches, *Chem. Rev.* 97, 1515–66.

Desvergne J.-P. and Czarnik A. W. (Eds) **(1997)** *Chemosensors of Ion and Molecule Recognition*, NATO ASI series, Kluwer Academic Publishers, Dordrecht.

Fabbrizzi L. and Poggi A. **(1995)** Sensors and Switches from Supramolecular Chemistry *Chem. Soc. Rev.* 24, 197–202.

Fabbrizzi L., Lichelli M., Pallavicini P., Sacchi D., Taglietti A. **(1996)** Sensing of Transition Metals Through Fluorescence Quenching or Enhancement, *Analyst* 121, 1763–8.

Fabbrizzi L., Lichelli M., Poggi A., Rabaioli G., Taglietti A. **(2001)** Fluorometric Detection of Anion Activity and Temperature Changes, in: Valeur B. and Brochon J. C. (Eds), *New Trends in Fluorescence Spectroscopy. Applications to Chemical and Life Sciences*, Springer-Verlag, Berlin, pp. 209–27.

Fernandez-Gutierrez A. and Muñoz de la Peña A. **(1985)** Determination of Inorganic Substances by Luminescence Methods, in: Schulman S. G. (Ed.), *Molecular Luminescence Spectroscopy, Methods and Applications: Part 1*, Wiley & Sons, New York, pp. 371–546 and references therein (more than 1400).

Fuh M.-R. S., Burgess L. W. and Christian G. D. **(1991)** Fiber-Optic Chemical Fluorosensors, in: Warner I. M. and McGown L. B. (Eds), *Advances in Multidimensional Luminescence*, Vol. 1, Jai Press, Greenwich, pp. 111–29.

Grynkiewicz G., Poenie M. and Tsien R. Y. **(1985)** A New Generation of Ca^{2+} Indicators with Greatly Improved Fluorescence Properties, *J. Biol. Chem.* 260, 3440–50.

Hartmann W. K., Mortellaro M. A., Nocera D. G. and Pikramenou Z. **(1997)** Chemosensing of Monocyclic and Bicyclic Aromatic Hydrocarbons by Supramolecular Active Sites, in: Desvergne J.-P. and Czarnik A. W. (Eds), *Chemosensors of Ion and Molecule Recognition*, NATO ASI series, Kluwer Academic Publishers, Dordrecht, pp. 159–76.

Haugland R. P., *Handbook of Fluorescent Probes and Research Chemicals*, 6th edn, Molecular Probes, Inc., Eugene, OR, USA.

Janata J. **(1992)** Ion Optodes, *Anal. Chem.* 64, A921–7.

Jayaraman S., Biwersi J. and Verkman A. S. **(1999)** Synthesis and Characterization of Dual-Wavelength Cl^--Sensitive Fluorescent Indicators for Ratio Imaging, *Am. J. Physiol.* 276, C747–57.

Keefe M. H., Benkstein K. D. and Hupp J. T. **(2000)** Luminescent Sensor Molecules Based on Coordinated Metals: A Review of Recent Developments, *Coord. Chem. Rev.* 205, 201–8.

Lakowicz J. R. (Ed.) **(1994)** *Probe Design and Chemical Sensing*, Topics in Fluorescence Spectroscopy, Vol. 4, Plenum Press, New York.

Löhr H.-G. and Vögtle F. **(1985)** Chromo- and Fluororoionophores. A New Class of Dye Reagents, *Acc. Chem. Res.* 18, 65–72.

Martin M. M., Plaza P., Meyer Y. H., Badaoui F., Bourson J., Lefèvre J. P. and Valeur B. **(1996)** Steady-State and Picosecond Spectroscopy of Li^+ and Ca^{2+} Complexes with a Crowned Merocyanine, *J. Phys. Chem.* 100, 6879–88.

Pina F., Bernardo M. A. and Garcia-España E. **(2000)** Fluorescent Chemosensors Containing Polyamine Receptors, *Eur. J. Inorg. Chem.* 205, 59–83.

Prodi L., Bolletta F., Montalti M. and Zaccheroni N. **(2000)** Luminescent Chemosensors for Transition Metal Ions, *Coord. Chem. Rev.* 205, 2143–57.

Rettig W., Rurack K. and Sczepan M. **(2001)** From Cyanines to Styryl Bases – Photophysical Properties, Photochemical Mechanisms, and Cation Sensing Abilities of Charged and Neutral Polymethinic Dyes, in: Valeur B. and Brochon J. C. (Eds), *New Trends in Fluorescence Spectroscopy.*

Applications to Chemical and Life Sciences, Springer-Verlag, Berlin, pp. 125–55.

Robertson A. and Shinkai S. **(2000)** Cooperative Binding in Selective Sensors, Catalysts and Actuators, *Coord. Chem. Rev.* 205, 157–99.

Seitz W. R. **(1984)** Chemical Sensors Based on Fiber Optics, *Anal. Chem.* 56, A16–34.

Shinkai S. **(1997)** Aqueous Sugar Sensing by Boronic-Acid-Based Artificial Receptors, in: Desvergne J.-P. and Czarnik A. W. (Eds), *Chemosensors of Ion and Molecule Recognition*, NATO ASI series, Kluwer Academic Publishers, Dordrecht, pp. 37–59.

Shinkai S. and Robertson A. **(2001)** The Design of Molecular Artificial Sugar Sensing Systems, in: Valeur B. and Brochon J. C. (Eds), *New Trends in Fluorescence Spectroscopy. Applications to Chemical and Life Sciences*, Springer-Verlag, Berlin, pp. 173–85.

Slavik J. **(1994)** Fluorescent Probes in Cellular and Molecular Biology, CRC Press, Boca Raton.

Tsien R. Y. **(1989a)** Fluorescent Probes of Cell Signalling, *Ann. Rev. Neurosci.* 12, 227–53.

Tsien R. Y. **(1989b)** Fluorescent Indicators of Ion Concentration, *Methods Cell Biol.* 30, 127–56.

Ueno A., Ikeda H. and Wang J. **(1997)** Signal Transduction in Chemosensors of Modified Cyclodextrins, in: Desvergne J.-P. and Czarnik A. W. (Eds), *Chemosensors of Ion and Molecule Recognition*, NATO ASI Series, Kluwer Academic Publishers, Dordrecht, pp. 105–19.

Valeur B. **(1994)** Principles of Fluorescent Probes Design for Ion Recognition, in: J. R. Lakowicz (Ed.), *Probe Design and Chemical Sensing, Topics in Fluorescence Spectroscopy*, Vol. 4, Plenum, New York, pp. 21–48.

Valeur B. and Leray I. **(2000)** Design Principles of Fluorescent Molecular Sensors for Cation Recognition, *Coord. Chem. Rev.* 205, 3–40.

Valeur B. and Leray I. **(2001)** PCT (Photoinduced Charge Transfer) Fluorescent Molecular Sensors for Cation Recognition, in: Valeur B. and Brochon J. C. (Eds), *New Trends in Fluorescence Spectroscopy. Applications to Chemical and Life Sciences*, Springer-Verlag, Berlin, pp. 187–207.

Valeur B., Badaoui F., Bardez E., Bourson J., Boutin P., Chatelain A., Devol I., Larrey B., Lefèvre J. P. and Soulet A. **(1997)** Cation-Responsive Fluorescent Sensors. Understanding of Structural and Environmental Effects, in: Desvergne J.-P. and Czarnik A. W. (Eds), *Chemosensors of Ion and Molecule Recognition*, NATO ASI Series, Kluwer Academic Publishers, Dordrecht, pp. 195–220.

Whitaker J. E., Haugland R. P. and Prendergast F. G. **(1991)** Spectral and Photophysical Studies of Benzo[*c*]xanthene Dyes: Dual Emission pH Sensors, *Anal. Biochem.* 194, 330–44.

Wolfbeis O. S. **(1988)** Fiber Optical Fluorosensors in Analytical and Clinical Chemistry, in: Schulman S. G. (Ed.), *Molecular Luminescence Spectroscopy. Methods and Applications: Part 2*, John Wiley & Sons, New York, pp. 129–281.

Wolfbeis O. S. (Ed.) **(1991)** *Fiber Optic Chemical Sensors and Biosensors*, Vols. I–II, CRC Press, Boca Raton, FL.

Wolfbeis O. S., Fürlinger E., Kroneis H. and Marsoner H. **(1983)** Fluorimetric Analysis. 1. A Study on Fluorescent Indicators for Measuring Near Neutral ('Physiological') pH Values, *Fresenius Z. Anal. Chem.* 314, 119–24.

Wolfbeis O. S., Reisfeld R. and Oehme I. **(1996)** Sol–Gels and Chemical Sensors, *Struct. Bonding* 85, 51–98.

11
Advanced techniques in fluorescence spectroscopy

Dans tous les états de la vie & de la société, on a si souvent occasion d'admirer le jeu merveilleux de la lumière, l'importance et la réalité des secours que nous procurent les instruments d'optique, pour étendre notre vue, et pour suppléer à ses défauts, ...

M. l'Abbé de la Caille, 1766

[*In all walks of life and on all levels of society, one has so many opportunities to admire the marvellous play of light, the importance and the effectiveness of the help provided by optical instruments to extend our sight, and to compensate for its defects, ...*]

11.1
Time-resolved fluorescence in the femtosecond time range: fluorescence up-conversion technique

Chapter 6 described the current techniques employed in time-resolved fluorescence spectrocopy. The time resolution of these techniques ranges from a few picoseconds (streak cameras) to a few hundreds of picoseconds (single-photon timing with flash lamp excitation). The time resolution can be greatly improved by using the fluorescence up-conversion technique.

A schematic diagram illustrating this technique is shown in Figure 11.1. A laser source provides an ultra-short pulse (duration of 50–100 fs) at wavelength λ_1 corresponding to frequency ω_1. The second harmonic (frequency ω_2) is generated in a nonlinear crystal and separated from the fundamental light by a dichroic beam splitter. The *probe pulse* at frequency ω_1 passes through an optical delay line, and the *excitation pulse* at frequency ω_2 is focused on the sample. The incoherent fluorescence (frequency ω_{fl}) emitted by the sample is collected and mixed with the

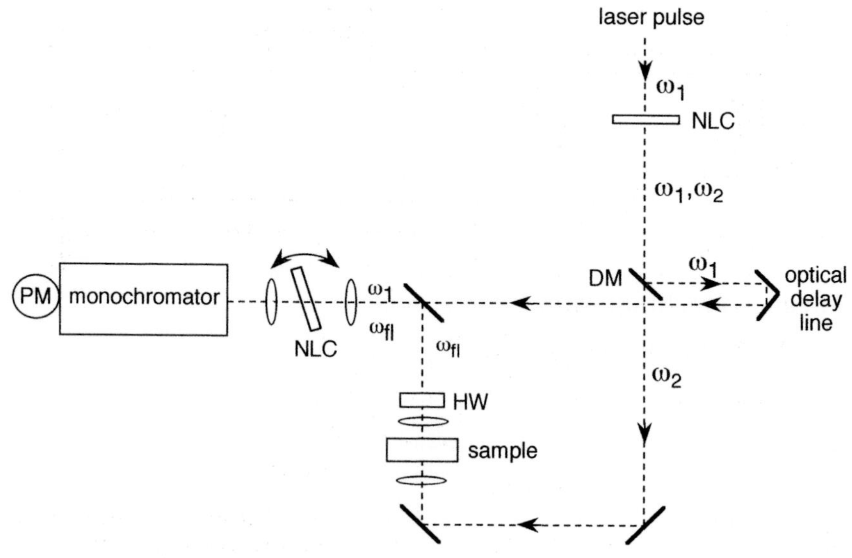

Fig. 11.1. Principles of fluorescence up-conversion. NLC: nonlinear crystal; DM: dichroic mirror; HW: half-wave plate; PM: photomultiplier.

probe pulse ω_1 in a nonlinear crystal. This frequency mixing generates light at the sum frequency, $\omega_{\text{sum}} = \omega_1 + \omega_{\text{fl}}$[1], *provided that there is a spatial and temporal coincidence of the ω_1 and ω_{fl} beams.* The greater the delay time between the probe pulse ω_1 and the fluorescence beam ω_{fl}, the smaller the fluorescence intensity[2]. The fluorescence intensity versus time is thus obtained by varying the delay time (ajustable thanks to the optical delay line), and by measuring the intensity of the sum frequency light for each delay time.

Currently used nonlinear optical crystals are potassium dihydrogen phosphate (KDP) and barium borate (BBO). Compared to KDP, the advantages of BBO are its transparency in the UV and its larger quantum efficiency of up-conversion by a factor of 4–6. For a given position of the crystal, only a narrow band of the fluorescence spectrum is up-converted. Therefore, if the full fluorescence spectrum is of interest, the crystal must be rotated at a series of angles. An example of experimental set-up is presented in Figure 11.2. The fwhm of the response is 210 fs.

Various ultrafast phenomena occuring in the femtosecond time-scale in the condensed phase have been studied by fluorescence up-conversion (for a review, see Mialocq and Gustavsson, 2001). As already mentioned in Chapter 7 (Box 7.1),

1) Such a conversion to higher frequencies is at the origin of the term 'up-conversion'.

2) The intensity of the sum frequency light I_{sum} at a given delay time τ between the probe pulse ω_1 and the fluorescence beam ω_{fl} is proportional to the correlation function of the fluorescence intensity with the intensity of the probe pulse ω_1:

$$I_{\text{sum}}(\tau) \approx \int_{-\infty}^{\infty} I_{\text{fl}}(t) I_1(t-\tau)\,\mathrm{d}t$$

where I_{fl} is the intensity of fluorescence and I_1 is the intensity of the probe pulse. This equation holds in the limit of small depletion of the fluorescence (conversion efficiency $< 10\%$).

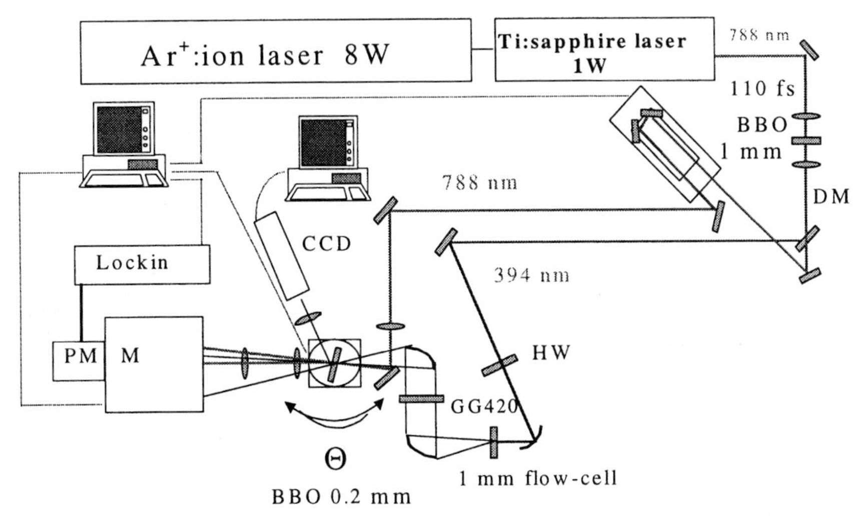

Fig. 11.2. Fluorescence up-conversion instrument. DM: dichroic mirror; HW: half-wave plate; GG420: Schott filter; CCD: video camera for the visual superposition of the beams in the BBO crystal; M: monochromator; PM: photomultiplier connected to a lock-in photon counter (reproduced with permission from Mialocq and Gustavsson, 2001).

fluorescence up-conversion is the method of choice to study solvation dynamics. Moreover, this technique is well suited to the investigation of photoinduced intramolecular processes (e.g. charge transfer, proton transfer) and intermolecular processes (e.g. electron transfer). Applications to the study of light-driven biological processes have also been reported (e.g. photoactive yellow protein, blue fluorescent protein (mutant of the green fluorescent protein)).

11.2 Advanced fluorescence microscopy

Fluorescence microscopy is principally used for the investigation of living cells and tissues by biologists and physiologists, but it is also a powerful tool to study chemical systems such as colloids, liquid crystals, polymer blends, photodegradation of naturally occurring polymers, dyeing of fibers and measurement of the glass transition temperature (Davidson, 1996). This section will focus on the improvements in conventional fluorescence microscopy and to the development of time-resolved fluorescence microscopy.

11.2.1 Improvements in conventional fluorescence microscopy

A conventional fluorescence microscope differs from a standard microscope by the light source (mercury or xenon lamp), which produces UV–visible light. The excitation wavelength is selected by an interference filter or a monochromator. Observation of the fluorescence is made by eye, photographic film or CCD (charge-

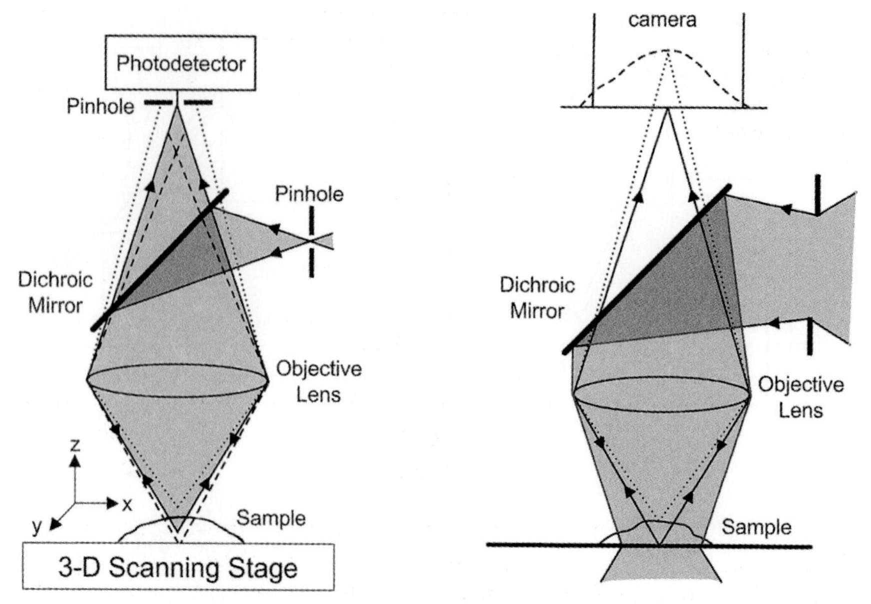

Fig. 11.3. Principle of confocal microscopy (left) compared with conventional microscopy (right).

coupled device) camera. The depth of field of a conventional fluorescence microscope is 2–3 μm and the maximal resolution is approximately equal to half the wavelength of the radiation used (i.e. 0.2–0.3 μm for visible radiation).

For samples thicker than the depth of field, the images are blurred by out-of-focus fluorescence. Corrections using a computer are possible, but other techniques are generally preferred such as *confocal microscopy* and *two-photon excitation microscopy*. It is possible to overcome the optical diffraction limit in *near-field scanning optical microscopy (NSOM)*.

11.2.1.1 **Confocal fluorescence microscopy**

In a confocal microscope, invented in the mid-1950s, a focused spot of light scans the specimen. The fluorescence emitted by the specimen is separated from the incident beam by a dichroic mirror and is focused by the objective lens through a pinhole aperture to a photomultiplier. Fluorescence from out-of-focus planes above and below the specimen strikes the wall of the aperture and cannot pass through the pinhole (Figure 11.3).

The principle is somewhat similar to the reading of a compact disk: a focused laser beam is reflected by the microscopic pits (embedded inside a plastic layer) towards a small photodiode so that scratches and dust have no effect. Scanning is achieved by rotation of the disk. A laser is also often used in confocal fluorescence microscopy, but scanning is achieved using vibrating mirrors or a rotating disk containing multiple pinholes in a spiral arrangement (Nipkow disk). In laser scanning confocal microscopy, images are stored on a computer and displayed on a monitor.

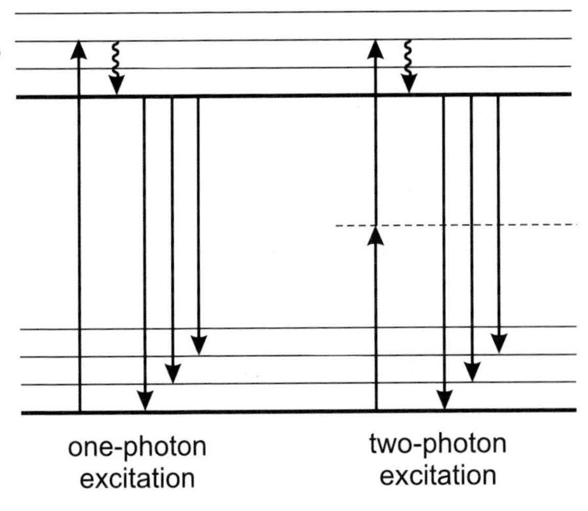

Fig. 11.4. Schematic of two-photon excitation compared to one-photon excitation. The dashed line represents the virtual state that mediates the absorption.

One of the features of confocal microscopy is that it can produce optical slices of defined thickness through thick specimens. Using a lens of high numerical aperture, thickness of the confocal sections can reach a theoretical limit of about 0.5 μm. Therefore, by moving the specimen up and down, a three-dimensional (3-D) image can be recorded.

It should be noted that, because confocal microscopy collects only a fraction of the total fluorescence emitted by a sample, the excitation energy required to image this fluorescence must be higher than in conventional fluorescence microscopy. Therefore, the amount of photobleaching per detected photon is higher. Photobleaching should be minimized by using stable fluorophores and by operating the confocal microscope at low laser power, high detector sensitivity, and maximum objective numerical aperture.

Confocal fluorescence microscopy has been extensively used in cell biology. Single living cells can indeed be studied by this technique: visualization of organelles, distribution of electrical potential, pH imaging, Ca^{2+} imaging, etc. (Lemasters, 1996). Interesting applications in chemistry have also been reported in the fields of colloids, liquid crystals and polymer blends.

Confocal fluorescence microscopy can be combined with time-domain and frequency-domain techniques to produce lifetime imaging (see Section 11.2.2.3).

11.2.1.2 Two-photon excitation fluorescence microscopy

In conventional fluorescence spectroscopy, a fluorophore is excited by absorption of one photon whose energy corresponds to the energy difference between the ground state and the excited state. Excitation is also possible by the simultaneous absorption of two photons of lower energy (i.e. of longer wavelength) via a short-lived virtual state (Figure 11.4)[3]. For instance, absorption of two photons in the red can

3) Combination of the energies of two photons to cause the transition to an excited state was predicted in 1931 but observed only 30 years later when the required large intensities from lasers became available.

excite a molecule that absorbs in the UV. Two-photon excitation is a nonlinear process; there is a quadratic dependence of absorption on excitation light intensity.

When a single laser is used, the two photons are of identical wavelength, and the technique is called *two-photon excitation fluorescence microscopy*. When the photons are of different wavelengths λ_1 and λ_2 (so that $1/\lambda_1 + 1/\lambda_2 = 1/\lambda_e$), the technique is called *two-color excitation fluorescence microscopy*.

The probability of two-photon absorption depends on both spatial and temporal overlap of the incident photons (the photons must arrive within 10^{-18} s). The cross-sections for two-photon absorption are small, typically 10^{-50} cm^4 s photon^{-1} molecule^{-1} for rhodamine B. Consequently, only fluorophores located in a region of very large photon flux can be excited. Mode-locked, high-peak power lasers like titanium–sapphire lasers can provide enough intensity for two-photon excitation in microscopy.

Because the excitation intensity varies as the square of the distance from the focal plane, the probability of two-photon absorption outside the focal region falls off with the fourth power of the distance along the z optical axis. Excitation of fluorophores can occur only at the point of focus. Using an objective with a numerical aperture of 1.25 and an excitation beam at 780 nm, over 80% of total fluorescence intensity is confined to within 1 μm of the focal plane. The excitation volume is of the order of 0.1–1 femtoliter. Compared to conventional fluorometers, this represents a reduction by a factor of 10^{10} of the excitation volume.

Two-photon excitation provides intrinsic 3-D resolution in laser scanning fluorescence microscopy. The 3-D sectioning effect is comparable to that of confocal microscopy, but it offers two advantages with respect to the latter: because the illumination is concentrated in both time and space, there is *no out-of-focus photobleaching*, and the *excitation beam is not attenuated by out-of-focus absorption*, which results in increased penetration depth of the excitation light.

The advantage of two-color excitation over two-photon excitation is not an improvement in imaging resolution, but the easier observation of microscopic objects through highly scattering media. In fact, in two-color excitation, scattering decreases the in-focus fluorescence but only minimally increases the unwanted fluorescence background, in contrast to two-photon excitation.

11.2.1.3 **Near-field scanning optical microscopy (NSOM)**

The resolution of a conventional microcope is limited by the classical phenomena of interference and diffraction. The limit is approximately $\lambda/2$, λ being the wavelength. This limit can be overcome by using a sub-wavelength light source and by placing the sample very close to this source (i.e. in the near field). The relevant domain is *near-field optics* (as opposed to far-field conventional optics), which has been applied to microscopy, spectroscopy and optical sensors. In particular, *near-field scanning optical microscopy* (NSOM) has proved to be a powerful tool in physical, chemical and life sciences (Dunn, 1999).

The idea of near-field optics to bypass the diffraction limit was described in three visionary papers published by Synge in the period 1928–32. Synge's idea is illus-

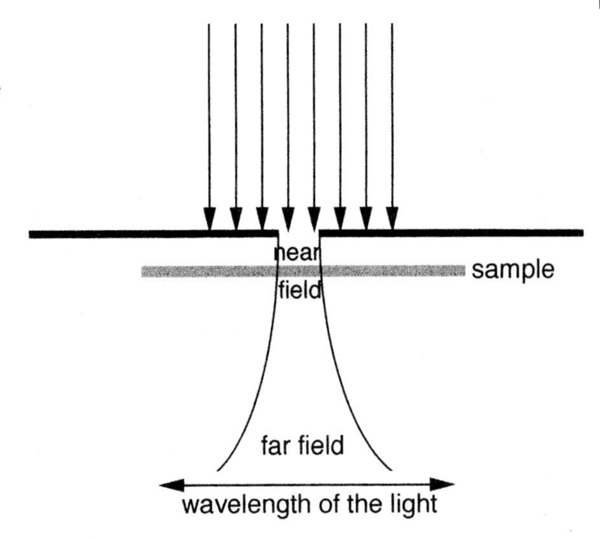

Fig. 11.5. Principle of near-field optics according to Synge's idea to overcome the diffraction limit.

trated in Figure 11.5. An incident light passes through a sub-wavelength hole in an opaque screen. The surface of the sample is positioned in close proximity to the hole so that the emerging light is forced to interact with it. The hole acts as a sub-wavelength-sized light probe that can be used to image a specimen before the light is diffracted out. The first measurement using this idea was reported half a century later, and today NSOM is used in many fields. In addition to its very high spatial resolution at the sub-micrometer level, this technique has an outstanding sensitivity that permits single-molecule measurements (see Section 11.4.2).

Most NSOM instruments are built around an inverted fluorescence microscope that offers the advantage of providing normal images, allowing the region to be studied to be located with the higher resolution NSOM mode (Figure 11.6). A laser beam passes through a single mode optical fiber whose end is fashioned into a near-field tip. The tip is held in a z-piezo head. An x–y piezo stage on which the sample is mounted permits scanning of the sample. Light from the tip excites the sample whose emitted fluorescence is collected from below by an objective with a high numerical aperture and detected through a filter (to remove residual laser excitation light) and a detector (e.g. avalanche photodiode or OMA (optical multi-channel analyzer)). This mode is called the *illumination mode*. Alternatively, in the *collection mode*, the sample is illuminated from the far field, and fluorescence is collected by the NSOM tip.

The systems that scan the piezos and record the image are similar to those used in atomic force microscopy.

The NSOM tip is obtained by heating and pulling a single-mode optical fiber down to a fine point. A reflective metal coating (aluminum, silver or gold) is deposited by vacuum evaporative techniques in order to prevent light from escaping.

Precise positioning of the tip within nanometers of the sample surface is required to obtain high-resolution images. This can be achieved by a feedback

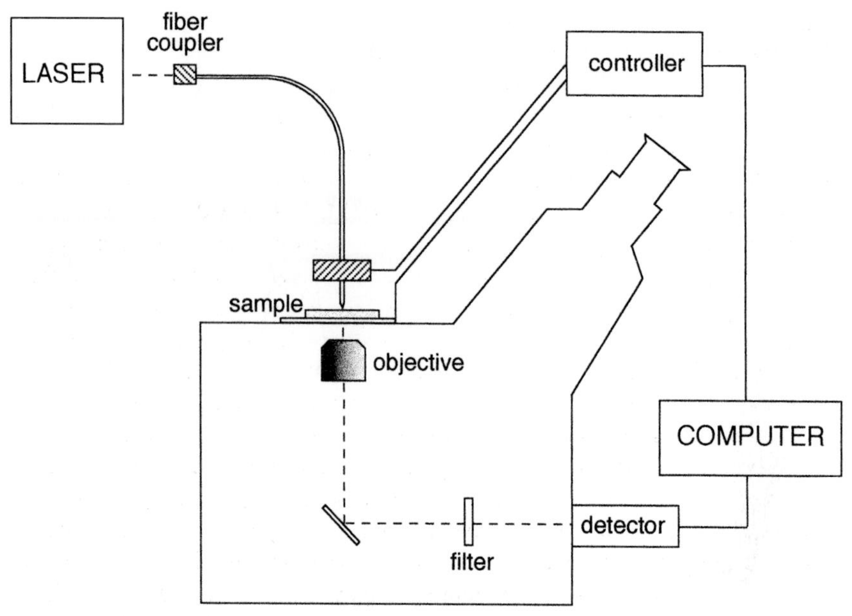

Fig. 11.6. Schematic of an NSOM instrument built around an inverted fluorescence microscope and operating according to the illumination mode.

mechanism that is generally based on the *shear-force method*: the tip is dithered laterally at one of its resonating frequencies with an amplitude of about 2–5 nm. As the tip comes within the van der Waals force field of the sample, the shear forces dampen the amplitude of the tip vibration. This amplitude can be monitored and used to generate a feedback signal to control the distance between the sample and the tip during imaging.

NSOM is a remarkable tool for the analysis of thin films such as electroluminescent polymers (e.g. poly(*p*-phenylene vinylene), poly(*p*-pyridyl vinylene)), J-aggregates, liquid crystals and Langmuir–Blodgett films. Moreover, NSOM can provide new insights into complex biological systems owing to the higher resolution than in confocal microscopy, with the additional capability of force mapping of the surface topography, and the advantage of reduced photobleaching. Photosynthetic systems, protein localization, chromosome mapping and membrane microstructure are examples of systems that have been successfully investigated by NSOM. Imaging of fixed biological samples in aqueous environments is in fact possible, but the study of unfixed cells is problematic.

Discrimination of species in complex samples can be made via lifetime measurements using the single-photon timing method coupled to NSOM.

Two-photon excitation in NSOM has been shown to be possible with uncoated fiber tips in shared aperture arrangement. This represents an interesting extension of the technique to applications requiring UV light through two-photon excitation.

11.2.2
Fluorescence lifetime imaging microscopy (FLIM)

Fluorescence lifetime imaging uses differences in the excited-state lifetime of fluorophores as a contrast mechanism for imaging. As emphasized in several chapters of this book, the excited-state lifetime of a fluorophore is sensitive to its microenvironment. Therefore, imaging of the lifetime provides complementary information on local physical parameters (e.g. microviscosity) and chemical parameters (e.g. pH, ion concentration), in addition to information obtained from steady-state characteristics (fluorescence spectra, excitation spectra and polarization). FLIM is an outstanding tool for the study of single cells with the possibility of coupling multi-parameter imaging of cellular structures with spectral information. Discrimination of autofluorescence of living cells from true fluorescence is possible on the basis of distinct lifetime differences. Various applications have been reported: calcium (or other chemical) imaging; membrane fluidity, transport and fusion; imaging using RET (resonance energy transfer) for quantifying the distance between two species labelled with two different fluorophores; DNA sequencing; clinical imaging (e.g. use of antibodies and nucleic acids labeled with fluorophores for quantitative measurements of multiple disease markers in individual cells of patient specimens; etc.).

FLIM has been developed using either time-domain or frequency-domain methods (Herman et al., 1997).

11.2.2.1 Time-domain FLIM

In principle, lifetime imaging is possible by combination of the single-photon timing technique with scanning techniques. However, the long measurement time required for collecting photons at each point is problematic.

Alternatively, a gated microchannel plate (MCP) image intensifier (operating at a maximal frequency of 10 kHz) can be used in conjunction with a slow-scan cooled CCD camera for digital recording (Wang et al., 1992). Laser picosecond pulses are used to illuminate the entire field of view via an optical fiber and a lens of large numerical aperture (Figure 11.7).

The principle of lifetime measurement by the gated image intensifier is illustrated in Figure 11.8. At time t_d after excitation by a light pulse, a sampling gate pulse (duration ΔT) is applied to the photocathode of the image intensifier. Fluorescence can thus be detected at various delay times (multigate detection). For a single exponential decay of the form $\alpha \exp(-t/\tau)$, two delay times t_1 and t_2 are sufficient. From corresponding fluorescence signals D_1 and D_2 given by

$$D_1 = \int_{t_1}^{t_1+\Delta T} \alpha \exp(-t/\tau)\,\mathrm{d}t \qquad (11.1)$$

$$D_2 = \int_{t_2}^{t_2+\Delta T} \alpha \exp(-t/\tau)\,\mathrm{d}t \qquad (11.2)$$

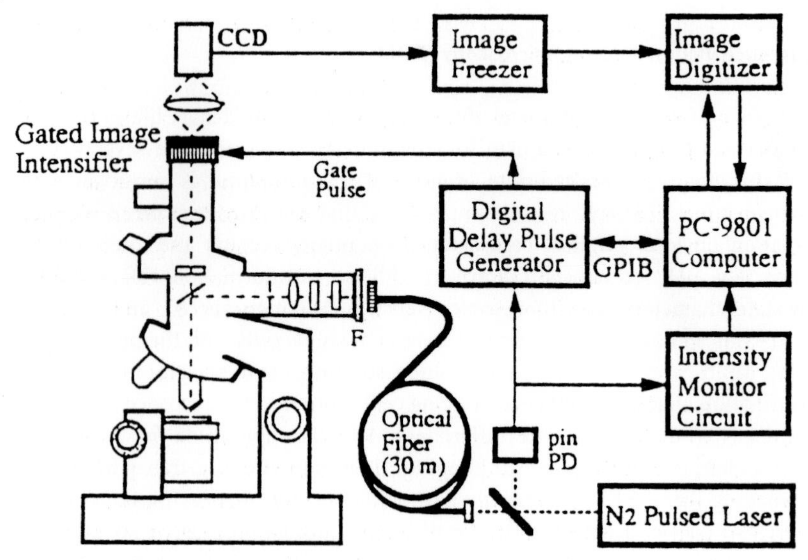

Fig. 11.7. A fluorescence lifetime microscope using a gated image intensifier (reproduced with permission from Wang et al., 1992).

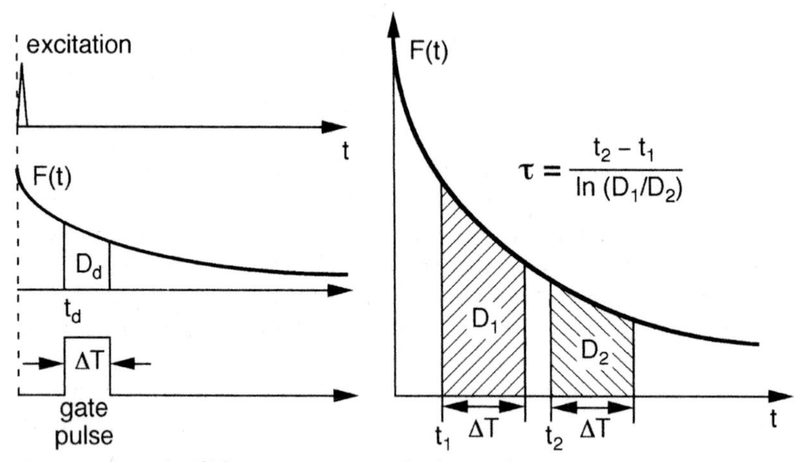

Fig. 11.8. Principle of lifetime measurement by the gated image intensifier (adapted from Wang et al., 1992).

the lifetime can be calculated by means of the following expression:

$$\tau = \frac{t_2 - t_1}{\ln(D_1/D_2)} \tag{11.3}$$

By this procedure, which requires calculation from only four parameters (D_1, D_2, t_1 and t_2), lifetime images can be obtained very quickly. Time-resolved images can be accumulated on the CCD chip by repeated pulse excitation. The images are then read out to a computer.

The time resolution of the image intensifier is about 3 ns (minimal gate width), which may not be sufficient for fast decaying probes. Moreover, a pixel-by-pixel deconvolution, if necessary, would require excessively long computation times.

A much better time resolution, together with space resolution, can be obtained by new imaging detectors consisting of a microchannel plate photomultiplier (MCP) in which the disk anode is replaced by a 'coded' anode (Kemnitz, 2001). Using a Ti–sapphire laser as excitation source and the single-photon timing method of detection, the time resolution is <10 ps. The space resolution is 100 μm (250×250 channels).

11.2.2.2 **Frequency-domain FLIM**

In frequency-domain FLIM, the optics and detection system (MCP image intensifier and slow scan CCD camera) are similar to that of time-domain FLIM, except for the light source, which consists of a CW laser and an acousto-optical modulator instead of a pulsed laser. The principle of lifetime measurement is the same as that described in Chapter 6 (Section 6.2.3.1). The phase shift and modulation depth are measured relative to a known fluorescence standard or to scattering of the excitation light. There are two possible modes of detection: heterodyne and homodyne detection.

Heterodyne detection is achieved by modulating the high-voltage amplification of the image intensifier at frequency $f + \Delta f$, while fluorescence is modulated at frequency f. The high-frequency fluorescence signal is then transformed into a signal modulated at frequency Δf (a few tens of Hz), which contains the same phase and modulation information as the original high-frequency signal. The phase and modulation of the Δf signal are determined simultaneously at each pixel of the CCD camera. To do so, it is advantageous to use a 'boxcar' method of measurement by gating the cathode of the image intensifier for a time Δt repetitively and synchronously with the Δf frequency master (Figure 11.9) (Gadella et al., 1993). Several phase delays settings are selected by systematically delaying the position of the gate. Separate images are recorded and accumulated on the CCD camera.

In the homodyne mode of detection, the modulation frequency of the excitation light is the same as that of the image intensifier. An example of data is shown in Box 11.1.

In the case of a single exponential decay, the lifetime can be rapidly calculated by either the phase shift Φ or the modulation ratio M by means of Eqs (6.25) and (6.26) established in Chapter 6 (Section 6.2.3):

$$\tau_\Phi = \frac{1}{\omega} \tan^{-1} \Phi \tag{11.4}$$

$$\tau_M = \frac{1}{\omega}\left(\frac{1}{M^2} - 1\right)^{1/2} \tag{11.5}$$

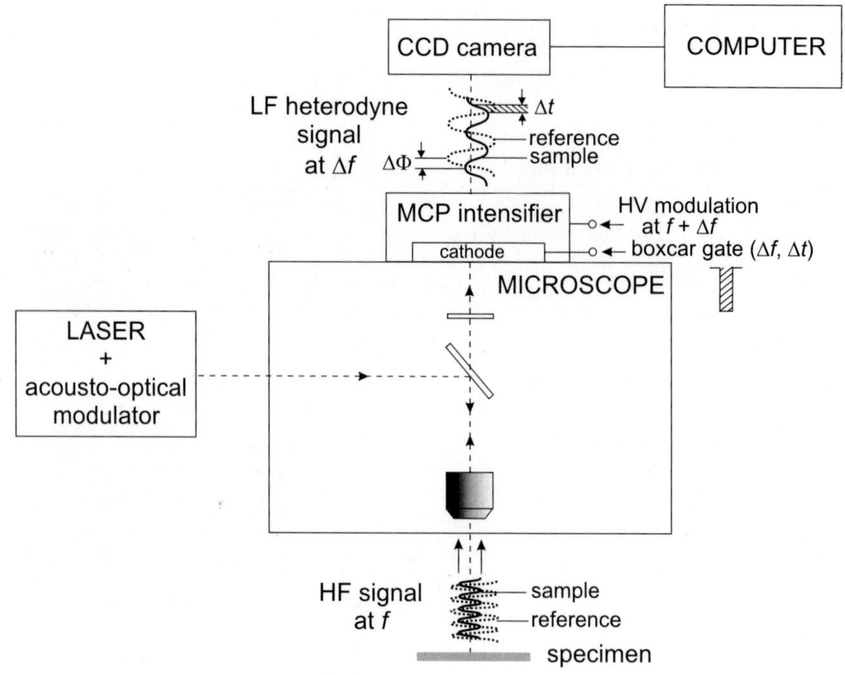

Fig. 11.9. Schematic of phase measurement using a 'boxcar' method (adapted from Gadella et al., 1993).

If the values calculated in these two ways are identical, the fluorescence decay is indeed a single exponential. Otherwise, for a multi-component decay, $\tau_\Phi < \tau_M$. In this case, several series of images have to be acquired at different frequencies (at least 5 for a triple exponential decay because three lifetimes and two fractional amplitudes are to be determined), which is a challenging computational problem.

11.2.2.3 **Confocal FLIM (CFLIM)**

As shown in Section 11.2.1.1, more details can be obtained by confocal fluorescence microscopy than by conventional fluorescence microscopy. In principle, the extension of conventional FLIM to confocal FLIM using either time- or frequency-domain methods is possible. However, the time-domain method based on single-photon timing requires expensive lasers with high repetition rates to acquire an image in a reasonable time, because each pixel requires many photon events to generate a decay curve. In contrast, the frequency-domain method using an inexpensive CW laser coupled with an acoustooptic modulator is well suited to confocal FLIM.

11.2.2.4 **Two-photon FLIM**

As described above, two-photon excitation microscopy provides several advantages (reduced photobleaching, deeper penetration into the specimen). A fluorescence microscope combining two-photon excitation and fluorescence time-resolved

Box 11.1 An example of FLIM data[a)]

A great deal of information is available when lifetimes are imaged. It is indispensable for the user to have as much information as possible presented in images and plots that convey multiple parameters simultaneously and conveniently, especially when the images are available in real time. The example presented in this box illustrates how to achieve this.

The data in Figure B11.1.1 were acquired with a frequency-domain FLIM instrument working in homodyne mode. In this mode, the modulation frequency of the intensifier (40 MHz) is identical to that of the excitation light (40 MHz). The phase and modulation are calculated from a series of images taken at different phase delays between the excitation light and the intensifier.

Figure B11.1.1 represents a 3T3 cell stained with BODIPY FL C5-ceramide (from Molecular Probes), a specific stain for the Golgi apparatus The color coding for the lifetimes is from 0 to 5 ns. The lifetime is coded in color (right upper) and this color-coded lifetime information is mapped onto the intensity surface (upper left) to give the combined lifetime/intensity plot (lower right). The final combined image shows intensity contours (in white), and a lit intensity surface is employed to accentuate the information in a three-dimensional form.

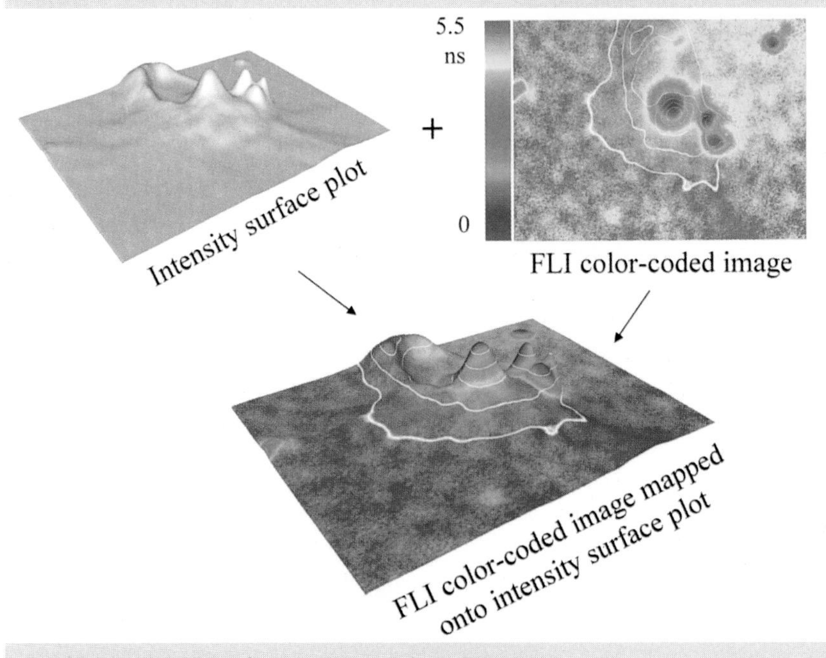

Fig. B11.1.1. FLIM data of a 3T3 cell (stained with BODIPY FL C5-ceramide) obtained by a frequency-domain FLIM instrument (see text)[a)].

a) Courtesy of Robert Clegg, Department of Physics, University of Illinois at Urbana-Champaign, USA.

imaging has been developed using a Ti:sapphire laser and the frequency-domain method with heterodyne detection (So et al., 1996). The Ti:sapphire laser is operated at a fixed frequency of 80 MHz. Because the pulse width is very narrow (150 fs), the train pulse has a harmonic content of up to 220 GHz. However, it is desirable to have modulation frequencies below 80 MHz in order to measure longer decay times. For this purpose, an acousto-optic modulator is placed in the beam path of the Ti:sapphire laser and provides modulation frequencies from 30 to 120 MHz. These frequencies are mixed with the 80 MHz repetition frequency of the laser and its high harmonics which generates new frequencies (sums and differences). Finally, frequencies are available in the range of kilohertz to gigahertz.

11.3 Fluorescence correlation spectroscopy

In fluorescence correlation spectroscopy (FCS), the temporal fluctuations of the fluorescence intensity are recorded and analyzed in order to determine physical or chemical parameters such as translational diffusion coefficients, flow rates, chemical kinetic rate constants, rotational diffusion coefficients, molecular weights and aggregation. The principles of FCS for the determination of translational and rotational diffusion and chemical reactions were first described in the early 1970s. But it is only in the early 1990s that progress in instrumentation (confocal excitation, photon detection and correlation) generated renewed interest in FCS.

11.3.1 Conceptual basis and instrumentation

Fluctuations in fluorescence intensity in a small open region (in general created by a focused laser beam) arise from the motion of fluorescent species in and out of this region via translational diffusion or flow. Fluctuations can also arise from chemical reactions accompanied by a change in fluorescence intensity: association and dissociation of a complex, conformational transitions, photochemical reactions (Figure 11.10) (Thompson, 1991).

The fluctuations $\delta I(t)$ of the fluorescence around the mean value $\langle I \rangle$, defined as

$$\delta I(t) = I(t) - \langle I \rangle \tag{11.6}$$

are analyzed in the form of an autocorrelation function $G(\tau)$ which relates the fluorescence intensity $I(t)$ at time t to the fluorescence intensity $I(t+\tau)$ at time $t+\tau$:

$$G(\tau) = \frac{\langle I(t).I(t+\tau)\rangle}{\langle I(t)\rangle^2} = \frac{\langle[\langle I \rangle + \delta I(t)][\langle I \rangle + \delta I(t+\tau)]\rangle}{\langle I \rangle^2} = 1 + \frac{\langle \delta I(t) \delta I(t+\tau)\rangle}{\langle I \rangle^2} \tag{11.7}$$

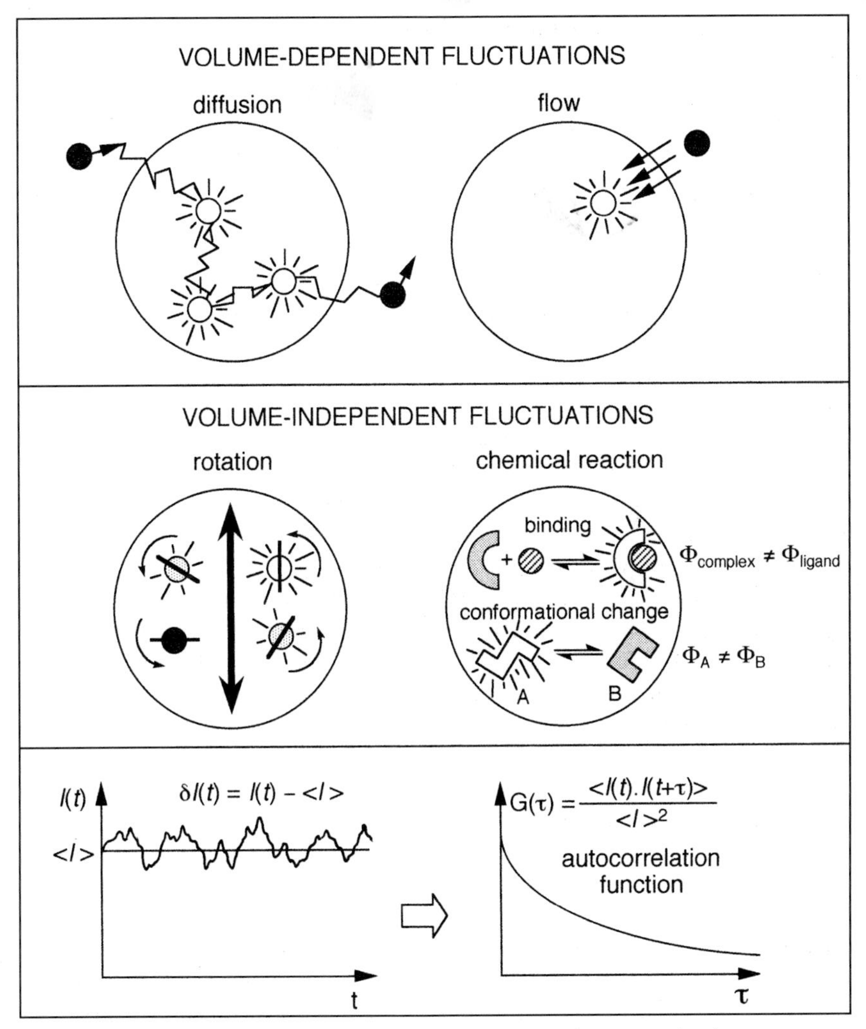

Fig. 11.10. Schematic illustration of fluorescence correlation spectroscopy. The autocorrelation function characterises the fluctuations of the fluorescence intensity: its decay time expresses the average duration of a fluctuation, and its extrapolated time-zero value represents the magnitude of the fluctuation. In the case of rotational diffusion, the double arrow represents the direction of transmission of the polarizer.

The FCS autocorrelation functions[4] contain two types of information (Thompson, 1991):

4) In several papers, only the time-dependent part of the autocorrelation function is considered, and the definition is then

$$G(\tau) = \frac{\langle \delta I(t)\delta I(t+\tau)\rangle}{\langle I\rangle^2}$$

- the magnitude of the fluctuation signal characterized by $G(0)$, i.e. the value of $G(\tau)$ at $\tau = 0$;
- a kinetic information provided by the rate and shape of the temporal decay of $G(\tau)$. The decay rate represents the average duration of the fluctuation signal.

$G(0)$ depends on the average number of molecules $\bar{N}$ inside the excitation volume. The larger this number, the smaller the value of $G(0)$; more precisely, $G(0) - 1$ is inversely proportional to $\bar{N}$. Therefore, the sensitivity of FCS increases with decreasing fluorophore concentration. It is worth introducing the volume V_T of the fluorophore territory, which is the reciprocal of the concentration, and to compare it to the excitation volume (sample volume element) V_S. If $V_S < V_T$, the fluctuations are large, whereas if $V_S \gg V_T$, we have large average intensities. Typical V_S values in a confocal microscope are 0.2–10 fL (femtoliters), and the typical working concentrations range from 10^{-9} M to 10^{-15} M (1 femtomol L^{-1}). At such low concentrations, we can expect to detect single molecules (see Section 11.4.3).

$G(0)$ can yield information on molecular weights (e.g. labeled proteins and nucleic acids) and molecular aggregation. In these applications, the laser beam is focused through a microscope objective to a small spot on the specimen; the latter is laterally translated through the beam by a computer-controlled microstepping stage, at a speed higher than the rate of translational diffusion of the species under study. This technique is called *scanning-FCS*.

$G(\tau)$ decays with correlation time because the fluctuation is more and more uncorrelated as the temporal separation increases. The rate and shape of the temporal decay of $G(\tau)$ depend on the transport and/or kinetic processes that are responsible for fluctuations in fluorescence intensity. Analysis of $G(\tau)$ thus yields information on translational diffusion, flow, rotational mobility and chemical kinetics. When translational diffusion is the cause of the fluctuations, the phenomenon depends on the excitation volume, which in turn depends on the objective magnification. The larger the volume, the longer the diffusion time, i.e. the residence time of the fluorophore in the excitation volume. On the contrary, the fluctuations are not volume-dependent in the case of chemical processes or rotational diffusion (Figure 11.10). Chemical reactions can be studied only when the involved fluorescent species have different fluorescence quantum yields.

Most FCS instruments have been designed around optical microscopes. A typical optical configuration is shown in Figure 11.11. The emission is detected through a pinhole conjugated with the image plane of the excitation volume. The size of the latter depends on the magnification and pinhole size. The detector is a photomultiplier (or an avalanche photodiode) operating in the analog mode, or more often in single-photon counting mode, and is connected to an amplifier/discriminator. The autocorrelation function can be instantaneously obtained from the analysis of the fluorescence intensity fluctuations by a fast correlator. For the determination of rotational mobility, polarizers are introduced in the excitation and/or emission path.

Two-photon FCS has been successfully developed; it offers the advantage of 3-D resolution (see Section 11.2.1.2) (Chen et al., 2001).

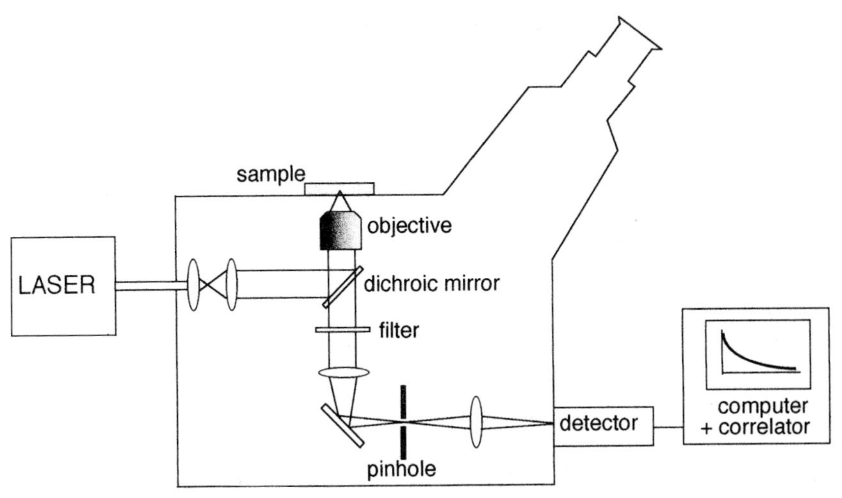

Fig. 11.11. An optical microscope adapted for fluorescence correlation measurements.

11.3.2
Determination of translational diffusion coefficients[5)]

For a single fluorescent species undergoing Brownian motion with a translational diffusion coefficient D_t (see Chapter 8, Section 8.1), the autocorrelation function, in the case of Gaussian intensity distribution in the x, y plane and infinite dimension in the z-direction, is given by

$$G(\tau) = 1 + \frac{1}{\bar{N}}\left(\frac{1}{1 + 4D_t\tau/\omega^2}\right) = 1 + \frac{1}{\bar{N}}\left(\frac{1}{1 + \tau/\tau_D}\right) \tag{11.8}$$

where $\tau_D = \omega^2/4D_t$ is the characteristic time for diffusion, and ω is the distance from the center of the illuminated area in the x, y plane at which the detected fluorescence has dropped by a factor e^2.

For a finite volume element with a Gaussian intensity distribution in three dimensions, the autocorrelation can be written as

$$G(\tau) = 1 + \frac{1}{\bar{N}}\left(\frac{1}{1 + 4D_t\tau/\omega_1^2}\right)\left(\frac{1}{1 + 4D_t\tau/\omega_2^2}\right)^{1/2} \tag{11.9}$$

where ω_1 and ω_2 are the distances from the center of the excitation volume in the radial and axial direction, respectively, at which the detected fluorescence has dropped by a factor e^2.

5) Translational diffusion can also be studied by fluorescence recovery after photobleaching (FRAP). This technique will not be described in this chapter. For a comparison with FCS, see Elson (1985) and Petersen and Elson (1986).

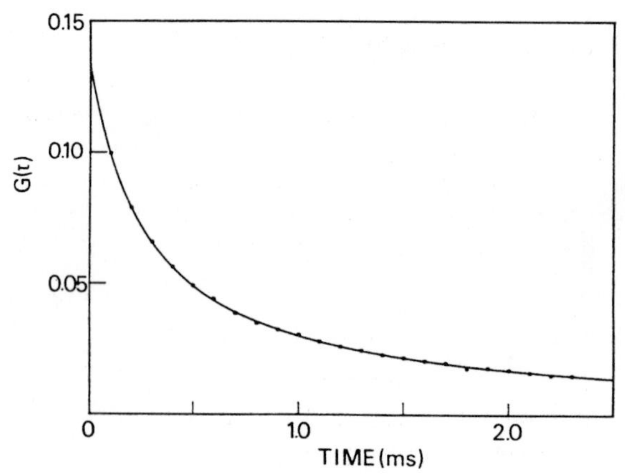

Fig. 11.12. Autocorrelation function for rhodamine 6G 10^{-9} M in ethanol. The best fit (solid line) yields $D_t \cong 3 \times 10^{-6}$ $cm^2\ s^{-1}$ (reproduced with permission from Thompson, 1991).

Translational diffusion coefficients of fluorophores like rhodamine 6G have been determined by FCS and reasonable values of $\sim 3 \times 10^{-6}$ $cm^2\ s^{-1}$ were found (Figure 11.12). Tests with latex beads showed good agreement with known values.

Applications to fluorescent or fluorescently labeled proteins and nucleic acids, and to fluorescent lipid probes in phospholipid bilayers, have been reported. In the latter case, the diffusion coefficients measured above the chain melting temperature were found to be $\approx 10^{-7}$ $cm^2\ s^{-1}$, which is in agreement with values obtained by other techniques.

Translational diffusion times of micelles can be measured by FCS, which allows calculation of the aggregation number (see Box 11.2).

Application of FCS to the determination of diffusion coefficients of atoms in the gas phase has also been proposed.

11.3.3 Chemical kinetic studies

When translational diffusion and chemical reactions are coupled, information can be obtained on the kinetic rate constants. Expressions for the autocorrelation function in the case of unimolecular and bimolecular reactions between states of different quantum yields have been obtained. In a general form, these expressions contain a large number of terms that reflect different combinations of diffusion and reaction mechanisms.

In the case of complex formation, i.e. association–dissociation kinetics, there are two limiting cases of interest:

1. $\tau_{chem} \ll \tau_D = \omega^2/4D_t$: the chemical relaxation time is much smaller than the characteristic diffusion time so that the equilibrium is reached during diffusion through the excitation volume. Then, it suffices to replace the diffusion coefficient appearing in Eq. (11.8) by the weighted average $\langle D_t \rangle$ of the diffusion coefficients of the free and bound ligand:

Box 11.2 Determination of the size of micelles by FCS

The translational diffusion coefficient of micelles loaded with a fluorophore can be determined from the autocorrelation function by means of Eqs (11.8) or (11.9). The hydrodynamic radius can then be calculated using the Stokes–Einstein relation (see Chapter 8, Section 8.1):

$$r_h = \frac{kT}{6\pi\eta D_t}$$

where k is the Boltzmann constant, T the temperature and η the viscosity (Pa s). Assuming that the micelles are spherical, the aggregation number, i.e. the number of surfactant molecules per micelle, is given by

$$\bar{n} = \frac{4\pi\rho r_h^3 N_a}{3m}$$

where ρ is the mean density of the micelle (g cm^{-3}), m is the molecular mass of a surfactant molecule (g mol^{-1}) and N_a is Avogadro's number.

Figure B11.2.1 shows the normalized autocorrelation functions of various micelles loaded with octadecyl rhodamine B chloride (ODRB) at pH 7 (PBS buffer)[a]. The differences in size of the micelles are clearly reflected by the differences in diffusion times τ_D. The translational diffusion coefficients are reported in Table B11.2.1, together with the hydrodynamic radii and the aggregation numbers.

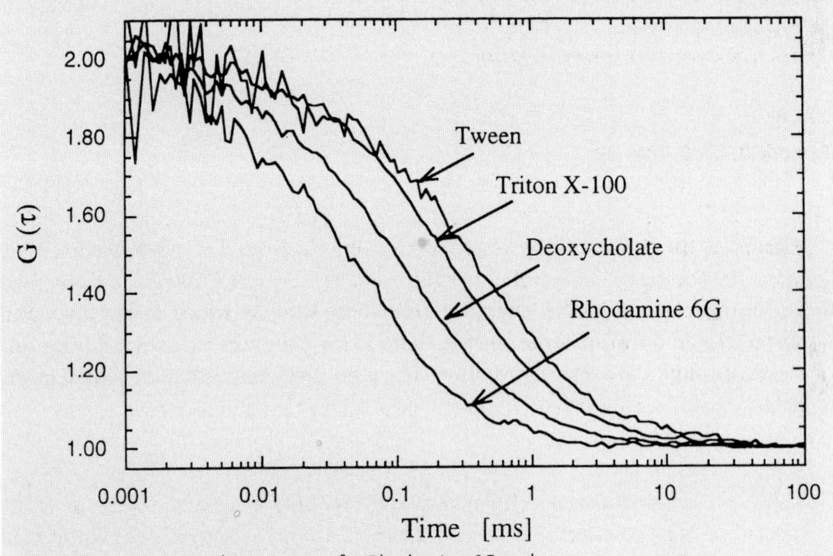

Fig. B11.2.1. Autocorrelation curves for Rhodamine 6G and various micelles loaded with ODBR (reproduced with permission from Hink and Visser[a]).

Tab. B11.2.1. Translational diffusion coefficients, hydrodynamic volumes and aggregation numbers of various micelles loaded with ODRB

	D_t (10^{-11} m^2 s^{-11})	r_h (nm)	$\bar{n}$
CTAB	6.3 ± 0.3	3.7 ± 0.2	319 ± 41
Deoxycholate	11 ± 1	2.2 ± 0.2	50 ± 13
SDS	5.9 ± 0.3	3.7 ± 0.2	357 ± 51
Triton X-100	5.5 ± 0.4	3.5 ± 0.2	92 ± 19
Tween 80	4.2 ± 0.4	5.6 ± 0.2	250 ± 68

The diffusion coefficients obtained with another fluorophore (NBD derivative) were slightly different. The values of the aggregations numbers were found to be often overestimated because incorporation of the fluorescent probe may require extra surfactant molecules. However, the relative size differences between the micelles are in good agreement with the values reported in the literature. In addition to the size of micelles, FCS can give information on the size distribution.

a) Hink M. and Visser A. J. W. G. (1999) in: Rettig W. et al. (Eds), *Applied Fluorescence in Chemistry, Biology and Medicine*, Springer-Verlag, Berlin, pp. 101–18.

$$G(\tau) = 1 + \frac{1}{\overline{N}}\left(\frac{1}{1 + 4\langle D_t\rangle\tau/\omega^2}\right) \tag{11.10}$$

with

$$\langle D_t\rangle = \alpha D_t^{\text{free}} + (1-\alpha)D_t^{\text{bound}} \tag{11.11}$$

where α is the fraction of free ligand.

2. $\tau_{\text{chem}} \gg \tau_D = \omega^2/4D_t$: the chemical relaxation time is much larger than the characteristic diffusion time so that there is no chemical exchange during diffusion through the excitation volume. The autocorrelation function is then given by

$$G(\tau) = 1 + \frac{1}{\overline{N}}\left[\frac{\alpha}{1 + 4D_t^{\text{free}}\tau/\omega^2} + \frac{1-\alpha}{1 + 4D_t^{\text{bound}}\tau/\omega^2}\right] \tag{11.12}$$

In both cases, the fractions of free and bound ligands can be determined provided that the diffusion coefficients of these species are known.

Triplet state kinetics can also be studied by FCS (Widengren et al., 1995). In fact, with dyes such as fluoresceins and rhodamines, additional fluctuations in fluorescence are observed when increasing excitation intensities as the molecules enter and leave their triplet states. The time-dependent part of the autocorrelation function is given by

$$G_{\mathrm{T}}(\tau) = G_{\mathrm{D}}(\tau)\left[1 + \frac{{}^{3}\overline{N}}{1 - {}^{3}\overline{N}} \exp(-t/\tau_{\mathrm{T}})\right] \tag{11.13}$$

where $G_{\mathrm{D}}(\tau)$ is the time-dependent part of the autocorrelation function for translational diffusion (i.e. $G(\tau) - 1$ obtained from Eqs 11.8 or 11.9), ${}^{3}\overline{N}$ is the average fraction of fluorophores within the sample volume element that are in their triplet state, and τ_{T} is the relaxation time from the triplet state. FCS is a convenient method for the determination of triplet parameters of fluorophores in solution. Because these parameters are sensitive to the fluorophore environment, FCS can be used for probing molecular microenvironments by monitoring triplet states.

Photoinduced electron transfer and photoisomerization can also be studied by FCS.

11.3.4
Determination of rotational diffusion coefficients

When the excitation light is polarized and/or if the emitted fluorescence is detected through a polarizer, rotational motion of a fluorophore causes fluctuations in fluorescence intensity. We will consider only the case where the fluorescence decay, the rotational motion and the translational diffusion are well separated in time. In other words, the relevant parameters are such that $\tau_{\mathrm{S}} \ll \tau_{\mathrm{c}} \ll \tau_{\mathrm{D}}$, where τ_{S} is the lifetime of the singlet excited state, τ_{c} is the rotational correlation time (defined as $1/6D_{\mathrm{r}}$ where D_{r} is the rotational diffusion coefficient; see Chapter 5, Section 5.6.1), and τ_{D} is the diffusion time defined above. Then, the normalized autocorrelation function can be written as (Rigler et al., 1993)

$$G(\tau) = 1 + \frac{1}{\overline{N}}\left[\frac{1}{1 + 4D_{\mathrm{t}}\tau/\omega^{2}} + \frac{4}{5}\exp(-t/\tau_{\mathrm{c}}) - \frac{9}{5}\exp(-t/\tau_{\mathrm{S}})\right] \tag{11.14}$$

This relation shows that *the rotational correlation time is uncoupled from the excited-state lifetime*, in contrast to classical steady-state or time-resolved fluorescence polarization measurements (see Chapter 5). The important consequence is the possibility of observing slow rotations with fluorophores of short lifetime. This is the case for biological macromolecules labeled with fluorophores (e.g. rhodamine) whose lifetime is of a few nanoseconds.

Note that the translational diffusion time decreases when the beam radius is decreased but not the rotational correlation time.

11.4 Single-molecule fluorescence spectroscopy

11.4.1 General remarks

Since the pioneering work of Moerner and Kador in 1989 on doped crystals, single-molecule detection has considerably expanded because it opens up new opportunities in analytical, material and biological sciences. For instance, sensitive detection is a major issue for the study of devices operating at a molecular level, and for observation of single biological macromolecules (tagged proteins or DNA). In bulk measurements, the properties of individual molecules are hidden in ensemble averages, whereas observation at the single molecule level provides new insights into physical, chemical and biological phenomena. New applications have emerged in analytical chemistry, biotechnology and the pharmaceutical industries.

The challenge in single-molecule detection is more a matter of background reduction than sensitive detection. In fact, single-photon detection techniques have long been used in spectroscopy, but the major difficulty is detecting fluorescence photons on top of background photons arising from Rayleigh and Raman scattering and from fluorescence impurities.

While a fluorescent molecule transits in a focused laser beam (during a few ms), it undergoes cycles of photon absorption and emission so that its presence is signaled by a burst of emitted photons, which allows us to distinguish the signal from background (Figure 11.13).

The point is now to estimate the maximum number of photons that can be detected from a burst. The maximum rate at which a molecule can emit is roughly the reciprocal of the excited-state lifetime. Therefore, the maximum number of photons emitted in a burst is approximately equal to the transit time divided by the excited-state lifetime. For a transit time of 1 ms and a lifetime of 1 ns, the maximum number is 10^6. However, photobleaching limits this number to about 10^5 photons for the most stable fluorescent molecules. The detection efficiency of specially designed optical systems with high numerical aperture being about 1%, we cannot expect to detect more than 1000 photons per burst. The background can be minimized by careful clean-up of the solvent and by using small excitation volumes (≈ 1 pL in hydrodynamically focused sample streams, ≈ 1 fL in confocal excitation and detection with one- and two-photon excitation, and even smaller volumes with near-field excitation).

11.4.2 Single-molecule detection in flowing solutions

A small sample volume in a flowing solution can be obtained by introducing the sample from a capillary tube inserted into a flow cell (Ambrose et al., 1999) (Figure 11.14). This tube is surrounded by a rapidly flowing sheath fluid so that the sample stream is focused as it exits the capillary. The sample stream resulting from such

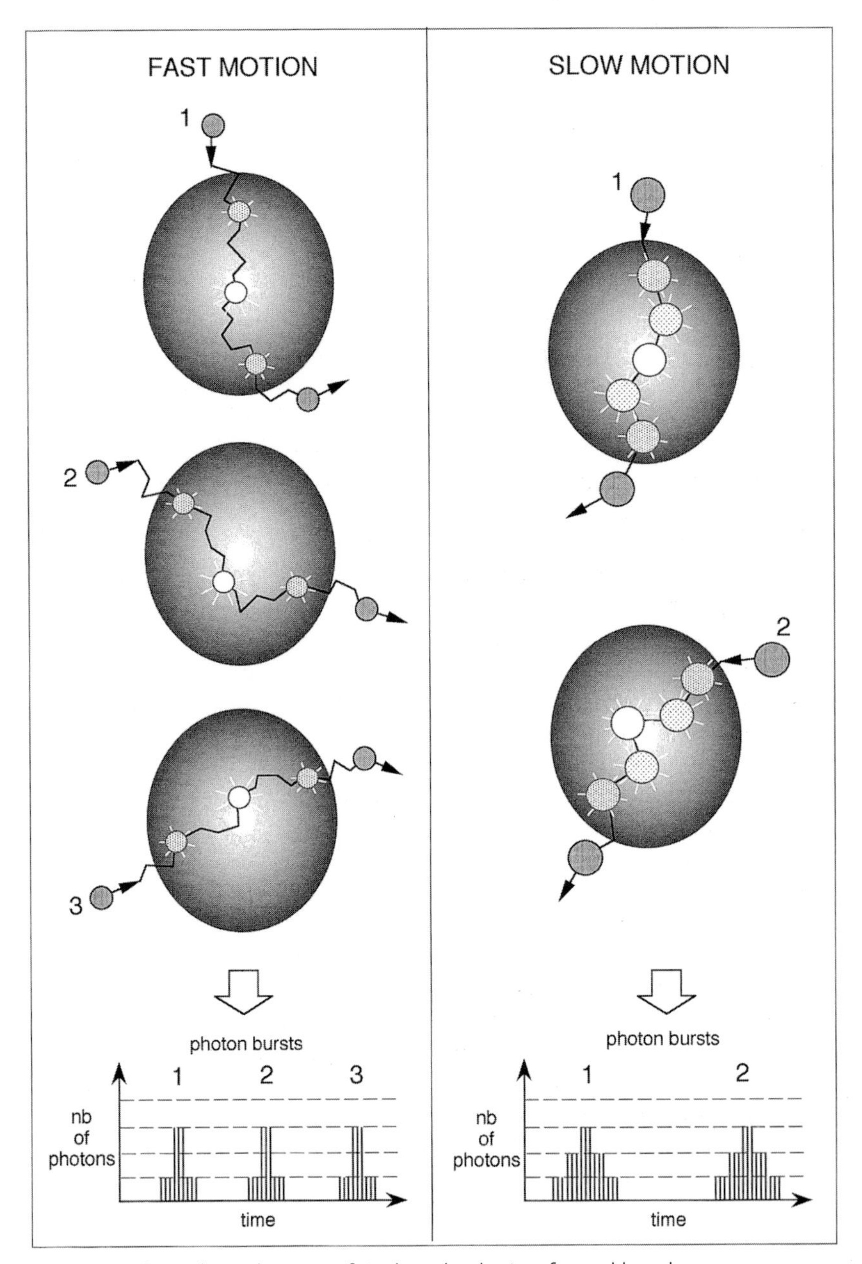

Fig. 11.13. Photon burst detection of single molecules in a focused laser beam.

hydrodynamic focusing has a diameter ranging from 1 to 20 μm. The laser beam is focused to a diameter of 10 μm. Typical sample volume elements are 1–10 pL.

The background resulting from Raman and Rayleigh scattering can be drastically reduced using a pulsed laser and the single-photon timing technique (see Chapter

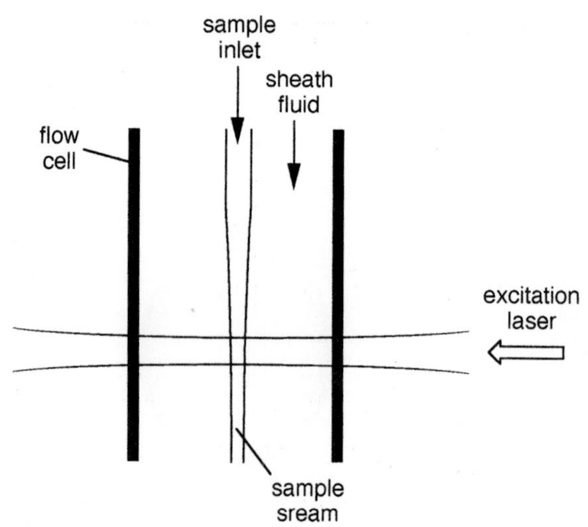

Fig. 11.14. Principle of single-molecule detection in flowing solutions (adapted from Goodwin et al., 1996, *Acc. Chem. Res.* **29**, 607).

6, Section 6.2.2) as follows: because the excited-state lifetime is generally of a few nanoseconds, a single-channel analyzer is used in conjunction with a time-to-amplitude converter to process only the photons that are detected at times longer than 1 ns after the excitation pulse.

The single-photon timing technique has the additional advantage of providing an estimate of the excited-state lifetime from the histogram of arrival times of the photons on the detector. From the fluorescence decay of a single molecule of Rhodamine 110 (Figure 11.15), an excited-state lifetime of 3.9 ± 0.6 ns is estimated. Lifetime measurement is of interest for identification of single molecules.

Steady-state and time-resolved emission anisotropy measurements also allows distinction of single molecules on the basis of their rotational correlation time.

As far as applications are concerned, DNA sequencing has received much attention (see Box 11.3).

11.4.3
Single-molecule detection using advanced fluorescence microscopy techniques

Confocal microscopes (see Section 11.2.1.1) are well suited to the detection of single molecules. A photon burst is emitted when the molecule diffuses through the excitation volume (0.1–1 fL). An example is given in Figure 11.16.

Single-molecule detection in confocal spectroscopy is characterized by an excellent signal-to-noise ratio, but the detection efficiency is in general very low because the excitation volume is very small with respect to the whole sample volume, and most molecules do not pass through the excitation volume. Moreover, the same molecule may re-enter this volume several times, which complicates data interpretation. Better detection efficiencies can be obtained by using microcapillaries and microstructures to force the molecules to enter the excitation volume. A nice example of the application of single-molecule detection with confocal microscopy is

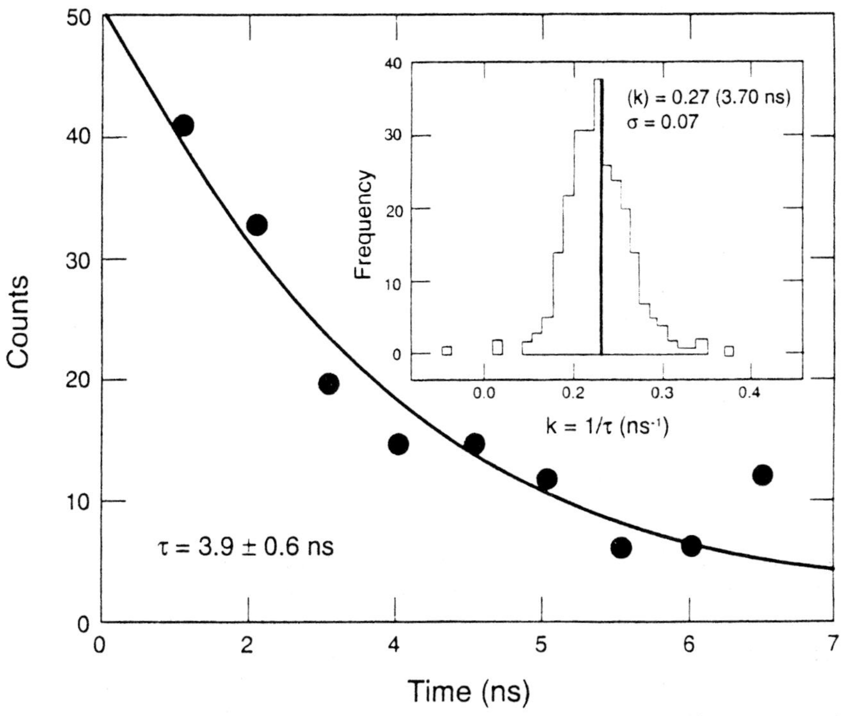

Fig. 11.15. Fluorescence decay of a single molecule of rhodamine 110. The estimated lifetime is 3.9 ± 0.6 ns. Repetitive measurements on several hundreds of single molecules lead to a value of 3.7 ± 0.1 ns, in excellent agreement with the lifetime measured in bulk solutions (3.8 ± 0.1 ns) (adapted from Wilkerson et al., 1993, *Appl. Phys. Lett.* **62**, 2030).

the observation of the dynamics of a single fluorescent molecule interacting with guanine on a DNA strand.

Under the same optical configuration, FCS (Fluorescence Correlation Spectroscopy) measurements (see Section 11.3) can be carried out on samples at the single-molecule level under conditions where the average number of fluorescent molecules in the excitation volume is less than 1. It should be noted that at low fluorophore concentrations, the time required to obtain satisfactory statistics for the fluctuations may become problematic in practical applications (e.g. for a concentration of 1 fM, a fluorophore crosses a confocal excitation volume every 15 min).

Several interesting applications to DNA molecules have been reported (e.g. hybridization, replication, detection of single point mutations, etc.). Single-molecule FCS was also used to study proteins and enzymes (e.g. Green Fluorescent Proteins (GFP), interactions of proteins with carbohydrates, conformational change of H^+-ATPase upon binding to nucleotides, etc.). Finally, single-molecule FCS is the method of choice for drug screening (Buehler et al., 2001).

Two-photon excitation (TPE) fluorescence microscopy (Section 11.2.1.2) can be applied to the detection of single molecules in solution. By comparison with one-

Box 11.3 DNA sequencing using single-molecule detection of fluorescently labeled nucleotides

After complete sequencing of the human genome using Sanger's enzymatic chain termination method and automated DNA sequencing machines, it is of interest to develop alternative methods that are more efficient and more accurate in order to understand the function of each gene and the corresponding health implications. This requires investigation of genetic variations in different

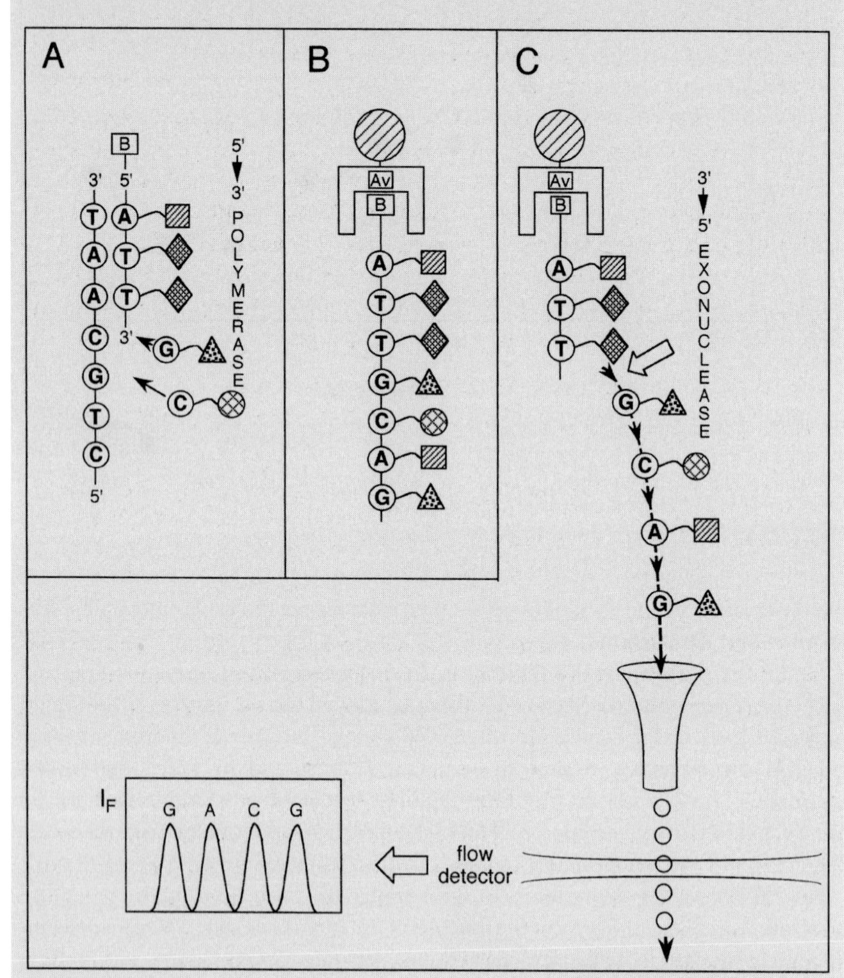

Fig. B11.3.1. Principle of a flow-based method to sequence single fragments of DNA. A: synthesis of the complementary strand with nucleotides labeled with fluorophores. B: attachment of this strand to a microsphere by an avidin (Av)–biotin (B) bond and suspension in a flowing sample stream. C: sequential cleavage by an exonuclease and detection (adapted from Keller et al.[c]).

cell types, individuals and organisms[a]. Fluorescence-based detection of single molecules is one of these methods.

In the early 1990s, Keller and coworkers proposed a very clever method to sequence single fragments of DNA[b,c]. The principle is illustrated in Figure B11.3.1. A DNA strand is replicated by a polymerase using nucleotides linked to a fluorophore via a linker arm. The fluorescently tagged DNA strand is attached to a support (e.g. a latex bead) and suspended in a flowing sample stream. The DNA bases are then sequentially cleaved by an exonuclease enzyme. The released labeled nucleotides are detected and identified by their fluorescence signature. The DNA sequence can thus be determined by the order in which the labeled nucleotides pass through the laser beam. This method has the potential for reading long DNA sequences ($\approx 10^4$ bases) in contrast to gel-based techniques ($< 10^3$ bases). The rate can reach several hundred bases per second.

When the nucleotides are labeled with different fluorophores, they are identified by their spectral characteristics. Alternatively, the same fluorophore can be used and distinction is made on the basis of different lifetimes as a result of different interactions between the nucleotide and the fluorophore.

The use of sub-micrometer channels and detection by confocal fluorescence microscopy is an interesting alternative, which should allow precise control of the movement of single molecules by electrokinetic or electro-osmotic forces[a].

a) Neumann M., Herten D.-P. and Sauer M. (2001) in: Valeur B. and Brochon J. C. (Eds), *New Trends in Fluorescence Spectroscopy. Applications to Chemical and Life Sciences*, Springer-Verlag, Berlin, pp. 303–29.

b) Ambrose W. P., Goodwin P. M., Jett J. H., Johnson M. E., Martin J. C., Marrone B. L., Schecker J. A., Wilkerson C. W. and Keller R. A. (1993) *Ber. Bunsenges. Phys. Chem.* **97**, 1535.

c) Keller R. A., Ambrose W. P., Goodwin P. M., Jette J. H., Martin J. C. and Wu M. (1996) *Appl. Spectrosc.* **50**, 12A.

photon confocal detection, single-molecule detection by TPE is more sensitive because TPE's ability to suppress background is better, and the background is smaller than in the UV. The advantages of TPE have been exploited in FCS (Chen et al., 2001).

Single molecules can be detected by NSOM (Near-field Scanning Optical Microscopy; see Section 11.2.1.3) with the advantages of (i) higher spatial resolution over far-field techniques, (ii) reduced photobleaching, (iii) simultaneous information of the surrounding of the molecule obtained from force mapping (Dunn, 1999), (iv) possible information on the orientation of the fluorophore transition moment. However, lifetime measurements on single molecules are perturbed by the nearby metal-coated tip. The effect of the latter on spectral measurements is generally negligible at room temperature and such measurements can reveal new insights into sample properties.

Finally, the choice between far-field and near-field techniques largely depends on the application. When spatial resolution is not critical, far-field techniques are preferred, especially for studying the photophysical properties of single molecules in

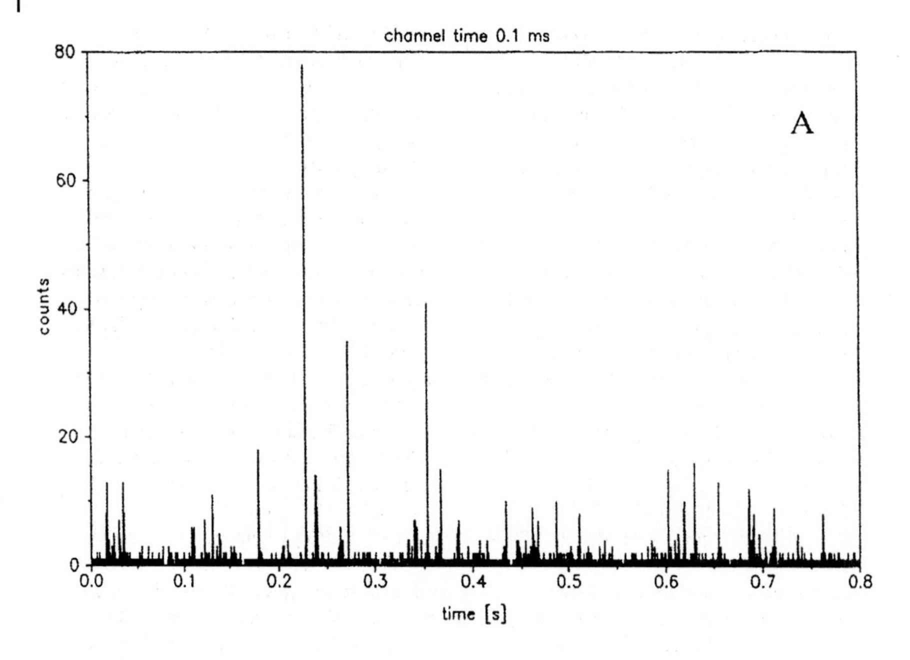

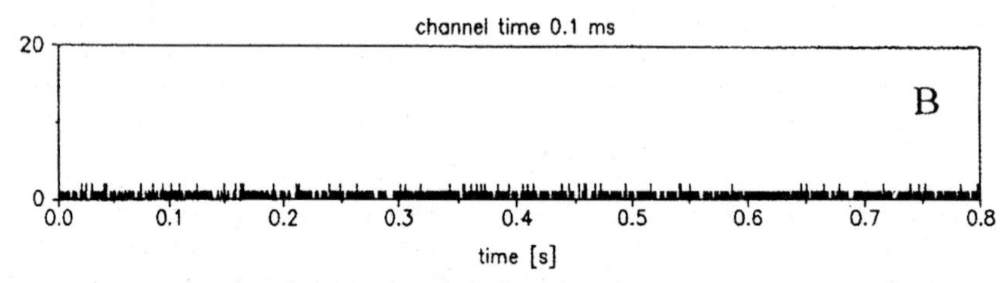

Fig. 11.16. Detection of single molecules of Rhodamine 6G by confocal fluorescence microscopy. A: solution of Rhodamine 6G 2×10^{-12} M in water; B: pure water (reproduced with permission from Mets and Rigler, 1994, *J. Fluorescence* **4**, 259).

samples where coverage can be controlled. In fact, the advantages of these techniques are the enhanced signal-to-noise ratio and the higher speed of data acquisition.

11.5 Bibliography

Ambrose W. P., Goodwin P. M., Jett J. H., Van Horden A., Werner J. H. and Keller R. A. **(1999)** Single Molecule Fluorescence Spectroscopy at Ambient Temperature, *Chem. Rev.* 99, 2929–56.

Bashé T., Moerner W. E., Orrit M. and Wild U. P. (Eds) **(1996)** *Single-Molecule Optical Detection, Imaging and Spectroscopy*, VCH, Weinheim.

Buehler C., Stoeckli K. and Auer M. **(2001)**

The Integration of Single Molecule Detection Technologies into Miniaturized Drug Screening: Current Status and Future Perspectives, in: Valeur B. and Brochon J. C. (Eds), *New Trends in Fluorescence Spectroscopy. Applications to Chemical and Life Sciences*, Springer-Verlag, Berlin, pp. 331–79.

Chen Y., Müller J. D., Eid J. S. and Gratton E. **(2001)** Two-Photon Fluorescence Fluctuation Spectroscopy, in: Valeur B. and Brochon J. C. (Eds) *New Trends in Fluorescence Spectroscopy. Applications to Chemical and Life Sciences*, Springer-Verlag, Berlin, pp. 276–96.

Davidson R. S. **(1996)** Application of Fluorescence Microscopy to a Study of Chemical Problems, *Chem. Soc. Rev.* 241–53.

Dunn R. C. **(1999)** Near-Field Scanning Optical Microscopy, *Chem. Rev.* 99, 2891–927.

Ehrenberg M. and Rigler R. **(1976)** Fluorescence Correlation Spectroscopy Applied to Rotational Diffusion of Macromolecules, *Quart. Rev. Biophys.* 9, 69–81.

Elson E. L. **(1985)** Fluorescence Correlation Spectroscopy and Photobleaching Recovery, *Ann. Rev. Phys. Chem.* 36, 379–406.

Gadella T. W. J., Jovin T. M. and Clegg R. M. **(1993)** Fluorescence Lifetime Imaging Microscopy (FLIM): Spatial Resolution of Microstructures on the Nanosecond Time Scale, *Biophys. Chem.* 48, 221–39.

Goodwin P. M., Ambrose W. P. and Keller R. A. **(1996)** Single-Molecule Detection in Liquids by Laser-Induced Fluorescence, *Acc. Chem. Res.* 29, 607–13.

Keller R. A., Ambrose W. P., Goodwin P. M., Jette J. H., Martin J. C. and Wu M. **(1996)** Single-Molecule Fluorescence Analysis in Solution, *Appl. Spectrosc.* 50, A12–32.

Kotyk A. (Ed.) **(1999)** *Fluorescence Microscopy and Fluorescent Probes* (Proceedings of the Third Conference held in Prague on June 20–23, 1999), Espero Publishing, Prague.

Herman B., Wang X. F., Wodnicki P., Perisamy A., Mahajan N., Berry G. and Gordon G. **(1999)** Fluorescence Lifetime Imaging Microscopy, in: Rettig W. et al. (Eds), *Applied Fluorescence in Chemistry, Biology and Medicine*, Springer-Verlag, Berlin, pp. 491–507.

Hink M. and Visser A. J. W. G. **(1999)** Characterization of Membrane Mimetic Systems with Fluorescence, in: Rettig W. et al. (Eds), *Applied Fluorescence in Chemistry, Biology and Medicine*, Springer-Verlag, Berlin, pp. 101–18.

Lakowicz J. R. and Szymacinski H. **(1996)** Imaging Applications of Time-Resolved Fluorescence Spectroscopy, in: Wang X. F. and Herman B. (Eds), *Fluorescence Imaging Spectroscopy and Microscopy*, Chemical Analysis Series, Vol. 137, John Wiley & Sons, New York, pp. 273–311.

Lemasters J. J. **(1996)** Confocal Microscopy of Single Living Cells, in: Wang X. F. and Herman B. (Eds), *Fluorescence Imaging Spectroscopy and Microscopy*, Chemical Analysis Series, Vol. 137, John Wiley & Sons, New York, pp. 157–77.

Madge D. **(1976)** Chemical Kinetics and Applications of Fluorescence Correlation Spectroscopy, *Quart. Rev. Biophys.* 9, 35–47.

Mialocq J.-C. and Gustavsson T. **(2001)** Investigation of Femtosecond Chemical Reactivity by Means of Fluorescence Up-Conversion, in: Valeur B. and Brochon J. C. (Eds), *New Trends in Fluorescence Spectroscopy. Applications to Chemical and Life Sciences*, Springer-Verlag, Berlin, pp. 61–80.

Neumann M., Herten D.-P. and Sauer M. **(2001)** New Techniques for DNA Sequencing Based on Diode Laser Excitation and Time-Resolved Fluorescence Detection, in: Valeur B. and Brochon J. C. (Eds), *New Trends in Fluorescence Spectroscopy. Applications to Chemical and Life Sciences*, Springer-Verlag, Berlin, pp. 303–29.

Petersen N. O. and Elson E. L. **(1986)** Measurements of Diffusion and Chemical Kinetics by Fluorescence Photobleaching Recovery and Fluorescence Correlation Spectroscopy, *Methods Enzymol.* 130, 454–84.

Piston D. W. **(1996)** Two-photon Excitation Microscopy, in: Wang X. F. and Herman B. (Eds), *Fluorescence Imaging Spectroscopy and Microscopy*, Chemical Analysis Series, Vol. 137, John Wiley & Sons, New York, pp. 253–72.

Rigler R., Widengren J. and Mets Ü. **(1993)** Interactions and Kinetics of Single

Molecules as Observed by Fluorescence Correlation Spectroscopy, in: WOLFBEIS O. S. (Ed.), *Fluorescence Spectroscopy. New Methods and Applications*, Springer-Verlag, Berlin, pp. 15–24.

SO P. T. C., FRENCH T., YU W. M., BERLAND K. M., DONG C. Y. and GRATTON E. **(1996)** Two-Photon Fluorescence Microscopy: Time-Resolved and Intensity Imaging, in: WANG X. F. and HERMAN B. (Eds), *Fluorescence Imaging Spectroscopy and Microscopy*, Chemical Analysis Series, Vol. 137, John Wiley & Sons, New York, pp. 351–74.

TAN W. and KOPELMAN R. **(1996)** Nanoscale Imaging and Sensing by Near-Field Optics, in: WANG X. F. and HERMAN B. (Eds), *Fluorescence Imaging Spectroscopy and Microscopy*, Chemical Analysis Series, Vol. 137, John Wiley & Sons, New York, pp. 407–75.

THOMPSON N. L. **(1991)** Fluorescence Correlation Spectroscopy, in: LAKOWICZ J. R. (Ed.), *Topics in Fluorescence Spectroscopy, Vol. 1: Techniques*, Plenum Press, New York, pp. 337–78.

WANG X. F., PERIASAMY A. and HERMAN B. **(1992)** Fluorescence Lifetime Imaging Microscopy (FLIM): Instrumentation and Applications, *Crit. Rev. Anal. Chem.* 23, 369–95.

WANG X. F., PERIASAMY A., WODNICKI P., GORDON G. W. and HERMAN B. **(1996)** Time-Resolved Fluorescence Lifetime Imaging Microscopy: Instrumentation and Biomedical Applications, in: WANG X. F. and HERMAN B. (Eds), *Fluorescence Imaging Spectroscopy and Microscopy*, Chemical Analysis Series, Vol. 137, John Wiley & Sons, New York, pp. 313–50.

WEBB W. W. **(1976)** Applications of Fluorescence Correlation Spectroscopy, *Quart. Rev. Biophys.* 9, 49–68.

WIDENGREN J., METS Ü. and RIGLER R. **(1995)** Fluorescence Correlation Spectroscopy of Triplet States in Solution: A Theoretical and Experimental Study, *J. Phys. Chem.* 99, 13368–79.

Epilogue

Dans la phase initiale de la démarche (...), le scientifique fonctionne par l'imagination, comme l'artiste. Après seulement, quand interviennent l'épreuve critique et l'expérimentation, la science se sépare de l'art ...

[In the initial phase of the process (...), the scientist works through the imagination, as does the artist. Only afterwards, when critical testing and experimentation come into play, does science diverge from art ...]

F. Jacob, 1997

Index